Translational Bioinformatics

Volume 16

Series editor
Xiangdong Wang, MD, Ph.D.
Zhongshan Hospital Institute of Clinical Science, Fudan University Shanghai Medical College, Shang Hai, China

Aims and Scope

The Book Series in Translational Bioinformatics is a powerful and integrative resource for understanding and translating discoveries and advances of genomic, transcriptomic, proteomic and bioinformatic technologies into the study of human diseases. The Series represents leading global opinions on the translation of bioinformatics sciences into both the clinical setting and descriptions to medical informatics. It presents the critical evidence to further understand the molecular mechanisms underlying organ or cell dysfunctions in human diseases, the results of genomic, transcriptomic, proteomic and bioinformatic studies from human tissues dedicated to the discovery and validation of diagnostic and prognostic disease biomarkers, essential information on the identification and validation of novel drug targets and the application of tissue genomics, transcriptomics, proteomics and bioinformatics in drug efficacy and toxicity in clinical research.

The Book Series in Translational Bioinformatics focuses on outstanding articles/chapters presenting significant recent works in genomic, transcriptomic, proteomic and bioinformatic profiles related to human organ or cell dysfunctions and clinical findings. The Series includes bioinformatics-driven molecular and cellular disease mechanisms, the understanding of human diseases and the improvement of patient prognoses. Additionally, it provides practical and useful study insights into and protocols of design and methodology.

Series Description

Translational bioinformatics is defined as the development of storage-related, analytic, and interpretive methods to optimize the transformation of increasingly voluminous biomedical data, and genomic data in particular, into proactive, predictive, preventive, and participatory health. Translational bioinformatics includes research on the development of novel techniques for the integration of biological and clinical data and the evolution of clinical informatics methodology to encompass biological observations. The end product of translational bioinformatics is the newly found knowledge from these integrative efforts that can be disseminated to a variety of stakeholders including biomedical scientists, clinicians, and patients. Issues related to database management, administration, or policy will be coordinated through the clinical research informatics domain. Analytic, storage-related, and interpretive methods should be used to improve predictions, early diagnostics, severity monitoring, therapeutic effects, and the prognosis of human diseases.

Recently Published and Forthcoming Volumes

Transcriptomics and Gene Regulation
Editor: Jiaqian Wu
Volume 9

Application of Clinical Bioinformatics
Editors: Xiangdong Wang, Christian Baumgartner, Denis C. Shields, Hong-Wen Deng, Jacques S Beckmann
Volume 11

Pediatric Biomedical Informatics Computer Applications in Pediatric Research
Editor: John J. Hutton
Volume 10

Genomic Approach to Asthma
Editors: Xiangdong Wang, Zhihong Chen
Volume 12

More information about this series at http://www.springer.com/series/11057

Hong Jiang • Ming Liu
Editors

Heart Genomics

Springer

Editors
Hong Jiang
Zhongshan Hospital
Fudan University
Shanghai, China

Ming Liu
Zhongshan Hospital
Fudan University
Shanghai, China

ISSN 2213-2775 ISSN 2213-2783 (electronic)
Translational Bioinformatics
ISBN 978-981-13-1428-5 ISBN 978-981-13-1429-2 (eBook)
https://doi.org/10.1007/978-981-13-1429-2

Library of Congress Control Number: 2018956134

This Springer imprint is published by the registered company Springer Nature Singapore Pte Ltd.
The registered company address is: 152 Beach Road, #21-01/04 Gateway East, Singapore 189721, Singapore

Contents

Chapter 1
Genome-Wide Association Studies of Hypertension and Several Other Cardiovascular Diseases

Yan Wang and Ji-Guang Wang

Abstract Genome-wide association studies (GWAS) have greatly expanded our understanding on the genetic architecture of cardiovascular disease in the past decade. It has revealed hundreds of suggestive genetic loci that replicated known biological candidate genes and indicated existence of previously unsuspected new biology relevant to cardiovascular disorders. These data have been used successfully to create genetic risk scores that may improve risk prediction and susceptive individual identification. Furthermore, the GWAS-identified novel pathways may augur a new era of novel drug development and stratification of patients. In this chapter, we will briefly review the literature on the candidate genes and signals discovered by GWAS in hypertension and coronary artery disease and discuss their implications in clinical medicine.

Keywords Genome-wide association study · Genetic risk score · Hypertension · Coronary artery disease

1 Introduction

The large majority of heart disease is polygenetic, which is the result of combination of multiple common genetic variants and environmental factors. Unraveling the genetic basis of heart disease has evolved slowly through linkage studies, because of the much smaller effect size attributable to common modifying variants in complex disorders (Watkins and Farrall 2006). Chip-based microarray technology for assaying over one million inter-individual genetic variants provides the foundation of genome-wide association studies (GWAS), which are defined as any studies of common genetic variation across the entire human genome designed to identify genetic associations with observable traits (Pearson and Manolio 2008). The common variants usually refer to those with a prevalence of at least 5% in population. A stringent genome-wide significance threshold of $P < 5e\text{-}8$ is routinely used as a

Y. Wang · J.-G. Wang (✉)
The Shanghai Institute of Hypertension, Ruijin Hospital, Shanghai Jiao Tong University School of Medicine, Shanghai, China

H. Jiang, M. Liu (eds.), *Heart Genomics*, Translational Bioinformatics 16,
https://doi.org/10.1007/978-981-13-1429-2_1

correction for multiple testing, which is based on the estimation of approximately one million independent SNPs in a population (Pe'er et al. 2008).

The past decade has witnessed substantial advances in understanding the genetic basis of heart diseases through GWAS. Furthermore, the HapMap- and 1000 Genomes-based meta-analysis including hundreds of thousands subjects has expanded our understanding on genetic architecture of heart disease to a great extent. In the GWAS catalog (www.ebi.ac.uk/gwas/), there are hundreds of loci that have been reported showing association with more than ten heart diseases or traits. The major achievements are listed in Table 1.1.

In this chapter, we will give a brief survey of the achievements of GWAS in blood pressure (BP) and coronary arterial disease (CAD), the progress in risk prediction at individual level, and the novel pathophysiological pathways and drug discoveries.

2 GWAS Significant Loci

2.1 Blood Pressure

BP is a quantitative trait, normally distributed in the general population, whereas the 30–50% of the variation of BP is determined by the inherited genetic factors (Miall and Oldham 1963; Levy et al. 2007). Hypertension ranks as the leading cause of morbidity and mortality worldwide, contributing to CAD, atrial fibrillation, and heart failure, so that the knowledge about the genetics of hypertension or BP is an important factor in the diagnosis, control, and treatment to all of these heart diseases.

Until the middle of 2017, GWAS has identified and replicated genetic variants of modest or weak effect on BP over 200 loci, and the strongest SNPs associated with different BP traits are summarized in Tables 1.2a and 1.2b. However, the attempts to identify genetic variants associated with BP have been challenging and of relatively low yield in the early phase. In 2007, the first GWAS (Wellcome Trust Case Control Consortium, WTCCC) (2007) adopted a case-control study design using 3000 shared controls and 14,000 cases (2000 for hypertension) of European ancestry to study 7 complex diseases simultaneously. Hypertension was the only disease without any significant results. The first GWAS of quantitative BP phenotypes was conducted by the Framingham Heart Study, which included 1400 family subjects and found no significant results either (Levy et al. 2007). These two studies let researchers realize the complexity of the genetic mechanisms underlying BP regulation and the need for much larger sample size to look for genes associated with BP/hypertension.

To enhance statistical power, international collaborations were established among studies and are organized in consortia. Furthermore, detecting the association with BP as a continuous variable rather than as a case-control study also had

Table 1.1 Genome-wide association studies of hypertension and several other cardiovascular diseases and abnormalities

Study	Initial sample size	Replication sample size	Major ethnic group	Top locus
Blood pressure				
Levy et al. (2009)	29,136	34,433	European	SH2B3
Newton-Cheh et al. (2009)	34,433	113,236	European	CNNM2
Kato et al. (2011)	19,608	30,765	Asian	HECTD4
Wain et al. (2011)	74,064	48,607	European	ATXN2
Ehret et al. (2011)	69,395	133,361	European	CSK
Warren et al. (2017)	140,886	190,318	European	JAG1
Coronary artery disease				
Samani et al. (2007)	4,864	2,519	European	CDKN2B-AS1 – DMRTA1
Helgadottir et al. (2007)	8,335	9,289	European	CDKN2B-AS1 – DMRTA1
Tregouet et al. (2009)	4,864	14,398	European	LPA
Erdmann et al. (2009)	2,520	38,253	European	MRAS
Reilly et al. (2011)	2,723	17,053	European	ADAMTS7
C4D Consortium et al. (2011)	30,482	40,593	European, Asian	PHACTR1
Schunkert et al. (2011)	86,995	56,682	European	CDKN2B-AS1
Lu et al. (2012)	6,534	26,932	Asian	CDKN2B-AS1
Nikpay et al. (2015)	187,599	None	European, Asian	CDKN2B-AS1
Nelson et al. (2017)	63,731	276,868	European,	CDKN2B-AS1
Congestive heart failure				
Smith et al. (2010)	23,821	None	European, African	LRIG3
Dilated cardiomyopathy				
Villard et al. (2011)	2,287	2,467	European	TSPB7
Atrial fibrillation				
Gudbjartsson et al. 2009 (2009)	36,137	8,855	European	ZFHX3
Ellinor et al. (2010)	14,179	4,771	European	KCNN3
Ellinor et al. (2012)	59,133	5,381	European	PITX2 – MIR297
Low et al. (2017)	36,792	257,896	Asian, European	PITX2 – MIR297
Electrocardiographic abnormalities				
PR interval				
Pfeufer et al. (2010)	28,517	None	European	SCN10A

(continued)

Table 1.1 (continued)

Study	Initial sample size	Replication sample size	Major ethnic group	Top locus
Chambers et al. (2010)	6,543	6,243	Asian	SCN10A
QRS duration				
Sotoodehnia et al. (2010)	40407	7,170	European	SCN10A
QT interval				
Arking et al. (2006)	200	4,451	European	NOS1AP, LOC105371475
Newton-Cheh et al. (2009)	13,685	15,854	European	LOC107985450 – NOS1AP
Arking et al. (2014)	70,389	33,316	European	LOC107985450 – NOS1AP
Heart rate				
den Hoed et al. (2013)	92,355	88,823	European	MYH6
Nolte et al. (2017)	26,523	38,558	European, Hispanic, African American	NDUFA11
Dyslipidemia				
Saxena et al. (2007)	5,217	None	European	APOC1 – APOC1P1
Kooner et al. (2008)	2,011	10,536	European, Mexican, Asian	MLXIPL
Kathiresan et al. (2008)	19,840	20,623	European	HERPUD1 – CETP
Sandhu et al. (2008)	11,685	4,979	European	CELSR2 – PSRC1
Teslovich et al. (2010)	96,598	None	European	ZPR1
Kim et al. (2011)	12,545	30,395	Asian	ZPR1
Willer et al. (2013)	94,595	93,982	European	LDLR
Ko et al. (2014)	3,323	6,017	Mexican	ZPR1
Surakka et al. (2015)	62,166	None	European	APOE

success. In 2009, two large-scale meta-analyses of GWAS from the Global Blood Pressure Genetics (Global BPgen) (Newton-Cheh et al. 2009) and Cohorts for Heart and Aging Research in Genomic Epidemiology (CHARGE) (Levy et al. 2009) consortia identified associations withstanding correction for multiple testing (genome-wide significance). Each of the studies contained nearly 30,000 subjects at the discovery phase and found 8 genomic loci, among which 3 loci overlapped. In 2011, the International Consortium of Blood Pressure (ICBP) (2011) used a multistage design in 200,000 individuals of European descent, replicated the previous 13 loci effectively, and discovered 16 new loci significant at the genome-wide level. Then ICBP analyzed two derived BP phenotypes: mean arterial pressure (MAP) and

Table 1.2a Significant genetic loci for blood pressure and hypertension reported in genome-wide association studies in Europeans

Chr	Strongest SNP	Position	EA	EAF	OR or BETA	*P* value	*N*	Closest gene	Trait	References
1	rs1886773	1319056	A	0.03	−0.63	1.90E-10	233,789	NADK, CPSF3L	PP	Warren et al. (2017)
1	rs14057	6623180	A	0.35	−0.3	2.50E-10	329,584	RNF207	SBP	Warren et al. (2017)
1	rs9662255	9381890	A	0.43	−0.21	1.90E-10	310,618	chr1mb9	PP	Warren et al. (2017)
1	rs880315	10736809	C	0.35	0.64	9.10E-16	133,125	CASZ1	SBP	Wain et al. (2017)
1	rs17367504	11802721	G	0.15	−0.53	2.68E-18	149,714	MTHFR, NPPB	DBP	Wain et al. (2017)
1	rs12744757	11846764	T	0.06	−0.7	7.54E-23	359,786	NPPA-AS1, NPPA	SBP	Wain et al. (2017)
1	rs1042010	15467418	A	0.19	−0.41	3.03E-13	345,219	CELA2A	SBP	Wain et al. (2017)
1	rs1048238	16015154	T	0.57	0.37	8.09E-07	140,299	HSPB7	SBP	Wain et al. (2017)
1	rs6686889	24703979	T	0.25	0.19	3.60E-09	322,575	chr1mb25	DBP	Warren et al. (2017)
1	rs871524	37945773	A	0.33	0.23	1.27E-11	304,987	INPP5B	PP	Wain et al. (2017)
1	rs4360494	37990219	C	0.55	0.28	3.70E-16	282,851	SF3A3	PP	Warren et al. (2017)
1	rs112557609	56111252	A	0.35	0.23	6.80E-12	325,952	RP4-710M16.1, PPAP2B	PP	Warren et al. (2017)
1	rs3889199	59188070	A	0.71	0.35	1.80E-24	329,486	FGGY	PP	Warren et al. (2017)
1	rs61767086	88600899	G	0.14	−0.41	1.90E-09	136,914	PKN2-AS1	PP	Wain et al. (2017)
1	rs10922502	88894475	A	0.62	−0.38	2.20E-15	323,666	GTF2B	SBP	Warren et al. (2017)
1	rs12129649	112688881	T	0.06	0.55	5.93E-19	343,172	MOV10	DBP	Wain et al. (2017)
1	rs11466111	115286557	T	0.02	0.65	5.08E-10	319,490	chr1mb115	DBP	Warren et al. (2017)
1	rs12405515	172388301	T	0.56	−0.17	1.40E-09	328,543	DNM3	DBP	Warren et al. (2017)
1	rs12408022	217545447	T	0.26	0.2	2.40E-10	320,983	GPATCH2	DBP	Warren et al. (2017)
1	rs10916082	227064925	A	0.73	−0.18	8.40E-09	327,636	CDC42BPA	DBP	Warren et al. (2017)
1	rs2760061	228003374	A	0.47	0.23	2.10E-16	312,761	WNT3A	DBP	Warren et al. (2017)
1	rs953492	243307890	A	0.46	0.22	7.40E-16	325,253	SDCCAG8	DBP	Warren et al. (2017)

(continued)

Table 1.2a (continued)

Chr	Strongest SNP	Position	EA	EAF	OR or BETA	*P* value	*N*	Closest gene	Trait	References
2	rs2289081	20682080	C	0.36	−0.22	5.50E-12	329,140	C2orf43	PP	Warren et al. (2017)
2	rs55701159	24916727	T	0.89	0.29	7.20E-11	321,052	ADCY3	DBP	Warren et al. (2017)
2	rs1275988	26691496	T	0.61	−0.58	2.35E-15	149,285	KCNK3	SBP	Wain et al. (2017)
2	rs7562	28412873	T	0.52	0.26	1.90E-08	319,942	FOSL2	SBP	Warren et al. (2017)
2	rs13420463	37290423	A	0.77	0.36	7.00E-11	330,307	PRKD3	SBP	Warren et al. (2017)
2	rs4952611	40340603	T	0.58	−0.16	4.00E-08	309,395	SLC8A1	DBP	Warren et al. (2017)
2	rs76326501	42940738	A	0.91	0.42	3.60E-18	318,127	AC016735.1	DBP	Warren et al. (2017)
2	rs72876037	42967456	T	0.12	−0.53	1.65E-14	319,468	HAAO, ZFP36L2	SBP	Wain et al. (2017)
2	rs11690961	46136197	A	0.88	0.34	3.90E-12	327,847	PRKCE	PP	Warren et al. (2017)
2	rs74181299	65056838	T	0.62	0.23	9.60E-13	324,224	CEP68	PP	Warren et al. (2017)
2	rs11689667	85264242	T	0.54	0.18	1.70E-08	330,634	TCF7L1	PP	Warren et al. (2017)
2	rs2579519	96009418	T	0.63	−0.2	4.80E-12	311,557	GPAT2, FAHD2CP	DBP	Warren et al. (2017)
2	rs62167177	145181135	T	0.14	−0.26	1.04E-09	289,338	TEX41	DBP	Warren et al. (2017)
2	rs1446468	164106976	C	0.55	0.54	2.98E-13	141,331	FIGN, GRB14	SBP	Wain et al. (2017)
2	rs79146658	178921341	T	0.91	−0.31	2.40E-10	321,318	CCDC141	DBP	Warren et al. (2017)
2	rs13407401	179850979	A	0.29	−0.43	4.30E-08	145,919	ZNF385B	SBP	Wain et al. (2017)
2	rs144073138	182410033	G	0.06	0.35	1.90E-08	295,411	PDE1A	DBP	Warren et al. (2017)
2	rs7592578	190574865	T	0.19	−0.24	9.50E-12	304,672	TMEM194B	DBP	Warren et al. (2017)
2	rs55780018	207661416	T	0.54	−0.39	5.90E-16	304,567	METTL21A, AC079767.3	SBP	Warren et al. (2017)
2	rs1250259	215435759	A	0.74	−0.31	8.70E-19	325,485	FN1	PP	Warren et al. (2017)
2	rs10932679	216787868	T	0.19	−0.23	1.70E-08	371,085	LOC101928278	PP	Wain et al. (2017)
2	rs1063281	217804009	T	0.6	−0.2	1.30E-12	315,354	TNS1	DBP	Warren et al. (2017)
3	rs142892876	1110752	T	0.001	10.69	5.00E-08	19,546	CNTN6	DBP	Nagy et al. (2017)
3	rs11716531	27415717	A	0.24	0.21	5.72E-14	373,386	SLC4A7	DBP	Wain et al. (2017)

3	rs13082711	27496418	C	0.23	0.33	2.72E-10	146,340	SLC4A7	DBP	Wain et al. (2017)
3	rs36022378	49876272	T	0.8	−0.2	4.70E-09	319,983	CAMKV, ACTBP13	DBP	Warren et al. (2017)
3	rs743757	50438947	C	0.14	0.25	2.40E-10	328,836	CACNA2D2	DBP	Warren et al. (2017)
3	rs9827472	56692618	T	0.37	−0.18	4.30E-10	323,058	FAM208A	DBP	Warren et al. (2017)
3	rs9834975	122398816	T	0.44	0.15	1.14E-08	330,670	chr3mb122	DBP	Warren et al. (2017)
3	rs62270945	128483046	T	0.03	0.61	1.80E-09	279,925	GATA2	PP	Warren et al. (2017)
3	rs9859176	134281183	T	0.4	0.32	1.30E-11	322,428	RYK	SBP	Warren et al. (2017)
3	rs143112823	154990178	A	0.09	−0.4	1.40E-14	297,343	RP11-439C8.2	DBP	Warren et al. (2017)
3	rs419076	169383098	T	0.47	0.41	1.78E-13	193,725	MECOM	SBP	International Consortium for Blood Pressure Genome-Wide Association S et al. (2011)
3	rs12374077	185599886	C	0.35	0.16	9.20E-09	327,513	SENP2	DBP	Warren et al. (2017)
3	rs528908640	193644588	C	0.0005	14.67	2.00E-08	19,546	OPA1	DBP	Nagy et al. (2017)
3	rs9815354	41912651	A	0.17	0.4	7.12E-11	135,381	ULK4	DBP	Wain et al. (2017)
4	rs871606	53933078	C	0.1	−0.5	1.53E-10	144,494	CHIC2	PP	Wain et al. (2017)
4	rs17004869	80283879	T	0.05	0.72	1.02E-09	278,498	FGF5	SBP	Warren et al. (2017)
4	rs2014912	85794517	C	0.85	−0.42	3.86E-10	143,933	ARHGAP24	PP	Wain et al. (2017)
4	rs13107325	102267552	T	0.05	−0.68	2.28E-17	150,920	SLC39A8	DBP	International Consortium for Blood Pressure Genome-Wide Association S et al. (2011)
4	rs13112725	105990585	C	0.77	0.41	3.11E-16	357,230	NPNT, TBCK	SBP	Wain et al. (2017)
4	rs33966350	110510288	A	0.01	1.66	2.10E-11	216,630	ENPEP	SBP	Warren et al. (2017)
4	rs66887589	119588124	T	0.52	−0.22	3.40E-15	324,397	PDE5A	DBP	Warren et al. (2017)
4	rs78049276	147506351	C	0.13	0.27	1.04E-08	321,315	chr4mb148	PP	Warren et al. (2017)
4	rs13139571	155724361	C	0.76	0.26	2.17E-10	185,393	GUCY1A3, GUCY1B3	DBP	International Consortium for Blood Pressure Genome-Wide Association S et al. (2011)

(continued)

Table 1.2a (continued)

Chr	Strongest SNP	Position	EA	EAF	OR or BETA	*P* value	*N*	Closest gene	Trait	References
4	rs1566497	168795997	A	0.42	0.24	1.90E-13	320,948	PALLD	PP	Warren et al. (2017)
4	rs17059668	173663512	C	0.92	−0.33	2.80E-08	313,277	chr4mb174	PP	Warren et al. (2017)
5	rs1173771	32814922	G	0.6	0.5	1.79E-16	158,664	C5orf23, NPR3	SBP	International Consortium for Blood Pressure Genome-Wide Association S et al. (2011)
5	rs10078021	75742606	T	0.63	−0.16	1.30E-08	314,172	POC5	DBP	Warren et al. (2017)
5	rs10057188	78541966	A	0.46	−0.21	6.70E-11	325,985	LHFPL2	PP	Warren et al. (2017)
5	rs10059921	88218698	T	0.08	−0.53	4.00E-09	298,543	TMEM161B	SBP	Warren et al. (2017)
5	rs337100	123210816	A	0.41	−0.28	4.61E-09	149,200	PRDM6 – SUMO1P5	PP	Wain et al. (2017)
5	rs6595838	128532506	A	0.3	0.34	7.60E-12	328,401	FBN2	SBP	Warren et al. (2017)
5	rs31864	158793185	A	0.55	0.21	5.50E-11	326,557	EBF1	PP	Warren et al. (2017)
5	rs72812846	173950633	A	0.28	−0.21	2.20E-11	312,601	CPEB4	DBP	Warren et al. (2017)
6	rs6911827	22130372	T	0.45	0.3	2.00E-10	326,471	CASC15	SBP	Warren et al. (2017)
6	rs1800562	26092913	A	0.06	0.39	8.69E-17	373,770	HFE	DBP	Wain et al. (2017)
6	rs151168737	31638615	A	0.46	0.25	3.60E-20	315,660	PRRC2A – BAG6	DBP	Wain et al. (2017)
6	rs805303	31648589	G	0.61	0.38	1.49E-11	201,745	BAT2, BAT5	SBP	International Consortium for Blood Pressure Genome-Wide Association S et al. (2011)
6	rs185819	32082290	C	0.51	0.37	1.04E-17	364,144	TNXB	SBP	Wain et al. (2017)
6	rs78648104	50715296	T	0.92	−0.48	1.30E-08	305,426	TFAP2D	SBP	Warren et al. (2017)
6	rs13205180	51967696	T	0.49	0.17	7.00E-10	325,419	PKHD1	DBP	Warren et al. (2017)
6	rs9372498	118251323	A	0.08	0.33	1.80E-11	330,625	SLC35F1	DBP	Warren et al. (2017)
6	rs11154027	121460244	T	0.47	0.21	1.10E-10	316,708	GJA1	PP	Warren et al. (2017)
6	rs13209747	126794309	T	0.45	0.27	1.12E-09	146,105	RSPO3	DBP	Wain et al. (2017)
6	rs17080102	150683634	C	0.07	−0.57	1.04E-10	147,869	PLEKHG1	DBP	Wain et al. (2017)

6	rs36083386	152076778	I	0.11	0.44	1.50E-18	323,303	ESR1	PP	Warren et al. (2017)
6	rs449789	159278093	C	0.14	0.36	2.40E-15	325,584	FNDC1	PP	Warren et al. (2017)
6	rs147212971	165764963	T	0.06	−0.36	1.60E-09	296,010	PDE10A	DBP	Warren et al. (2017)
6	rs1322639	169187008	A	0.78	0.32	4.80E-17	319,866	THBS2	PP	Warren et al. (2017)
7	rs6461992	27181212	G	0.93	0.91	8.49E-24	311,598	HOXA3	SBP	Warren et al. (2017)
7	rs1859168	27202740	C	0.92	0.44	1.85E-20	373,643	HOTTIP	DBP	Wain et al. (2017)
7	rs76206723	40408372	A	0.1	−0.35	7.40E-12	328,162	SUGCT	PP	Warren et al. (2017)
7	rs10260816	45970501	G	0.43	−0.3	5.31E-10	143,371	IGFBP3	PP	Wain et al. (2017)
7	rs17477177	106771412	C	0.21	0.55	4.14E-22	149,296	PIK3CG	PP	Wain et al. (2017)
7	rs13238550	131374297	A	0.4	0.33	1.90E-12	325,647	MKLN1	SBP	Warren et al. (2017)
7	rs1011018	139763465	A	0.2	-0.33	1.50E-08	325,110	HIPK2	SBP	(Warren et al. 2017)
7	rs3918226	150993088	T	0.08	0.67	2.27E-13	118,604	NOS3	DBP	(Wain et al. 2017)
8	rs9693857	9409607	T	0.45	−0.34	2.40E-15	374,178	LOC105379231	SBP	Wain et al. (2017)
8	rs1821002	10782555	G	0.6	−0.42	4.26E-19	323,712	BLK, GATA4	SBP	Warren et al. (2017)
8	rs10107145	10900703	G	0.53	−0.36	1.44E-17	374,413	XKR6	SBP	Wain et al. (2017)
8	rs6557876	26043159	T	0.25	−0.37	2.85E-14	369,457	EBF2	SBP	Wain et al. (2017)
8	rs2978456	42467247	T	0.55	−0.19	1.20E-08	304,964	SLC20A2	PP	Warren et al. (2017)
8	rs2978098	100664447	A	0.54	0.17	1.50E-09	324,424	SNX31	DBP	Warren et al. (2017)
8	rs35783704	104954030	A	0.11	−0.41	7.08E-09	349,452	LRP12, ZFPM2	SBP	Wain et al. (2017)
8	rs2071518	119423572	T	0.17	0.3	1.56E-08	149,046	NOV	PP	Wain et al. (2011)
8	rs894344	134600502	A	0.6	−0.26	3.20E-08	329,834	ZFAT	SBP	Warren et al. (2017)
8	rs4454254	140049929	A	0.63	−0.26	5.10E-16	330,022	TRAPPC9	PP	Warren et al. (2017)
8	rs62524579	142979538	A	0.53	−0.18	3.80E-09	268,645	RP11-273G15.2	DBP	Warren et al. (2017)
8	rs4076877	143820544	T	0.05	−0.42	3.57E-08	252,983	chr8mb144	DBP	Warren et al. (2017)
9	rs872256	2496480	NA	NA	0.1	9.00E-09	8,423	SMARCA2, VLDLR	SBP	Parmar et al. (2016)
9	rs4364717	21801531	A	0.55	−0.18	1.30E-10	327,173	MTAP	DBP	Warren et al. (2017)
9	rs568998724	107040156	A	0.0007	13.49	3.00E-08	19,546	LOC340512	DBP	Nagy et al. (2017)

(continued)

Table 1.2a (continued)

Chr	Strongest SNP	Position	EA	EAF	OR or BETA	*P* value	*N*	Closest gene	Trait	References
9	rs72765298	125138717	T	0.87	−0.37	2.70E−14	316,271	SCAI	PP	Warren et al. (2017)
10	rs9337951	30028144	A	0.34	0.28	2.50E-15	299,646	KIAA1462	PP	Warren et al. (2017)
10	rs10826995	31793730	T	0.71	−0.21	1.10E-09	327,373	ARHGAP12, ZEB1	PP	Warren et al. (2017)
10	rs4590817	61707795	G	0.83	−0.44	3.52E-13	146,522	C10orf107	MAP	Wain et al. (2011)
10	rs932764	94136183	G	0.44	0.48	7.10E-16	160,885	PLCE1	SBP	International Consortium for Blood Pressure Genome-Wide Association S et al. (2011)
10	rs112184198	100844757	A	0.1	−0.66	3.60E-18	323,791	PAX2	SBP	Warren et al. (2017)
10	rs11191548	103086421	T	0.91	1.1	6.90E-26	161,709	NT5C2, CYP17A1	SBP	International Consortium for Blood Pressure Genome-Wide Association S et al. (2011)
10	rs10787517	114055047	A	0.62	0.44	3.31E-18	272,852	ADRB1 – RNU6-709P	SBP	Wain et al. (2017)
11	rs2649044	9742422	T	0.55	0.2	1.12E-14	343,652	SWAP70	DBP	Wain et al. (2017)
11	rs7129220	10328991	G	0.89	−0.62	2.97E-12	182,871	ADM	SBP	International Consortium for Blood Pressure Genome-Wide Association S et al. (2011)
11	rs177542	16901107	A	0.5	0.24	4.18E-08	143,257	PLEKHA7	DBP	Wain et al. (2017)
11	rs11030119	27706555	A	0.31	−0.16	2.90E-08	330,002	BDNF	DBP	Warren et al. (2017)
11	rs11442819	45186591	I	0.11	−0.28	7.10E-09	326,483	PRDM11	PP	Warren et al. (2017)
11	rs751984	61510774	C	0.12	−0.41	1.98E-09	140,821	LRRC10B	DBP	Wain et al. (2017)
11	rs4980532	63913247	T	0.56	0.3	1.53E-10	147,416	RCOR2	PP	Wain et al. (2017)
11	rs67330701	69312240	T	0.09	−0.37	2.10E-12	276,760	MYEOV	DBP	Warren et al. (2017)
11	rs2289125	89491285	A	0.21	−0.38	9.10E-22	307,682	NOX4	PP	Warren et al. (2017)
11	rs633185	100722807	G	0.28	−0.56	1.21E-17	160,461	TMEM133, FLJ32810	SBP	International Consortium for Blood Pressure Genome-Wide Association S et al. (2011)

11	rs8258	117412960	T	0.38	0.24	2.90E-13	327,038	CEP164	PP	Warren et al. (2017)
11	rs11222084	130403335	T	0.35	0.35	2.26E-12	139,748	ADAMTS8	PP	Wain et al. (2017)
12	rs10770612	20077705	A	0.8	0.31	6.90E−15	311,586	PDE3A	PP	Warren et al. (2017)
12	rs73099903	53046995	T	0.07	0.51	1.95E-10	343,318	LOC283335	SBP	Wain et al. (2017)
12	rs7312464	65980467	G	0.52	0.21	5.35E-10	289,978	chr12mb66	PP	Warren et al. (2017)
12	rs17249754	89666809	A	0.16	−0.9	6.59E-20	146,304	ATP2B1	SBP	Wain et al. (2017)
12	rs139236208	94486966	A	0.1	−0.36	1.60E-10	291,244	CCDC41	PP	Warren et al. (2017)
12	rs3184504	111446804	T	0.47	0.45	3.59E-25	120,633	SH2B3	DBP	International Consortium for Blood Pressure Genome-Wide Association S et al. (2011)
12	rs10850519	115490635	C	0.3	−0.21	5.10E-13	327,837	TBX5, TBX3	DBP	Warren et al. (2017)
13	rs9549328	112981842	T	0.23	0.32	1.50E-08	313,787	MCF2L	SBP	Warren et al. (2017)
14	rs12050260	23291885	T	0.35	0.19	2.60E-08	304,390	MYH6	PP	Warren et al. (2017)
14	rs8904	35402011	A	0.38	0.31	1.31E-12	365,195	NFKBIA	SBP	Wain et al. (2017)
14	rs9888615	52910822	T	0.29	−0.32	3.50E-10	326,235	FERMT2	SBP	Warren et al. (2017)
14	rs8016306	63461828	A	0.8	0.34	3.70E-09	329,869	PPP2R5E	SBP	Warren et al. (2017)
14	rs12434998	93989208	C	0.37	−0.19	1.64E-08	315,683	chr14mb94	PP	Warren et al. (2017)
14	rs9323988	98121293	T	0.63	−0.21	4.10E-11	327,551	RP11-61O1.1	PP	Warren et al. (2017)
15	rs1036477	48622729	G	0.1	−0.42	3.81E-08	149,967	FBN1	PP	Wain et al. (2017)
15	rs7178615	66576734	A	0.37	−0.18	2.60E-10	318,076	RP11-321F6.1	DBP	Warren et al. (2017)
15	rs1563894	68343437	NA	NA	0.09	3.00E-08	10,090	ITGA11	SBP	Parmar et al. (2016)
15	rs117539635	69390577	G	0.03	−0.65	2.15E-08	196,789	chr15mb69	DBP	Warren et al. (2017)
15	rs1378942	74785026	C	0.35	0.42	2.69E-26	163,115	CYP1A1, ULK3	DBP	International Consortium for Blood Pressure Genome-Wide Association S et al. (2011)
15	rs6495122	74833304	A	0.42	0.4	1.84E-10	NA	CSK, ULK3	DBP	Levy et al. (2009)
15	rs187680191	76002970	T	0.0006	18.58	3.00E-09	19,546	NRG4	DBP	Nagy et al. (2017)

(continued)

Table 1.2a (continued)

Chr	Strongest SNP	Position	EA	EAF	OR or BETA	*P* value	*N*	Closest gene	Trait	References
15	rs62012628	78777658	T	0.29	−0.24	5.10E-12	244,143	ADAMTS7	DBP	Warren et al. (2017)
15	rs11634851	80736624	G	0.46	0.32	5.38E-14	374,208	FAM108C1	SBP	Wain et al. (2017)
15	rs2521501	90894158	T	0.31	0.65	5.20E-19	127,208	FURIN, FES	SBP	International Consortium for Blood Pressure Genome-Wide Association S et al. (2011)
15	rs12906962	94768842	T	0.68	−0.22	5.60E-14	319,952	chr15mb95	DBP	Warren et al. (2017)
15	rs4984496	96092669	G	0.67	−0.18	2.86E-09	307,660	chr15mb96	DBP	Warren et al. (2017)
16	rs12921187	4893018	T	0.43	−0.17	2.50E-10	326,469	PPL	DBP	Warren et al. (2017)
16	rs13333226	20354332	A	0.81	0.85	1.50E-13	39,706	UMOD	HT	Padmanabhan et al. (2010)
16	rs72799341	30925422	A	0.24	0.19	5.80E-09	324,502	FBXL19	DBP	Warren et al. (2017)
16	rs56249585	65231799	T	0.53	0.18	8.98E-09	318,082	chr16mb65	PP	Warren et al. (2017)
16	rs141767645	69869262	T	0.02	1.3	1.90E-08	174,313	NFAT5	PP	Warren et al. (2017)
16	rs200337503	70084312	I	0.03	1.31	6.88E-13	193,318	NFAT5	PP	Warren et al. (2017)
16	rs117006983	70721707	A	0.01	0.99	4.10E-12	250,766	VAC14	PP	Warren et al. (2017)
16	rs11643209	75297146	T	0.42	−0.34	1.80E-12	309,242	CFDP1	SBP	Warren et al. (2017)
16	rs8059962	81540592	T	0.42	−0.17	1.30E-09	319,839	CMIP	DBP	Warren et al. (2017)
16	rs7500448	83012185	A	0.75	0.33	1.10E-19	321,958	CDH13	PP	Warren et al. (2017)
17	rs12941318	1430304	T	0.49	−0.27	2.50E-08	299,739	CRK	SBP	Warren et al. (2017)
17	rs7226020	6570508	T	0.56	−0.26	2.30E-14	303,389	KIAA0753	PP	Warren et al. (2017)
17	rs5417	7281743	A	0.57	0.21	1.10E-13	319,299	TP53, SLC2A4	DBP	Warren et al. (2017)
17	rs138643143	42557849	A	0.07	0.5	3.30E-09	229,161	KCNH4, HSD17B1	PP	Warren et al. (2017)
17	rs62080325	43983263	A	0.66	−0.19	4.00E-08	315,689	PYY	PP	Warren et al. (2017)
17	rs12946454	45130754	T	0.28	0.57	1.00E-08	7,7690	PLCD3	SBP	Newton-Cheh et al. (2009a)
17	rs551011992	45851264	G	0.24	−0.33	3.32E-08	275,778	GOSR2	SBP	Warren et al. (2017)
17	rs17608766	46935905	C	0.14	0.55	1.97E-15	137,066	GOSR2	PP	Wain et al. (2017)

17	rs79917357	48747312	A	0.17	0.34	5.33E-09	336,863	LOC105371811 – LOC105371812	SBP	Wain et al. (2017)
17	rs585736	48796910	A	0.03	0.61	2.50E-11	301,845	HOXB7	PP	Warren et al. (2017)
17	rs12940887	49325445	T	0.37	0.26	1.18E-08	144,603	ZNF652	DBP	(Wain et al. (2017)
17	rs16948048	49363104	G	0.39	0.31	8.24E+04	82,441	PHB, ZNF652	DBP	Newton-Cheh et al. (2009a)
17	rs740698	62689790	T	0.56	−0.23	3.10E-12	311,450	MRC2	PP	Warren et al. (2017)
17	rs4459609	63471587	A	0.61	−0.2	2.03E-14	361,581	CYB561, ACE	DBP	Wain et al. (2017)
17	rs4308	63482264	A	0.37	0.21	6.80E-14	319,394	ACE	DBP	Warren et al. (2017)
17	rs2467099	75952964	T	0.22	−0.31	3.30E-08	326,401	ACOX1	SBP	Warren et al. (2017)
17	rs57927100	77321218	G	0.26	−0.31	4.04E−10	347,188	SPET9	SBP	Wain et al. (2017)
18	rs7236548	45517785	A	0.18	0.35	2.00E-18	330,075	SLC14A2	PP	Warren et al. (2017)
18	rs745821	50616484	T	0.76	0.19	1.40E-09	330,954	MAPK4	DBP	Warren et al. (2017)
19	rs36047283	7255690	G	0.11	−0.8	3.25E-26	281,588	INSR	SBP	Wain et al. (2017)
19	rs2116941	10223767	C	0.81	−0.22	2.77E-08	321,960	chr19mb10	PP	Warren et al. (2017)
19	rs77279095	11415506	A	0.04	0.86	7.26E-09	224,215	RGL3	SBP	Warren et al. (2017)
19	rs62104477	29804084	T	0.33	0.18	1.20E-09	320,347	CCNE1	DBP	Warren et al. (2017)
19	rs9710247	40254542	G	0.45	0.16	1.61E-09	308,028	AKT2	DBP	Wain et al. (2017)
20	rs6108168	8645624	A	0.25	−0.21	1.10E-11	327,368	PLCB1	DBP	Warren et al. (2017)
20	rs1327235	10988382	G	0.46	0.3	1.41E-15	158,454	JAG1	DBP	International Consortium for Blood Pressure Genome-Wide Association S et al. (2011)
20	rs6081613	19485263	A	0.28	0.26	1.60E-13	315,546	SLC24A3	PP	Warren et al. (2017)
20	rs80346118	48794612	A	0.15	−0.27	1.10E-12	327,614	PREX1	DBP	Warren et al. (2017)
20	rs6015450	59176062	G	0.12	0.9	3.87E-23	159,190	EDN3, GNAS	SBP	International Consortium for Blood Pressure Genome-Wide Association S et al. (2011)
21	rs12627651	43340723	A	0.29	0.32	2.40E-09	121,169	CRYAA, SIK1	DBP	Wain et al. (2017)
22	rs4819852	20000644	A	0.29	0.26	1.43E-13	367,460	ARVCF	PP	Wain et al. (2017)
22	rs73161324	41642782	T	0.05	0.5	2.80E-11	267,722	XRCC6	PP	Warren et al. (2017)

Chr chromosome, *EA* effect allele, *EAF* effect allele frequency, *SBP* systolic blood pressure, *DBP* diastolic blood pressure, *PP* pulse pressure, *MAP* mean arterial pressure, *HT* hypertension, *NA* no available

Table 1.2b Significant genetic loci for blood pressure and hypertension reported in genome-wide association studies in Asians and Africans

Chr	Strongest SNP	Position	EA	EAF	OR or BETA	*P* value	*N*	Closest gene	Trait	References
Asians										
1	rs880315	10736809	C	0.65	0.56	3.05E-10	32,611	CASZ1	DBP	Kato et al. (2011)
1	rs10745332	112646431	A	0.82	0.96	2.52E-09	46,269	MOV10	SBP	Lu et al. (2015)
2	rs1344653	19531084	A	0.53	−0.27	7.79E-12	220,853	OSR1	PP	Kato et al. (2015)
2	rs1275988	26691496	T	0.53	−0.37	4.95E-21	236,311	KCNK3	OH	Kato et al. (2015)
2	rs7604423	43155602	C	0.66	−0.21	2.40E-08	217,072	NR	DBP	Kato et al. (2015)
2	rs6736587	81628601	C	0.16	−1.38	5.3E-08	NA	CTNNA2	OH	Li et al. (2017)
2	rs16849225	164050310	C	0.61	0.75	3.45E-11	49,511	GRB14, FIGN	SBP	Kato et al. (2011)
3	rs820430	27507409	A	0.32	0.76	1.36E-12	79,318	SLC4A7	SBP	Lu et al. (2015)
3	rs9810888	53601568	G	0.39	0.39	4.0E-12	77,555	CACNA1D	DBP	Lu et al. (2015)
4	rs1902859	80236549	C	0.41	1.34	1.76E-22	45,856	FGF5	SBP	Lu et al. (2015)
4	rs2014912	85794517	T	0.16	0.62	5.37E-17	242,456	ARHGAP24	SBP	Kato et al. (2015)
4	rs6825911	110460482	C	0.51	0.39	8.96E-09	49,511	ENPEP	DBP	Kato et al. (2011)
4	rs13143871	155698052	T	0.80	0.96	5.16E-08	45,737	GUCY1A3	SBP	Lu et al. (2015)
5	rs1173766	32804422	C	0.60	0.63	1.95E-08	49,970	NPR3	SBP	Kato et al. (2011)
5	rs13359291	123140763	A	0.31	0.53	8.88E-16	229,584	PRDM6	SBP	Kato et al. (2015)
5	rs9687065	149011577	A	0.78	0.26	7.36E-11	259,216	ABLIM3, SH3TC2	DBP	Kato et al. (2015)
6	rs2021783	32077074	C	0.79	0.49	2.18E-12	78,911	CYP21A2	DBP	Lu et al. (2015)
6	rs1563788	43340625	T	0.31	0.51	2.22E-16	220,757	ZNF318	SBP	Kato et al. (2015)
6	rs1474698	56199399	T	0.62	−0.20	4.52E-09	274,981	NR	PP	Kato et al. (2015)
7	rs2107595	19009765	A	0.23	0.31	3.91E-11	209,305	HDAC9	PP	Kato et al. (2015)
7	rs10260816	45970501	C	0.61	0.31	1.51E-14	207,070	IGFBP3	PP	Kato et al. (2015)
7	rs17477177	106771412	T	0.77	−0.53	3.69E-12	99,344	PIK3CG	PP	Kato et al. (2015)
10	rs4919669	102712218	A	0.43	−0.65	2.63E-08	NA	ARL3	SBP	Li et al. (2017)
10	rs284844	102794772	A	0.49	−0.75	1.05E-11	NA	WBP1L	SBP	Li et al. (2017)
10	rs4409766	102856906	T	0.71	1.24	6.08E-17	46,030	CYP17A1	SBP	Lu et al. (2015)

10	rs11191548	103086421	T	0.74	1.18	3.94E-17	41,315	CNNM2	SBP	Kato et al. (2011)
10	rs11191580	103146454	T	0.74	0.97	4.44E-15	NA	NT5C2	SBP	Li et al. (2017)
11	rs4757391	16281393	C	0.28	0.88	5.20E-9	46,336	SOX6	SBP	Lu et al. (2015)
11	rs751984	61510774	T	0.79	0.33	7.66E-12	233,082	LRRC10B, SYT7	MAP	Kato et al. (2015)
12	rs12579720	20020830	C	0.32	−0.32	2.2E-16	218,606	PDE3A	DBP	Kato et al. (2015)
12	rs17249754	89666809	G	0.64	1.17	7.72E-20	40,719	ATP2B1	SBP	Kato et al. (2011)
12	rs3184504	111446804	T	0.46	0.54	2.20E-10	35,342	SH2B3	DBP	Kato et al. (2015)
12	rs653178	111569952	T	0.50	−0.55	1.20E-08	35,342	ATXN2	DBP	Kato et al. (2015)
12	rs11066280	112379979	T	0.75	1.01	1.32E-35	46,957	HECTD4	DBP	Kato et al. (2011)
12	rs35444	115114632	A	0.75	0.52	9.62E-08	29,746	TBX3	DBP	Kato et al. (2011)
12	rs11067763	115760536	A	0.62	0.51	2E-18	79,651	MED13L	DBP	Lu et al. (2015)
17	rs2240736	61408032	T	0.66	0.35	2.20E−16	217,197	C17orf82, TBX2	MAP	Kato et al. (2015)
18	rs403814	6282594	A	NA	1.15	6.13E-09	NA	L3MBTL4	HT	Liu et al. (2016)
19	rs740406	2232222	A	0.87	−0.55	3.10E−15	193,219	PLEKHJ1, DOT1L	PP	Kato et al. (2015)
20	rs1887320	10985350	A	0.53	0.78	1.48E-08	46,123	JAG1	SBP	Lu et al. (2015)
Africans										
3	rs1717027	41946428	T	0.64	0.49	4.90E-13	29,322	ULK4	HT	Franceschini et al. (2013)
6	rs13209747	126794309	T	0.19	0.56	2.40E-11	28,708	RSPO3	DBP	Franceschini et al. (2013)
6	rs76987554	133759717	C	0.09	−1.69	2.28E-10	62,297	TARID, TCF21	SBP	Liang et al. (2017)
6	rs17080102	150683634	C	0.1	−0.74	1.90E-11	29,323	PLEKHG1	DBP	Franceschini et al. (2013)
7	rs6969780	27119517	C	0.3	0.46	3.27E-10	85,397	HOXA	DBP	Liang et al. (2017)
7	rs17428471	27298248	T	0.14	1.20	2.00E-12	29,326	HOXA, EVX1	SBP	Franceschini et al. (2013)
7	rs11563582	27312031	A	0.13	0.67	1.71E-11	84,816	EVX1, HOXA	DBP	Liang et al. (2017)
7	rs7801190	100860471	C	0.72	1.31	3.40E-08	10,771	SLC12A9	HT	Lettre et al. (2011)
8	rs7006531	94098516	G	0.85	−1.02	5.96E-12	72,887	CDH17	PP	Liang et al. (2017)
8	rs78192203	141364973	T	0.2	0.84	4.00E-11	62,681	GPR20	DBP	Liang et al. (2017)
9	rs115795127	83378986	T	0.89	0.17	1.13E-08	32,149	FRMD3	HT	Liang et al. (2017)

Chr chromosome, *EA* effect allele, *EAF* effect allele frequency, *SBP* systolic blood pressure, *DBP* diastolic blood pressure, *PP* pulse pressure, *MAP* mean arterial pressure, *HT* hypertension, *OH* orthostatic hypotension, *NA* no available pulse pressure (PP) (Wain et al. 2011). This study discovered four novel loci for PP and two novel loci for MAP. The latest GWAS among UK Biobank participants of European ancestry included ~140,000 subjects from single source at the discovery phase. With dense 1000 Genomes and UK10K imputation, they yielded a data set with ~9.8 million variants for the meta-analysis. With the help of major international consortia for parallel replication, they totally included GWAS data from 330,956 individuals and reported 107 significant loci, among which 24 were associated with SBP, 41 associated with DBP, and 42 with PP (Warren et al. 2017).

The first large meta-analysis of GWAS on BP traits among East Asians was conducted by Asian Genetic Epidemiology Network (AGEN) consortium (Kato et al. 2011). Our group was one of the eight groups participating in stage 1 meta-analysis, which included 19,608 individuals totally. After de novo genotyping in two stages of replication involving 10,518 and 20,247 East Asian samples, we confirmed seven loci previously identified in European population reported by the Global BPgen and CHARGE and additionally identified 6 novel loci as ST7L-CAPZA1, FIGN-GRB14, ENPEP, NPR3, TBX3, and ALDH2. In another meta-analysis of Chinese Han population, which involved 11,816 subjects in the discovery stage and 69,146 for the replication (Lu et al. 2015), they replicated 8 previously reported loci and 4 novel loci, including CASZ1, MOV10, FGF5, CYP17A1, SOX6, ATP2B1, ALDH2, JAG1, CACNA1D, CYP21A2, MED13L, and SLC4A7. These findings suggested the possible presence of allelic heterogeneity in BP regulation between European and Asian and provided new mechanistic insights into hypertension.

African-descent populations are the most ancestral and have smaller regions of linkage disequilibrium due to the accumulation of more recombination events in that group, so that the GWAS performed in Africans need more SNPs with better overall genomic coverage (Bush and Moore 2012). There are fewer loci reaching GWAS significance in African population than European or Asian. The largest GWAS performed in African origin population analyzed 21 GWAS comprised of 31,968 individuals of African ancestry, validated their results with additional 54,395 individuals from multiethnic studies (Liang et al. 2017), and found 9 loci with 11 independent variants which are for either SBP, DBP, hypertension, or combined traits.

2.2 *Coronary Artery Disease (CAD)*

The heritability of CAD has been estimated between 40% and 60% based on family and twin studies (Teslovich et al. 2010). The first robust locus associated with CAD identified by GWAS was published in 2007 (Wellcome Trust Case Control C 2007; Samani et al. 2007; McPherson et al. 2007), when three independent groups reported common variants at the 9p21 locus. It was associated with a ~30% increased risk of CAD per copy of the risk allele. Since then, progressively larger sample size meta-analyses of additional GWAS, mainly in European descent, have identified additional loci of smaller effect size. In 2015, the CARDIoGRAMplusC4D Consortium reported a GWAS meta-analysis of 185,000 CAD cases and controls, interrogating

6.7 million common variants as well as 2.7 million low-frequency variants (Nikpay et al. 2015). In this study, they confirmed most known CAD loci at that time and identified 10 novel loci. The majority of the significant loci with a frequency of greater than 5%, which indicated the genetic of CAD, was largely determined by the cumulative effect of multiple common risks of small effect size and strongly supported the common disease/common variant hypothesis. In 2017, Nelson et al. (2017) further meta-analyzed the results from 10,801 cases in the UK Biobank against 137,914 controls in combination with the CARDIoGRAMplusC4D 1000 Genomes and the MIGen/CARDIoGRAM Exome chip study. They found 12 new signals reaching genome-wide significance. They also reported that 304 independent variants meeting the 5% FDR threshold explained 21.2% of CAD heritability. Such finding highlighted the importance of the FDR approach in expansion of associated variants. The strongest SNPs associated with CAD which had reached genome-wide significance are listed in Tables 1.3a and 1.3b.

Although the majority of GWAS for CAD have been carried in European, there were still important studies performed in Asian and black populations with smaller sample size (Nikpay et al. 2015). Unlike a similar effect size of CAD risk alleles in East Asian and European populations, the effect size was apparently attenuated in South Asian and black populations. The lower level of LD in African genome might lead to the failure of tagging potentially shared casual variants (Marigorta and Navarro 2013).

3 Clinical Implications

It is undeniable that GWAS have achieved considerable success in exploring the genetic architecture of heart disease. However, it is important to point out that the significant variant is often merely tagged. The gene indicated by the GWAS signal would not be considered as the causal variant until the functional mechanism can be found. The ultimate goals of the study of the genetics of CAD or hypertension are to understand the pathophysiology mechanism and subsequently to establish risk prediction method and develop effective therapy strategies (Fig. 1.1).

The effect of the identified susceptibility variants of GWAS is individually small, but their effects are independent and additive, which can be calculated as genetic risk score (GRS) with an unweighted manner by adding risk allele numbers or with a weighted mean (McPherson and Tybjaerg-Hansen 2016). As DNA is stable over the course of the lifetime, genetic risk can be ascertained from birth; therefore, GRS may be particularly useful in risk prediction among younger patients in whom the cumulative impact of lifestyle factors is less pronounced.

The interpretation of GWAS often is complicated by the nature of significant loci, which is mostly located in the intergenic region. The bioinformatics approaches could narrow the list prioritizing a gene for subsequent functional study, including expression quantitative trait locus (Consortium 2011), genome-wide chromatin profiles of histone modifications, data on transcription factor-binding sites, chroma-

Table 1.3a Significant loci for coronary artery disease reported by genome-wide association studies in Europeans

Chr	Strongest SNP	Position	EA	EAF	OR or BETA	*P* value	*N*	Closestgene	References
1	rs11206510	55030366	T	0.85	1.08	2.34E-08	188,372	PCSK9	Nikpay et al. (2015)
1	rs17114036	56497149	A	0.91	1.17	3.81E-19	133,226	PPAP2B	Schunkert et al. (2011)
1	rs629301	109275684	NA	NA	NA	1.00E-49	NA	CELSR2	He et al. (2016)
1	rs11810571	151789832	G	0.79	1.06	4.34E-08	345,035	TDRKH	Nelson et al. (2017)
1	rs6689306	154423470	A	0.44	1.05	1.50E-09	NA	IL6R	Nelson et al. (2017)
2	rs16986953	19742712	A	0.07	1.11	4.80E-10	NA	AK097927	Nelson et al. (2017)
2	rs7568458	85561052	A	0.45	1.06	2.40E-13	NA	VAMP5, VAMP8, GGCX	Nelson et al. (2017)
2	rs17678683	144528992	G	0.09	1.10	3.00E-09	179,585	ZEB2, ACO74093.1	Nikpay et al. (2015)
2	rs1250229	215439661	T	0.26	1.07	2.88E-13	270,189	FN1	Nelson et al. (2017)
3	rs7623687	49411133	A	0.86	1.08	3.54E-10	333,836	RHOA	Nelson et al. (2017)
3	rs142695226	124756354	G	0.14	1.07	1.58E-09	345,188	UMPS, ITGB5	Nelson et al. (2017)
3	rs9818870	138403280	T	0.15	1.15	7.44E-13	40,773	MRAS	Erdmann et al. (2009)
3	rs12493885	154122077	C	0.87	1.07	3.23E-08	251,805	ARHGEF26	Nelson et al. (2017)
4	rs17087335	56972417	T	0.21	1.06	4.60E-08	NA	NOA1, REST	Nikpay et al. (2015)
4	rs10857147	80259918	T	0.27	1.05	5.81E-09	345,133	PRDM8, FGF5	Nelson et al. (2017)
4	rs7678555	119988346	C	0.28	1.05	1.35E-08	345,126	PDE5A, MAD2L1	Nelson et al. (2017)
4	rs6841581	147480038	A	0.15	1.07	4.60E-10	NA	EDNAA	Nelson et al. (2017)
6	rs6909752	22612400	A	0.33	1.05	2.26E-09	343,171	HDGFL1	Nelson et al. (2017)
6	rs17609940	35067023	G	0.75	1.07	1.36E-08	137,412	ANKS1A	Schunkert et al. (2011)
6	rs56336142	39166323	T	0.81	1.07	1.85E-08	188,349	KCNK5	Nikpay et al. (2015)
6	rs12190287	133893387	C	0.62	1.08	1.70E-12	130,888	TCF21	Schunkert et al. (2011)

(continued)

Table 1.3a (continued)

Chr	Strongest SNP	Position	EA	EAF	OR or BETA	*P* value	*N*	Closestgene	References
6	rs10455872	160589086	G	0.07	1.31	1.70E-49	NA	LPA	Nelson et al. (2017)
6	rs4252185	160702419	C	0.06	1.34	1.64E-32	NA	PLG	Nikpay et al. (2015)
7	rs2107595	19009765	A	0.18	1.08	3.40E-13	NA	HDAC9	Nelson et al. (2017)
7	rs11556924	130023656	C	0.62	1.09	9.18E-18	134,200	ZC3HC1	Schunkert et al. (2011)
7	rs3918226	150993088	T	0.07	1.13	1.60E-12	NA	NOS3	Nelson et al. (2017)
8	rs2954029	125478730	A	0.54	1.06	5.20E-13	NA	TRIB1	Nelson et al. (2017)
9	rs579459	133278724	C	0.21	1.10	4.08E-14	123,978	ABO	Schunkert et al. (2011)
10	rs2505083	30046193	C	0.39	1.06	6.85E-09	169,421	KIAA1462	Nikpay et al. (2015)
10	rs501120	44258419	T	0.87	1.33	9.46E-08	7,181	CXCL12	Samani et al. (2007)
10	rs2246942	89245129	G	0.35	1.08	3.50E-16	NA	LIPA	Nelson et al. (2017)
10	rs12413409	102959339	G	0.89	1.12	1.03E-09	129,741	NT5C2, CNNM2, CYP17A1	Schunkert et al. (2011)
11	rs10840293	9729649	A	0.55	1.05		NA	SWAP70	Nelson et al. (2017)
11	rs3993105	13281524	T	0.68	1.05	4.88E-08	340,699	ARNTL	Nelson et al. (2017)
11	rs974819	103789839	T	0.32	1.07	2.41E-09	64,881	PDGFD	Coronary Artery Disease Genetics C (2011)
11	rs964184	116778201	NA	NA	NA	2.00E-108	NA	ZPR1	He et al. (2016)
12	rs3184504	111446804	T	0.42	1.07	1.03E-09	NA	SH2B3	Nikpay et al. (2015)
12	rs11830157	117827636	G	0.36	1.12	2.12E-09	NA	KSR2	Nikpay et al. (2015)
12	rs2244608	120979185	G	0.35	1.05	7.90E-10	345,106	HFN1A	Nelson et al. (2017)
13	rs1924981	28448508	T	0.33	1.05	1.90E-07	NA	FLT1	Nelson et al. (2017)
14	rs2895811	99667605	C	0.43	1.07	1.14E-09	114,238	HHIPL1	Schunkert et al. (2011)

(continued)

Table 1.3a (continued)

Chr	Strongest SNP	Position	EA	EAF	OR or BETA	*P* value	*N*	Closestgene	References
15	rs56062135	67163292	C	0.79	1.07	4.52E-09	184,037	SMAD3	Nikpay et al. (2015)
15	rs7164479	78830712	T	0.58	1.07	6.40E-18	NA	ADAMTS7	Nelson et al. (2017)
15	rs8042271	89030987	G	0.90	1.10	3.68E-08	162,206	MFGE8, ABHD2	Nikpay et al. (2015)
15	rs17514846	90873320	A	0.44	1.05	3.10E-07	NA	FURIN, FES	Nikpay et al. (2015)
16	rs7500448	83012185	A	0.77	1.06	4.90E-10	337,399	CDH13	Nelson et al. (2017)
17	rs216172	2223210	C	0.37	1.07	1.15E-09	111,538	SMG6, SRR	Schunkert et al. (2011)
17	rs46522	48911235	T	0.53	1.06	1.81E-08	137,633	UBE2Z	Schunkert et al. (2011)
17	rs7212798	60936127	C	0.15	1.08	1.88E-08	188,377	BCAS3	Nikpay et al. (2015)
18	rs8082812	8522684	NA	NA	NA	5.00E-67	NA	THEMIS3P – AKR1B1P6	He et al. (2016)
18	rs663129	60171168	A	0.26	1.06	3.20E-08	NA	PMAIP1, MC4R	Nikpay et al. (2015)
19	rs6511720	11091630	G	0.88	1.14	7.90E-22	NA	LDLR	Nelson et al. (2017)
19	rs12976411	32391114	A	0.09	1.49	1.18E-14	NA	ZNF507, LOC400684	Nikpay et al. (2015)
19	rs8108632	41348629	T	0.48	1.05	4.13E-08	345,058	TGFB1	Nelson et al. (2017)
19	rs4420638	44919689	NA	NA	NA	1.00E-22	NA	APOC1 – APOC1P1	He et al. (2016)
19	rs1964272	45687010	G	0.52	1.05	2.52E-08	343,548	SNAPD2	Nelson et al. (2017)
21	rs28451064	34221526	A	0.12	1.14	2.60E-23	NA	KCNE2	Nelson et al. (2017)
21	rs460976	41463567	NA	NA	NA	1.00E-08	NA	MX1 – TMPRSS2	He et al. (2016)
22	rs180803	24262890	G	0.97	1.20	1.64E-10	174,076	POM121L9P, ADORA2A	Nikpay et al. (2015)

tin immune-precipitation sequencing (Park 2009), and others. Below we describe some examples of clinical implication of GWAS results in risk prediction and mechanism interpretation.

Table 1.3b Significant loci for coronary artery disease reported by genome-wide association studies in Asians

Chr	Strongest SNP	Position	EA	EAF	OR or BETA	*P* value	*N*	Closest gene	References
2	rs2123536	19745816	T	0.39	1.12	6.83E-11	33,466	TTC32, WDR35	Lu et al. (2012)
4	rs1842896	155590307	T	0.76	1.14	1.26E-11	33,466	GUCY1A3	Lu et al. (2012)
6	rs9349379	12903725	G	0.349	1.34	8.02E-10	4,496	PHACTR1	Hager et al. (2012)
6	rs9268402	32373576	G	0.59	1.16	2.77E-15	33,466	BTNL2, C6orf10	Lu et al. (2012)
6	rs12524865	133875536	C	0.61	1.11	1.87E-07	24,724	TCF21	Lu et al. (2012)
9	rs10738607	22088095	G	0.64	0.78	2.37E-08	1,237	CDKN2A, CDKN2B	Wakil et al. (2016)
12	rs7136259	89687411	T	0.39	1.11	5.68E-10	33,466	ATP2B1	Lu et al. (2012)
12	rs3782889	110912851	C	0.21	1.26	3.95E-14	13,742	MYL2	Lee et al. (2013)
12	rs671	111803962	A	0.23	1.43	1.6E-34	12,041	ALDH2	Takeuchi et al. (2012)
12	rs11066280	112379979	A	0.17	1.19	1.69E-11	24,741	C12orf51	Lu et al. (2012)

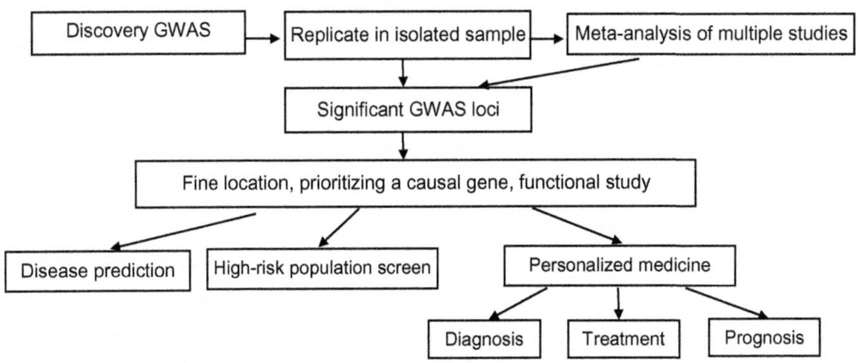

Fig. 1.1 Clinical implication from genome-wide association studies (GWAS)

3.1 Hypertension

With the expansion of significant loci of BP in GWAS, robust association for some biological candidate genes previously suspected to influence BP was detected. The latest GWAS by Warren et al. (2017) identified multiple loci involved in BP regulation, including angiotensin-converting enzyme, voltage-dependent calcium channel auxiliary subunit, metallo-endopeptidase/neutral endopeptidase, adrenergic β 2B receptor, and phosphodiesterase 5a. Moreover, in the pathway analysis, they also showed an enrichment of pathways associated with cardiovascular disease, including α-adrenergic pathway, CXCR4 chemokine signaling pathway, endothelin system, and angiotensin-receptor pathways.

In ICBP GWAS (Wain et al. 2011), the SBP and DBP differences between the top and bottom quintiles of GRS were 4.6 and 3.0 mmHg, respectively. Furthermore, the GRS generated by combination of 107 loci by Warren et al. (2017) showed that the group with lowest quintile GRS had a SBP approximately 9–10 mmHg lower than those with highest quintile GRS. Reductions in BP of such a magnitude might lead to significantly lower cardiovascular morbidity and mortality in hypertensive patients. GRS could therefore be useful in early life to assess the risk of hypertension and direct dietary or lifestyle modifications.

The GWAS findings had also provided clues for personalized prevention and treatment of hypertension. One good example was the uromodulin (UMOD) gene, which was identified in an extreme case-control design (Padmanabhan et al. 2010). The rs13333226-G in UMOD showed association with the lower risk of hypertension and reduced urinary UMOD excretion. Uromodulin is mainly expressed in the thick portion of the ascending limb of Henle, which indicated that it may participate in the BP regulation through effect on sodium hemostasis. Trudu et al. (2013) modeled the effect of UMOD in transgenic mice and demonstrated that uromodulin influenced BP through activation of the renal sodium cotransporter NKCC2. The subjects with UMOD risk variant increased UMOD excretion, greater salt sensitivity, hypertension, and greater BP response to loop diuretics. This finding presented an opportunity for hypertension precision medicine and new drug development.

The study of AGEN (Kato et al. 2011) found ethnicity-specific locus on 12q24.13, where acetaldehyde dehydrogenase (ALDH2) gene is located. ALDH2 was a key enzyme in the major pathway of alcohol metabolism, and glutamate-lysine substitution (rs671, E504K) produced an inactive subunit of ALDH2, resulting in an inability to metabolize acetaldehyde and a subsequent accumulation of acetaldehyde after alcohol intake (Takeshita et al. 1993). The rs671 was not polymorphic in Europeans, but it was close to rs3184504 at the SH2B3 locus, which was significantly associated with BP in European and had no polymorphism in East Asian. Such phenomenon indicated the undergone natural selection. Furthermore, rs671 displayed pleiotropic effects both on risk factors of cardiovascular disease and on CAD susceptibility. Interestingly, the locus showed deleterious effect on BP but protective effects on HDL cholesterol and LDL cholesterol, which resulted in a net reduction in CAD risk. Moreover, most of the associations between rs671 and each of the cardiovascular risk factors were influenced by the alcohol intake.

3.2 CAD

Since 2007, nearly 70 distinct genetic loci for CAD have been yielded depending on the progressively larger sample sizes. A minority of all such risk variants appears to modulate CAD risk by influencing classical risk factors like plasma lipids, diabetes, and hypertension, reinforcing key roles for these pathways in the development of CAD. The rest of the risk variants identified by GWAS are located in regulatory regions of unclear function.

In a large prospective cohort study with a median of 10.7 years of follow-up, Ripatti et al. (2010) found that individuals with a GRS in the top quintile derived from 13 multi-locus SNPs of CAD exhibited a 1.66fold increased risk adjusting for traditional risk factors (TRFs). However, the GRS did not improve the C index over the TRFs and family history or the net reclassification of risk categories. A recent survey on 55,685 subjects from three prospective studies and one cross-sectional study confirmed the association between GRS and incidence of CAD (Khera et al. 2016), with subjects in the top GRS quintiles having a 91% higher relative risk as compared to those in the bottom quintiles. They also reported that a favorable lifestyle was associated with a nearly 50% lower relative risk of coronary artery disease than was an unfavorable lifestyle.

9p21.3 locus was the first one identified by GWAS, which consisted of a cluster of 59 linked SNPs over 53,000 bp region (Chen et al. 2014). Among the CAD risk alleles, rs10811656 and rs10757278 were located in an enhancer element and disrupted a binding site for ATAT1 (Harismendy et al. 2011). This enhancer locus interacted with CDKN2A/CDKN2AB and IFNA21, which encoded interferon in human vascular endothelial cells and participated in the inflammatory process. It also suggested that this region probably encodes for a long noncoding RNA designated as ANRIL (antisense noncoding RNA in the INK4 locus). ANRIL influenced the expression of CDKN2A/CDKN2AB that was involved in regulating the cell cycle and cellular proliferation (Holdt et al. 2013). The meta-analysis also confirmed that 9p21.3 locus only increased the burden of atherosclerosis but not the risk of myocardial infarction, which indicated the stimulation effect of this locus in the atherosclerosis (Chan et al. 2013).

1p13 locus had been independently associated with CAD by several GWAS. Among a number of genes located in this region, SORT1 was found as the target one by eQTL and expression studies. SORT1 encoded sortilin, which was a multi-ligand type 1 receptor and expresses in many cell and tissue types (Petersen et al. 1997). The overexpression of Sort 1 might decrease VLDL secretion rate and increase plasma LDL turnover (Kathiresan et al. 2008). However, the knockout or knockdown of Sortl in different studies resulted in either increased or decreased VLDL secretion (Kjolby et al. 2010; Musunuru et al. 2010). Whether Sortl was the causal gene to CAD or just a signal found by GWAS was still needed to be clarified. The concrete molecular mechanism by which sortilin influenced the lipid metabolism and CAD risk had not been elucidated.

4 Perspectives

The GWAS findings have identified novel biological signals involved in human heart diseases. However, there are still certain challenges. The heritability of BP from family studies varies from 30% to 50%, but the collective effect of all BP loci identified through GWAS only explains ≈3.5% of BP heritability. The missing heritability is not unique to BP genetics but is universally observed in almost all common heart diseases. Heritability reflects the degree of phenotypic resemblance between relatives, which not only depends on genetic architecture contributing to the trait but also is the result of environmental factors and interactions within the genome. In GWAS for hypertension-related phenotypes, significant gene and environment interaction has been identified with alcohol consumption (Simino et al. 2013), body mass index, smoking (Sung et al. 2015), educational level, and other modifiable lifestyle factors. These findings may help identify the causal genetic loci that contribute to missing heritability. In addition, the gene-gene interaction (Slavin et al. 2011), rare variants (Lu et al. 2017), and epigenetic mechanisms will likely carry an unknown fraction of complex heart disease. Future research will focus on the application of variants identified in GWAS in the identification of individuals at risk of disease, guidance of clinical management decisions, and prediction of prognosis.

References

Arking DE, Pfeufer A, Post W, Kao WH, Newton-Cheh C, Ikeda M, West K, Kashuk C, Akyol M, Perz S, et al. A common genetic variant in the NOS1 regulator NOS1AP modulates cardiac repolarization. Nat Genet. 2006;38(6):644–51.

Arking DE, Pulit SL, Crotti L, van der Harst P, Munroe PB, Koopmann TT, Sotoodehnia N, Rossin EJ, Morley M, Wang X, et al. Genetic association study of QT interval highlights role for calcium signaling pathways in myocardial repolarization. Nat Genet. 2014;46(8):826–36.

Bush WS, Moore JH. Chapter 11: genome-wide association studies. PLoS Comput Biol. 2012;8(12):e1002822.

Chambers JC, Zhao J, Terracciano CM, Bezzina CR, Zhang W, Kaba R, Navaratnarajah M, Lotlikar A, Sehmi JS, Kooner MK, et al. Genetic variation in SCN10A influences cardiac conduction. Nat Genet. 2010;42(2):149–52.

Chan K, Patel RS, Newcombe P, Nelson CP, Qasim A, Epstein SE, Burnett S, Vaccarino VL, Zafari AM, Shah SH, et al. Association between the chromosome 9p21 locus and angiographic coronary artery disease burden: a collaborative meta-analysis. J Am Coll Cardiol. 2013;61(9):957–70.

Chen HH, Almontashiri NA, Antoine D, Stewart AF. Functional genomics of the 9p21.3 locus for atherosclerosis: clarity or confusion? Curr Cardiol Rep. 2014;16(7):502.

Consortium EP. A user's guide to the encyclopedia of DNA elements (ENCODE). PLoS Biol. 2011;9(4):e1001046.

Coronary Artery Disease Genetics C. A genome-wide association study in Europeans and south Asians identifies five new loci for coronary artery disease. Nat Genet. 2011;43(4):339–44.

Diabetes Genetics Initiative of Broad Institute of H, Mit LU, Novartis Institutes of BioMedical R, Saxena R, Voight BF, Lyssenko V, Burtt NP, de Bakker PI, Chen H, Roix JJ, et al.

Genome-wide association analysis identifies loci for type 2 diabetes and triglyceride levels. Science. 2007;316(5829):1331–6.

Ellinor PT, Lunetta KL, Albert CM, Glazer NL, Ritchie MD, Smith AV, Arking DE, Muller-Nurasyid M, Krijthe BP, Lubitz SA, et al. Meta-analysis identifies six new susceptibility loci for atrial fibrillation. Nat Genet. 2012;44(6):670–5.

Ellinor PT, Lunetta KL, Glazer NL, Pfeufer A, Alonso A, Chung MK, Sinner MF, de Bakker PI, Mueller M, Lubitz SA, et al. Common variants in KCNN3 are associated with lone atrial fibrillation. Nat Genet. 2010;42(3):240–4.

Erdmann J, Grosshennig A, Braund PS, Konig IR, Hengstenberg C, Hall AS, Linsel-Nitschke P, Kathiresan S, Wright B, Tregouet DA, et al. New susceptibility locus for coronary artery disease on chromosome 3q22.3. Nat Genet. 2009;41(3):280–2.

Franceschini N, Fox E, Zhang Z, Edwards TL, Nalls MA, Sung YJ, Tayo BO, Sun YV, Gottesman O, Adeyemo A, et al. Genome-wide association analysis of blood-pressure traits in African-ancestry individuals reveals common associated genes in African and non-African populations. Am J Hum Genet. 2013;93(3):545–54.

Gudbjartsson DF, Holm H, Gretarsdottir S, Thorleifsson G, Walters GB, Thorgeirsson G, Gulcher J, Mathiesen EB, Njolstad I, Nyrnes A, et al. A sequence variant in ZFHX3 on 16q22 associates with atrial fibrillation and ischemic stroke. Nat Genet. 2009;41(8):876–8.

Hager J, Kamatani Y, Cazier JB, Youhanna S, Ghassibe-Sabbagh M, Platt DE, Abchee AB, Romanos J, Khazen G, Othman R, et al. Genome-wide association study in a Lebanese cohort confirms PHACTR1 as a major determinant of coronary artery stenosis. PLoS One. 2012;7(6):e38663.

Harismendy O, Notani D, Song X, Rahim NG, Tanasa B, Heintzman N, Ren B, Fu XD, Topol EJ, Rosenfeld MG, et al. 9p21 DNA variants associated with coronary artery disease impair interferon-gamma signalling response. Nature. 2011;470(7333):264–8.

He L, Kernogitski Y, Kulminskaya I, Loika Y, Arbeev KG, Loiko E, Bagley O, Duan M, Yashkin A, Ukraintseva SV, et al. Pleiotropic meta-analyses of longitudinal studies discover novel genetic variants associated with age-related diseases. Front Genet. 2016;7:179.

Helgadottir A, Thorleifsson G, Manolescu A, Gretarsdottir S, Blondal T, Jonasdottir A, Jonasdottir A, Sigurdsson A, Baker A, Palsson A, et al. A common variant on chromosome 9p21 affects the risk of myocardial infarction. Science. 2007;316(5830):1491–3.

Holdt LM, Hoffmann S, Sass K, Langenberger D, Scholz M, Krohn K, Finstermeier K, Stahringer A, Wilfert W, Beutner F, et al. Alu elements in ANRIL non-coding RNA at chromosome 9p21 modulate atherogenic cell functions through trans-regulation of gene networks. PLoS Genet. 2013;9(7):e1003588.

International Consortium for Blood Pressure Genome-Wide Association S, Ehret GB, Munroe PB, Rice KM, Bochud M, Johnson AD, Chasman DI, Smith AV, Tobin MD, Verwoert GC, et al. Genetic variants in novel pathways influence blood pressure and cardiovascular disease risk. Nature. 2011;478(7367):103–9.

Kathiresan S, Melander O, Guiducci C, Surti A, Burtt NP, Rieder MJ, Cooper GM, Roos C, Voight BF, Havulinna AS, et al. Six new loci associated with blood low-density lipoprotein cholesterol, high-density lipoprotein cholesterol or triglycerides in humans. Nat Genet. 2008;40(2):189–97.

Kato N, Loh M, Takeuchi F, Verweij N, Wang X, Zhang W, Kelly TN, Saleheen D, Lehne B, Leach IM, et al. Trans-ancestry genome-wide association study identifies 12 genetic loci influencing blood pressure and implicates a role for DNA methylation. Nat Genet. 2015;47(11):1282–93.

Kato N, Takeuchi F, Tabara Y, Kelly TN, Go MJ, Sim X, Tay WT, Chen CH, Zhang Y, Yamamoto K, et al. Meta-analysis of genome-wide association studies identifies common variants associated with blood pressure variation in east Asians. Nat Genet. 2011;43(6):531–8.

Khera AV, Emdin CA, Drake I, Natarajan P, Bick AG, Cook NR, Chasman DI, Baber U, Mehran R, Rader DJ, et al. Genetic risk, adherence to a healthy lifestyle, and coronary disease. N Engl J Med. 2016;375(24):2349–58.

Kim YJ, Go MJ, Hu C, Hong CB, Kim YK, Lee JY, Hwang JY, Oh JH, Kim DJ, Kim NH, et al. Large-scale genome-wide association studies in East Asians identify new genetic loci influencing metabolic traits. Nat Genet. 2011;43(10):990–5.

Kjolby M, Andersen OM, Breiderhoff T, Fjorback AW, Pedersen KM, Madsen P, Jansen P, Heeren J, Willnow TE, Nykjaer A. Sort1, encoded by the cardiovascular risk locus 1p13.3, is a regulator of hepatic lipoprotein export. Cell Metab. 2010;12(3):213–23.

Ko A, Cantor RM, Weissglas-Volkov D, Nikkola E, Reddy PM, Sinsheimer JS, Pasaniuc B, Brown R, Alvarez M, Rodriguez A, et al. Amerindian-specific regions under positive selection harbour new lipid variants in Latinos. Nat Commun. 2014;5:3983.

Kooner JS, Chambers JC, Aguilar-Salinas CA, Hinds DA, Hyde CL, Warnes GR, Gomez Perez FJ, Frazer KA, Elliott P, Scott J, et al. Genome-wide scan identifies variation in MLXIPL associated with plasma triglycerides. Nat Genet. 2008;40(2):149–51.

Lee JY, Lee BS, Shin DJ, Woo Park K, Shin YA, Joong Kim K, Heo L, Young Lee J, Kyoung Kim Y, Jin Kim Y, et al. A genome-wide association study of a coronary artery disease risk variant. J Hum Genet. 2013;58(3):120–6.

Lettre G, Palmer CD, Young T, Ejebe KG, Allayee H, Benjamin EJ, Bennett F, Bowden DW, Chakravarti A, Dreisbach A, et al. Genome-wide association study of coronary heart disease and its risk factors in 8,090 African Americans: the NHLBI CARe Project. PLoS Genet. 2011;7(2):e1001300.

Levy D, Ehret GB, Rice K, Verwoert GC, Launer LJ, Dehghan A, Glazer NL, Morrison AC, Johnson AD, Aspelund T, et al. Genome-wide association study of blood pressure and hypertension. Nat Genet. 2009;41(6):677–87.

Levy D, Larson MG, Benjamin EJ, Newton-Cheh C, Wang TJ, Hwang SJ, Vasan RS, Mitchell GF. Framingham Heart Study 100K Project: genome-wide associations for blood pressure and arterial stiffness. BMC Med Genet. 2007;8(Suppl 1):S3.

Li C, Kim YK, Dorajoo R, Li H, Lee IT, Cheng CY, He M, Sheu WH, Guo X, Ganesh SK, et al. Genome-wide association study meta-analysis of long-term average blood pressure in east Asians. Circ Cardiovasc Genet. 2017;10(2):e001527.

Liang J, Le TH, Edwards DRV, Tayo BO, Gaulton KJ, Smith JA, Lu Y, Jensen RA, Chen G, Yanek LR, et al. Single-trait and multi-trait genome-wide association analyses identify novel loci for blood pressure in African-ancestry populations. PLoS Genet. 2017;13(5):e1006728.

Liu X, Hu C, Bao M, Li J, Liu X, Tan X, Zhou Y, Chen Y, Wu S, Chen S, et al. Genome wide association study identifies L3MBTL4 as a novel susceptibility gene for hypertension. Sci Rep. 2016;6:30811.

Low SK, Takahashi A, Ebana Y, Ozaki K, Christophersen IE, Ellinor PT, Consortium AF, Ogishima S, Yamamoto M, Satoh M, et al. Identification of six new genetic loci associated with atrial fibrillation in the Japanese population. Nat Genet. 2017;49(6):953–8.

Lu X, Peloso GM, Liu DJ, Wu Y, Zhang H, Zhou W, Li J, Tang CS, Dorajoo R, Li H, et al. Exome chip meta-analysis identifies novel loci and East Asian-specific coding variants that contribute to lipid levels and coronary artery disease. Nat Genet. 2017;49:1722.

Lu X, Wang L, Chen S, He L, Yang X, Shi Y, Cheng J, Zhang L, Gu CC, Huang J, et al. Genome-wide association study in Han Chinese identifies four new susceptibility loci for coronary artery disease. Nat Genet. 2012;44(8):890–4.

Lu X, Wang L, Lin X, Huang J, Charles Gu C, He M, Shen H, He J, Zhu J, Li H, et al. Genome-wide association study in Chinese identifies novel loci for blood pressure and hypertension. Hum Mol Genet. 2015;24(3):865–74.

Marigorta UM, Navarro A. High trans-ethnic replicability of GWAS results implies common causal variants. PLoS Genet. 2013;9(6):e1003566.

McPherson R, Pertsemlidis A, Kavaslar N, Stewart A, Roberts R, Cox DR, Hinds DA, Pennacchio LA, Tybjaerg-Hansen A, Folsom AR, et al. A common allele on chromosome 9 associated with coronary heart disease. Science. 2007;316(5830):1488–91.

McPherson R, Tybjaerg-Hansen A. Genetics of coronary artery disease. Circ Res. 2016;118(4):564–78.

Miall WE, Oldham PD. The hereditary factor in arterial blood-pressure. Br Med J. 1963;1(5323):75–80.

Musunuru K, Strong A, Frank-Kamenetsky M, Lee NE, Ahfeldt T, Sachs KV, Li X, Li H, Kuperwasser N, Ruda VM, et al. From noncoding variant to phenotype via SORT1 at the 1p13 cholesterol locus. Nature. 2010;466(7307):714–9.

den Hoed M, Eijgelsheim M, Esko T, Brundel BJ, Peal DS, Evans DM, Nolte IM, Segre AV, Holm H, Handsaker RE, et al. Identification of heart rate-associated loci and their effects on cardiac conduction and rhythm disorders. Nat Genet. 2013;45(6):621–31.

Nagy R, Boutin TS, Marten J, Huffman JE, Kerr SM, Campbell A, Evenden L, Gibson J, Amador C, Howard DM, et al. Exploration of haplotype research consortium imputation for genome-wide association studies in 20,032 Generation Scotland participants. Genome Med. 2017;9(1):23.

Nelson CP, Goel A, Butterworth AS, Kanoni S, Webb TR, Marouli E, Zeng L, Ntalla I, Lai FY, Hopewell JC, et al. Association analyses based on false discovery rate implicate new loci for coronary artery disease. Nat Genet. 2017;49(9):1385–91.

Newton-Cheh C, Eijgelsheim M, Rice KM, de Bakker PI, Yin X, Estrada K, Bis JC, Marciante K, Rivadeneira F, Noseworthy PA, et al. Common variants at ten loci influence QT interval duration in the QTGEN study. Nat Genet. 2009;41(4):399–406.

Newton-Cheh C, Johnson T, Gateva V, Tobin MD, Bochud M, Coin L, Najjar SS, Zhao JH, Heath SC, Eyheramendy S, et al. Genome-wide association study identifies eight loci associated with blood pressure. Nat Genet. 2009;41(6):666–76.

Nikpay M, Goel A, Won HH, Hall LM, Willenborg C, Kanoni S, Saleheen D, Kyriakou T, Nelson CP, Hopewell JC, et al. A comprehensive 1,000 Genomes-based genome-wide association meta-analysis of coronary artery disease. Nat Genet. 2015;47(10):1121–30.

Nolte IM, Munoz ML, Tragante V, Amare AT, Jansen R, Vaez A, von der Heyde B, Avery CL, Bis JC, Dierckx B, et al. Genetic loci associated with heart rate variability and their effects on cardiac disease risk. Nat Commun. 2017;8:15805.

Padmanabhan S, Melander O, Johnson T, Di Blasio AM, Lee WK, Gentilini D, Hastie CE, Menni C, Monti MC, Delles C, et al. Genome-wide association study of blood pressure extremes identifies variant near UMOD associated with hypertension. PLoS Genet. 2010;6(10):e1001177.

Park PJ. ChIP-seq: advantages and challenges of a maturing technology. Nat Rev Genet. 2009;10(10):669–80.

Parmar PG, Taal HR, Timpson NJ, Thiering E, Lehtimaki T, Marinelli M, Lind PA, Howe LD, Verwoert G, Aalto V, et al. International genome-wide association study consortium identifies novel loci associated with blood pressure in children and adolescents. Circ Cardiovasc Genet. 2016;9(3):266–78.

Pearson TA, Manolio TA. How to interpret a genome-wide association study. JAMA. 2008;299(11):1335–44.

Pe'er I, Yelensky R, Altshuler D, Daly MJ. Estimation of the multiple testing burden for genome-wide association studies of nearly all common variants. Genet Epidemiol. 2008;32(4):381–5.

Petersen CM, Nielsen MS, Nykjaer A, Jacobsen L, Tommerup N, Rasmussen HH, Roigaard H, Gliemann J, Madsen P, Moestrup SK. Molecular identification of a novel candidate sorting receptor purified from human brain by receptor-associated protein affinity chromatography. J Biol Chem. 1997;272(6):3599–605.

Pfeufer A, van Noord C, Marciante KD, Arking DE, Larson MG, Smith AV, Tarasov KV, Muller M, Sotoodehnia N, Sinner MF, et al. Genome-wide association study of PR interval. Nat Genet. 2010;42(2):153–9.

Reilly MP, Li M, He J, Ferguson JF, Stylianou IM, Mehta NN, Burnett MS, Devaney JM, Knouff CW, Thompson JR, et al. Identification of ADAMTS7 as a novel locus for coronary atherosclerosis and association of ABO with myocardial infarction in the presence of coronary atherosclerosis: two genome-wide association studies. Lancet. 2011;377(9763):383–92.

Ripatti S, Tikkanen E, Orho-Melander M, Havulinna AS, Silander K, Sharma A, Guiducci C, Perola M, Jula A, Sinisalo J, et al. A multilocus genetic risk score for coronary heart disease: case-control and prospective cohort analyses. Lancet. 2010;376(9750):1393–400.

Samani NJ, Erdmann J, Hall AS, Hengstenberg C, Mangino M, Mayer B, Dixon RJ, Meitinger T, Braund P, Wichmann HE, et al. Genomewide association analysis of coronary artery disease. N Engl J Med. 2007;357(5):443–53.

Sandhu MS, Waterworth DM, Debenham SL, Wheeler E, Papadakis K, Zhao JH, Song K, Yuan X, Johnson T, Ashford S, et al. LDL-cholesterol concentrations: a genome-wide association study. Lancet. 2008;371(9611):483–91.

Schunkert H, Konig IR, Kathiresan S, Reilly MP, Assimes TL, Holm H, Preuss M, Stewart AF, Barbalic M, Gieger C, et al. Large-scale association analysis identifies 13 new susceptibility loci for coronary artery disease. Nat Genet. 2011;43(4):333–8.

Simino J, Sung YJ, Kume R, Schwander K, Rao DC. Gene-alcohol interactions identify several novel blood pressure loci including a promising locus near SLC16A9. Front Genet. 2013;4:277.

Slavin TP, Feng T, Schnell A, Zhu X, Elston RC. Two-marker association tests yield new disease associations for coronary artery disease and hypertension. Hum Genet. 2011;130(6):725–33.

Smith NL, Felix JF, Morrison AC, Demissie S, Glazer NL, Loehr LR, Cupples LA, Dehghan A, Lumley T, Rosamond WD, et al. Association of genome-wide variation with the risk of incident heart failure in adults of European and African ancestry: a prospective meta-analysis from the cohorts for heart and aging research in genomic epidemiology (CHARGE) consortium. Circ Cardiovasc Genet. 2010;3(3):256–66.

Sotoodehnia N, Isaacs A, de Bakker PI, Dorr M, Newton-Cheh C, Nolte IM, van der Harst P, Muller M, Eijgelsheim M, Alonso A, et al. Common variants in 22 loci are associated with QRS duration and cardiac ventricular conduction. Nat Genet. 2010;42(12):1068–76.

Sung YJ, de Las Fuentes L, Schwander KL, Simino J, Rao DC. Gene-smoking interactions identify several novel blood pressure loci in the Framingham Heart Study. Am J Hypertens. 2015;28(3):343–54.

Surakka I, Horikoshi M, Magi R, Sarin AP, Mahajan A, Lagou V, Marullo L, Ferreira T, Miraglio B, Timonen S, et al. The impact of low-frequency and rare variants on lipid levels. Nat Genet. 2015;47(6):589–97.

Takeshita T, Morimoto K, Mao XQ, Hashimoto T, Furuyama J. Phenotypic differences in low km aldehyde dehydrogenase in Japanese workers. Lancet. 1993;341(8848):837–8.

Takeuchi F, Yokota M, Yamamoto K, Nakashima E, Katsuya T, Asano H, Isono M, Nabika T, Sugiyama T, Fujioka A, et al. Genome-wide association study of coronary artery disease in the Japanese. Eur J Hum Genet. 2012;20(3):333–40.

Teslovich TM, Musunuru K, Smith AV, Edmondson AC, Stylianou IM, Koseki M, Pirruccello JP, Ripatti S, Chasman DI, Willer CJ, et al. Biological, clinical and population relevance of 95 loci for blood lipids. Nature. 2010;466(7307):707–13.

Tregouet DA, Konig IR, Erdmann J, Munteanu A, Braund PS, Hall AS, Grosshennig A, Linsel-Nitschke P, Perret C, DeSuremain M, et al. Genome-wide haplotype association study identifies the SLC22A3-LPAL2-LPA gene cluster as a risk locus for coronary artery disease. Nat Genet. 2009;41(3):283–5.

Trudu M, Janas S, Lanzani C, Debaix H, Schaeffer C, Ikehata M, Citterio L, Demaretz S, Trevisani F, Ristagno G, et al. Common noncoding UMOD gene variants induce salt-sensitive hypertension and kidney damage by increasing uromodulin expression. Nat Med. 2013;19(12):1655–60.

Villard E, Perret C, Gary F, Proust C, Dilanian G, Hengstenberg C, Ruppert V, Arbustini E, Wichter T, Germain M, et al. A genome-wide association study identifies two loci associated with heart failure due to dilated cardiomyopathy. Eur Heart J. 2011;32(9):1065–76.

Wain LV, Vaez A, Jansen R, Joehanes R, van der Most PJ, Erzurumluoglu AM, O'Reilly PF, Cabrera CP, Warren HR, Rose LM, et al. Novel blood pressure locus and gene discovery using genome-wide association study and expression data sets from blood and the kidney. Hypertension. 2017;70:e4-e19.

Wain LV, Verwoert GC, O'Reilly PF, Shi G, Johnson T, Johnson AD, Bochud M, Rice KM, Henneman P, Smith AV, et al. Genome-wide association study identifies six new loci influencing pulse pressure and mean arterial pressure. Nat Genet. 2011;43(10):1005–11.

Wakil SM, Ram R, Muiya NP, Mehta M, Andres E, Mazhar N, Baz B, Hagos S, Alshahid M, Meyer BF, et al. A genome-wide association study reveals susceptibility loci for myocardial infarction/coronary artery disease in Saudi Arabs. Atherosclerosis. 2016;245:62–70.

Warren HR, Evangelou E, Cabrera CP, Gao H, Ren M, Mifsud B, Ntalla I, Surendran P, Liu C, Cook JP, et al. Genome-wide association analysis identifies novel blood pressure loci and offers biological insights into cardiovascular risk. Nat Genet. 2017;49(3):403–15.

Watkins H, Farrall M. Genetic susceptibility to coronary artery disease: from promise to progress. Nat Rev Genet. 2006;7(3):163–73.

Wellcome Trust Case Control C. Genome-wide association study of 14,000 cases of seven common diseases and 3,000 shared controls. Nature. 2007;447(7145):661–78.

Willer CJ, Schmidt EM, Sengupta S, Peloso GM, Gustafsson S, Kanoni S, Ganna A, Chen J, Buchkovich ML, Mora S, et al. Discovery and refinement of loci associated with lipid levels. Nat Genet. 2013;45(11):1274–83.

Chapter 2
Cardiac Transcriptome Profile in Heart Diseases

Lili Hao, Shiyu Chen, Jing Ma, Deyong Xiao, and Duan Ma

Abstract Heart development is a complex process. It requires morphological and functional changes of multiple cell types that must organize into a complex structure. This process necessitates elaborate control of multiple transcripts' expression in a temporal and spatial manner. New sequencing technologies, combined with bioinformatics and computational tools, have allowed the scientific community to appreciate the great complexity of the cardiac transcriptome. These will promote the understanding of the complex molecular mechanisms in heart developmental processes. In this chapter, we briefly introduce several major technology platforms of cardiac transcriptome profile and compare the differences between them to realize their better applications in cardiac transcriptomics. Since various cardiac cells play their unique functions in the heart, we describe the expression profile in cardiac development and different conditions by summarizing the expression changes and biomarkers in different cardiac cells. Finally, we review the transcriptomics-based biomarkers or candidate transcripts in several heart diseases including coronary artery disease, congestive heart failure, and common congenital heart disease.

Keywords Cardiac transcriptome profile · RNA sequencing · Noncoding RNAs · Cardiac cells · Heart diseases · Biomarkers

---

Lili Hao and Shiyu Chen are contributed equally with all other contributors.

L. Hao · D. Xiao · D. Ma (✉)
Key Laboratory of Metabolism and Molecular Medicine, Ministry of Education, Department of Biochemistry and Molecular Biology, Collaborative Innovation Center of Genetics and Development, Institutes of Biomedical Sciences, School of Basic Medical Sciences, Fudan University, Shanghai, China
e-mail: duanma@fudan.edu.cn

S. Chen
Institutes of Biomedical Sciences, School of Basic Medical Sciences, Shanghai Medical College of Fudan University, Shanghai, China

J. Ma
Shanghai Key Lab of Birth Defect, Children's Hospital of Fudan University, Shanghai, China

H. Jiang, M. Liu (eds.), *Heart Genomics*, Translational Bioinformatics 16,
https://doi.org/10.1007/978-981-13-1429-2_2

1 Introduction

Heart development requires complex cellular differentiation involving morphological and functional changes during the embryonic and postnatal periods. This process is marked by distinct changes in gene expression profiles in both cardiomyocyte and non-myocyte components that must organize into an intact structure. Studies have shown that the establishment of the heart necessitates elaborate control of gene expression patterns in a temporal and spatial manner (Miquerol and Kelly 2013; Rana et al. 2013). And the process relies on exquisite cardiac transcription factor network. Several molecules in the network with inducing or inhibiting properties have been found involved in each of the different stages of mouse heart development. At the beginning, during early cardiac development, the precursors for two specialized cell types (cardiomyocytes and endothelial cells) emerge from a population of mesodermal progenitor cells marked by *Mesp1* expression or *Flk1/Pdgfra*. Then, multipotent cardiac progenitor cells (CPCs) are marked by *Isl-1*, *Nkx2–5*, or *BAF60c* expression. The early action of BMP2 and FGF signaling plays crucial roles in initial migration of CPCs and the further cardiac field formation. Then, NKX2–5 regulates the formation of the tubular heart, and during this process, Wnt/β-catenin signaling (WNT-3A, WNT-8), ISL-1 and FGF10 are necessary for correct formation of the second heart field and the expression of *Shh*, *Bmp4*, and *Bmp7*. In addition, differentiation of cells within the heart is regulated by a variety of transcription factors. For example, NKX2–5 interacts with the T-box family of transcription factors and is involved in the activation or repression of gene expression required for the differentiation of numerous cardiac structures (Espinozalewis and Wang 2012). Faulty regulation of transcriptional programs results in both congenital heart diseases and adult cardiac diseases that have a developmental basis. In fact, many of the same transcriptional regulators involved in cardiac development also have important roles in cardiac hypertrophy, pathological remodeling, and heart failure (Gao and Wang 2014).

While the binding of sequence-specific transcription factors to gene regulatory elements is the major way to regulate heart development, it has become increasingly clear that additional mechanisms contribute to the cardiac regulatory network. New sequencing technologies, combined with bioinformatics and computational tools, have uncovered the great number of noncoding RNAs. These molecules, including microRNAs (miRNAs) and long noncoding RNAs (lncRNAs), have emerged as powerful regulators of cellular and tissue function and constitute potential therapeutic targets. Many studies have revealed that the noncoding RNAs functioned in cardiovascular development and pathophysiology of cardiovascular diseases. They could regulate and interact with the cardiac transcriptional network, and their functions may represent key distinctions that give rise to the more complex substructures of the mammalian heart (Scheuermann and Boyer 2013).

The distinct expression pattern of various molecules (mRNA, miRNA, and lncRNA) in the heart can be analyzed by cardiac transcriptome profile. The cardiac transcriptome profile is the study of the whole set of RNA transcripts expressed in the cardiac-specific tissues or cells via high-throughput techniques. And microarray analysis, array RT-qPCR, and RNA sequencing are major techniques for transcriptome profiling (Matkovich 2016). In the field of cardiac transcriptomics, many exciting profiling results have already been reported. And distinct gene expression levels and pathways have been associated with heart development (differentiation and remodeling) and different heart conditions (coronary artery disease/atherosclerosis, congestive heart failure, and common congenital heart disease). The cardiac regulatory networks are constantly updated and enriched, and lots of biomarkers have been identified, which promoted the understanding of the complex molecular mechanisms in developmental processes and potentially allowed stratified healthcare and therapeutic intervention.

In this chapter, we briefly introduce several major technology platforms of cardiac transcriptome profile and compare the differences between them. Then, we describe the expression profile in cardiac development and different conditions by summarizing the expression changes and biomarkers in different cardiac cells. Finally, we review the transcriptomics-based biomarkers or candidate transcripts in several heart diseases including coronary artery disease, congestive heart failure, and common congenital heart disease.

2 Technology Platforms

2.1 Methods for Cardiac Transcriptome

The transcriptome is the total of all RNA molecules in one cell or a population of cells. Sometimes it refers to all RNAs including mRNAs and noncoding RNAs but sometimes just mRNA. Unlike the genome that is generally immobilized on a given cell line (excluding mutations), the transcriptome will change as external environmental changes. The transcriptome reflects the genes that are actively being expressed at any given time, except for mRNA degradation and other degradation phenomena, because it contains all mRNA transcripts in a cell. The accurate quantification of transcripts levels is crucial for transcriptome study. There are three general methods in current use for examining the transcriptome, including RNA sequencing (RNA-seq), microarrays, and reverse transcription followed by quantitative polymerase chain reaction (RT-qPCR).

2.1.1 RNA Sequencing

RNA-seq is a technique that uses high-throughput sequencing technologies to determine the presence and quantity of transcriptome, such as mRNA, noncoding RNA, microRNA, and so on. In vivo, DNA sequences are transcribed and spliced to produce mature mRNA or other RNAs, which are extracted from the organism. Then

RNA is fragmented and reverse transcribed into ds-cDNA fragments in vitro. The ds-cDNA is sequenced using high-throughput, short-read sequencing methods (Hoeijmakers et al. 2013). This technology is a convenient and accurate tool for quantitative analysis of gene expression, gene discovery, differential use of promoters, and splicing variation of all genes expressed in a single tissue. The depth of information acquired by sequencing requires accurate assessment of RNA intactness and DNA contamination.

Theoretically, RNA-seq can accurately recognize and quantify all RNA species, small or large, at low or high abundance. However, RNA-seq is a complex, multistep process including reverse transcription, amplification, fragmentation, purification, adaptor ligation, and sequencing. Any improper operations of these steps will lead to incorrect results. What's more, RNA-seq intrinsic biases (such as GC bias and nucleotide composition bias) and transcriptome complexity can also make data defective.

RNA-seq can obtain a tremendous amount of data, which is a precisely double-edged sword. The rich data contains a lot of valuable information, but such rich data need complicated analysis of bioinformatics. Thus the analysis of RNA-seq data is the key of RNA-seq (Wang et al. 2009).

2.1.2 Microarray Analysis

A microarray is a multiplex gene chip, which is a technology of molecular hybridization based on the principle of DNA double helix. There are tens of thousands of gene units as a ligand on its solid substrate (usually a glass slide or silicon thin-film cell) to screen genes. The genes to be tested are amplified and labeled by fluorescence to produce a recognizable fluorescence emission or spectral shift during the screening process. This fluorescence signal is detected by fluorescence microscope when tested genes are hybridized with genes of the gene chip. Microarray can quickly analyze a large number of genetic information, so that biomedical workers can study and collect gene expression and mutation information. This type of experiment allows to determine relative levels of mRNA abundance in a set of tissues or cell populations for thousands of genes simultaneously.

Because microarray is a high-throughput detection technology, it can detect tens of thousands of DNA fragments at the same time. Thus this technology can greatly improve the efficiency of detection, lower the cost of detection, and ensure the quality of detection.

In the study of gene expression analysis based on microarray, such an experiment requires computational and statistical analysis techniques, which directly impacts the accuracy of results and the reliability of experiments.

The database analysis includes three parts: the processing of gene image to obtain the gene expression level of one single experiment, the gene expression matrix obtained by the integration of multiple experiments, and knowledge mining according to the gene expression matrix. Two crucial technologies are involved: normalization and clustering analysis.

2.1.3 Array RT-qPCR

RT-qPCR, a variant of polymerase chain reaction (PCR), is a technique commonly used in molecular biology to detect RNA expression. In RT-qPCR, the RNA template is first converted into a complementary DNA (cDNA) using a reverse transcriptase. The cDNA is then used as a template for exponential amplification in PCR. The RNA template can be total RNA, mRNA, or an RNA product that is transcribed in vitro. No matter what kind of RNA is used, the key is to ensure that there is no RNA enzyme and genomic DNA contamination.

The use of RT-qPCR alters the study of gene expression in several ways for the detection of RNA transcripts: (1) making it possible to detect transcripts of all genes, (2) using sample amplification and eliminating the need for large amounts of starting materials, and (3) improving the tolerance for RNA degradation.

In practice, RT-qPCR is often divided into one-step RT-qPCR and two-step RT-qPCR. The one-step RT-qPCR reaction is more rapid, more sensitive, simpler, and easier to operate with low contamination rates. Moreover it reduces the mismatch rate of the PCR reaction due to the proofreading activity of the polymerase. In the two-step RT-qPCR, the reaction has more accuracy and relatively little random error. And it is easier to preserve cDNA since it is reversely transcribed into cDNA in the first step. In addition, the whole process of the two-step method saves more money than the one-step method.

RT-qPCR is a very sensitive technique that can detect a very low copy number of RNAs. It is used to detect or quantify expression results. In addition, RT-qPCR can be used to detect differences in gene expression or to clone cDNA without constructing cDNA library.

Despite its major advantages, RT-qPCR also has some drawbacks. In multi-cycle PCR cycles, the exponential growth of reverse-transcribed cDNA results in inaccurate endpoint quantification because it is hard to stay linear. To provide accurate detection and quantification of RNA in samples, RT-qPCR uses fluorescence-based modifications to detect amplification of products in every cycle of PCR. The great sensitivity of the technique can be a double-edged sword because even the smallest DNA contamination can lead to bad results. A simple way to solve this problem is to add anchors, or tags, to the 5′ end of the gene-specific primers. In addition, planning and designing quantification studies can be technically challenging because of a lot of sources of variation including template concentration and amplification efficiency.

2.1.4 Evaluation and Comparison

Gene detection in RNA-seq, unlike microarray, is not dependent on probe design, but relies on short-read nucleotide mapping which can attain exceedingly high resolution. What's more, the RNA-seq can cover a larger dynamic range than microarray probe hybridization. In spite of the advantages of RNA-seq, microarray technology is still widely used because of lower costs and wider availability.

However, the gene expression measurements obtained by microarray and RNA-seq are subject to various measurement errors. RT-qPCR provides more accurate quantification of gene expression levels than microarray and RNA-seq, while it has limited throughput capacity.

The strengths and weaknesses of RT-qPCR, RNA-seq, and microarray can be summarized as follows. In terms of throughput capacity, RNA-seq is the highest, microarray is the second, and RT-qPCR is the lowest, whereas in terms of accuracy, the order from the highest to the lowest is RT-qPCR, RNA-seq, and microarray.

Of these three techniques for cardiac transcriptome, RNA-seq has overtaken the other two earlier technologies in popularity as a primary assay format, because of its higher throughput, independence from predesigned probes and primers, and continuously decreasing cost. RNA-seq is gradually becoming a method of measuring organism transcriptome, although the older microarray technology and RT-qPCR technology are still being used (Matkovich 2016).

2.2 *Analysis of Transcriptome Data*

A large amount of tissue-specific transcriptome databases have been constructed and annotated to identify differently expressed in different cell populations. Analysis of transcriptome data can be divided into three categories, including genome mapping, transcriptome mapping, and reference-free assembly. The reference-free assembly is mainly used to analyze species that do not have reference genomes and gene annotations. When a genome sequence is available for the studied organism, it is possible to identify transcripts by mapping RNA-seq reads onto the genome. However, for organisms that don't have sequenced genomes, quantification would be achieved by first assembling reads de novo into contigs and then mapping these contigs onto the transcriptome (Yang and Kim 2015).

2.2.1 De Novo Transcriptome Analysis

There are several steps in de novo transcriptome analysis: (1) the processing of raw data, (2) statistical analysis of contig and unigene, (3) functional annotation of unigene (NR, GO, KEGG), (4) analysis of SNPs and SSR, and (5) differential expression of analysis.

2.2.2 Transcriptome Analysis with Reference Genome

There are several steps in transcriptome analysis with reference genome: the processing of raw data, short-read mapping to reference genome, functional annotation of genes, structural analysis of genes, analysis of variable splicing, differential expression of analysis, and prediction of new genes.

At each of these steps, specific checks should be applied to monitor the quality of the data. Quality control for the raw reads involves the analysis of sequence quality, GC content, the presence of adaptors, overrepresented k-mers, and duplicated reads in order to detect sequencing errors, PCR artifacts, or contaminations. One important mapping quality parameter is the percentage of mapped reads, and the other is the uniformity of read coverage on exons and the mapped strand. Once actual transcript quantification values have been calculated, they should be checked for GC content and gene length biases so that correcting normalization methods can be applied if necessary. In addition, it is also crucial to assess the global quality of the RNA-seq dataset by checking on the reproducibility among replicates and for possible batch effects (Conesa et al. 2016).

There are many tools to estimate gene expression, each producing different results. Comparing results of these tools, though Cufflinks, RSEM, and IsoEM might not correlate as well as HTSeq with RT-qPCR measurements, they have expression values with higher accuracy. RNA-seq has become an important method for transcriptome analysis, but the technology and tools are evolving. However, the results obtained from different tools are still unsatisfactory and affected by parameter settings, especially when genes are expressed at low levels (Chandramohan et al. 2013).

Many RNA-seq studies used insufficient biological replicates, resulting in low statistical power and inefficient use of sequencing resources. Sequencing less reads and performing more biological replication are effective strategies to increase power and accuracy in large-scale differential expression RNA-seq studies and provide new insights into efficient experiment design of RNA-seq studies.

2.3 *Intact Hearts vs. Isolated Cells in Cardiac Disease Studies*

The majority of transcriptome analysis of hearts are proceeded on intact tissue samples. However, the results of RNA profiles are not just from cardiac myocytes but also from cardiac fibroblasts and other cell types present in the sample (Matkovich 2016).

Interest in single-cell whole-transcriptome analysis is growing rapidly, especially for profiling rare or heterogeneous populations of cells. Single-cell transcriptomes include about 3000–8000 expressed genes, which are far fewer than that counted in the transcriptomes of corresponding pooled populations (Conesa et al. 2016). Single-cell RNA-seq (scRNA-seq) is one of the newest and most active fields of RNA-seq with its unique set of opportunities and challenges, which promises to evaluate transcriptional differences between individual cells of the same type and also distinguish cells of different origin that neighbor each other in the same tissue.

There are two main strategies for achieving single-cell sequencing currently. The first one, as most people think, is to isolate individual cells and build a library of them and eventually perform sequence. We can do it with flow cytometry (including microfluidic chips) or laser capture microdissection (LCM). However, the flux of this method is very low because of cost. To overcome this difficulty, a second strategy has been adopted in recent years: barcode-based single-cell recognition. Add unique DNA sequences to each cell so that when sequencing, the sequence carrying the same barcode is considered to be from a cell. Thus, hundreds of individual cells can be measured in a single database. However, there is about a 10% chance that two single cells will be mistaken for the same one.

Single-cell RNA-sequencing data provides a unique view of transcriptome function. However, careful analysis is required in order to use single-cell RNA-sequencing measurements for this purpose. Technical variation must be considered in single-cell RNA-sequencing studies of expression variation. For a subset of genes, biological variability within each cell type appears to be regulated in order to perform dynamic functions, rather than solely molecular noise.

At present, they can measure only one functional genomic data type at a time in the same single cell, but there is an expectation that in the future we will be able to recover the transcriptome of a single cell simultaneously with additional functional data.

3 Developmental Transcriptome Profiles of Cardiac Cells

Formation and maturation of the mammalian heart involves distinct changes in gene expression profiles in both cardiomyocyte and non-myocyte components that must organize into an intact structure. Deconvoluting the multiple cell populations within the heart helps in understanding how these relate to cellular differentiation and maturation and contributes to identifying novel pathways involved in the generation of heart disease. Earlier, the roles of key developmental regulators were revealed by conventional gene deletion at tissue-level resolution. Recent advances in single-cell genome-wide RNA sequencing (RNA-seq) enable us to understand the gene expression depending on anatomical location, spatial context, developmental stage, and cell type. The transcriptional data of distinct cardiac lineages provided insights into lineage-specific markers and subpopulations that contribute to normal and aberrant cardiac development and will facilitate future efforts to identify transcriptional perturbations that indicate the onset of heart diseases that have a developmental basis (Delaughter et al. 2016; Li et al. 2016).

3.1 Developmental Transcriptional Profiles of Cardiomyocytes

Cardiomyocytes (CMs) are major cell type in the heart and are required for chamber contraction and efficient blood flow throughout the body. They possess the property of automatic or spontaneous depolarization. The crucial role of the CMs in ensuring proper heart function is maintained throughout life. Injury to the cardiac muscle often leads to heart failure due to the loss of a large number of CMs. The limited intrinsic capacity of CMs regenerates the damaged tissue, making it one of the leading causes of morbidity and mortality worldwide. Thus, maintaining CM number and function has emerged as a focus for the prevention and treatment of cardiac diseases. The transcriptomes of CMs display significant heterogeneity in their function with regard to anatomical location and development stages as well as various conditions (pluripotent stem cell [PSC] differentiation and cardiac myocyte remodeling). In this section, we will introduce the spatial and temporal characteristics of myocardial expression profiles and related changes in CM differentiation (Später et al. 2014).

3.1.1 The Developmental Spatial Transcriptomes of CMs

The transcriptomes of CMs vary among different parts of the heart; it can be reflected by cardiac myocyte subtypes and their distinct anatomical locations. Specialized cardiac myocyte subtypes include atrial and ventricular CMs and cardiac conduction cells such as those of the sinoatrial node (SAN), the atrioventricular node (AVN), the HIS bundle, the left and right bundle branches (LBB and RBB), and Purkinje fibers. And the cross-species molecular markers of those myocyte subtypes have been identified which helped in characterizing their functional properties (Table 2.1).

3.1.2 The Developmental Temporal Transcriptomes of CMs

The temporal evolution of CM transcriptomes is demonstrated by gene expression changes between the stages of CM maturation and the differentiation process of pluripotent stem cell-derived CM (PSC-CM).

The multilayered atlas of gene expression and regulation, temporal changes of gene expression, pathways, and upstream transcriptional regulators from one stage to the next during mouse development have been demonstrated. Pathway enrichment analysis was applied to determine how the genes are involved in CM function (Uosaki et al. 2015; Delaughter et al. 2016):

1. The cardiac muscle contraction pathway and cardiovascular disease pathways, including dilated cardiomyopathy, hypertrophic cardiomyopathy, arrhythmogenic right ventricular cardiomyopathy, and viral myocarditis, were enriched in the upregulated group of genes up to the neonatal stage, but none of them were enriched in upregulated genes from neonate to adult. These pathways contain structural genes (e.g., sarcomere, desmosome, and sarcoglycan genes) in different compositions whose expressions are the biomarkers of matured CMs and are necessities in CM function.
2. The fatty acid metabolism pathway was enriched in the upregulated group of genes during embryonic development. And that PPAR (a master regulator of

Table 2.1 The location as well as functional and molecular properties of cardiac myocytes

Cardiac myocyte	Location	Molecular markers	Function
Atrial	Right and left atria	MYPC3, MLRA, CXA5, SARCO, ANF, CXA1, SMPX, HEY1	Contractibility
Ventricular	Right and left ventricle	MYPC3, MYRV, IRX4, CXA1, HEY2, SERCA	Contractibility
SA nodal	Right atrial wall, entrance of superior vena cava	HCN4, TBX18, TBX3, SHOX2, LBH	Generate spontaneous action potentials leading to the depolarization and contraction of the atria
AV nodal	Base of interatrial septum, right side	HCN4, CXG3, TBX3	Converge electrical impulses
HIS bundle, LBB, RBB	Top, left, and right of ventricular septum	HCN4, CXA5, TBX3 (proximal part of LBB+RBB)	Spread electrical impulses, leading to a depolarization of the syncytium of CMs, thereby ensuring orchestrated contraction of the chambers
Purkinje fiber cell	Subendocardial, ventricular wall	HCN4, CXA5, sMHC, SCN1A, NFM, ANF, MYPC3	

From reference Später et al. (2014)
The expression of some markers is enriched but not specific
SERCA sarcoplasmic/endoplasmic reticulum calcium ATPase, *sMHC* slow tonic myosin heavy chain

fatty acid metabolisms) pathway was the only pathway enriched in upregulated genes throughout the maturation, while carbohydrate metabolism pathways, including glycolysis, pentose phosphate pathway, fructose and mannose metabolism, and galactose metabolisms, were enriched in the group of downregulated genes at the neonate stage. These pathway changes revealed how metabolic changes are accomplished in vivo. Embryonic CMs use glycolysis to generate energy, but adult CMs use fatty acid oxidation.

3. The cell cycle pathway was enriched in downregulated genes from the earlier stages, and more cell cycle-related pathways were enriched in downregulated at the adult stage.

What's more, the atlas of temporal transcriptional regulations was well correlated with transcriptional and pathway changes, which can be seen in Table 2.2.

The pluripotent stem cells (PSCs) have great prospects in regenerative medicine, disease modeling, and drug discovery because of their strong potential for self-renewing and multi-lineage differentiation. Now the methods that are inducing PSCs to differentiate into CMs in vitro have been reported, and PSC-derived CMs (PSC-CMs) are among the most desired and studied cell types as live CMs are rarely ever obtained from patients (Kattman et al. 2011). Although the strong applied

Table 2.2 Temporal changes of gene expression, pathways, and upstream transcriptional regulators between the stages of CM maturation

Pathway changes	Development stages	Regulators	Gene expression
Cardiac muscle contraction pathway (up)	Up to the neonatal stage	Activated: MEF2C, GATA4, TBX5, MYOCD, HIF1A HOPX, AND NKX2-5	*myl9*, *myh6*, *myh7*, *csrp2*, *myl2*, *tnni3*, and *ttn*↑
Fatty acid metabolism pathway (up)	Through maturation	Activated: PPAR	*fabp3*, *fabp4*↑
Carbohydrate metabolism pathways (down)	Neonate stage	Inactivated: ARNT	*Pgam1*↓
Cell cycle pathway (down)	From the mid-embryonic stages	Activated: RB1, RBL1, CDKN2A, TP53 Inactivated: CCND1, E2F1–3, MYC	*Prc1*, *ccna2*, *cdca8*, *cdca3*, *top2a*, *ccnb2*, *mki67*, *ccnb1*↓

MEF2C MADS box transcription enhancer factor 2, *GATA4* GATA-binding protein 4, *TBX5* T-box protein 5, *MYOCD* myocardin, *HIF1A* hypoxia-inducible factor 1 alpha subunit, *HOPX* HOP homeobox, *NKX2–5* NK2 homeobox 5, *PPAR* peroxisome proliferator-activated receptor, *ARNT* aryl hydrocarbon receptor nuclear translocator, *RB1* RB transcriptional corepressor 1, *RBL1* RB transcriptional corepressor like 1, *CDKN2A* cyclin-dependent kinase inhibitor 2A, *TP53* tumor protein P53, *CCND1* cyclin D1, *E2F1–3* E2F transcription factor 1–3, *MYC* MYC proto-oncogene, bHLH transcription factor, *myl9*, *myl2* encode myosin light chain 2, 9, *myh6*, *myh7* encode myosin heavy chain 6, 7, *csrp2* encodes cysteine and glycine-rich protein 2, *tnni3* encodes *troponin i3*, *ttn* encodes titin, *fabp3*, *fabp4* encode fatty acid-binding protein 3, 4, *pgam1* encodes phosphoglycerate mutase 1, *prc1* encodes protein regulator of cytokinesis 1, *ccna2*, *ccnb1*, *ccnb2*, encode cyclin a2, b1, b2, *cdca3*, *cdca8* encode cell division cycle associated 3, 8, *top2a* encodes DNA topoisomerase II alpha, *mki67* encodes marker of proliferation Ki-67

value, proper maturation of PSC-CMs remains a critical barrier for recapitulating the adult phenotype. The early maturation stage of PSCs resembled that of CMs, which then are unexpectedly arrested at the late embryonic stage after 20 days of culture (Uosaki et al. 2015). Global gene expression pattern was close to late embryonic or neonatal hearts, while many upstream transcriptional regulators displayed aberrant action or inaction. The aberrant regulators mainly include PPARs, PPAR gamma coactivator 1A (PPARGC1A), CCAAT/enhancer-binding protein A/B (CEBPA/B), and beta-catenin (CTNNB1). (1) PPARs are fatty acid metabolism regulator and the CM-specific deletion of it caused cardiomyopathy and hypertrophy. Studies showed the PPARs displayed lower activities in PSC-CMs. (2) PPARGC1A/B, coactivators of PPARs, and other nuclear receptors regulate mitochondrial biogenesis in the heart and had higher activities in PSC-CMs. (3) CEBPB as the inhibitor of CM proliferation and hypertrophy had higher activities in PSC-CMs. (4) The higher activity of CTNNB1 found in PSC-CMs could increase CM proliferation and cause sarcomere disarray, suggesting the alterations of CM maturation. Getting clear of the aberrant molecules changes in PSC-CMs helps the improvement of the application of PSC-CMs.

3.1.3 The Noncoding RNAs in CMs

Several miRNAs are involved in CM proliferation and maturation. (1) MicroRNA-29a was found involved in rat CM proliferation during postnatal development. The inhibition of microRNA-29a induced CMs to proliferate, accelerated the G1/S and G2/M transition, and upregulated the cell cycle gene expression. Cyclin D2 (CCND2) was identified as a direct target of microRNA-29a (Cao et al. 2013). (2) Two miRNAs (hsa-miR-590 and hsa-miR-199a) were shown to promote cell cycle reentry of adult CMs ex vivo and to promote CM proliferation in both neonatal and adult animals (Eulalio et al. 2012). (3) Let-7 was both required and sufficient for maturation, but not for early differentiation of CMs, which were demonstrated by gain- and loss-of-function analyses in embryonic stem cell-derived CMs (hESC-CMs). It can augment metabolic energetics in maturing CMs. Let-7-driven CM maturation could be a result of downregulation of the phosphoinositide 3-kinase (PI3K)/AKT protein kinase/insulin pathway and an upregulation of fatty acid metabolism (Kuppusamy et al. 2015).

Several lines of evidence indicate that lncRNAs can also regulate CM metabolism (*CARL*, *ANRIL*) (Bochenek et al. 2013; Long et al. 2014), hypertrophy (*MHRT*, *CHRF*) (Pei et al. 2014; Wang et al. 2014a, b), differentiation, and proliferation (*BVHT*, *FENDRR*) (Grote et al. 2013; Klattenhoff et al. 2013).

3.2 Developmental Transcriptional Profiles of Endothelial Cells

The endothelium is a continuous layer of cells that provides a functional and structural barrier between the blood and the vessel. The endothelial cells (ECs) serve many functions, such as regulation of tone, coagulation, and fibrinolysis, and it secretes several substances to ensure proper functioning. For example, the ECs release tissue-type plasminogen activator (t-PA) and urokinase when participates in fibrinolysis. Dysfunction and/or injury of endothelial cells precedes the development of arteriosclerosis and plays an important role in its progression. Various expression profiling have been performed for ECs, and this may be attributed to the distinct response of the EC under the different conditions.

3.2.1 The Developmental Temporal Transcriptomes of ECs

Studies have demonstrated that the transcriptional profiles of ventricular ECs showed a distinct temporal progression with heart development by single-cell RNA sequencing. During heart looping and chamber formation, high expression levels of *Hapln1* (hyaluronan and proteoglycan link protein 1) and *Hmga1* (high-mobility group AT-hook protein) were identified. Among these, HAPLN1 is an important regulator of developmental extracellular matrix (ECM) interactions, and HMGA1 regulates chromatin organization and promotes endothelial-mesenchymal transformation (EMT). These findings are consistent with matrix remodeling that participates in early ventricular trabeculation. Later in development, ECs showed increased expression of *Fabp4* (fatty acid-binding protein 4), a molecule that is involved in cardiac metabolic shifts from glycolysis to oxidative phosphorylation. And after birth, the EC transcriptional program remained unchanged (Delaughter et al. 2016).

3.2.2 The Function-Related Gene Expression in ECs

Endothelial cells have many functions, and specific gene expressions are required for ECs to function. (1) Coagulation formation and fibrinolysis: ECs do not express tissue factor normally, but when exposed to thrombin, cytokines, or lipopolysaccharides, they will synthesize and express tissue factor on their surface to initiate blood coagulation (Oeth et al. 1997; Michiels 2003). (2) Participation in inflammatory reactions: Endothelial cells have the potential to synthesize and secrete cytokines and growth factors, such as interleukins, colony-stimulating factors, and chemokines, after being stimulated by lipopolysaccharide, tumor necrosis factor-β (TNF-β), or IL-1 and recruit inflammatory cells to sites of tissue injury or infection (Krishnaswamy et al. 1999).

3.2.3 Gene Expression in ECs that Participate in the Pathogenesis of Atherosclerosis

When atherogenesis occurs, vascular endothelial cells are activated, and then the adhesion molecules such as vascular adhesion molecule-1 (VCAM-1), endothelial-leukocyte adhesion molecule (E-selectin), and intercellular adhesion molecule 1 (ICAM-1) are expressed. The VCAM-1 protein mediates the adhesion of lymphocytes, monocytes, eosinophils, and basophils to vascular endothelium. It also functions in leukocyte-endothelial cell signal transduction and may play a role in the development of atherosclerosis and rheumatoid arthritis. E-selectin functions in recruiting leukocytes to the injury site during inflammation. The local release of cytokines IL-1 and TNF by damaged cells induces the overexpression of E-selectin on endothelial cells of nearby blood vessels. ICAM-1 is a ligand for LFA-1 (integrin), a receptor found on leukocytes. The ICAM-1/LFA-1 will mediate the leukocyte attachment to vascular endothelial surfaces. Normally ICAM are kept in low concentrations in the membranes of leukocytes and endothelial cells. The high expression of ICAM needs to be induced by cytokines or other pro-inflammatory mediators (Delaughter et al. 2016).

In addition, expression profiling has documented the changes in the expression of groups of genes by many common agents including turbulent flow (upregulation of VCAM-1 and E-selectin), infectious agents such as CP (upregulation of IL-1, MCP-1, IL-8, FGF and PDGF-β), and toxic agents like homocysteine, oxidized LDL (upregulation of RXR-β), and nicotine (upregulation of proteins in the inositol phospholipid pathway, namely, phosphatidylinositol phosphate kinase and diacylglycerol kinase). These may provide further medical strategies and targets for reducing the inflammatory response which ultimately result in atherosclerosis.

3.2.4 The Noncoding RNAs in ECs

Recent studies have identified miRNAs as key regulators of endothelial cell function. Tissue-specific Dicer deletion revealed the global requirement of miRNAs in angiogenesis. It was reported that conditional deletion of Dicer in endothelial cells impaired embryonic and postnatal angiogenesis (Chen and Wang 2012). From miRNA profiling of human endothelial cells, miR-126 and the miR-17-92, miR-23-27-24, and miR-222-221 clusters are major conserved families of miRNAs expressed in endothelial cells, and miR-126 is the best known endothelial cell-specific "angiomiR" that modulates angiogenic signaling and vascular integrity (Santoro and Nicoli 2013). It is expressed in the vascular tree of vertebrate throughout development. Deletion of miR-126 in mice results in vascular developmental defects as well as diminished angiogenesis and increased mortality after coronary ligation. In a zebrafish study, miR-126 regulated the endothelial cell response to vascular endothelial growth factor by targeting Spred-1 and PIK3R2, both regulators of the vascular endothelial growth factor pathway.

LncRNAs in ECs are involved in the regulation of migration and differentiation. For example, *MALAT1* regulates endothelial cell function and vessel growth by targeting various cell cycle regulators such as *CCNA2*, *CCNB1/2*, *CDKN1A*, and *CDKN1B* (Michalik et al. 2014).

3.3 Developmental Transcriptional Profiles of Fibroblasts

Cardiac fibroblasts (CFs) comprise about 2/3 of the cardiac cell types. Because of their mesenchymal origin, CFs have abilities to adhere to plastic and to secrete extracellular matrix (ECM). CFs are the stromal elements of the heart where they play roles in scaffolding support, inflammation, cardiac remodeling, and fibrosis. Studies demonstrating the characteristics of CFs have increased awareness of their potential contribution to multiple aspects of myocardial function and pathophysiology that must be rectified to achieve successful therapy (Weber et al. 2013).

3.3.1 The Function-Related Gene Expression in CFs

CFs play an essential role in cardiac remodeling and fibrosis. Functions of CFs associated with these conditions are cell proliferation, cell death by apoptosis, autophagy, migration, adhesion, secretion of extracellular matrix (ECM) protein, collagen gel contraction, CF differentiation to cardiac myofibroblasts (CMF), and finally cytokine and growth factor synthesis. And increasing attention has been gained for CF functions as inflammatory supporting cells in the process of inflammation. Several reports demonstrated the presence of different classes of receptors, which are produced by CFs, which are the most important regulators of those functions. (1) β2-adrenergic receptors (β2-AR) are the only adrenergic receptors expressed by CFs. The activation of β2-AR increases levels of cAMP which modulates the activity of both PKA and EPAC proteins, so the CFs can participate in many cellular responses, such as proliferation, differentiation, ECM protein secretion, migration, and ion transport. In addition, the stimulation of β2-AR also activates ERK1/2, a critical signaling pathway for cell growth and protein synthesis. What's more, CFs, but not CMs, are involved in the inflammation after cardiac injury by serving as the predominant source of IL-6 after an increase in cAMP levels induced by β2-AR activation. (2) Angiotensin receptors. CFs express both AT1R and AT2R, and AT1R activation induces cell proliferation and increases ECM protein synthesis. Hypertension or cardiac infarct promotes the increase in CF AT1R expression. And CFs are the main cardiac cells responsible for TNF- α and TGF- β synthesis and secretion induced by the AT1R ligands, Ang II, associated with cardiac fibrosis development. (3) Kinin receptors. CF can express both B1R and B2R, and B1R is increased in pathological conditions such as ischemia, atheromatous disease, or exposure to inflammatory cytokines. The exact function of those two receptors in CFs remains unclear (Díaz-Araya et al. 2015).

A number of studies have suggested that CFs can also communicate with immune cells, such as lymphocytes, monocytes, and macrophages, to modulate their behavior and orchestrate heart injury and repair process (Lech and Anders 2013). For example, at the early stages, CFs are able to recruit leukocytes via the release of chemokines including MCP-1, macrophage inflammatory protein (MIP)-1, and RANTES. Besides, CFs can express adhesion molecules (ICAM-1 and VCAM-1) on their surface to achieve the adhesion with neutrophils and B lymphocytes when they receive pro-inflammatory stimuli (Díaz-Araya et al. 2015).

CFs share many generic properties and functions with mesenchymal stem cells (MSCs). Studies showed that a high proportion of adult mouse cardiac fibroblasts expressed defined MSC markers including stem cell antigen 1 (SCA1) and CD140α. CD140α can interact with ECM components, reflecting the capacity to adhere and interact to the microenvironment. And the principal role of SCA1+ mesenchymal cells in the adult heart is to provide support to the myocardium in an injury scenario. These implicate CF functions as mesenchymal support and myofibroblast formation, representing a putative source of progenitors that can differentiate into multiple cell types in certain disease states.

The current results showed that many cardiogenic factors like *Tcf21*, *Tbx5*, *Tbx20*, *Gata4*, *Gata5*, *Gata6*, *Tie1*, *Hand2*, *Mef2c*, *Nkx2-5*, and *Wt1* genes were all significantly upregulated in cardiac fibroblasts compared with tail fibroblasts. This indicates that at least a subpopulation of cardiac fibroblasts prepare for transdifferentiation by expression of cardiogenic factors. The results will prompt novel strategies to target specific cardiac cell populations with enhanced transdifferentiative capacities (Furtado et al. 2014).

3.3.2 The Noncoding RNAs in CFs

In the heart, noncoding RNAs, especially miRNAs, have been shown to be key players in CF biology. And direct fibroblast-specific alterations of miRNAs in fibrosis and cardiac remodeling have been demonstrated. For example, miR-21 expression is enriched in CFs and directly drives fibrosis. MiR-21 targets a negative regulator of the ERK-MAP kinase pathway, sprouty homology 1 (Spry1), to drive fibroblast growth factor (FGF) secretion and fibroblast survival. MiR-21 also targets PTEN whose downregulation derepresses the matrix metalloproteinase 2 (MMP2) and TGF-beta receptor III (Bang et al. 2014). Studies also showed that fibroblast exosomal-derived miR-21_3p (miR-21*) as a potent paracrine-acting RNA molecule induces CM hypertrophy by targeting sorbin and SH3 domain-containing protein 2 (SORBS2) and PDZ and LIM domain 5 (PDLIM5) (Bang et al. 2014). Other prominent fibroblast-enriched and fibrosis-associated microRNAs are represented by the three members of the miR-29 family (miR-29a, miR-29b, miR-29c), which target ECM components (MMPs, collagens, elastin, fibrillin). The downregulation of miR-29 thus likely contributes to cardiac fibrosis through increasing ECM production (Rooij et al. 2008). Besides, miR-15 in fibroblasts targets a number of members within the TGF- β signaling cascade, such as TGF- β1, p38, endoglin, SMAD3, and SMAD7, and therefore negatively regulates ECM production. MiR-30 is another example of a microRNA that is highly expressed in cardiac fibroblasts and downregulated in the failing heart in humans and rodent models. Mechanistically, miR-30 directly targets connective tissue growth factor (CTGF) to reduce the production of collagens.

Several lncRNAs specifically influencing CF activity have been reported. Mhrt (myosin heavy-chain-associated RNA transcripts) is the antagonist of Brg1 (part of the Brg1–HDAC–Parp chromatin repressor complex). Under cardiac stress conditions, Mhrt expression is inhibited and Brg1 levels increase to an extent outcompeting Mhrt1. As a consequence, Brg1 can repress Myh6 (and Mhrt) leading to a pathological switch from the Myh6 to the Myh7 isoform which ultimately contributes to heart failure (Pei et al. 2014). What's more, lncRNAs control telomere structure (*TERRA*) and are associated with senescence (*SAL-RNA153*, *H19*) (Ratajczak 2012; Abdelmohsen et al. 2013).

4 Transcriptomics-Based Biomarkers in Typical Heart Disease

Heart disease can be divided into two main types: congenital heart defects (CHDs) and acquired heart disease. According to the data published by the WHO in 2008, the number of deaths due to cardiovascular disease (CVD) totals 17.3 million a year, representing 30% of all global deaths. The studies of heart disease transcriptomes have led to the identification of unique subsets of transcripts associated with specific diseases and disease processes. These will provide insights into the molecular mechanism of cardiovascular disease and may generate new diagnostic and therapeutic markers.

4.1 Coronary Artery Disease (CAD)

CAD is the most common type of heart disease, which is generally due to the accumulation of atherosclerotic plaques (atherosclerosis) in the walls of the coronary arteries. Atherosclerosis leads to symptoms of stable angina if no clinically apparent plaque rupture or significant breach of the arterial wall occurs, but when the arterial wall integrity is lost and thrombosis has occurred, unstable angina, acute myocardial infarction (MI), or sudden cardiac death will happen. Over the past 10–15 years, developments in expression profiling have opened new areas for the discovery of important factors in the pathogenesis of CAD. Messenger RNA and noncoding RNAs are differentially expressed at each stage of a patient's progression through the natural history of CAD, including evaluation of risk factors, prediction and detection of coronary artery disease, and acute coronary syndromes. In this section, we mainly review the transcriptomics-based biomarkers helping to the identification of CAD and detection of MI.

4.1.1 Traditional Biomarkers of MI

The traditional biomarkers of MI include troponin T (cTnT), troponin I (cTnI), creatine kinase-MB isoenzyme (CK-MB), and myoglobin (Mb). (1) Troponin. Cardiac troponins regulate muscle contraction through interaction with actin and myosin. The troponin–tropomyosin complex is composed of three subunits: troponins T, I, and C. And only T (cTnT) and I (cTnI) are expressed as cardiac muscle-specific isoforms and are encoded by different genes in striated muscle. Troponin T anchors the troponin complex to tropomyosin on thin myofibril filaments. Troponin I inhibits the binding of myosin with actin. Both troponins I and T display biphasic release profiles in acute MI, and the time to peak depends on reperfusion of the infarct-related coronary artery. Troponin T peaks within 24–96 h, whereas troponin I peaks earlier, at approximately 12–24 h. Once released from disintegrating myocytes into the blood, cTnT and cTnI can be measured by antibody-dependent tools, providing the potential for developing highly specific bioassays for myocardial injury (Mueller et al. 2013; Maznyczka et al. 2015). (2) Creatine kinase-MB isoenzyme. $CK\text{-}MB_{mass}$, $CK\text{-}MB_{activity}$, and total CK are more specific than myoglobin, but may not be detectable for 4–8 h in the bloodstream following MI. In acute MI, $CK\text{-}MB_{mass}$ levels rise to twice the upper limit of normal at 6 h and peak at approximately 24 h after cardiac injury. Thus, the sensitivity of $CK\text{-}MB_{mass}$ for acute MI is only 50% when measured early at the time of presentation. Quantitative measurements of $CK\text{-}MB_{mass}$ have largely replaced $CK_{activity}$ and $CK\text{-}MB_{activity}$ as measures of myocardial injury. As with myoglobin, plasma concentrations of $CK\text{-}MB_{mass}$ may be elevated in patients with skeletal muscle injury in the absence of myocardial injury. Nevertheless, because it is more specific than CK or myoglobin alone, CK-MB_{mass} has become the predominant laboratory marker for diagnosing acute MI. Many hospitals have begun to collect CK-MB specimens at frequent intervals (e.g., 0, 3, 6, 9, and 12 h) to allow more timely and sensitive detection of myocardial injury. This strategy is frequently used in emergency departments and chest pain observation units (CPU) emphasizing early triage and risk stratification. (3) Myoglobin has a lower molecular weight than CK-MB or the cardiac troponin molecules and is released more rapidly during acute MI, reaching twice-normal values within 2 h and peaking within 4 h. Myoglobin is found in large amounts in both skeletal and cardiac muscle, and because of this lack of specificity, it is not an ideal marker of cardiac injury. Myoglobin is also rapidly cleared compared with other markers limiting its diagnostic usefulness in patients presenting after 8–12 h of symptoms (Hudson et al. 1999). Early diagnosis or rule out of acute MI might possibly reduce morbidity and mortality, as well as facilitate early discharge from emergency departments.

4.1.2 Message RNAs Implicated in the Prediction and Detection of MI

One of the most clinically useful functions of gene expression analysis is to identify CAD in patients who present with symptoms concerning for angina but has not yet been identified. In the PREDICT study, the gene expression profiles in various samples (peripheral blood mononuclear cells [PBMCs] or whole blood) from those patients have been analyzed. And the expression algorithm for prediction of angiographic CAD involved genes in several different pathways including neutrophil activation, apoptosis, cell necrosis, natural killer cell activation, and both the adaptive and innate immune responses (Rosenberg et al. 2010) (Table 2.3).

Several studies of platelet and whole blood mRNA in patients with MI versus patients with stable CAD have identified a number of differentially expressed genes which are candidates for further investigation. The identified target mRNAs were clustered in several pathways, including inflammatory pathways involving PPAR, IL-6, IL-10, natural killer cell, and glucocorticoid receptor signaling; cholesterol metabolism involving liver X and retinoic acid receptor activation; and the extrinsic pathway of the coagulation system (Table 2.3). These findings reflect the intimate relationships between cholesterol deposition, intimal proliferation and inflammation, and acute thrombosis involved in the natural history of CAD (Friede et al. 2015). These data support ongoing efforts to incorporate biomarkers based on expression profiling to determine prognosis and response to therapy.

4.1.3 MiRNAs and LncRNAs as Diagnostic Tools in CAD

It is evident that miRNAs are involved in the development of atherosclerosis and on CAD progression. The measurement of their levels in the peripheral blood (circulating miRNAs) can serve diagnostic or prognostic purposes in cases of stable CAD or

Table 2.3 mRNAs implicated in the prediction and detection of MI

Sample	Pathway	mRNA	References
Prediction of future adverse cardiac events			
PMBCs	Metabolism	*ALOX5, MGST1, HK3*	Wingrove et al. (2008)
	Cytokines	*CSF2R, CSF3R, IL1RN*	
	Adaptive immune responses	*VSIG4*	
	Intracellular processes	*VCAN, CAPG*	
Whole blood	Neutrophil activation	*IL18RAP, TNFAIP6, CASP5, IL8RB, TNFRSF10C, TLR4, KCNE3, S100A12, CLC4E, RPL28, NCF4, AQP9*	Rosenberg et al. (2010) and Taurino et al. (2010)
	Natural killer cell activation	*SLAMF7, KLRC4, TMC8, CD3D*	
	Metabolism	*NDUFB3, ATP5C1, ATP51, COX7B, COX7C, NDUFA4*	
	Adaptive immune responses	*SPIB, CD79B, TMC8, CD3D*	
	Intracellular processes	*UQCRQ*	
Diagnosis of MI			
Platelets	Adaptive immune response	*CD69, MYOZ3, TCF3*	Healy et al. (2006) and Calverley et al. (2010)
	Inflammation	*MRP14*	
	Signal processing	*AGAP3, KLHL22, CELF4, MACROD1*	
	Cytokines	*IL28RA*	
PMBCs	Adaptive immune response	*VSIG4, MYBL1, FCGR1A*	Kiliszek et al. (2012)
	Innate immune response	*KLRC2, KLRC4, CR1*	
	Signal processing	*STAB1, RNASE1, RNASE2, PPARG*	
	Cytokines	*SOCS3*	

From reference Friede et al. (2015)

even in acute coronary syndromes. The role of miRNAs as biomarkers for stable coronary artery disease and MI is summarized in Table 2.4.

The number of lncRNAs studied in heart disease is still very limited. But several different transcripts have been studied as potential biomarkers for coronary artery disease:

1. *ANRIL* was found lower in patients with acute MI compared to healthy volunteers. *ANRIL* is highly expressed in atherosclerotic plaques and might be a novel component of inflammatory NF-κB pathway (TNF-α–NF-κB– ANRIL/YY1–IL6/8). *ANRIL* acts as an antisense regulator to the *CDKN2B-CDKN2A* gene cluster at the 9p21 locus. It can also interact with polycomb repressive complex-1 (PRC1) and polycomb repressive complex-2 (PRC2), leading to epigenetic silencing of other genes in this cluster.

Table 2.4 MiRNAs implicated in stable coronary artery disease and MI

Sample	MicroRNAs	Regulation and expression	Patients/controls	References
MicroRNAs as biomarkers of atherosclerosis and stable coronary artery disease				
Plasma	miR-133, -208a	Upregulation. Cardiac muscle-enriched microRNAs	67 CAD/31 healthy controls	Fichtlscherer et al. (2010)
	miR-126, -17, -92a, -145, -155	Downregulation. miR-145 is enriched in smooth muscle	67 CAD/31 healthy controls	Fichtlscherer et al. (2010)
	miR-1, -122, -126, -133a/b, -199a	Upregulation	53 CAD (34 SA, 19 UA/20 healthy controls	D'Alessandra et al. (2013)
	miR-337-5p, 485-3p	Upregulation	53 CAD/20 healthy controls	D'Alessandra et al. (2013)
Serum	miR-27b, -130a, -210	Upregulation	104 arteriosclerosis obliterans/105 controls	Li et al. (2011)
	miR-208, -499	Upregulation in patients with angina pectoris compared to patients with AMI	117 AMI/182 angina pectoris/100 controls	Li et al. (2013a, b, c)
PBMCs	miR-135a	Upregulation	50 CAD (25 SA, 25 UA)/20 controls	Hoekstra et al. (2010)
	miR-146a/b	Upregulation	66 CAD/33 controls	Aad et al. (2010)
	miR-147	Downregulation	50 CAD (25 SA, 25 UA)/20 controls	Hoekstra et al. (2010)
	miR-181a	Downregulation	81 CAD/44 controls	Aad et al. (2010)
Platelets	miR-340*, -624*	Upregulation	12 premature CAD/12 healthy controls	Sondermeijer et al. (2011)
	miR-1280	Downregulation	12 premature CAD/12 healthy controls	Sondermeijer et al. (2011)
Whole blood	miR-19a, -484, -155, -222, -145, -29a, -378, -342, -181d,-150, -30e-5p	Downregulation	15 CAD/20 healthy controls	Weber et al. (2011)
EPCs	miR-221, -222	Upregulation	4 CAD/4 controls	Minami et al. (2009)

(continued)

Table 2.4 (continued)

Sample	MicroRNAs	Regulation and expression	Patients/controls	References
MicroRNAs as biomarkers in myocardial infraction				
Plasma	miR-208a	Upregulation. cardiac-specific miRNA	33 AMI/63controls (30 healthy, 16 non-AMI coronary heart disease, 17 other cardiovascular diseases)	Wang et al. (2010)
	miR-208b, -499	Upregulation	32 AMI/36 controls with atypical chest pain and no cardiac disease	Corsten et al. (2010)
	miR-1, -133, -499-5p	Upregulation both in human and mice	33 acute STEMI/17 healthy controls	D'Alessandra et al. (2010)
	miR-208b, -499, -320a	Upregulation with no incremental to TnT diagnostic value	224 AMI/931 controls with atypical chest pain and no AMI	Devaux et al. (2015)
	miR-1, -499 -21	Upregulation	106 ACS/226 controls	Oerlemans et al. (2012)
	miR-92a	Upregulation	82 STEMI/116 SA	Wang et al. (2011)
	miR-1, -21 -133a, -423-5p, -499-5p	Upregulation	92 Non-STEMI/180 controls (81 acute heart failure; 99 healthy controls)	Olivieri et al. (2013)
	miR-122, -375	Downregulation	25 STEMI/17 healthy controls	D'Alessandra et al. (2010)
	miR-1	Upregulation	56 AMI/28 controls	Li et al. (2014)
	miR-21-5p, 361-5p, 519e-5p	Upregulation (MiR-21-5p and 361-5p); downregulation (519e-5p)	17 AMI/28 healthy controls	Wang et al. (2014)
Serum	miR-1, -134, -186, -208, -223, -499	Upregulation	117 AMI/282 controls (182 angina/100 healthy controls)	Li et al. (2013a)
PBMCs	miR-21, 146a	Upregulation	31 ACS/31 controls (15 SA; 16 chest pain syndrome)	Yao et al. (2011)
	miR-155	Downregulation	31 ACS/31 controls (15 SA; 16 chest pain syndrome)	Yao et al. (2011)
Whole blood	miR-1291, -663b, -30c, -145	Upregulation	20 AMI/20 controls	Meder et al. (2011)

From reference Economou et al. (2015)
CAD coronary artery disease, *AMI* acute myocardial infarction, *vs.* versus, *PBMCs* peripheral blood mononuclear cells, *PE* pulmonary embolism, *STEMI* ST-elevation myocardial infarction, *TnT* troponin T, *ACS* acute coronary syndrome

2. *KCNQ1OT1* is the antisense to the KCNQ1 gene, an epigenetic regulator of many targets, known for its pathogenic role in Beckwith–Wiedemann syndrome. Patients with ST-elevation myocardial infarction (STEMI) had lower levels of KCNQ1OT1 than patients with non-STEMI (NSTEMI). It is a good predictive marker to distinguish between STEMI and NSTEMI.
3. *HIF1a-AS1* and *-AS2* are also antisense transcripts modulating *HIF1A*, a pro-angiogenic and anti-apoptotic gene upregulated in many heart diseases. The expression level of *HIF1a-AS2* is higher in MI patients, and *HIF1a-AS2* levels varied based on time of presentation after onset of chest pain.

Moreover, altered expression of *MIAT* by the SNP has been linked to genetic susceptibility for MI and might act as a competing endogenous RNA for various targets. And CAD could be distinguished with high sensitivity from other CVD by higher plasma levels of *CoroMarker (*Busch et al. 2016*)*.

As the science of mRNA and noncoding RNA research progresses over the next decade, more investigation is needed to consolidate the discovery of differential mRNA and noncoding RNA expression into validated predictive models for prediction, detection, and prognostication of CAD. Ultimately, profiling patients using this genomic information is an extremely promising development for clinicians who strive to improve outcomes and provide additional information for their patients with CAD.

4.2 Congestive Heart Failure

Congestive heart failure (CHF) includes conditions other than artery disease in which the heart cannot pump enough blood throughout the body. The damage of pumping capacity may result in a weakened or stiffened heart muscle or valve defects. CHF often occurs as the end stage of cardiac disease, referred to as sustained pathological hypertrophy and ischemic cardiomyopathy (ICM) due to severe CAD. It is a leading cause of morbidity and mortality in the world. And the prevalence of HF is expected to continue to rise as early detection therapies for myocardial infarction (MI), valvular diseases, and arrhythmias improve, thereby allowing patients to survive longer. The development of transcriptomics-based biomarkers will play an increasing role in the early diagnosis, therapeutic monitoring, and management of heart failure patients in the future.

4.2.1 Blood-Based Biomarkers in CHF

Although the high prevalence of CHF, the diagnosis remains difficult as none of the signs and symptoms are specific or particularly sensitive. Due to this, history and physical examination may not be sufficient to reach the diagnosis. Diagnostic aids in the form of blood-based biomarkers have been sought, and these will play an

Table 2.5 Classification of blood-based biomarkers in heart failure

Marker source	Biomarker
Inflammation	C-reactive protein (CRP), pentraxin 3, myeloperoxidase, uric acid/xanthine oxidase, tumor necrosis factor (TNF), interleukin-6 (IL-6), osteoprotegerin, monocyte chemoattractant factor, FasR, CD40, P-selectin, E-selectin, L-selectin, intercellular adhesion molecule 1 (ICAM-1), vascular cell adhesion molecule 1 (VCAM-1), vascular endothelial growth factor (VEGF), placenta growth factor (PlGF), hepatic growth factor (HGF), endothelial growth factor (EGF)
Extracellular matrix turnover and remodeling	Matrix metalloproteinases and tissue inhibitors of matrix metalloproteinases, procollagen type III aminopropeptide (PIIINP), procollagen type I carboxy-terminal peptide (PIP), procollagen type I amino-terminal peptide (PINP), type I collagen telopeptide (ICTP), basement membrane laminin (BML), tenascin C, osteopontin, interleukin-33/ST2, galectin-3 (gal-3)
Biochemical strain	Natriuretic peptides, growth differentiation factor (GDF-15)
Neurohormonal activation	Aldosterone, ADH/copeptin, adrenomedullin, apelin, endothelin, relaxin, urotensin
Nutrition and metabolism	Albumin, ghrelin, leptin, adiponectin, resistin
Cardiomyocyte injury	Cardiac troponins, heart fatty acid-binding protein (HFBP), myosin light chains (MLC)

From reference Liquori et al. (2014)

increasing role in advancing CHF therapies and interventions. Much research for the past decade has examined numerous possible biomarkers, and these can be broken down into various biologic pathways and physiologic processes including inflammation, remodeling, strain, neurohormonal activation, metabolism, and cardiac myocyte injury (Table 2.5). The biomarkers could reflect distinct pathophysiological changes in the cardiac myocytes seen in patients with HF such as myocyte stress (natriuretic peptides and midregional natriuretic peptides), myocyte injury (cardiac troponins), myocyte inflammation/ischemia (GDF-15), renal dysfunction (NGAL), HF (procalcitonin), neurohormone (copeptin), and myocyte fibrosis (ST2, galectin-3) (Choudhary et al. 2013). Next we will briefly describe biomarkers that have seen extensive research in both the experimental and clinical settings in patients with HF.

1. *B-type natriuretic peptide (BNP)* was identified both as a cardiac hormone and biomarker of HF (Yasue et al. 1994). The BNP will be released primarily from the ventricular myocardium when the left ventricle experiences dysfunction with intracardiac pressure rising. The elevated filling pressure in the ventricle activates compensatory pathways, which work to restore cardiac output. As the heart continues to activate these pathways in an attempt to maintain blood flow, the result is ultimately ventricular remodeling and worsening HF (Green et al. 2009). Thus, BNP appears to be a useful marker of the function and hemodynamic status of the ventricle. In addition, BNP has also been related to changes in pulmonary capillary wedge pressure. Preliminary results of the Rapid Emergency

Department Heart Failure Outpatient Trial (REDHOT) study indicated that BNP significantly reflected the severity of CHF, in contrast to physicians' perceived severity (Maisel et al. 2004).

2. *Galectin-3 (gal-3)* is a β-galactosidase-binding lectin that is expressed and secreted by macrophages. Increased gal-3 levels are associated with fibroblast proliferation and activation of inflammatory response that eventually leads to cardiac hypertrophy, fibrosis, remodeling, and ventricular dysfunction. Multiple studies have showed that the gal-3 level was significantly elevated in HF patients compared to those without HF. Importantly, gal-3 was a superior predictor of death or HF hospitalization. What's more, gal-3 levels correlated with right ventricular failure and were found to be independently predictive of 4-year mortality (Lok et al. 2013).
3. *ST2* is a member of the Toll-like/IL-1 receptor superfamily. ST2 has both a transmembrane (ST2L) and a soluble (sST2) form with the functional ligand for each being IL-33. Interaction of IL-33 with ST2L appears to activate the NF-κB and MAPK signaling pathways to prevent apoptosis. An important strength in the clinical value of ST2 is its independence from natriuretic peptides, which allows for both unique and complementary predictive information. Several studies have shown that, while ST2 alone is an effective marker for HF, its predictive value is independent of natriuretic peptide values and its prognostic usefulness increases when used in conjunction with natriuretic peptides (Lok et al. 2013).

4.2.2 Expression Profiling of Human Cardiac Tissues in CHF

In addition to the clinical HF biomarkers, there are many studies focus on the expression profiling of human cardiac tissues. These analyses have been employed to identify genes whose expression is linked to CHF; identified genes may become the new candidate biomarkers for the HF prediction and treatment. And they will contribute to the development of the heart failure phenotype and represent compensatory mechanisms to sustain cardiac function in failing human hearts.

Several studies have shown that in the heart tissue of end-stage dilated cardiomyopathy (DCM) patients, ANP (atrial natriuretic factor and brain natriuretic peptide) and numerous sarcomeric and cytoskeletal proteins (e.g., cardiac troponin, tropomyosin), stress response proteins (e.g. HSP-40, HSP-70), and transcription/translation regulators (e.g., CCAAT box-binding factor, eIF-1AY) are found to be increased, and genes classified as cell-signaling channels and mediators, particularly those involved in calcium pathways (calcium/calmodulin-dependent kinase, inositol 1,4,5-trisphosphate receptor, and sarcoplasmic reticulum Ca 2^{+-}ATPase [SERCA]), were decreased (Barrans et al. 2002; Hwang et al. 2002). Besides, the pro-apoptotic shift in the TNF-α signaling pathway has also been found in human idiopathic DCM hearts that have progressed to CHF, which can be reflected by decreased expression of TNF-α- and NF-kB-induced anti-apoptotic genes such as growth arrest and DNA damage-inducible (GADD) 45-b, FLICE inhibitory protein (FLIP), and TNF-induced protein 3 (A20) genes (Steenbergen et al. 2003).

As for the treatment, the use of a left ventricular assist device (LVAD) can decrease cardiac mass and myocyte size and has the potential to improve contractile function. Studies demonstrated that the LVAD support resulted in downregulation of cytokines genes and upregulation of genes encoding a large proportion of transcription factors, related to cell growth/apoptosis/DNA repair, cell structure proteins, metabolism, and cell signaling/communication. Among those profiles, the most significantly upregulated gene was the G-protein-coupled receptor APJ and the specific receptor for apelin. The apelin is localized primarily in the endothelium of the coronary arteries after the mechanical off-loading by the LVAD, acting as mediators of cardiovascular function, fluid homeostasis, and adipocyte endocrine secretion (Chen et al. 2003).

In summary, gene expression changes in CHF can reveal the underlying molecular mechanisms regulating hypertrophy, proliferation, and apoptosis of CMs. These studies reflected the dysregulation of various molecular pathways involved in CHF (Archacki and Wang 2004).

4.2.3 CHF Related Noncoding RNAs

The role of miRNA as biomarkers in HF has been studied extensively for a long time, and the first preclinical trials on therapeutic applications come from this field. Due to the heterogeneous clinical phenotypes of HF, the power of the miRNA biomarkers was often insufficient.

High circulating levels of miRNA-21, miRNA-27a, miRNA-29a, miRNA-155, and miRNA-210 and low levels of miRNA-1, miRNA-142, and miRNA-150, singly or in different combinations, have been found to correlate with the established protein biomarker for HF and BNP. For instance, lower level of miRNA-210 in EDTA serum was correlated to BNP. And lower level of miRNA-1 reflected inverse correlation with NT-proBNP. Although the roles of miRNA-1, miRNA-21, miRNA-29a, miRNA-155, and miRNA-210 have been more or less defined from coherent in vitro and tissue studies, little is known about the other candidates. Evidence revealed that some candidate levels did however correlate well with disease severity. For example, a lower expression of miRNA-150 was consistently associated with worse left ventricular function. The miRNA-133a expression levels were associated with a worse outcome for left ventricular remodeling. It can be seen from all research above that there still a long way to go for miRNA as biomarkers for disease or severity predictors in heart failure (Busch et al. 2016).

Several lncRNAs that showed conservation of both sequence and function between mouse and human are implicated in the failing heart maladaptive remodeling process, which is considered an important target for therapy. (1) *CHRF* (cardiac hypertrophy-related factor) was the first lncRNA reported to function in HF, promoting cardiac hypertrophy. By acting as the sponge to miR-489, it regulates its downstream target, MYD88, which is the key gene in activating cardiac hypertrophy. (2) *Chaer* (cardiac hypertrophy-associated epigenetic regulator) is also required for cardiac hypertrophy. It inhibits histone lysine methylation at the promoter regions of pro-hypertrophic genes by interacting with PRC2, so that these genes could express. (3) *CHAST* (cardiac hypertrophy-associated transcript) plays a role in promoting hypertrophy. Induced overexpression of *Chast* in mice led to CM hypertrophy, while its suppression attenuated remodeling and hypertrophy without signs of toxicity. This is one of the most prominent examples of the strong therapeutic potential of lncRNAs for cardiac remodeling, showing that manipulation of a specific lncRNA can improve cardiac function (Gomes et al. 2017).

4.3 *Congenital Heart Disease*

Congenital heart diseases (CHDs) are usually occur at birth and are characterized by structural abnormalities of the heart and/or the great vessels, such as atrial or ventricular septation defects and transposition of the arteries, and by functional abnormalities, such as electrical conduction abnormalities or cardiomyopathies. CHDs are caused by polygenetic perturbations of highly conserved molecular mechanisms in heart development and are often linked to mutations in core cardiac transcriptional factors. CHD is the most common birth defects, but the deterministic roadmap of CHD pathogenesis does not currently exist. The broad goals of CHD research are to find common mechanisms/pathways to predict and diagnose cardiac defects and ultimately engineer preemptive therapeutic strategies, and the most basic and essential step is exploring transcriptome profile of CHD.

4.3.1 Expression Profiling of CHD

Cardiogenic transcriptional landscape defining the time-dependent expression of cardiac ontogeny and prioritizing regulatory networks at the interface between health and disease has shown that the top enriched pathways associated with non-syndromic CHD genes and their partners (*Foxh1*, *Acvr2b*, *Notch1*, *Nkx2–5*, *Gata4*, *Gata6*, *Vegfa*, *Hand2*, and *Actc1*) expressed at different developmental stages are epithelial-to-mesenchymal transition (EMT), BMP signaling, NF-AT signaling, TGF β-dependent EMT, and Notch signaling. Among these, the genes involved in the regulation of EMT are distributed throughout early to late stages of heart development, while the BMP pathway is mainly expressed during early embryonic stages, followed by activation of the NF-AT pathway in later stages of cardiogenesis. What's more, the genes involved in syndromic CHD are mainly related to the MAPK family, which affect many different pathways and are distinct from the non-syndromic CHD.

Only a few distinct gene expression profiles based on comparative transcriptome analysis and associated with CHDs including tetralogy of Fallot (TOF), ventricular

Table 2.6 Gene changes in congenital heart disease

CHDS	Pathways	Gene	References
VSD	Ribosomal proteins	Downregulation of S11, L18A, L36, LP0, L31, and MRPS7	Kaynak et al. (2003)
	Cell proliferation, differentiation, and apoptosis	Downregulation of AMD1, RIPK3, EGLN1, SIAHBP1, and ARVCF	
	Ion channel	SLC26A8, SLC16A5, SLC4A7, KCNS2, and KCNN3	
ASD	Cardiac-specific transcriptional factors	Downregulation of GATA4 and NKX2–5	Wang et al. (2016)
	Extracellular signal molecules	Downregulation of VEGFA and BMP10	
	Cardiac sarcomeric proteins	Downregulation of MYL2, MYL3, MYH7, TNNT1, and TNNT3	
TOF	Cell cycle and cardiac development	Upregulation of SNIP, A2BP1, and KIAA1437 and downregulation of STK33, BRDG1, and TEKT2	Kaynak et al. (2003)
	Ribosomal proteins	Upregulation of S6, l37a, S3A, S14, and L13A	
RVH	Stress response, cell proliferation, and metabolism	Upregulation of ADD2	Kaynak et al. (2003)

septal defect (VSD), atrial septal defect (ASD), and right ventricular hypertrophy (RVH) were reported (Table 2.6). The results showed that genes coding cardiac-specific transcriptional factors; regulators of cell proliferation, differentiation, and apoptosis; ribosomal proteins; and sarcomeric proteins displayed significant changes in various CHDs. These findings provided evidences of understanding pathogenic mechanism of CHDs and resources for validation of CHD genomic studies.

4.3.2 MicroRNAs in CHD

Recent evidences suggest a role in the dysregulation of miRNAs in CHD. Misexpression of these miRNAs has been shown to cause structural abnormalities in the heart and vasculature, thus furthering our understanding of the processes which result in CHD. Table 2.7 summarized the miRNAs so far implicated in development of CHD. These advances have provided new insight into the signaling pathways responsible for CHD. Furthermore, this new appreciation for miRNAs in the development of CHD has uncovered their potential for new therapeutic targets where modulated miR activity may reduce the burden of disease.

Table 2.7 MiRNAs in congenital heart disease

Congenital heart defect	Origin	MicroRNA	Annotation	References
VSD, chamber dilatation	Mice	Downregulation of miR-133a-1/miR-1-2; miR-133a-2/miR-1-1	Myocyte enhancer factor (MEF)-2-dependant enhancers	Liu et al. (2007), and Catalucci et al. (2008)
VSD	Human cardiac tissue	Downregulation of miR-1-1/miR-181c	MiR-1-1 targets BMPR2, Hand2 and GJA1; miR-181c targets SOX9	Li et al. (2013)
VSD	Mouse embryos	Downregulation of miR-92		Catalucci et al. (2008)
VSD	Mice	Downregulation of miR-17-92 cluster		Ventura et al. (2008)
VSD	Human maternal blood	Upregulation of 19b, 29c	Clinical biomarkers	Zhu and Department (2013)
VSD	Human plasma	Downregulation of let-7e-5p, miR-155-5p, miR-222-3p, miR-379-5p, miR-409-3p, miR-433, miR-487b	NOTCH1, HAND1, ZFPM2, and GATA3 were predicted as targets of let-7e-5p, miR-222-3p, and miR-433	Li et al. (2013)
Cardiac septation, morphogenesis, valve formation	Fetal human heart samples	Downregulation of miR-196	HOXB8-Shh signaling	Goddeeris et al. (2008)
Down syndrome	Human DNA	upregulation of miR-99a, let-7c, miR-125b-b, miR-155, miR-802		Latronico et al. (2008)
TOF	Human maternal blood	Downregulation of miR-19b, miR-22, miR-29c, miR -375, miR -421	Inverse correlation between the expression of miR-421 and SOX4	O'Brien et al. (2012), and Zhu and Department (2013)
BAV	Human aortic valves; porcine valvular interstitial cells	Downregulation of miR-26a, miR-95, miR-30b and miR-141		Nigam et al. (2010)

From reference Smith et al. (2015)
List of microRNAs implicated in congenital heart disease (CHD)
miR microRNA, *VSD* ventricular septal defect, *TOF* tetralogy of Fallot, *BAV* bicuspid aortic valve

References

Aad G, Abbott B, Abdallah J, Abdelalim AA, Abdesselam A, Abdinov O, Abi B, Abramowicz M, Abreu H, Acerbi H. Observation of a centrality-dependent dijet asymmetry in lead-lead collisions at root s(NN)=2.76 TeV with the ATLAS detector at the LHC. Phys Rev Lett. 2010;105(25):252303.

Abdelmohsen K, Panda A, Kang MJ, Xu J, Selimyan R, Yoon JH, et al. Senescence-associated lncRNAs: senescence-associated long noncoding RNAs. Aging Cell,12,5(2013-07-14). 2013;12(5):890.

Archacki S, Wang Q. Expression profiling of cardiovascular disease. Hum Genomics. 2004;1(5):1–16.

Bang C, Batkai S, Dangwal S, Gupta S, Foinquinos A, Holzmann A, Just A, Remke J, Zimmer K, Zeug A. Cardiac fibroblast–derived microRNA passenger strand-enriched exosomes mediate cardiomyocyte hypertrophy. J Clin Invest. 2014;124(5):2136.

Barrans JD, Allen PD, Stamatiou D, Dzau VJ, Liew CC. Global gene expression profiling of end-stage dilated cardiomyopathy using a human cardiovascular-based cDNA microarray. Am J Pathol. 2002;160(6):2035–43.

Bochenek G, Häsler R, El Mokhtari NE, König IR, Loos BG, Jepsen S, Rosenstiel P, Schreiber S, Schaefer AS. The large non-coding RNA ANRIL, which is associated with atherosclerosis, periodontitis and several forms of cancer, regulates ADIPOR1, VAMP3 and C11ORF10. Hum Mol Genet. 2013;22(22):4516.

Busch A, Eken SM, Maegdefessel L. Prospective and therapeutic screening value of non-coding RNA as biomarkers in cardiovascular disease. Ann Transl Med. 2016;4(12):236.

Calverley DC, Casserly IP, Choudhury QG, Phang TL, Gao B. Platelet gene expression as a biomarker risk stratification tool in acute myocardial infarction: a pilot investigation. Clin Med Insights: Blood Disord,2010,3(2010-07-29). 2010;2010(3):9–15.

Cao X, Wang J, Wang Z, Du J, Yuan X, Huang W, Meng J, Gu H, Nie Y, Ji B. MicroRNA profiling during rat ventricular maturation: a role for miR-29a in regulating cardiomyocyte cell cycle re-entry. FEBS Lett. 2013;587(10):1548–55.

Catalucci D, Latronico MV, Condorelli G. MicroRNAs control gene expression: importance for cardiac development and pathophysiology. Ann N Y Acad Sci. 2008;1123(1):20.

Chandramohan R, Wu PY, Phan JH, Wang MD. Benchmarking RNA-Seq quantification tools. Conf Proc IEEE Eng Med Biol Soc. 2013;2013(2013):647–50.

Chen J, Wang DZ. microRNAs in cardiovascular development. J Mol Cell Cardiol. 2012;52(5):949.

Chen Y, Park S, Li Y, Missov E, Hou M, Han X, Hall JL, Miller LW, Bache RJ. Alterations of gene expression in failing myocardium following left ventricular assist device support. Physiol Genomics. 2003;14(3):251.

Choudhary R, Iqbal N, Khusro F, Higginbotham E, Green E, Maisel A. Heart failure biomarkers. J Cardiovasc Transl Res. 2013;6(4):471.

Conesa A, Madrigal P, Tarazona S, Gomez-Cabrero D, Cervera A, Mcpherson A, Szcześniak MW, Gaffney DJ, Elo LL, Zhang X. A survey of best practices for RNA-seq data analysis. Genome Biol. 2016;17(1):13.

Corsten MF, Dennert R, Jochems S, Kuznetsova T, Devaux Y, Hofstra L, Wagner DR, Staessen JA, Heymans S, Schroen B. Circulating MicroRNA-208b and MicroRNA-499 reflect myocardial damage in cardiovascular disease. Circ Cardiovasc Genet. 2010;3(6):499.

D'alessandra Y, Devanna P, Limana F, Straino S, Di CA, Brambilla PG, Rubino M, Carena MC, Spazzafumo L, De SM. Circulating microRNAs are new and sensitive biomarkers of myocardial infarction. Eur Heart J. 2010;31(22):2765.

D'alessandra Y, Carena MC, Spazzafumo L, Martinelli F, Bassetti B, Devanna P, Rubino M, Marenzi G, Colombo GI, Achilli F. Diagnostic potential of plasmatic microRNA signatures in stable and unstable angina. PLoS One. 2013;8(11):e80345.

Delaughter DM, Bick AG, Wakimoto H, Mckean D, Gorham JM, Kathiriya IS, Hinson JT, Homsy J, Gray J, Pu W. Single-cell resolution of temporal gene expression during heart development. Dev Cell. 2016;39(4):480.

Devaux Y, Mueller M, Haaf P, Goretti E, Twerenbold R, Zangrando J, Vausort M, Reichlin T, Wildi K, Moehring B. Diagnostic and prognostic value of circulating microRNAs in patients with acute chest pain. J Intern Med. 2015;277(2):260–71.

Díaz-Araya G, Vivar R, Humeres C, Boza P, Bolivar S, Muñoz C. Cardiac fibroblasts as sentinel cells in cardiac tissue: receptors, signaling pathways and cellular functions. Pharmacol Res. 2015;101:30.

Economou EK, Oikonomou E, Siasos G, Papageorgiou N, Tsalamandris S, Mourouzis K, Papaioanou S, Tousoulis D. The role of microRNAs in coronary artery disease: from pathophysiology to diagnosis and treatment. Atherosclerosis. 2015;241(2):624.

Espinozalewis RA, Wang DZ. MicroRNAs in heart development. Curr Top Dev Biol. 2012;100(2):279.

Eulalio A, Mano M, Dal FM, Zentilin L, Sinagra G, Zacchigna S, Giacca M. Functional screening identifies miRNAs inducing cardiac regeneration. Nature. 2012;492(7429):376.

Fichtlscherer S, De Rosa S, Fox H, Schwietz T, Fischer A, Liebetrau C, Weber M, Hamm CW, Röxe T, Müller-Ardogan M. Circulating microRNAs in patients with coronary artery disease. Circ Res. 2010;107(5):677.

Friede KA, Ginsburg GS, Voora D. Gene expression signatures and the spectrum of coronary artery disease. J Cardiovasc Transl Res. 2015;8(6):339.

Furtado MB, Costa MW, Pranoto EA, Salimova E, Pinto AR, Lam NT, Park A, Snider P, Chandran A, Harvey RP. Cardiogenic genes expressed in cardiac fibroblasts contribute to heart development and repair. Circ Res. 2014;114(9):1422–34.

Gao C, Wang Y. Transcriptome complexity in cardiac development and diseases: an expanding universe between genome and phenome. Circ J Off J Jpn Circ Soc. 2014;78(5):1038–47.

Goddeeris MM, Rho S, Petiet A, Davenport CL, Johnson GA, Meyers EN, Klingensmith J. Intracardiac septation requires hedgehog-dependent cellular contributions from outside the heart. Development. 2008;135(10):1887–95.

Gomes CPC, Spencer H, Ford KL, Michel LYM, Baker AH, Emanueli C, Balligand JL, Devaux Y. The function and therapeutic potential of long non-coding RNAs in cardiovascular development and disease. Mol Ther Nucleic Acids. 2017;8(C):494.

Green SM, Green JA, Jr JJ. Natriuretic peptide testing for heart failure therapy guidance in the inpatient and outpatient setting. Am J Ther. 2009;16(2):171.

Grote P, Wittler L, Hendrix D, Koch F, Währisch S, Beisaw A, Macura K, Bläss G, Kellis M, Werber M. The tissue-specific lncRNA Fendrr is an essential regulator of heart and body wall development in the mouse. Dev Cell. 2013;24(2):206–14.

Healy AM, Pickard MD, Pradhan AD, Wang Y, Chen Z, Croce K, Sakuma M, Shi C, Zago AC, Garasic J. Platelet expression profiling and clinical validation of myeloid-related protein-14 as a novel determinant of cardiovascular events. Circulation. 2006;113(19):2278.

Hoeijmakers WA, Bártfai R, Stunnenberg HG. Transcriptome analysis using RNA-Seq. Methods Mol Biol. 2013;923(923):221.

Hoekstra M, Ca VDL, Halvorsen B, Gullestad L, Kuiper J, Aukrust P, Van Berkel TJ, Biessen EA. The peripheral blood mononuclear cell microRNA signature of coronary artery disease. Biochem Biophys Res Commun. 2010;394(3):792–7.

Hudson MP, Christenson RH, Newby LK, Kaplan AL, Ohman EM. Cardiac markers: point of care testing. Clin Chim Acta. 1999;284(2):223–37.

Hwang JJ, Allen PD, Tseng GC, Lam CW, Fananapazir L, Dzau VJ, Liew CC. Microarray gene expression profiles in dilated and hypertrophic cardiomyopathic end-stage heart failure. Physiol Genomics. 2002;10(1):31–44.

Kattman SJ, Witty AD, Gagliardi M, Dubois NC, Niapour M, Hotta A, Ellis J, Keller G. Stage-specific optimization of activin/nodal and BMP signaling promotes cardiac differentiation of mouse and human pluripotent stem cell lines. Cell Stem Cell. 2011;8(2):228.

Kaynak B, von Heydebreck A, Mebus S, Seelow D, Henig S, Vogel J, Sperling HP, Pregla R, Alexi-Meskeshvili V, Hetzer R. Genome-wide array analysis of normal and malformed human hearts. Circulation. 2003;107(19):2467.

Kiliszek M, Burzynska B, Michalak M, Gora M, Winkler A, Maciejak A, Leszczynska A, Gajda E, Kochanowski J, Opolski G. Altered gene expression pattern in peripheral blood mononuclear cells in patients with acute myocardial infarction. PLoS One. 2012;7(11):e50054.

Klattenhoff CA, Scheuermann JC, Surface LE, Bradley RK, Fields PA, Steinhauser ML, Ding H, Butty VL, Torrey L, Haas S. Braveheart, a long noncoding RNA required for cardiovascular lineage commitment. Cell. 2013;152(3):570–83.

Krishnaswamy G, Kelley J, Yerra L, Smith JK, Chi DS. Human endothelium as a source of multifunctional cytokines: molecular regulation and possible role in human disease. J Interferon Cytokine Res Off J Int Soc Interferon Cytokine Res. 1999;19(2):91.

Kuppusamy KT, Jones DC, Sperber H, Madan A, Fischer KA, Rodriguez ML, Pabon L, Zhu WZ, Tulloch NL, Yang X. Let-7 family of microRNA is required for maturation and adult-like metabolism in stem cell-derived cardiomyocytes. Proc Natl Acad Sci U S A. 2015;112(21):2785–94.

Latronico MV, Catalucci D, Condorelli G. MicroRNA and cardiac pathologies. Physiol Genomics. 2008;34(3):239–42.

Lech M, Anders HJ. Macrophages and fibrosis: how resident and infiltrating mononuclear phagocytes orchestrate all phases of tissue injury and repair. Biochim Biophys Acta. 2013;1832(7):989–97.

Li T, Cao H, Zhuang J, Wan J, Guan M, Yu B, Li X, Zhang W. Identification of miR-130a, miR-27b and miR-210 as serum biomarkers for atherosclerosis obliterans. Clin Chim Acta. 2011;412(1–2):66–70.

Li C, Fang Z, Jiang T, Zhang Q, Liu C, Zhang C, Xiang Y. Serum microRNAs profile from genome-wide serves as a fingerprint for diagnosis of acute myocardial infarction and angina pectoris. BMC Med Genet. 2013a;6(1):1–9.

Li D, Ji L, Liu L, Liu Y, Hou H, Yu K, Sun Q, Zhao Z. Characterization of circulating microRNA expression in patients with a ventricular septal defect. PLoS One. 2013b;9(8):e106318.

Li J, Cao Y, Ma XJ, Wang HJ, Ma D, Huang GY. Roles of miR-1-1 and miR-181c in ventricular septal defects. Int J Cardiol. 2013c;168(2):1441.

Li LM, Cai WB, Ye Q, Liu JM, Li X, Liao XX. Comparison of plasma microRNA-1 and cardiac troponin T in early diagnosis of patients with acute myocardial infarction. World J Emerg Med. 2014;5(3):182.

Li G, Xu A, Sim S, Priest JR, Tian X, Khan T, Quertermous T, Zhou B, Tsao PS, Quake SR. Transcriptomic profiling maps anatomically patterned subpopulations among single embryonic cardiac cells. Dev Cell. 2016;39(4):491.

Liquori ME, Christenson RH, Collinson PO, Defilippi CR. Cardiac biomarkers in heart failure. Clin Biochem. 2014;47(6):327–37.

Liu N, Williams AH, Kim Y, McAnally J, Bezprozvannaya S, Sutherland LB, Richardson JA, Bassel-Duby R, Olson EN. An intragenic MEF2-dependent enhancer directs muscle-specific expression of microRNAs 1 and 133. Proc Natl Acad Sci U S A. 2007;104(52):20844.

Lok DJ, Lok SI, Badings E, Lipsic E, Wijngaarden JV, Boer RAD, Veldhuisen DJV, Meer PVD. Galectin-3 is an independent marker for ventricular remodeling and mortality in patients with chronic heart failure. Clin Res Cardiol. 2013;102(2):103–10.

Long B, Liu CY, Liu F, Wang K, Zhou LY, Li PF, Zhou QY, Fan YYCARL. lncRNA inhibits anoxia-induced mitochondrial fission and apoptosis in cardiomyocytes by impairing miR-539-dependent PHB2 downregulation. Nat Commun. 2014;5(5):3596.

Maisel A, Hollander JE, Guss D, Mccullough P, Nowak R, Green G, Saltzberg M, Ellison SR, Bhalla MA, Bhalla V. Primary results of the Rapid Emergency Department Heart Failure Outpatient Trial (REDHOT). A multicenter study of B-type natriuretic peptide levels, emergency department decision making, and outcomes in patients presenting with shortness of breath. J Am Coll Cardiol. 2004;44(6):1328–33.

Matkovich SJ. Transcriptome analysis in heart failure. Curr Opin Cardiol. 2016;31(3):242.

Maznyczka A, Kaier T, Marber M. Troponins and other biomarkers in the early diagnosis of acute myocardial infarction. Postgrad Med J. 2015;91(1076):322–30.

Meder B, Keller A, Vogel B, Haas J, Sedaghat-Hamedani F, Kayvanpour E, Just S, Borries A, Rudloff J, Leidinger P. MicroRNA signatures in total peripheral blood as novel biomarkers for acute myocardial infarction. Basic Res Cardiol. 2011;106(1):13–23.

Michalik KM, You X, Manavski Y, Doddaballapur A, Zörnig M, Braun T, John D, Ponomareva Y, Chen W, Uchida S. Long noncoding RNA MALAT1 regulates endothelial cell function and vessel growthnovelty and significance. Circ Res. 2014;114(9):1389.

Michiels C. Endothelial cell functions. J Cell Physiol. 2003;196(3):430–43.

Minami Y, Satoh M, Maesawa C, Takahashi Y, Tabuchi T, Itoh T, Nakamura M. Effect of atorvastatin on microRNA 221/222 expression in endothelial progenitor cells obtained from patients with coronary artery disease. Eur J Clin Investig. 2009;39(5):359–67.

Miquerol L, Kelly RG. Organogenesis of the vertebrate heart. Wires Dev Biol. 2013;2(1):17–29.

Mueller M, Vafaie M, Biener M, Giannitsis E, Katus HA. Cardiac troponin T: from diagnosis of myocardial infarction to cardiovascular risk prediction. Circ J Off J Jpn Circ Soc. 2013;77(7):1653–61.

Nigam V, Sievers HH, Jensen BC, Sier HA, Simpson PC, Srivastava D, Mohamed SA. Altered micrornas in bicuspid aortic valve: a comparison between stenotic and insufficient valves. J Heart Valve Dis. 2010;19(4):459.

O'Brien J Jr, Kibiryeva N, Zhou XG, Marshall JA, Lofland GK, Artman M, Chen J, Bittel DC. Noncoding RNA expression in myocardium from infants with tetralogy of Fallot. Circ Cardiovasc Genet. 2012;5(3):279.

Oerlemans MI, Mosterd A, Dekker MS, De Vrey EA, Van MA, Pasterkamp G, Doevendans PA, Hoes AW, Sluijter JP. Early assessment of acute coronary syndromes in the emergency department: the potential diagnostic value of circulating microRNAs. EMBO Mol Med. 2012;4(11):1176–85.

Oeth P, Parry GC, Mackman N. Regulation of the tissue factor gene in human monocytic cells. Role of AP-1, NF-kappa B/Rel, and Sp1 proteins in uninduced and lipopolysaccharide-induced expression. Thromb Haemost. 1997;78(1):747–54.

Olivieri F, Antonicelli R, Lorenzi M, D'alessandra Y, Lazzarini R, Santini G, Spazzafumo L, Lisa R, La SL, Galeazzi R. Diagnostic potential of circulating miR-499-5p in elderly patients with acute non ST-elevation myocardial infarction. Int J Cardiol. 2013;167(2):531–6.

Pei H, Wei L, Lin CH, Jin Y, Shang C, Nuernberg ST, Kai JK, Xu W, Lin CY, Lin CJ. A long noncoding RNA protects the heart from pathological hypertrophy. Nature. 2014;514(7520):102–6.

Rana MS, Christoffels VM, Moorman AF. A molecular and genetic outline of cardiac morphogenesis. Acta Physiol. 2013;207(4):588–615.

Ratajczak MZ. Igf2-H19, an imprinted tandem gene, is an important regulator of embryonic development, a guardian of proliferation of adult pluripotent stem cells, a regulator of longevity, and a 'passkey' to cancerogenesis. Folia Histochem Cytobiol. 2012;50(2):171–9.

Rooij EV, Sutherland LB, Thatcher JE, Dimaio JM, Naseem RH, Marshall WS, Hill JA, Olson EN. Dysregulation of microRNAs after myocardial infarction reveals a role of miR-29 in cardiac fibrosis. Proc Natl Acad Sci U S A. 2008;105(35):13027.

Rosenberg S, Elashoff MR, Beineke P, Daniels SE, Wingrove JA, Tingley WG, Sager PT, Kraus WE, Newby LK, Schwartz RS. Multi-center validation of the diagnostic accuracy of a blood-based gene expression test for assessment of obstructive coronary artery disease in non-diabetic patients. Ann Intern Med. 2010;153(7):425–34.

Santoro MM, Nicoli S. miRNAs in endothelial cell signaling: the endomiRNAs. Exp Cell Res. 2013;319(9):1324–30.

Scheuermann JC, Boyer LA. Getting to the heart of the matter: long non-coding RNAs in cardiac development and disease. EMBO J. 2013;32(13):1805–16.

Smith T, Rajakaruna C, Caputo M, Emanueli C. MicroRNAs in congenital heart disease. Ann Transl Med. 2015;3(21):333.

Sondermeijer BM, Bakker A, Halliani A, Ronde MWJD, Marquart AA, Tijsen AJ, Mulders TA, Kok MGM, Battjes S, Maiwald S. Platelets in patients with premature coronary artery disease exhibit upregulation of miRNA340* and miRNA624*. PLoS One. 2011;6(10):e25946.
Später D, Hansson EM, Zangi L, Chien KR. How to make a cardiomyocyte. Development. 2014;141(23):4418–31.
Steenbergen C, Afshari CA, Petranka JG, Collins J, Martin K, Bennett L, Haugen A, Bushel P, Murphy E. Alterations in apoptotic signaling in human idiopathic cardiomyopathic hearts in failure. Am J Physiol Heart Circ Physiol. 2003;284(1):H268.
Taurino C, Miller WH, Mcbride MW, Mcclure JD, Khanin R, Moreno MU, Dymott JA, Delles C, Dominiczak AF. Gene expression profiling in whole blood of patients with coronary artery disease. Clin Sci. 2010;119(8):335.
Uosaki H, Cahan P, Lee D, Wang S, Miyamoto M, Fernandez L, Kass D, Kwon C. Transcriptional landscape of cardiomyocyte maturation. Cell Rep. 2015;13(8):1705–16.
Ventura A, Young AG, Winslow MM, Lintault L, Meissner A, Erkeland SJ, Newman J, Bronson RT, Crowley D, Stone JR. Targeted deletion reveals essential and overlapping functions of the miR-17 through 92 family of miRNA clusters. Cell. 2008;132(5):875–86.
Wang Z, Gerstein M, Snyder M. RNA-Seq: a revolutionary tool for transcriptomics. Nat Rev Genet. 2009;10(1):57.
Wang GK, Zhu JQ, Zhang JT, Li Q, Li Y, He J, Qin YW, Jing Q. Circulating microRNA: a novel potential biomarker for early diagnosis of acute myocardial infarction in humans. Eur Heart J. 2010;31(6):659.
Wang H, Lin YZ, Lu HM, Zhou Y, Luo C, Lu SH, Zhao L, Liu F. Circulating microRNA-92a in patients with ST-segment elevation myocardial infarction. Zhongguo Wei Zhong Bing Ji Jiu Yi Xue = Chin Crit Care Med = Zhongguo Weizhongbing Jijiuyixue. 2011;23(12):718.
Wang F, Long G, Zhao C, Li H, Chaugai S, Wang Y, Chen C, Wang DW. Atherosclerosis-related circulating miRNAs as novel and sensitive predictors for acute myocardial infarction. PLoS One. 2014a;9(9):e105734.
Wang K, Liu F, Zhou LY, Long B, Yuan SM, Wang Y, Liu CY, Sun T, Zhang XJ, Li PF. The long noncoding RNA CHRF regulates cardiac hypertrophy by targeting miR-489. Circ Res. 2014b;114(9):1377–88.
Wang W, Niu Z, Wang Y, Li Y, Zou H, Yang L, Meng M, Wei C, Li Q, Duan L. Comparative transcriptome analysis of atrial septal defect identifies dysregulated genes during heart septum morphogenesis. Gene. 2016;575(2):303–12.
Weber M, Baker MB, Patel RS, Quyyumi AA, Bao G, Searles CD. MicroRNA expression profile in CAD patients and the impact of ACEI/ARB. Cardiol Res Pract. 2011;2011(6):532915.
Weber KT, Sun Y, Bhattacharya SK, Ahokas RA, Gerling IC. Myofibroblast-mediated mechanisms of pathological remodelling of the heart. Nat Rev Cardiol. 2013;10(1):15.
Wingrove JA, Daniels SE, Sehnert AJ, Tingley W, Elashoff MR, Rosenberg S, Buellesfeld L, Grube E, Newby LK, Ginsburg GS. Correlation of peripheral-blood gene expression with the extent of coronary artery stenosis. Circ Cardiovasc Genet. 2008;1(1):31.
Yang IS, Kim S. Analysis of whole transcriptome sequencing data: workflow and software. Genomics Inform. 2015;13(4):119.
Yao R, Ma Y, Du Y, Liao M, Li H, Liang W, Yuan J, Ma Z, Yu X, Xiao H. The altered expression of inflammation-related microRNAs with microRNA-155 expression correlates with Th17 differentiation in patients with acute coronary syndrome. Cell Mol Immunol. 2011;8(6):486.
Yasue H, Yoshimura M, Sumida H, Kikuta K, Kugiyama K, Jougasaki M, Ogawa H, Okumura K, Mukoyama M, Nakao K. Localization and mechanism of secretion of B-type natriuretic peptide in comparison with those of A-type natriuretic peptide in normal subjects and patients with heart failure. Circulation. 1994;90(1):195–203.
Zhu S, Department C. Identification of maternal serum microRNAs as novel noninvasive biomarkers for prenatal detection of fetal congenital heart defects. Clin Chim Acta. 2013;424:66–72.

Chapter 3
The Emerging Role of Epigenetics

Lu Qian Wang, Kailash Singh, Aung Moe Zaw, and Billy Kwok Chong Chow

Abstract Epigenetics is one of the most rapidly expanding fields in biology over the past decades. Epigenetic mechanisms, including DNA methylation, histone modifications, and RNA-associated editing, lead to the heritable silencing/activation of genes without changes in DNA sequence. The critical role of epigenetic modifications has been demonstrated in normal and disease development in humans. With the advent of next-generation sequencing, the technological breakthrough makes it possible to unveil the genome-wide mapping of epigenetic changes. Here, we give a comprehensive overview of epigenetic mechanisms and focus on the recent progress of epigenetic modifications involved in the pathogenesis or progression of human diseases, in particular, cardiovascular diseases and cancers. In addition, some current epigenetic therapies including the inhibitors of DNA methyltransferases and histone deacetylases that have shown promising therapeutic effects will be also discussed.

Keywords Epigenetics · DNA methylation · Histone modification · miRNAs and lncRNAs

Abbreviations

3′ UTR	3′ untranslated region
5-aza-dC	5-aza-2′-deoxycytidine
CADs	Coronary artery diseases
ChIP	Chromatin immunoprecipitation
CVDs	Cardiovascular diseases
DCM	Dilated cardiomyopathies

L. Q. Wang · K. Singh · A. M. Zaw · B. K. C. Chow (✉)
School of Biological Sciences, The University of Hong Kong, Hong Kong, China
e-mail: wanglucy@hku.hk; kailash@connect.hku.hk; moezaw@connect.hku.hk; bkcc@hku.hk

H. Jiang, M. Liu (eds.), *Heart Genomics*, Translational Bioinformatics 16,
https://doi.org/10.1007/978-981-13-1429-2_3

DNMTs	DNA methyltransferases
eNOS	Endothelial nitric oxide synthase
HATs	Histone acetyltransferases
HDACs	Histone deacetylases
HF	Heart failure
iNOS	Inducible nitric oxide synthase
lncRNAs	Long noncoding RNAs
MBDs	Methyl-CpG-binding proteins
miRNAs	microRNAs
NGS	Next-generation sequencing
NO	Nitric oxide
PAH	Pulmonary arterial hypertension
SMCs	Smooth muscle cells
TSGs	Tumor suppressor genes

1 Introduction

1.1 Cardiovascular Diseases

Cardiovascular diseases (CVDs) are the worldwide leading cause of human morbidity and mortality. CVDs encompass a series of pathological conditions, such as hypertension, coronary artery diseases (CADs), myocardial infarction (MI), and chronic heart failure (CHF). CHF is the final stage of most CVDs; however, recent pharmacotherapy on CHF management is limited to diuretics, renin-angiotensin-aldosterone (RAA) inhibitors, and vasodilators, targeting blood pressure, vasodilation, cardiac contractility, and hypertrophy (Swedberg et al. 2005). Despite efforts in revitalizing therapeutic management over decades, mortality remains unacceptably high with 50% of survival rate 5 years after diagnosis (Roger 2013). Moreover, after the process of thrombosis and coronary atherosclerosis, CAD could evolve into myocardial ischemia by plaque rupture or erosion (Ambrose et al. 1988). CAD is another leading cause of morbidity and mortality affecting the cardiovascular systems in developed countries. Therefore, in addition to the known mechanisms and therapies, there is clearly an unmet need for the exploration of better therapeutic targets for CVDs.

1.2 Epigenetics

The word "epigenesis" was coined by Conrad Waddington in the twentieth century to link the fields of developmental biology and genetics, which is broadly defined as how genotypes give rise to phenotypes during development (Waddington 1952). Several decades later, the term "epigenetics" was proposed as the study of

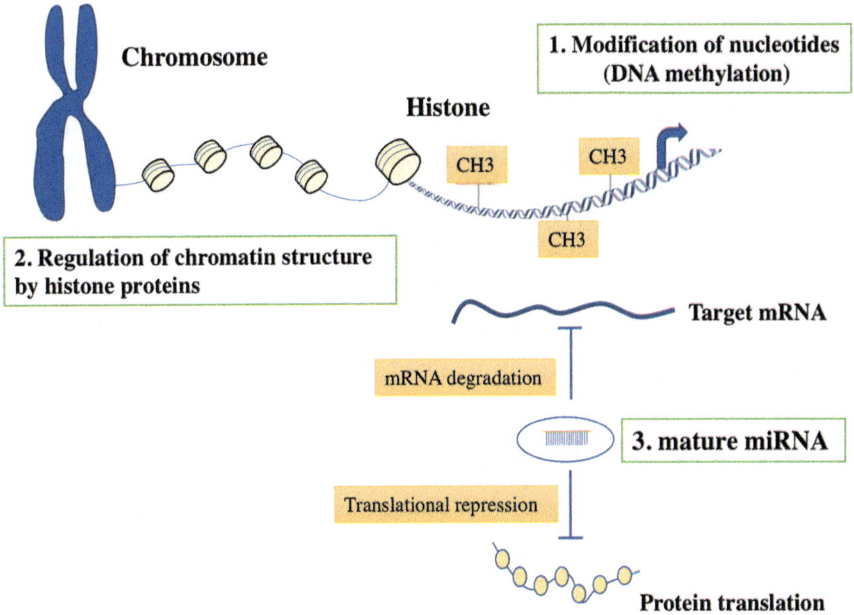

Fig. 3.1 Mechanisms of epigenetic modifications: DNA methylation, histone modification, and RNA-associated silencing

mitotically and meiotically heritable changes in gene function that are not coded in DNA sequence itself (Riggs 1975; Razin and Riggs 1980). Epigenetics explores the mechanisms that make cells adapt quickly to environmental changes and build a bridge between the environment and genes. The mechanisms of heritable epigenetics have been extensively studied in different organisms and encompassed an array of molecular modifications to both DNA and chromatin, including DNA methylation, histone modification, and RNA-associated silencing, which are used to initiate and sustain epigenetic silencing (Egger et al. 2004) (Fig. 3.1). These modulators could modify the binding of transcription activators and repressors with respective gene promoters and/or change the function and conformation of chromatin, further changing the expression and function of some key genes implicated in specific signaling pathways (Gluckman et al. 2009). In general, compared with the short and flexible effect of histone modifications, DNA methylation seems to work as long-term silencers of gene expression, in which the cross talk between them exists in different mechanisms (Goldberg et al. 2007). Furthermore, the function of epigenetics has been extensively evaluated in human cancers, and recent studies have started to explore the research field of epigenetic role implicated in the development and progression of CVDs (Turunen et al. 2009; Loscalzo and Handy 2014).

1.3 DNA Methylation

DNA methylation was proposed as the first crucial epigenetic mechanism for the switching and heritability of gene activities (Griffith and Mahler 1969; Holliday and Pugh 1975). As the most widely studied epigenetic modification, DNA methylation is defined as the addition of a methyl group (-CH3) to the carbon 5 position of cytosine ring in a CpG dinucleotide by DNA methyltransferases (DNMTs) (Jones 2001; Robertson 2005). The DNA methylation machinery is made up of two parts, DNMTs, which establish and maintain DNA methylation patterns, and the methyl-CpG-binding proteins (MBDs) family, which are candidates for the readout of DNA methylation (Robertson 2005).

MBD family is composed of five well-characterized proteins, including MeCP2, MBD1, MBD2, MBD3, and MBD4, of which, at least MBD1, MBD2, and MeCP2 could directly repress transcription (Ng et al. 1999; Ballestar and Esteller 2005). Methylation of the promoter CpG island results in the recruitment and binding of MBDs and transcription repressors including histone deacetylases (HDACs) and Sin3A, which further leads to the blockage of transcription initiation (Jones et al. 1998; Ng, Zhang et al. 1999). On the contrary, unmethylated promoters commonly lack MBD proteins with the exception of MBD1 (Lopez-Serra et al. 2006). Upon the treatment of demethylating agent, the promoter CpG islands of cancer cells will be hypomethylated and accompanied by MBD protein release and gene re-expression, suggesting the important association of MBD proteins and methylation status.

DNMT1, DNMT3a, and DNMT3b have the transcription repressor domains which can recruit HDACs and/or other corepressor proteins to DNA in a way similar to MBDs (Rountree et al. 2000; Bachman et al. 2001). Of the three DNMTs, DNMT1 is required for maintaining patterns of methylation, while DNMT3a and DNMT3b are responsible for de novo methylation and mammalian development (Okano et al. 1999; Branco et al. 2008). The maintenance DNMT1 is most abundant in the cells, which is needed to methylate hemimethylated sites that are generated during semiconservative DNA replication (Jones and Liang 2009; Portela and Esteller 2010). Different from DNMT1, DNMT3A and DNMT3B are highly expressed in embryonic stem cells and downregulated in differential cells, which might serve to establish the pattern of methylation during embryonic development (Esteller 2007). Methylation gives some explanations to phenomena in many cellular processes, such as gene imprinting or X chromosome inactivation in female mammals, both of which can be stably maintained (Reik and Walter 2001; Huynh and Lee 2005).

In normal tissues, DNA methylation occurs in 80% of repetitive DNA regions, including tandem repeats (such as satellite DNA) and dispersed repeats (such as long interspersed transposable elements (LINEs), short interspersed transposable elements (SINEs), and endogenous retroviruses) (Yoder et al. 1997; Robertson 2005). CpG islands are small concentrated regions of CpG-rich dinucleotides with the length of 0.5 to several Kb, which show elevated G+C base composition (Bird

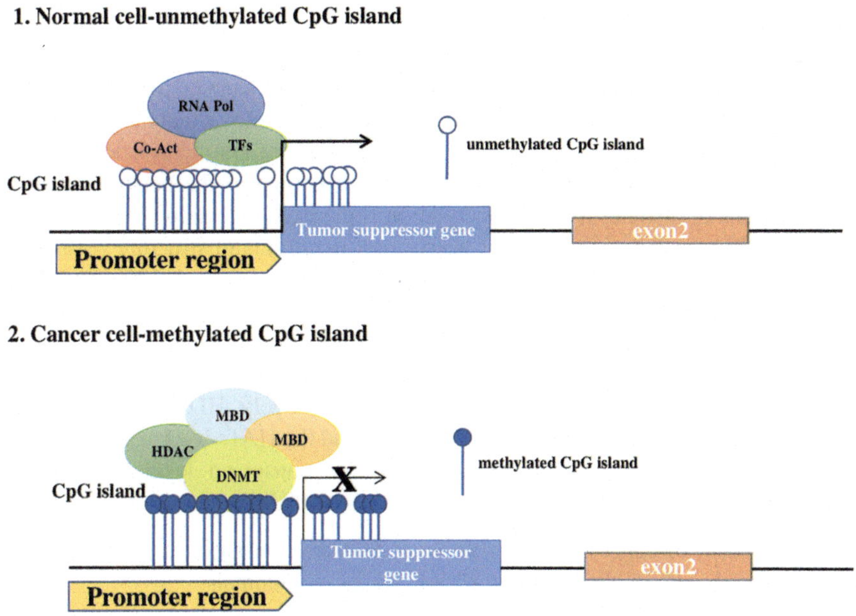

Fig. 3.2 DNA methylation patterns in normal and cancer cells

1986). The frequency of CpG dinucleotides in human genome is lower than expected; however, approximately 70% of human gene promoters contain CpG islands, whereas some CpG islands are found within the body of genes or even 3′ region (Saxonov et al. 2006). The majority of these promoters are associated with housekeeping genes, as well as half of tissue-specific genes and development regulatory genes (Bird 1986; Larsen et al. 1992).

DNA methylation of these promoter CpG islands correlates with transcriptional repression in mammalian cells (Goll and Bestor 2005) (Fig. 3.2). Normally, the CpG islands associated with the promoters are usually unmethylated, leading to the transcription of the gene. The importance of DNA methylation for inhibition of gene expression and transcriptional silencing is apparent in the exceptions of gene imprinting or X chromosome inactivation in females (Herman and Baylin 2003). The fully methylated CpG islands leading to global gene silencing on X chromosomes and the silenced alleles of some imprinted genes are considered as a form to distinguish maternal and parental determination which are programmed in condition that only one allele of the gene is expressed in normal tissues (Antequera and Bird 1993; Bird 2002; Reik and Lewis 2005). Genomic proper patterns of methylation in mammals are essential for gene regulation and chromatin organization during embryogenesis (Goll and Bestor 2005).

1.4 Histone Modifications

Chromatin is the complex of DNA and proteins that are packaged within the nucleus of eukaryotic cells. Histones are a family of those small proteins that facilitate DNA to condense into chromatin. Importantly, the acetylation and methylation of conserved lysine residues in histone tail domain are known as the posttranslational modifications of histones. Histone acetylation is the best-understood histone modification. Of note, the critical role of acetylation in chromatin structure regulation and gene expression has been recently demonstrated (Cheung et al. 2000; Yang and Seto 2007). Most well-characterized acetylation substrates are nucleus-localized transcription factors and its coregulatory machinery (Blander and Guarente 2004; Yang and Seto 2008). These transcription repressors and activators recruit HDACs and histone acetyltransferase (HAT), respectively. Largely, HDACs lead to transcriptional inhibition by removing acetyl group, and the HATs work to increase the transcription by adding the acetyl group to the conserved lysine residue on N-terminal tail of the histone proteins (Brown et al. 2000).

HDAC Recent evidence suggests that many HDACs consist of highly conserved domain and are subdivided into four major classes in mammals.

1. Class I HDACs (HDAC1, 2, 3, and 8) comprise of domain surrounded by short NH2 and COOH termini and are widely expressed.
2. Class II HDACs contain one or two catalytic sites and are further subdivided into two subclasses, including subclass IIa (HDAC4, 5, 7, and 9) and subclass IIb (HDAC6 and 10) (Yang and Seto 2008).
3. Class III HDACs are sirtuins (SIRT1-SIRT7).
4. Class IV is known to be homologs of Hda1 and Rpd3 protein of yeast and consists of a solitary membrane HDAC11 (Majumdar et al. 2012; Wang et al. 2014).

HAT There are two subcategories of HATs including nuclear and cytoplasmic, which are designated as type A HATs and type B HATs, respectively (Thiagarajan et al. 2016).

Type A is transcription related and further divided into five families.

1. GNAT-related acetyltransferase family represented by general control nonderepressible (GCN5), p300/CBP-associated factor (PCAF), and elongator complex protein 3 (ELP3).
2. p300/CBP family represented by p300 and CBP.
3. MYST family consists of MYST1 (HMOF, males absent on the first), MYST2 (HBO1, histone acetyltransferase binding to ORC), MYST3 (MOZ, monocytic leukemia zinc finger), MYST4 (MORF, monocytic leukemia zinc finger protein-related factor), and TIP60 (tat interacting protein 60 kDa).
4. Basal TF family: TFIIIC (transcription factor IIIC), TAF1.
5. Nuclear receptor cofactor (NRCF) family, steroid receptor coactivator (SRC), ACTR/NCOA3 (nuclear receptor coactivator 3).

Type B HATs such as HAT1, HAT2, Hat B3.1, and Rtt109 are relatively characterized.

Briefly, histone acetylation can modify gene transcription at both global acetylation level and promoter-specific acetylation level (Gottesfeld and Forbes 1997; Vaissière et al. 2008). More importantly, there is evidence showing the intimate cross talk between histone modification and DNA methylation in the process of gene silencing. The loss of histone acetylation may initiate gene silencing, which is followed by the induction of DNA methylation (Daujat et al. 2002; Vaissière et al. 2008).

1.5 Noncoding RNAs

Recently, the discovery of the noncoding RNAs (ncRNAs) provided further evidence to posttranslational mechanisms defined as epigenetic changes. It has become apparent that noncoding portion of the genome is of great importance for normal development, differentiation, and multiple diseases (Friedman and Jones 2009; Esteller 2011). Those ncRNAs involved in epigenetic processes could be categorized into the short ncRNAs (<30 nucleotides (nt)) and the long noncoding RNAs (lncRNAs) (>200 nts). The three major classes of short ncRNAs are microRNAs (miRNAs), short interfering RNAs (siRNAs), and piwi-interacting RNAs (piRNAs) (Kowalczyk et al. 2012). Both major groups are found to implicate in a wide range of cellular processes, heterochromatin formation, histone modification, and transcriptional silencing (Grewal and Moazed 2003; Moazed 2009). Here, we mainly focus on the epigenetic role of miRNAs and lncRNAs.

miRNAs belong to a class of single-stranded, small noncoding RNAs of 19–25 nts in length. miRNAs are transcribed from long RNA precursors (pri-miRNAs) by polymerase II and stabilized by 5′ capping and 3′ polyadenylation (Bartel 2009) (Fig. 3.3). The hairpin secondary structure of pri-miRNA is recognized and processed by the microprocessor complex (RNase III enzyme Drosha and DGCR8/Pasha) to release 60–80 nts pre-miRNA (Lee et al. 2003; Denli et al. 2004; Gregory et al. 2004). A pre-miRNA of the hairpin or stem-loop structure is exported into cytoplasm by Exportin 5 and further cleaved by another RNase III enzyme Dicer to generate the mature miRNA (Han et al. 2004; Kim 2005). Mature miRNAs result in translational repression or mRNA degradation of the target protein-coding genes through imperfect base pairing with the 3′ untranslated region (3′ UTR) of targeted mRNAs (Bartel 2004).

On the other hand, with the advent of high-throughput genome studies, a new class of lncRNAs measuring longer than 200 nts in length without evident protein-coding function was discovered (Martin et al. 2012). As a broad definition, lncRNA is encompassed by different classes of RNA transcripts, such as small nucleolar RNA (snoRNA), enhancer RNAs, and intergenic transcripts (Batista and Chang 2013). lncRNAs could regulate gene expression by modification of accessibility of gene promoters, which is further facilitated to tune the gene expression when

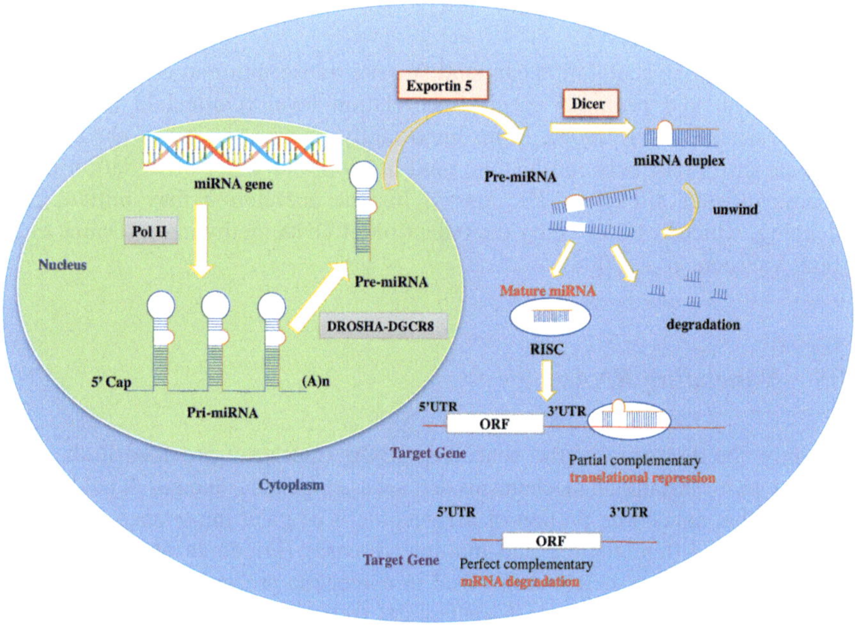

Fig. 3.3 miRNA biogenesis and function

adapting to the changes of environmental conditions or silence a gene during the developmental program (Batista and Chang 2013; Kornienko et al. 2013). Multiple studies have shown that the expression of miRNAs and lncRNAs is more tissue or cell specific than that of protein-coding genes (Lagos-Quintana et al. 2002; Ravasi et al. 2006; Djebali et al. 2012).

2 Epigenetics and Diseases

Gene expression patterns in different cell types should be appropriately maintained responding to environmental and developmental changes, whereas inappropriate expression might result in diseases. The important association of epigenetics with diseases is addressed by the increasing number of human diseases that are known to happen when the epigenetic information is not correctly maintained. It has become clear that aberrant regulation of DNA methylation, miRNAs, and lncRNAs contributes to the initiation or progression of human diseases (Robertson 2005; Calin and Croce 2006; Brunner et al. 2012). Further interests in this field come from the evidence of the interaction and cross talk among DNA methylation, histone modification, and small RNAs, as well as the developing pharmacological ways of reversing epigenetic abnormalities (Brown and Strathdee 2002; Tsai et al. 2010).

2.1 *DNA Methylation and Diseases*

2.1.1 DNA Methylation and Cancer

A large group of diseases are linked to the abnormalities of DNA methylation, showing the aberrant DNA methylation patterns and the pathological impact of their dysregulation in repeat instability. Among all diseases, the best-known studied disease is cancer. The association of methylation and cancer was firstly reported in 1983, which has been shown that cancer cells are hypomethylated compared with normal counterparts (Feinberg and Vogelstein 1983). Loss of methylation is an early and frequent event in carcinogenesis, which is generally more pronounced with cancer severity and metastatic potential in different tumor types (Widschwendter et al. 2004). For instance, age-dependent hypomethylation of specific DNA sequence was found to precede diploidy loss in a subset of gastrointestinal cancers (Suzuki et al. 2006). On the other hand, hypomethylation in cancer cells may result in the activation of oncogenes, further leading to the genomic rearrangements and instabilities (Herman and Baylin 2003). The synuclein-γ gene seems to be a target of oncogenic effect, demethylation of which serves as a molecular indicator of metastasis in a large group of human cancers, including breast cancer, ovarian cancer, and hepatocellular cancer (Gupta et al. 2003; Zhao et al. 2006).

In contrast to oncogene, biallelic loss of function in tumor suppressor genes (TSGs) is responsible for tumorigenesis as described by the Knudson's hypothesis. This hypothesis was derived from families with hereditary cancers, in which children who suffer an inactivating germ line mutation of one allele and later a second somatic mutation of the other allele have the chance to predispose to cancer formation (Knudson 1997). Of note, essential TSGs are found to be mapped to frequent regions of homozygous deletion or loss of heterozygosity (LOH) in some human cancers, such as CDKN2A mutation in melanoma, RB1 mutation in retinoblastoma, and TP53 inactivation in Li-Fraumeni syndrome (Malkin 1993; Monzon et al. 1998; Thiagalingam et al. 2002; Wang and Chim 2015). Given that DNA methylation results in gene silencing, promoter methylation of TSGs may work as a second hit in addition to gene mutation or deletion, to fulfill the Knudson's hypothesis in carcinogenesis (Wang and Chim 2015). Global DNA hypomethylation but gene-specific hypermethylation of promoter-associated CpG islands is a hallmark of human cancers (Esteller 2007). Interestingly, promoter CpG island methylation of TSGs has been noted in a variety of human cancers. In addition to gene silencing and loss of classic tumor suppressor function, it could also affect different cellular signaling pathways with tumor-specific profile and miRNAs with growth inhibitory role (Esteller 2007; Wang and Chim 2015). For instance, a diverse group of TSGs has been reported to be methylated in solid tumors and hematological cancers, suggesting the implication of hypermethylation in the carcinogenesis, including CDKN2A/CDKN2B regulating cell cycle in ependymal tumors, SFRP family (SFRP1/SFRP2/SFRP3/SFRP5) regulating WNT signaling in colorectal cancer and acute myeloid leukemia, and SOCS1/SOCS3/SHP1 regulating the JAK/STAT signaling pathways in hepatocellular carcinoma and multiple myeloma (Yoshikawa et al. 2001; Rousseau et al. 2003; Chim et al. 2004; Suzuki et al. 2004; Niwa et al. 2005; Valencia et al. 2009).

The genome-wide DNA hypomethylation is a consistent finding in human tumors; however, the importance and the detailed mechanism of this change remain an open question. Several animal models are used for studying the effects of changes in DNA hypomethylation implicated in carcinogenesis. Lin et al. investigated the effect of conditional inactivation of de novo methyltransferase DNMT3b in $Apc^{Min/+}$ mice. The results support the role of DNMT3b in the transition stage between microadenoma formation and macroscopic colonic tumor growth, and suggest the link of DNA hypomethylation and decreased progression of microadenoma to macroscopic tumor growth (Lin et al. 2006). On the other hand, several studies show the important function of maintenance DNA methyltransferase DNMT1 involved in tumor formation. DNA hypomethylation in the transgenic $Apc^{Min/+}$ mice containing a mutant DNMT1 is shown to promote early formation of hepatocellular adenomas and carcinomas but suppress overal progression of tumorigenesis, indicating the opposite effects of DNA hypomethylation in different cell types (Yamada et al. 2005). Mice carrying a hypomorphic allele of DNMT1 ($DNMT1^{chip/-}$) developed thymic lymphomas with the generation of intracisternal A particle (IAP) somatic insertion in the middle of the oncogenic Notch1 genomic locus, demonstrating that DNA hypomethylation could induce tumors accompanied by the activation of endogenous retroviral elements and thus the chromosomal instabilities (Howard et al. 2008). Therefore, the different outcomes of animal models may vary from different types of cells, tissues, and tumors that are studied, the background of the animals, as well as the specific DNA sequences and chromosomal variations (Song et al. 2005; Yamada et al. 2005; Illingworth et al. 2008).

2.1.2 DNA Methylation and Atherosclerosis

The role of methylation involved in CVDs remains largely unknown. It has been noted that loss of DNA methylation is present in human atherosclerotic lesions, whereas hypermethylation of several genes is implicated in the pathogenesis of atherosclerosis, including extracellular superoxide dismutase, estrogen receptor-α (ER-α), endothelial nitric oxide synthase (eNOS), and 15-lipoxygenase (Turunen et al. 2009). Hiltunen et al. firstly demonstrated that a global genomic hypomethylation with enhanced transcriptional activity was shown during atherogenesis in advanced human, mouse, and rabbit atherosclerotic lesions, further affecting the cellular proliferation and gene expression (Hiltunen et al. 2002). Interestingly, it has been noted that in mice model during early atherosclerosis, local DNA hypermethylation and hypomethylation occurred in normohomocysteinemic mice, and the atherogenic lipoprotein profiles could influence and promote DNA hypermethylation in cultured human macrophages (Lund et al. 2004; Zaina et al. 2005). This suggested that nutritional factors could modify DNA methylation patterns that are likely to be independent of vitamin or homocysteine levels, and moreover, DNA methylation profiles may serve as early markers of atherosclerosis (Zaina et al. 2005).

Indeed, DNA methylation status could accelerate the process of atherosclerosis through upregulating atherosclerosis-susceptible genes and downregulating atherosclerosis-protective genes (Stenvinkel et al. 2007). The epigenetic changes occur in smooth muscle cells (SMCs), which constitute most of the cellular mass in the arterial wall. For instance, methylation of ER-α promoter was found in SMCs instead of endothelial cells (Post et al. 1999). ER was noted to play the role of active proliferating atheromatous SMCs, and meanwhile, it was the first gene that was well characterized for the association between aging of vascular system and the methylation of CpG islands in normal tissues in vitro (Toyota and Issa 1999; Turunen et al. 2009). Moreover, DNA methylation was also investigated to change monocarboxylate transporters (MCTs) by suppressing MCT3 expression, leading to the regulation of SMC proliferation and the development of atherosclerosis (Zhu et al. 2005). Other genes affected by aging are insulin-like growth factor 2 (IGF-2), N33, MYOD1, and cMYC (Toyota and Issa 1999; Hiltunen and Ylä-Herttuala 2003). Furthermore, inducible nitric oxide synthase (iNOS) promoter was heavily methylated in a variety of normal primary endothelial cells and vascular smooth muscle cells, whereas eNOS is hypomethylated in normal endothelium and controlled by cell-specific methylation or histone modifications (Chan et al. 2005; Fish et al. 2005). However, iNOS is expressed in endothelial cells in atherosclerotic lesions which raises the possibility that modifications in epigenetic mechanisms may lead to the changes (Shirodkar and Marsden 2011). Therefore, recent studies showed that aberrant DNA methylation occurs during atherogenesis and contributes to the lesion development and cell proliferation.

2.1.3 DNA Methylation with Cardiovascular Risk and Diseases

DNA methylation was shown to have the correlation with cardiovascular risk and mortality. The association of repetitive element LINE-1 hypomethylation and cardiovascular events, such as ischemic heart diseases and stroke, in cross-sectional and longitudinal analyses of the population has revealed that DNA hypomethylation might predict the diagnosis of CVDs before the clinical onset (Baccarelli et al. 2010). This finding showed that DNA methylation could be an effective marker that predicts the transgenerational cardiovascular risks in early life of adult diseases (Baccarelli et al. 2010). Moreover, DNA methylation was also measured and analyzed in peripheral blood leukocytes from the group of chronic kidney diseases (CKDs). It showed that inflamed patients with CKDs had global hypermethylation in peripheral blood lymphocyte DNA, suggesting the correlation of DNA methylation with increased cardiovascular mortality (Stenvinkel et al. 2007).

With regard to the DNA methylation involved in the heart failure (HF) or cardiomyopathy, there are two interesting studies supporting the essential role of DNA methylation implicated in the pathogenesis of cardiomyopathies. The global methylation profile within the CpG islands of promoters and within gene bodies was examined between human cardiac tissues from end-stage cardiomyopathic patients who had taken cardiac transplantation and normal controls. This finding showed three angiogenesis-related genes that were differentially methylated, including angiomotin-like 2 (AMOLT2), Rho GTPase-activating protein 24 (ARHGAP24), and platelet endothelial cell adhesion molecule (PECAM1) (Movassagh et al. 2010). Of note, hypermethylation in the 5′ region of PECAM1 in dilated hearts was correlated with its lower expression, and hypomethylation within the gene body of AMOLT2 was associated with reduced gene expression, whereas hypermethylation within the gene body of ARHGAP24 was correlated with increased gene expression (Movassagh et al. 2010). Therefore, the above study showed that aberrant gene expression implicated in the pathogenesis of HF and a part of the etiologies can be ascribed to the differential DNA methylation. In addition, Haas et al. also reported that an altered methylation pattern is related to dilated cardiomyopathies (DCM) compared with normal controls (Haas et al. 2013). In particular, both lymphocyte antigen 75 (LY75) and adenosine receptors A2A (ADORA2A) showed the significant downregulation in the myocardium of DCM which was mediated by promoter hypermethylation; and meanwhile, in vivo studies of LY75 and ADORA2A in zebra fish demonstrated the function role in adaptive or maladaptive pathways implicated in CVDs (Haas et al. 2013).

Nitric oxide (NO) bioavailability is implicated in the pathogenesis and progression of CVDs (Napoli and Ignarro 2009). Interestingly, it has been shown that the synthesis of NO from the precursor L-arginine could be affected by the complex process of arginine methylation pathways upon the inflammation and oxidative stress (Moncada and Higgs 1993; Rochette et al. 2013). All methylated arginine metabolites could inhibit NO synthesis in an indirect way through competitive blockade of arginine transport in CHF (Closs et al. 1997; Usui et al. 1998). Tang et al. investigated that accumulated plasma levels of methylated arginine metabolites (but not methyl-lysine) were correlated with the presence of left ventricular diastolic dysfunction in chronic systolic heart failure patients (Wilson Tang et al. 2008). Meanwhile, the close association between the methylated arginine metabolites and N-terminal pro-brain natriuretic peptide (NT-proBNP) is also confirmed in this study (Dückelmann et al. 2007; Wilson Tang et al. 2008).

2.2 *Histone Modification with CVDs*

2.2.1 Histone Modification with Hypoxia and Shear Stress

Hypoxia is majorly responsible for the endothelial phenotype, and the global transcription activity is reported to decline under hypoxic conditions. An ancient eukaryotic response to hypoxia is known as hypoxia-inducible factor-1 (HIF-1) transcription paradigm, and it has also been found that hypoxia induces a global drop in H3K9 acetylation via HDAC upregulation (Gluckman et al. 2008; Wadhwa et al. 2009). Across different cell lines, it has been observed that hypoxic conditions mediate an increase in the global H3K9 methylation, owing to a rise in G9 histone methyltransferase expression (Waterland and Michels 2007). Moreover, eNOS is well recognized to be of significant importance in regulating the chromatin structure in a cell-specific manner, and it is noteworthy that eNOS proximal promoter histones undergo H3 and H4 acetylation (Waterland and Jirtle 2003).

Vascular endothelium is known to endure a constant shear stress, owing to the physical force of circulation. Shear stress response element (SSRE) is a transcription factor regulating the expression of a few essential genes which are responsible for adapting to the circulatory pressure changes, such as Krüppel-like factor 2 (KLF2), VEGF receptor 2 (VEGFR2), and eNOS (Bogdarina et al. 2007; Heijmans et al. 2008; Tobi et al. 2009). Notably, it has been evidenced that the chromatin-based mechanisms, in particular global acetylation of H3 and H4 and phosphorylation of serine 10 on histone H3 (H3S10), contribute to the transcriptional regulation of these genes (El-Maarri et al. 2007).

2.2.2 Histone Modifications with CAD

As mentioned, the main cause of CADs is atherosclerosis-mediated arterial damage. It is also known that acetylation status is linked to the occurrence of atherosclerosis. HAT modulation is derived from CBP via thrombin signaling through mitogen-activated protein kinase (MAPK) pathway (Reddy and Natarajan 2011). SMCs, serving as the most abundant cell type in the arterial wall, are known to be involved in all the stages of lesion formation in atherosclerosis (Doran et al. 2008). Serum response factor (SRF) and SRF cofactor have the ability to activate CArG elements of SMC gene transcription, the binding of which is associated with methylation and acetylation to histone H3 and H4 residues (Manabe and Owens 2001; Cao et al. 2005; McDonald et al. 2006). In addition, HDAC inhibitors have shown to exacerbate the neointimal lesions in low-density lipoprotein (LDL) receptor-deficient ($Ldlr^{-/-}$) mice, which are known to be atherosclerosis-prone (Choi et al. 2005). Myocardial ischemia and reperfusion injury are also heavily influenced by the HDAC activity, in particular in the case of MI (Granger et al. 2008).

2.2.3 Histone Modifications and HF

HF is regarded as the final stage of all CVDs, and the common causes include ischemic heart disease, hypertrophy, and cardiomyopathy (McMurray and Pfeffer 2005). There are many studies presenting the epigenetic regulation involved in cardiomyocytes. The expression of both class I and class II HDACs has been shown to be associated with cardiac hypertrophy (Kee and Kook 2011). In particular, class I HDACs are commonly considered as injurious to cardiac function (Montgomery et al. 2007; Xie and Hill 2013). It is known that upon the treatment of phenylephrine (PE), the hypertrophic response of cardiomyocytes could be induced with enhanced CBP/p300 transcription activity (Gusterson et al. 2002). Ectopic overexpression of

CBP/p300 is shown to stimulate cardiac growth in decompensated HF (Yanazume et al. 2003). On the global level, trimethylation of histone H3 on lysine-4 (K4TM) or lysine-9 (K9TM) markers has been demonstrated to be linked with the development of HF in a rat model by high-throughput pyrosequencing performed with chromatin immunoprecipitation (ChIP) (Kaneda et al. 2009). Furthermore, the evidence showing the increase of histone acetylation only in the complication-free diabetic subjects in vivo and in vitro may signify that histone acetylation functions as a protective mechanism in the heart (Chen et al. 2009a).

2.2.4 Histone Modifications with Hypertension and Pulmonary Arterial Hypertension (PAH)

Recently, some studies have shown that core histone acetylation is responsible for hypertension and PAH (Mu et al. 2011; Yang et al. 2012; Lee et al. 2013). Environment-induced gene expression alteration in PAH is shown to be caused by certain epigenetic mechanisms (Kim et al. 2011). Therefore, the regulators of acetylation could be used as potential drug targets for treating hypertension and PAH. Valproic acid, a HDAC inhibitor, has shown to prevent the development of hypertension through diminishing transcriptional activity of mineralocorticoid receptor (MR) via increasing its acetylation (Lee et al. 2013). Moreover, HDAC inhibitor-treated spontaneously hypertensive rats (SHR) have successfully decreased vascular inflammation along with blood pressure (Usui et al. 2012). Increased activity of class I HDACs could enhance the pro-inflammatory phenotype of pulmonary adventitial fibroblasts from chronically hypoxic hypertensive calves (PH-Fibs), and therefore, the inhibition of HDACs was able to attenuate PH-Fibs functional activity to induce monocyte migration and adhesion (Li et al. 2011).

2.3 *RNA Silencing with CVDs*

2.3.1 miRNAs and Cardiac Aging

Aging is a predominant risk factor for CVDs, and the physiological cardiovascular aging is mainly linked with left ventricular hypertrophy, enhanced cardiac fibrosis, diastolic dysfunction, and valvular degeneration (Dai et al. 2012). A variety of miRNAs are found to be differently expressed during the process of cardiac aging with cell-type-specific characteristic (de Lucia et al. 2017). Importantly, miR-21 was the first example of miRNA which was well illustrated to play some role implicated in cardiac function. It has been demonstrated that miR-21 was upregulated selectively in fibroblasts of the failing heart compared with non-failing hearts, activating ERK-MAPK signaling by the inhibition of sprouty homologue 1 (SPRY1) (Thum et al. 2008). Pressure overload of the left ventricle was induced in mice by transverse aortic constriction (TAC), which was followed by the treatment of antagomir-21 or a control oligonucleotide. Results showed that silencing of miR-21 could reverse the expression of SPRY1 with the reduction of MAPK activity. Meanwhile, the interstitial fibrosis, cardiac hypertrophy, and dysfunction were repressed 3 weeks after in vivo silencing of miR-21 (Thum et al. 2008). Moreover, the myocardial and plasma levels of miR-21 were significantly higher in aortic stenosis (AS) patients compared with normal donors and correlated with echocardiographic mean transvalvular gradients and myocardial collagen expression (Villar et al. 2013). Furthermore, in the investigation of how myocardial ischemia affects miRNA expression, Roy et al. identified that phosphatase and tensin homologue (PTEN) is a direct target of miR-21 in isolated cardiac fibroblasts and miR-21 could modulate the expression of matrix metalloprotease-2 (MMP-2) through PTEN pathway in the infarct zone (Roy et al. 2009). A decreased expression of PTEN in infarct zone was correlated with enhanced MMP-2 expression (Roy et al. 2009).

miR-34a was another representative miRNA showing an important role in cardiovascular aging (Boon et al. 2013). Boon et al. found that the expression of miR-34a was higher in the heart tissues of C57B16 aging mice compared with young mice. In vivo silencing of miR-34a with antisense antagomir Ant-34a in 18-month-old mice decreased the dead cells, and moreover, the age-induced reduction in cardiac function was preserved in miR-34a$^{-/-}$ mice with less cardiomyocyte cellular death and hypertrophy when compared with wild-type controls (Boon et al. 2013). Meanwhile, miR-34a was also assessed in the mouse model of acute myocardial infarction (AMI), and it has been shown that miR-34a was induced after AMI and inhibition of miR-34a by antagomir Ant-34a or LNA-based antimiRs was able to increase cardiac contractile function after AMI (Boon et al. 2013). In addition, phosphatase 1 nuclear targeting subunit (PNUTS), which could repress telomere erosion and DNA damage response, was demonstrated as a novel target gene of miR-34a, further identifying the pro-apoptotic function of miR-34a in cardiomyocytes implicated in the aging process of CVDs (Kim et al. 2009; Landsverk et al. 2010; Boon et al. 2013).

Furthermore, Jazbutyte et al. determined the miRNA expression profiles in neonatal, 4-week, 6-month, and 19-month healthy C57/B16N mice and found that miR-22 was the most prominent upregulated miRNA accompanied by the downregulation of its direct target gene osteoglycin (OGN) during cardiac aging (Jazbutyte et al. 2013). Interestingly, OGN has been shown to be implicated in the regulation of cardiac inflammation, collagen assembly, and cardiomyocyte hypertrophy (Petretto et al. 2008; Van Aelst et al. 2014). Functionally, overexpression of miR-22 was able to accelerate cell senescence and increase the migratory activity of cardiac fibroblasts (Jazbutyte et al. 2013).

2.3.2 miRNAs and HF

Several studies have reported the potential of circulating miRNAs for the diagnosis of HF. The miRNA signature of failing myocardium is comprised of the dysregulated miRNAs such as miR-1-1, miR-195, miR-199a-1, miR-199b, and miR-221

(Akat et al. 2014). miR-1 family has been investigated as the muscle-specific miRNA and one of the essential regulating miRNAs implicated in cardiogenesis and skeletal muscle development (Lagos-Quintana et al. 2002; Zhao et al. 2007). miR-1 is highly expressed in the heart muscles, and deletion of miR-1-2 could lead to the cardiac defects of left ventricular malfunction progressing to a DCM (Zhao et al. 2007; Rao et al. 2009). Moreover, miR-1 was shown to downregulate in the early stage of cardiac hypertrophy in the hearts of C57BL/6 mice with TAC (Sayed et al. 2007). Several important regulatory genes are demonstrated as the target genes of miR-1, such as insulin-like growth factor-1 (IGF-1), twinfilin-1, a cytoskeleton regulatory protein, and heart-type fatty acid-binding protein-3 (FABP3), which are implicated in the process of cell growth, cellular differentiation or cardiac cytoskeleton, and metabolism (Elia et al. 2009; Li et al. 2010; Varrone et al. 2013).

In acute heart failure (AHF), a subset of 7 circulating miRNAs (miR-18a-5p, miR-26b-5p, miR-27a-3p, miR-30e-5p, miR-106a-5p, miR-199a-3p, and miR-652-3p) were validated to differ mostly in healthy donor controls and AHF patients (Ovchinnikova et al. 2016). In particular, AHF patients were shown to get further downregulation of the above circulating miRNAs within 48 h after admission to hospital (Ovchinnikova et al. 2016). Voellenkle et al. also investigated the miRNA expression profile in peripheral blood derived from a population of CHF patients affected by ischemic and nonischemic DCMs (Voellenkle et al. 2010). Three circulating miRNAs, including miR-107, miR-139, and miR-142-5p, were found to significantly downregulate in all HF patients when compared with healthy control donors (Voellenkle et al. 2010). Interestingly, miR-125b and miR-497 were only shown to decrease in ischemic patients, whereas miR-142-3p and miR-29b were significantly enhanced in nonischemic patients (Voellenkle et al. 2010).

2.3.3 miRNAs as Potential Biomarkers

miRNAs can be easily detected by real-time polymerase chain reaction (qRT-PCR) or microarrays in body fluid, including blood, saliva, or urine, which provides the opportunities to use these small molecules as biomarkers for human diseases (Mitchell et al. 2008; Kroh et al. 2010; Gallo et al. 2012; Tijsen et al. 2012). Importantly, cell-free blood serum and plasma miRNAs are the most stable and suitable sources for clinical application such as the diagnosis of disease given that the process of sample collection is simple and reproducible as well as the expression of miRNAs in the blood is stable even in the long term (Chim et al. 2008; Mitchell et al. 2008). Interestingly, it has been found that genetic exchange of mRNA and extracellular miRNAs between cells could be accomplished through microvesicles (MVs) or exosomes (Hunter et al. 2008). These microvesicles and exosomes containing miRNAs could facilitate the communication between cells or organs in the body. Exosomes are a population of extracellular vesicles (50–100 nm in diameter). They are released from cells in the process of inward budding of endosomal membrane that are fused with plasma membrane. Compared with exosomes, MVs are larger-sized vesicles (50–1000 nm in diameter) that are released from cells via outward budding (Raposo and Stoorvogel 2013; Lawson et al. 2016).

In CVDs, it has been noted that exosomes are manipulated by cardiovascular cells for intercellular communication as paracrine signaling mediators. Normally, the passenger strands of miRNAs, called star miRNAs (miRNA*), are considered to have no important biological function since they undergo intracellular degradation (Bartel 2004). Surprisingly, cardiac fibroblasts could secrete exosomes that are enriched with several miRNA passenger strands, in particular miR-21-3p, which has been identified to take the essential role to induce the cardiomyocyte hypertrophy (Bang et al. 2014). Moreover, the inhibition against miR-21-3p in mice with Ang II-induced cardiac hypertrophy could also reduce the pathological development of cardiac hypertrophy. Accordingly, it showed that cardiac fibroblast-derived miR-21-3p could serve as an important paracrine signaling effector in the process of fibroblast-derived cardiomyocyte hypertrophy, suggesting the potential therapeutic potential in HF patients in the future (Bang et al. 2014). Furthermore, Hergenreider et al. reported that endothelial-derived miR-143/145-containing vesicles could induce an atheroprotective SMC phenotype (Hergenreider et al. 2012). The shear-responsive transcription factor KLF2, which is a crucial mediator of endothelial gene expression patterns induced by atheroprotective flow, could directly bind to the promoter of miR-143/145 cluster and induce its upregulation (Dekker et al. 2006). In addition, extracellular vesicles generated from KLF2-transduced mouse endothelial cells could protect the mice from atherosclerotic lesion formation in the aorta of $ApoE^{-/-}$ mice, further demonstrating the protective role of KLF2 in miR-143/145-dependent manner (Hergenreider et al. 2012). Altogether, the atheroprotective stimuli could initiate the cross talk between endothelial cells and SMCs by the miRNAs-extracellular vesicle mechanisms, implying the promising therapy for atherosclerosis.

3 The Environment, Epigenetics, and CVD Risk

Today, approximately 7.6 billion people around the world from diverse natural environments live in their own lifestyle in different societies. Of note, lifestyle factors, society, and natural environment not only influence on daily living but also on the genome function through the dynamic and plastic epigenetic mechanisms including DNA methylation, histone modification, or miRNA aberrant expression (Alegría-Torres et al. 2011). As the life expectancy increases, the prevalence of chronic diseases will increase. In particular, it is estimated that by 2030, 40.5% of the population in the USA may have some forms of CVDs (Heidenreich et al. 2011). Therefore, the better understanding and early identification of how the natural, social, and personal environments modify the outcome of CVDs are highlighted for our prevention and management.

3.1 Natural Environmental Exposures

In environmental studies, it is noted that air pollution, circadian rhythm, and sunlight exposure are important factors that proved to be linked with CVDs that are partly regulated by epigenetic modifications. Firstly, exposure to air pollutants has been associated with the increased risk of CVDs (Du et al. 2016). Evidence showed that exposure level to fine metal-rich particulates (particulate matter <2.5 μm in diameter [PM2.5]) was demonstrated to be inversely correlated with the methylation of mitochondrial DNA (mtDNA) in the blood and participants with higher level of mtDNA methylation are more susceptible to heart rate viability due to the effect of PM2.5 exposure (Byun et al. 2016). Moreover, it was also shown that older participants with higher toll-like receptor 2 (TLR2) methylation were more prone to adverse autonomic effect of PM2.5 exposure (Zhong et al. 2015). Dietary modulation by flavonoids might reduce the above effect through the attenuation of TLR2 methylation. In addition, cardiovascular events, such as MI, sudden cardiac death, and pulmonary embolism, are recognized to have circadian rhythmicity with the high incidence in the morning, suggesting that chronic circadian disruption may enhance the susceptibility to CVDs (Oishi 2009). The evidence demonstrated that CLOCK, a transcription factor for circadian function, has the intrinsic HAT activity, whereas SIRT1 was shown to function as a HDAC that counteracts CLOCK activity to manage circadian chromatin remodeling (Bellet and Sassone-Corsi 2010). It has been identified that vitamin D deficiency has contributed to cardiovascular events such as MI, stroke, and HF (Nemerovski et al. 2009). More importantly, adequate vitamin D supplementation or sunlight exposure are beneficial for the prevention of CVDs (Wimalawansa 2016). Interestingly, vitamin D could be interacted with epigenome on multiple levels, including the methylation alteration of several critical genes such as vitamin D receptor (VDR) and the enzyme 25-hydroxylase (CYP2R1), as well as the histone modification of VDR proteins through HATs or HDACs (Fetahu et al. 2014).

3.2 Personal Environment

Smoking, diet, alcohol consumption, and sleeping habits are important factors that may have impacts on CVDs. Prenatal exposure to tobacco smoke increases the risk of diseases later in child's life which is partly mediated by the hypomethylation of AluYb8 and hypermethylation of AXL and PTPRO in exposed children compared with controls (Breton et al. 2009). It suggested that lifelong effect of in utero exposures could be mediated by aberrant DNA methylation. Jiang et al. also found that persistent nicotine exposure selectively inhibited the expression of TBX5 and GATA4 by promoter hypermethylation, leading to cardiac function impairment with repressed left ventricular ejection fraction (Jiang et al. 2017). On the other hand, diet nutrition is another crucial factor influencing chronic disease susceptibility by the epigenetic mechanism. It is supported by the evidence that maternal diet habit during pregnancy could modify the DNA methylation pattern of specific genes, resulting in some phenotypic changes such as blood pressure, body weight, and coat color (Waterland and Jirtle 2003; Bogdarina et al. 2007). Calorie restriction (CR) is proved to be the most effective nutritional intervention to increase longevity and prevent chronic diseases such as CVDs, hypertension, and diabetes (Omodei and Fontana 2011). DNA methylation drift with age was demonstrated to be delayed by CR in mice and monkeys, showing that epigenetic drift is a good biomarker of life span (Maegawa et al. 2017). In addition, sleeping deprivation also increases the risk of CVDs, which may be mediated by the enhanced global methylation status that leads to systemic cell injury and uncompensated oxidative stress (Everson et al. 2005; Trivedi et al. 2017).

4 Technical Overview of Platform Used in Epigenetic Study

4.1 DNA Methylation with Regional and Genome-Wide Approach

In the genome, usually a proportion of potential target sequences could be methylated so that the distribution of methyl marks can carry and express epigenetic information through transcriptional silencing (Laird 2010). Both the genome-wide scale and regional DNA methylation profiles have critical implications for better understanding of how the specific loci of genome express under some condition and how the epigenetic aberration contributes to the pathogenesis of diseases (Robertson 2005). Recent advances in examination and quantification of DNA modifications facilitate the epigenetic changes to be connected to phenotypic consequences on an unprecedented scale (Plongthongkum et al. 2014).

With regard to the sequence-specific DNA methylation detection and analysis, methylation-dependent pre-treatment of DNA should be performed at first to reveal the presence or absence of methyl group at cytosine residues before DNA amplification or hybridization (Fraga and Esteller 2002; Laird 2003). There are three kinds of methods for methylation-dependent pre-treatments including bisulfite conversion, endonuclease digestion, and affinity enrichment (Laird 2010).

4.1.1 Bisulfite Conversion

The utility of sodium bisulfite on denatured genomic DNA is based on the function of bisulfite to selectively deaminate the unmethylated cytosine residue, providing the basis to distinguish the methylated and unmethylated CpG sites (Shapiro et al. 1970; Hayatsu 2008). The bisulfite treatment of DNA could deaminate unmethylated cytosine to uracil, leaving the methylated cytosine unchanged.

Therefore, bisulfite conversion enables the application of new techniques to detect the methylation status at a particular region (Dahl and Guldberg 2003; Laird 2003; Dupont et al. 2004). Bisulfite-treated genomic DNA was at first analyzed and demonstrated by Sanger sequencing of cloned polymerase chain reaction (PCR) amplification at the single base pair level (Frommer et al. 1992; Paul and Clark 1996). Methylation-specific PCR (MSP) can detect the DNA methylation status of specific genes with high sensitivity and specificity (Herman et al. 1996).

4.1.2 Endonuclease Digestion

Restriction enzyme was found in bacteria and archaea by working as a defense mechanism against invading viruses, which was now used as a powerful tool in molecular biology (Roberts et al. 2003). Each sequence-specific restriction enzyme was accompanied by DNMTs that could protect DNA from restriction defense system through the methylation of recognition site (Laird 2010). Predicted modifications in restriction enzyme sites following bisulfite reaction and amplification of the product by the PCR were used to assess the methylation of CpG sites (Sadri and Hornsby 1996). *Hpa*II and *Sma*I are the most widely used methylation-sensitive restriction enzymes for DNA methylation studies, which are not repressed by CpG methylation (Singer-Sam et al. 1990; Laird 2010).

4.1.3 Affinity Enrichment

The most recent way to enrich genomic methylated DNA is to do affinity purification and enrichment. One approach is to use MBDs with affinity for methylated CpG sites; alternatively, another way is to use monoclonal antibodies that are specific for 5-methylcytosine ($5^{me}C$, in the context of denatured DNA) to immunoprecipitate methylated DNA (Cross et al. 1994; Selker et al. 2003; Weber et al. 2005). Of note, the affinity of methylated DNA was firstly identified with the rat chromosomal protein MeCP2 (Cross et al. 1994). Meanwhile, the immunoprecipitation approach, which is followed by hybridization to a tiling array or microarray, is regarded as MeDIP or mDIP (Weber et al. 2005; Keshet et al. 2006; Weber et al. 2007). Affinity-based approaches are to measure the density of methylation in a selected region (Zilberman and Henikoff 2007). Thus, the target regions with spare methylation sites might be difficult to differentiate the methylated and unmethylated sites, especially for mammalian genomes (Weber et al. 2007; Zilberman and Henikoff 2007).

4.1.4 Genome-Wide Approaches for Methylation

Over the past years, a large amount of approaches for genome-wide methylation analysis have been reported. As a type of two-dimensional electrophoresis, restriction landmark genomic scanning (RLGS) was the first technique in the genome-wide assessment of both genomic and epigenetic modifications (Fischer and Lerman 1979; Rush and Plass 2002). RLGS uses methylation-sensitive restriction enzymes to create the "landmarks", and therefore, methylation difference could be detected by the presence of radiolabeled locus in unmethylated profile. In methylated profile, no radiolabeled RLGS fragment is identified because the methylation-sensitive enzyme fails to digest the DNA (Hatada et al. 1991; Rush and Plass 2002). RLGS analysis of different tumor samples has showed the relevant cancer-specific patterns of methylation, contributing to our understanding of the role of methylation implicated in carcinogenesis (Costello et al. 2000; Allegrucci et al. 2007). Nowadays with the advent of next-generation sequencing (NGS) techniques, the use of RLGS is replaced by the sequencing approaches. Sequence-based technique is more flexible and high-throughput to show the methylation patterns on a genome-wide scale within specific sequence since it can read allele-specific DNA methylation after restriction enzyme enrichment (Cokus et al. 2008; Laird 2010; Davey et al. 2011). However, it is restricted to the sequence library biases (Laird 2010).

4.2 Approaches for Histone Modification Analysis

Chromatin state could directly regulate transcription, and the cross talk between chromatin and transcription is dynamic in cellular process (Berger 2007). Importantly, histone modifications are crucial epigenetic mechanisms for eukaryotic gene regulation. ChIP assay serves as the primary powerful technique to investigate the defined genomic locations associated with histone modifications within cells or tissues (Spencer et al. 2003). In standard ChIP protocol, chromatin is released from lysed cells, digested into fragments and then selectively bound with specific antibodies against the protein of interest (Geng et al. 2011). The chromatin-antibody complex is removed, and then ChIP DNA could be further purified and analyzed by a series of approaches, such as conventional PCR, quantitative PCR, hybridization to array (ChIP-chip), or sequencing (Milne et al. 2009). Standard ChIP procedure is subjected to the multiple steps, the size of cell population, as well as the inefficient immunoprecipitation. The introduction of ChIP-chip enables the genome-wide profile of DNA-protein interaction (Ren et al. 2000). During recent years, ChIP coupled with high-throughput sequencing (ChIP-Seq) is increasingly used to map the high-resolution genome-scale views of protein-DNA interactions including transcription factor bindings, histone modification, DNA methylation, and nucleosome positioning (Park 2009; O'Geen et al. 2011). With wider coverage, less noise, single base pair resolution, and deeper sequencing, ChIP-Seq provides several advantages when compared to array-based ChIP-chip (Zhang et al. 2008;

Park 2009). In 2007, the first comprehensive genome-wide map produced by ChIP-Seq showed 20 histone methylation markers, histone variant H2A.Z, RNA polymerase II, and CTCF (Barski et al. 2007; Mikkelsen et al. 2007). As the cost of sequencing decreases and institutional support grows, ChIP-Seq is likely to work as the main approach for exploring histone modifications in the future (Park 2009).

4.3 Approaches for miRNA Studies

Over the past years, miRNAs have been extensively studied in cellular processes, development, human diseases, and in particular, human cancers (Bartel 2009). It was reported that approximately 2600 miRNAs regulate more than 60% of human genomes involved in the regulation of cellular growth and death (Friedman et al. 2009; Shu et al. 2017). Therefore, a number of measurement platforms have been developed to examine the relative miRNA expression in different biological samples for better understanding of miRNA roles implicated in human diseases (Shingara et al. 2005; Chen et al. 2008; Willenbrock et al. 2009; Mestdagh et al. 2014). For miRNA studies, three main techniques are included including qRT-PCR, hybridization (DNA microarrays), and small RNA sequencing (Kong et al. 2009). The choice of platform depends on the specific experimental goals. It has been found that TaqMan qPCRarray might show superior sensitivity and specificity when compared with microarray (Chen et al. 2009b). However, microarray and qPCR platforms are limited to the number of probes or assays available, which in turn is subjected to the update of miRBase (Chen et al. 2009b; Mestdagh et al. 2014). In addition, by comparison of microarray (Exiqon's LNA-based microarrays) and NGS (Illumina's GA-II sequencing platform), microarray expression analysis might offer both high specificity and sensitivity, while NGS appears the ability to discover new sequence variants (Willenbrock et al. 2009).

5 The Application of Epigenetic Therapy to Human Diseases

Given that the epigenetic etiology may contribute to the pathogenesis of human diseases, this encourages the development of new promising "epigenetic therapy." To date, several agents have been discovered that could modify the epigenetic information in a manner that is not specific to genes, including inhibitors of DNMTs and HDACs (Egger et al. 2004; Miyamoto and Ushijima 2005). Some of these agents are tested in clinical trials, while gene-specific approaches are under exploration (Lubbert 2000; Issa et al. 2004; Wagner et al. 2010).

5.1 *Demethylation by 5-Azacytidine, Decitabine, and Zebularine*

Since DNA methylation can be reversible silencing, the expression and function of TSGs, miRNAs, or lncRNAs that have undergone inactivation by DNA methylation will be restored by state-of-the-art epigenetic therapy (Wang and Chim 2015). DNMTs, especially DNMT1, are known to maintain CpG methylation (Robert et al. 2003). The DNMT1 inhibitor, 5-aza-2′-deoxycytidine (5-aza-dC, decitabine), forms covalent attachment with DNMT1, resulting in the degradation of DNMT1 and therefore DNA hypomethylation (Šorm et al. 1964; Christman 2002). Importantly, pharmacological grade 5-aza-dC has been identified to increase response rate and survival in patients with myelodysplastic syndrome (Silverman et al. 2002; Fenaux et al. 2009). Compared with 5-aza-dC used in high dose of 75–100 $mg/m^2/day$, low-dose treatment of 5–20 $mg/m^2/day$ for chronic myelogenous leukemia patients in phase I study appeared more effective (Kantarjian et al. 2003; Issa et al. 2004). The major adverse effect of 5-aza-dC was myelosuppression and toxic, and the administration of 5-aza-dC is restricted to the short half-life and cannot be given orally (Issa et al. 2004). To further improve the stability and efficacy of 5-aza-dC, some cytidine analogs are under development (Yang et al. 2010). For instance, zebularine is a chemically stable cytidine analog, which could inhibit DNMTs and be administered orally or intraperitoneally (Cheng et al. 2003; Yoo et al. 2008). Moreover, zebularine showed the preference of depleting DNMT1 and induction of cancer-related antigens in cancer cells rather than normal fibroblasts (Cheng et al. 2004). However, the drawback is that higher concentration of zebularine is needed to get similar effect of demethylation in cells when compared with 5-aza-dC (Cheng et al. 2004).

5.2 *Demethylation by Non-nucleoside Analogs*

Different from cytidine analogs, non-nucleoside DNMT inhibitors, such as hydralazine, procaine, and procainamide, have limited activity of demethylation and less cytotoxicity as they are not incorporated into DNA (Chuang et al. 2005). For instance, the vasodilator agent hydralazine was shown to induce demethylating activity in human breast adenocarcinoma cells (Segura-Pacheco et al. 2006). Moreover, the inhibitory function of hydralazine with DNMTs was rationalized through the interaction of similar pharmacophore as established by deoxycytidine (Singh et al. 2009). Another approach to inhibit DNMT1 is to use the short-chain oligodeoxynucleotides. MG98 is a 20 base pair antisense oligodeoxynucleotide, which specifically binds to the 3′ UTR of human DNMT1 mRNA (Amato 2007). Preclinical data showed that a patient with renal cell carcinoma treated at 80 mg/m^2 of MG98 showed partial response (Stewart et al. 2003). A number of phase I/II clinical trials are investigating the safety and efficacy of MG98 in several solid and hematopoietic cancers (Amato 2007).

5.3 *HDAC and HAT Inhibition*

Histones can be acetylated and deacetylated, and two opposing classes of enzymes are involved in determining the pattern of histone acetylation, HDACs and HATs. Several leukemogenic transcription factors depress the expression of specific genes through aberrant recruitment of HDAC, and consequently, HDAC-dependent abnormal transcription repression contributes to the oncogenic mechanism of acute promyelocytic leukemia and non-Hodgkin's lymphoma (Dhordain et al. 1998; Lin et al. 1998). Acetylation results in the relaxation of chromatin structure and transcriptional activation of many genes, which might serve as a powerful mediator of tumor response to chemotherapy or adaptation environment to changes such as hypoxia (Grunstein 1997; Cao et al. 2011).

5.3.1 HDAC Inhibitors

Importantly, a number of HDAC inhibitors are found to induce cellular growth, differentiation, and/or apoptosis in transformed cells in culture or in tumors (Yoshida et al. 1990; Kijima et al. 1993; Kim et al. 1999). In particular, the cell cycle kinase inhibitor, $p21^{WAF1}$, can be induced with the specific HDAC inhibitor trichostatin A (TSA) in p53-independent manner in NIH3T3 cells, which provides new insights to cancer therapy that could arrest tumor cells in p53-independent way (Xiao et al. 1999). Of note, as autophagy is found to be involved in the pathological cardiac remodeling, TSA treatment has been shown to reduce both load- and agonist-induced hypertrophic growth and depress the autophagy in Beclin 1 transgenic mice, suggesting the promising strategy of the HDAC inhibitor targeting autophagy (Cao et al. 2011).

To date, the HDAC inhibitors, vorinostat and romidepsin, have been used for the treatment of refractory cutaneous T cell lymphoma (CTCL) at clinic (Duvic et al. 2007; Piekarz et al. 2009). Approved by FDA for clinical application, either of them was demonstrated to have the single-agent clinical activity with significant and sustainable responses in CTCL patients, making an important therapeutic option for CTCL treatment (Olsen et al. 2007; Whittaker et al. 2010). At the same time, some newer compounds such as panobinostat, givinostat, and mocetinostat are currently being clinically assessed (Qiu et al. 2013; West and Johnstone 2014).

5.3.2 HAT Inhibitors

Different subtypes of HATs enable to catalyze the histone acetylation on specific lysine residues (Dekker and Haisma 2009). HAT families have been extensively studied including the GNAT family, the p300/CBP family, the MYST family, and the Rtt109 family (Hodawadekar and Marmorstein 2007; Nagy and Tora 2007). HATs are identified to implicate in several crucial cell proliferation pathways, and therefore disruption of theses enzymes leading to chromosomal translocations might be involved in the pathogenesis of leukemia (Sobulo et al. 1997; Perez-Campo et al. 2009; Chen et al. 2010). Therefore, the above findings make HAT an attractive target for cancer therapy. The first HAT inhibitor to be described for selectively targeting the HAT p300/CBP family was bisubstrate analog inhibitor (Lau et al. 2000). Spermidinyl-CoA-based HAT inhibitors are found to block DNA repair and serve as cancer-specific and radiosensitive drug (Bandyopadhyay et al. 2009). A novel inhibitor of p300 HAT activity, C646, is a relatively potent and cell-permeable small molecule, showing the ability to slow the cancer cell growth and impede intracellular histone acetylation (Bowers et al. 2010). Of note, in heart diseases, the balance of histone acetylation and deacetylation plays the role of regulating hypertrophic response of cardiomyocytes to phenylephrine (Gusterson et al. 2003). Inhibition of transcription coactivator p300 or CBP HAT reduced the cardiac hypertrophy, whereas activating them promoted hypertrophy in HAT-dependent manner (Gusterson et al. 2003).

5.4 Combined Therapies with Demethylating Drug and HDAC Inhibitors

Epigenetic drug, no matter demethylating drug or HDAC inhibitor, results in the reexpression and reactivation of tumor-suppressive genes, miRNAs, and lncRNAs, which play important roles involved in the cellular process and function (Wang and Chim 2015). Interestingly, these two inhibitors can be used together to improve its effectiveness. It has been firstly demonstrated that sequential exposure to 5-azacytidine and butyrate (HDAC inhibitor) had a strong synergistic effect to induce the β-adrenergic receptors in HeLa cells (Jahangeer et al. 1982). Moreover, the combination of 5-azacytidine and TSA (HDAC inhibitor) treatment on esophageal squamous cell carcinoma could lead to the functional reactivation of several TSGs (Yamashita et al. 2002). Moreover, DNMT1 contributed to the development of lung cancer in murine model, and importantly, simultaneous 5-azacytidine (DNMT inhibitor) and phenyl butyrate (HDAC inhibitor) treatment led to the re-expression of TSGs, indicating that the combined epigenetic therapy could be effective for preventing murine lung cancer (Belinsky et al. 2003). In addition, clinical trials are developed targeting two epigenetic modifications in cancer patients. It has shown that combined epigenetic therapy with low-dose 5-azacitidine and entinostat (HDAC inhibitor) led to the durable response in patients with recurrent metastatic non-small cell lung cancer (Juergens et al. 2011). In addition, epigenetic therapy can be used for sensitive drug-resistant tumors followed by chemotherapy or immunotherapy, showing a good response (Weber et al. 1994; Plumb et al. 2000).

Despite the advantages of epigenetic therapy, the therapeutic use is still very preliminary, and there are some concerns about the clinical application. One concern is regarding the nonspecific effect to normal cells, in which DNA methylation might enforce the inactivation of genes in normal cells or activation of oncogenes, leading to the potential mutagenicity (Jackson-Grusby et al. 1997; Liang et al. 2002; Sato et al. 2003; Egger et al. 2004). Moreover, both HDAC and methylation inhibitors are cytotoxic, which could induce the cellular apoptosis and/or cell cycle arrest by activating p21 or p53 (Karpf et al. 2001; Marks et al. 2003). Therefore, the clinical application of epigenetic therapy needs to be examined in patients to get a better understanding and response.

6 Future and Prospective

Epigenetic modifications have shown to contribute to several human diseases, including CVDs. Of them, DNA methylation is major epigenetic modification, which is more extensively studied compared with histone modification and miRNA alternations. In particular, DNA methylation of specific genes or miRNAs, as well as histone modifications and aberrant miRNA expression, have been noted in different subtypes of CVDs, suggesting the critical role of epigenetics involved in the pathogenesis of CVD. Interestingly, some environmental exposures could change the epigenetic markers, and therefore, it is promising to find epigenetic therapies to such reversible modifications. Some environmental factors including air pollution, smoking, aging, circadian rhythm, and diet nutrition, are found to be associated with epigenetic alternations in CVD patients when compared with those in normal controls (Dong et al. 2002; Oka et al. 2009; Baccarelli and Ghosh 2012). Recent studies have discovered how epigenetic mechanisms are involved in the pathogenesis of CVDs. As a result, there has been a rapid increase in epigenetic therapies for CVDs, such as the demethylating drug, HDAC inhibitors, or miRNA mimics/inhibitors. However, the difficulties of clinical application for CVDs cannot be underestimated. The dosage and adverse effect of drug, route of administration, and drug interactions, are crucial factors to be considered to avoid the toxicity. It is apparent that we are still at the beginning of understanding the substantial contribution of epigenetics to CVDs, and elucidating the complete epigenetic information map will be an exciting challenge for us and eventually result in a clearer direction to illustrating the diseases as well as the novel therapeutic concepts.

Acknowledgment This work was supported by Hong Kong Research Grants Council – General Research Fund (HKU 17127215; project title: Molecular Mechanisms Underlying the Progressive Development of Pulmonary Arterial Hypertension in Secretin Knockout Mice; PI: Professor BKC Chow, School of Biological Sciences; 1,100,302 HKD, ongoing).

References

Akat KM, et al. Comparative RNA-sequencing analysis of myocardial and circulating small RNAs in human heart failure and their utility as biomarkers. Proc Natl Acad Sci. 2014;111(30):11151–6.
Alegría-Torres JA, et al. Epigenetics and lifestyle. Epigenomics. 2011;3:267–77.

Allegrucci C, et al. Restriction landmark genome scanning identifies culture-induced DNA methylation instability in the human embryonic stem cell epigenome. Hum Mol Genet. 2007;16(10):1253–68.
Amato RJ. Inhibition of DNA methylation by antisense oligonucleotide MG98 as cancer therapy. Clin Genitourin Cancer. 2007;5(7):422–6.
Ambrose JA, et al. Angiographic progression of coronary artery disease and the development of myocardial infarction. J Am Coll Cardiol. 1988;12(1):56–62.
Antequera F, Bird A. CpG islands. DNA methylation. New York: Springer; 1993. p. 169–85.
Baccarelli A, Ghosh S. Environmental exposures, epigenetics and cardiovascular disease. Curr Opin Clin Nutr Metab Care. 2012;15(4):323.
Baccarelli A, et al. Ischemic heart disease and stroke in relation to blood DNA methylation. Epidimiology. 2010;21(6):819.
Bachman KE, et al. Dnmt3a and Dnmt3b are transcriptional repressors that exhibit unique localization properties to heterochromatin. J Biol Chem. 2001;276(34):32282–7.
Ballestar E, Esteller M. Methyl-CpG-binding proteins in cancer: blaming the DNA methylation messenger. Biochem Cell Biol. 2005;83(3):374–84.
Bandyopadhyay K, et al. Spermidinyl-CoA-based HAT inhibitors block DNA repair and provide cancer-specific chemo-and radiosensitization. Cell Cycle. 2009;8(17):2779–88.
Bang C, et al. Cardiac fibroblast–derived microRNA passenger strand-enriched exosomes mediate cardiomyocyte hypertrophy. J Clin Invest. 2014;124(5):2136.
Barski A, et al. High-resolution profiling of histone methylations in the human genome. Cell. 2007;129(4):823–37.
Bartel D. MicroRNAs: genomics, biogenesis, mechanism, and function. Cell. 2004;116(2):281–97.
Bartel DP. MicroRNAs: target recognition and regulatory functions. Cell. 2009;136(2):215–33.
Batista PJ, Chang HY. Long noncoding RNAs: cellular address codes in development and disease. Cell. 2013;152(6):1298–307.
Belinsky SA, et al. Inhibition of DNA methylation and histone deacetylation prevents murine lung cancer. Cancer Res. 2003;63(21):7089–93.
Bellet MM, Sassone-Corsi P. Mammalian circadian clock and metabolism–the epigenetic link. J Cell Sci. 2010;123(22):3837–48.
Berger SL. The complex language of chromatin regulation during transcription. Nature. 2007;447(7143):407–12.
Bird AP. CpG-rich islands and the function of DNA methylation. Nature. 1986;321(6067):209–13.
Bird A. DNA methylation patterns and epigenetic memory. Genes Dev. 2002;16(1):6–21.
Blander G, Guarente L. The Sir2 family of protein deacetylases. Annu Rev Biochem. 2004;73(1):417–35.
Bogdarina I, et al. Epigenetic modification of the renin-angiotensin system in the fetal programming of hypertension. Circ Res. 2007;100(4):520–6.
Boon RA, et al. MicroRNA-34a regulates cardiac ageing and function. Nature. 2013;495(7439):107–10.
Bowers EM, et al. Virtual ligand screening of the p300/CBP histone acetyltransferase: identification of a selective small molecule inhibitor. Chem Biol. 2010;17(5):471–82.
Branco MR, et al. Safeguarding parental identity: Dnmtl maintains imprints during epigenetic reprogramming in early embryogenesis. Genes Dev. 2008;22(12):1567–71.
Breton CV, et al. Prenatal tobacco smoke exposure affects global and gene-specific DNA methylation. Am J Respir Crit Care Med. 2009;180(5):462–7.
Brown R, Strathdee G. Epigenomics and epigenetic therapy of cancer. Trends Mol Med. 2002;8(4):S43–8.
Brown CE, et al. The many HATs of transcription coactivators. Trends Biochem Sci. 2000;25(1):15–9.
Brunner AL, et al. Transcriptional profiling of long non-coding RNAs and novel transcribed regions across a diverse panel of archived human cancers. Genome Biol. 2012;13(8):R75.

Byun HM, et al. Effects of air pollution and blood mitochondrial DNA methylation on markers of heart rate variability. J Am Heart Assoc. 2016;5(4):e003218.
Calin GA, Croce CM. MicroRNA signatures in human cancers. Nat Rev Cancer. 2006;6:857–66.
Cao D, et al. Modulation of smooth muscle gene expression by association of histone acetyltransferases and deacetylases with myocardin. Mol Cell Biol. 2005;25(1):364–76.
Cao DJ, et al. Histone deacetylase (HDAC) inhibitors attenuate cardiac hypertrophy by suppressing autophagy. Proc Natl Acad Sci. 2011;108(10):4123–8.
Chan GC, et al. Epigenetic basis for the transcriptional hyporesponsiveness of the human inducible nitric oxide synthase gene in vascular endothelial cells. J Immunol. 2005;175(6):3846–61.
Chen J, et al. Highly sensitive and specific microRNA expression profiling using BeadArray technology. Nucleic Acids Res. 2008;36(14):e87.
Chen SS, et al. Elevated plasma prostaglandins and acetylated histone in monocytes in type 1 diabetes patients. Diabet Med. 2009a;26(2):182–6.
Chen Y, et al. Reproducibility of quantitative RT-PCR array in miRNA expression profiling and comparison with microarray analysis. BMC Genomics. 2009b;10(1):407.
Chen J, et al. Leukaemogenesis: more than mutant genes. Nat Rev Cancer. 2010;10(1):23–36.
Cheng JC, et al. Inhibition of DNA methylation and reactivation of silenced genes by zebularine. J Natl Cancer Inst. 2003;95(5):399–409.
Cheng JC, et al. Preferential response of cancer cells to zebularine. Cancer Cell. 2004;6(2):151–8.
Cheung WL, et al. Acetylation and chromosomal functions. Curr Opin Cell Biol. 2000;12(3): 326–33.
Chim CS, et al. SOCS1 and SHP1 hypermethylation in multiple myeloma: implications for epigenetic activation of the Jak/STAT pathway. Blood. 2004;103(12):4630–5.
Chim SS, et al. Detection and characterization of placental microRNAs in maternal plasma. Clin Chem. 2008;54(3):482–90.
Choi J-H, et al. Trichostatin a exacerbates atherosclerosis in low density lipoprotein receptor–deficient mice. Arterioscler Thromb Vasc Biol. 2005;25(11):2404–9.
Christman JK. 5-Azacytidine and 5-aza-2′-deoxycytidine as inhibitors of DNA methylation: mechanistic studies and their implications for cancer therapy. Oncogene. 2002;21(35):5483–95.
Chuang JC, et al. Comparison of biological effects of non-nucleoside DNA methylation inhibitors versus 5-aza-2′-deoxycytidine. Mol Cancer Ther. 2005;4(10):1515–20.
Closs EI, et al. Interference of L-arginine analogues with L-arginine transport mediated by the y+ carrier hCAT-2B. Nitric Oxide. 1997;1(1):65–73.
Cokus SJ, et al. Shotgun bisulphite sequencing of the Arabidopsis genome reveals DNA methylation patterning. Nature. 2008;452(7184):215–9.
Costello JF, et al. Aberrant CpG-island methylation has non-random and tumour-type–specific patterns. Nat Genet. 2000;24(2):132.
Cross SH, et al. Purification of CpG islands using a methylated DNA binding column. Nat Genet. 1994;6(3):236–44.
Dahl C, Guldberg P. DNA methylation analysis techniques. Biogerontology. 2003;4(4):233–50.
Dai D-F, et al. Cardiac aging: from molecular mechanisms to significance in human health and disease. Antioxid Redox Signal. 2012;16(12):1492–526.
Daujat S, et al. Crosstalk between CARM1 methylation and CBP acetylation on histone H3. Curr Biol. 2002;12(24):2090–7.
Davey JW, et al. Genome-wide genetic marker discovery and genotyping using next-generation sequencing. Nat Rev Genet. 2011;12(7):499–510.
de Lucia C, et al. microRNA in cardiovascular aging and age-related cardiovascular diseases. Front Med. 2017;4(74):74.
Dekker FJ, Haisma HJ. Histone acetyl transferases as emerging drug targets. Drug Discov Today. 2009;14(19):942–8.
Dekker RJ, et al. KLF2 provokes a gene expression pattern that establishes functional quiescent differentiation of the endothelium. Blood. 2006;107(11):4354–63.

Denli AM, et al. Processing of primary microRNAs by the microprocessor complex. Nature. 2004;432(7014):231–5.
Dhordain P, et al. The LAZ3 (BCL-6) oncoprotein recruits a SMRT/mSIN3A/histone deacetylase containing complex to mediate transcriptional repression. Nucleic Acids Res. 1998;26(20):4645–51.
Djebali S, et al. Landscape of transcription in human cells. Nature. 2012;489(7414):101–8.
Dong C, et al. DNA methylation and atherosclerosis. J Nutr. 2002;132(8):2406S–9S.
Doran AC, et al. Role of smooth muscle cells in the initiation and early progression of atherosclerosis. Arterioscler Thromb Vasc Biol. 2008;28(5):812–9.
Du Y, et al. Air particulate matter and cardiovascular disease: the epidemiological, biomedical and clinical evidence. J Thorac Dis. 2016;8(1):E8.
Dückelmann C, et al. Asymmetric dimethylarginine enhances cardiovascular risk prediction in patients with chronic heart failure. Arterioscler Thromb Vasc Biol. 2007;27(9):2037–42.
Dupont J-M, et al. De novo quantitative bisulfite sequencing using the pyrosequencing technology. Anal Biochem. 2004;333(1):119–27.
Duvic M, et al. Phase 2 trial of oral vorinostat (suberoylanilide hydroxamic acid, SAHA) for refractory cutaneous T-cell lymphoma (CTCL). Blood. 2007;109(1):31–9.
Egger G, et al. Epigenetics in human disease and prospects for epigenetic therapy. Nature. 2004;429(6990):457–63.
Elia L, et al. Reciprocal regulation of microRNA-1 and insulin-like growth factor-1 signal transduction cascade in cardiac and skeletal muscle in physiological and pathological conditions. Circulation. 2009;120(23):2377–85.
El-Maarri O, et al. Gender specific differences in levels of DNA methylation at selected loci from human total blood: a tendency toward higher methylation levels in males. Hum Genet. 2007;122(5):505–14.
Esteller M. Epigenetic gene silencing in cancer: the DNA hypermethylome. Hum Mol Genet. 2007;16(R1):R50–9.
Esteller M. Non-coding RNAs in human disease. Nat Rev Genet. 2011;12(12):861–74.
Everson CA, et al. Antioxidant defense responses to sleep loss and sleep recovery. Am J Phys Regul Integr Comp Phys. 2005;288(2):R374–83.
Feinberg AP, Vogelstein B. Hypomethylation distinguishes genes of some human cancers from their normal counterparts. Nature. 1983;301(5895):89–92.
Fenaux P, et al. Efficacy of azacitidine compared with that of conventional care regimens in the treatment of higher-risk myelodysplastic syndromes: a randomised, open-label, phase III study. Lancet Oncol. 2009;10(3):223–32.
Fetahu IS, et al. Vitamin D and the epigenome. Front Physiol. 2014;5:164.
Fischer SG, Lerman LS. Length-independent separation of DNA restriction fragments in two-dimensional gel electrophoresis. Cell. 1979;16(1):191–200.
Fish JE, et al. The expression of endothelial nitric-oxide synthase is controlled by a cell-specific histone code. J Biol Chem. 2005;280(26):24824–38.
Fraga MF, Esteller M. DNA methylation: a profile of methods and applications. BioTechniques. 2002;33(3):632–49.
Friedman JM, Jones PA. MicroRNAs: critical mediators of differentiation, development and disease. Swiss Med Wkly. 2009;139(33–34):466–72.
Friedman RC, et al. Most mammalian mRNAs are conserved targets of microRNAs. Genome Res. 2009;19(1):92–105.
Frommer M, et al. A genomic sequencing protocol that yields a positive display of 5-methylcytosine residues in individual DNA strands. Proc Natl Acad Sci. 1992;89(5):1827–31.
Gallo A, et al. The majority of microRNAs detectable in serum and saliva is concentrated in exosomes. PLoS One. 2012;7(3):e30679.
Geng T, et al. Histone modification analysis by chromatin immunoprecipitation from a low number of cells on a microfluidic platform. Lab Chip. 2011;11(17):2842–8.

Gluckman PD, et al. Effect of in utero and early-life conditions on adult health and disease. N Engl J Med. 2008;359(1):61–73.
Gluckman PD, et al. Epigenetic mechanisms that underpin metabolic and cardiovascular diseases. Nat Rev Endocrinol. 2009;5(7):401–8.
Goldberg AD, et al. Epigenetics: a landscape takes shape. Cell. 2007;128(4):635–8.
Goll MG, Bestor TH. Eukaryotic cytosine methyltransferases. Annu Rev Biochem. 2005;74:481–514.
Gottesfeld JM, Forbes DJ. Mitotic repression of the transcriptional machinery. Trends Biochem Sci. 1997;22(6):197–202.
Granger A, et al. Histone deacetylase inhibition reduces myocardial ischemia-reperfusion injury in mice. FASEB J. 2008;22(10):3549–60.
Gregory RI, et al. The microprocessor complex mediates the genesis of microRNAs. Nature. 2004;432(7014):235–40.
Grewal SI, Moazed D. Heterochromatin and epigenetic control of gene expression. Science. 2003;301(5634):798–802.
Griffith J, Mahler H. DNA ticketing theory of memory. Nature. 1969;223:580–2.
Grunstein M. Histone acetylation in chromatin structure and transcription. Nature. 1997;389(6649):349–52.
Gupta A, et al. Hypomethylation of the synuclein γ gene CpG island promotes its aberrant expression in breast carcinoma and ovarian carcinoma. Cancer Res. 2003;63(3):664–73.
Gusterson R, et al. The transcriptional co-activators CBP and p300 are activated via phenylephrine through the p42/p44 MAPK cascade. J Biol Chem. 2002;277(4):2517–24.
Gusterson RJ, et al. The transcriptional co-activators CREB-binding protein (CBP) and p300 play a critical role in cardiac hypertrophy that is dependent on their histone acetyltransferase activity. J Biol Chem. 2003;278(9):6838–47.
Haas J, et al. Alterations in cardiac DNA methylation in human dilated cardiomyopathy. EMBO Mol Med. 2013;5(3):413–29.
Han J, et al. The Drosha-DGCR8 complex in primary microRNA processing. Genes Dev. 2004;18(24):3016–27.
Hatada I, et al. A genomic scanning method for higher organisms using restriction sites as landmarks. Proc Natl Acad Sci. 1991;88(21):9523–7.
Hayatsu H. Discovery of bisulfite-mediated cytosine conversion to uracil, the key reaction for DNA methylation analysis – a personal account. Proc Jpn Acad Ser B. 2008;84(8):321–30.
Heidenreich PA, et al. Forecasting the future of cardiovascular disease in the United States. Policy Statement Am Heart Assoc. 2011;123(8):933–44.
Heijmans BT, et al. Persistent epigenetic differences associated with prenatal exposure to famine in humans. Proc Natl Acad Sci U S A. 2008;105(44):17046–9.
Hergenreider E, et al. Atheroprotective communication between endothelial cells and smooth muscle cells through miRNAs. Nat Cell Biol. 2012;14(3):249–56.
Herman JG, Baylin SB. Gene silencing in cancer in association with promoter hypermethylation. N Engl J Med. 2003;349(21):2042–54.
Herman J, et al. Methylation-specific PCR: a novel PCR assay for methylation status of CpG islands. Proc Natl Acad Sci U S A. 1996;93:9821–6.
Hiltunen MO, Ylä-Herttuala S. DNA methylation, smooth muscle cells, and atherogenesis. Arterioscler Thromb Vasc Biol. 2003;23(10):1750–3.
Hiltunen MO, et al. DNA hypomethylation and methyltransferase expression in atherosclerotic lesions. Vasc Med. 2002;7(1):5–11.
Hodawadekar S, Marmorstein R. Chemistry of acetyl transfer by histone modifying enzymes: structure, mechanism and implications for effector design. Oncogene. 2007;26(37):5528–40.
Holliday R, Pugh JE. DNA modification mechanisms and gene activity during development. Science. 1975;187(4173):226–32.
Howard G, et al. Activation and transposition of endogenous retroviral elements in hypomethylation induced tumors in mice. Oncogene. 2008;27(3):404–8.

Hunter MP, et al. Detection of microRNA expression in human peripheral blood microvesicles. PLoS One. 2008;3(11):e3694.

Huynh KD, Lee JT. X-chromosome inactivation: a hypothesis linking ontogeny and phylogeny. Nat Rev Genet. 2005;6(5):410–8.

Illingworth R, et al. A novel CpG island set identifies tissue-specific methylation at developmental gene loci. PLoS Biol. 2008;6(1):e22.

Issa J-PJ, et al. Phase 1 study of low-dose prolonged exposure schedules of the hypomethylating agent 5-aza-2′-deoxycytidine (decitabine) in hematopoietic malignancies. Blood. 2004;103(5):1635–40.

Jackson-Grusby L, et al. Mutagenicity of 5-aza-2′-deoxycytidine is mediated by the mammalian DNA methyltransferase. Proc Natl Acad Sci. 1997;94(9):4681–5.

Jahangeer S, et al. β-Adrenergic receptor induction in HeLa cells: synergistic effect of 5-azacytidine and butyrate. Biochem Biophys Res Commun. 1982;108(4):1434–40.

Jazbutyte V, et al. MicroRNA-22 increases senescence and activates cardiac fibroblasts in the aging heart. Age. 2013;35(3):747–62.

Jiang X-Y, et al. Inhibition of gata4 and Tbx5 by nicotine-mediated DNA methylation in myocardial differentiation. Stem Cell Rep. 2017;8(2):290–304.

Jones PA. Cancer: death and methylation. Nature. 2001;409(6817):141–4.

Jones PA, Liang G. Rethinking how DNA methylation patterns are maintained. Nat Rev Genet. 2009;10(11):805–11.

Jones PL, et al. Methylated DNA and MeCP2 recruit histone deacetylase to repress transcription. Nat Genet. 1998;19(2):187–91.

Juergens RA, et al. Combination epigenetic therapy has efficacy in patients with refractory advanced non–small cell lung cancer. Cancer Discov. 2011;1(7):598–607.

Kaneda R, et al. Genome-wide histone methylation profile for heart failure. Genes Cells. 2009;14(1):69–77.

Kantarjian HM, et al. Results of decitabine (5-aza-2′ deoxycytidine) therapy in 130 patients with chronic myelogenous leukemia. Cancer. 2003;98(3):522–8.

Karpf AR, et al. Activation of the p53 DNA damage response pathway after inhibition of DNA methyltransferase by 5-aza-2′-deoxycytidine. Mol Pharmacol. 2001;59(4):751–7.

Kee HJ, Kook H. Roles and targets of class I and IIa histone deacetylases in cardiac hypertrophy. J Biomed Biotechnol. 2011;2011:928326.

Keshet I, et al. Evidence for an instructive mechanism of de novo methylation in cancer cells. Nat Genet. 2006;38(2):149–53.

Kijima M, et al. Trapoxin, an antitumor cyclic tetrapeptide, is an irreversible inhibitor of mammalian histone deacetylase. J Biol Chem. 1993;268(30):22429–35.

Kim VN. MicroRNA biogenesis: coordinated cropping and dicing. Nat Rev Mol Cell Biol. 2005;6(5):376–85.

Kim YB, et al. Oxamflatin is a novel antitumor compound that inhibits mammalian histone deacetylase. Oncogene. 1999;18(15):2461–70.

Kim H, et al. TRF2 functions as a protein hub and regulates telomere maintenance by recognizing specific peptide motifs. Nat Struct Mol Biol. 2009;16(4):372–9.

Kim GH, et al. Epigenetic mechanisms of pulmonary hypertension. Pulm Circ. 2011;1(3):347–56.

Knudson AG. Karnofsky memorial lecture. Hereditary cancer: theme and variations. J Clin Oncol. 1997;15(10):3280–7.

Kong W, et al. Strategies for profiling microRNA expression. J Cell Physiol. 2009;218(1):22–5.

Kornienko AE, et al. Gene regulation by the act of long non-coding RNA transcription. BMC Biol. 2013;11(1):59.

Kowalczyk MS, et al. Molecular biology: RNA discrimination. Nature. 2012;482(7385):310–1.

Kroh EM, et al. Analysis of circulating microRNA biomarkers in plasma and serum using quantitative reverse transcription-PCR (qRT-PCR). Methods. 2010;50(4):298–301.

Lagos-Quintana M, et al. Identification of tissue-specific microRNAs from mouse. Curr Biol. 2002;12(9):735–9.

Laird PW. The power and the promise of DNA methylation markers. Nat Rev Cancer. 2003;3(4):253.
Laird PW. Principles and challenges of genome-wide DNA methylation analysis. Nat Rev Genet. 2010;11(3):191–203.
Landsverk HB, et al. The protein phosphatase 1 regulator PNUTS is a new component of the DNA damage response. EMBO Rep. 2010;11(11):868–75.
Larsen F, et al. CpG islands as gene markers in the human genome. Genomics. 1992;13(4):1095–107.
Lau OD, et al. HATs off: selective synthetic inhibitors of the histone acetyltransferases p300 and PCAF. Mol Cell. 2000;5(3):589–95.
Lawson C, et al. Microvesicles and exosomes: new players in metabolic and cardiovascular disease. J Endocrinol. 2016;228(2):R57–71.
Lee Y, et al. The nuclear RNase III Drosha initiates microRNA processing. Nature. 2003;425(6956):415–9.
Lee HA, et al. Histone deacetylase inhibition attenuates transcriptional activity of mineralocorticoid receptor through its acetylation and prevents development of hypertension. Circ Res. 2013;112(7):1004–12.
Li Q, et al. Attenuation of microRNA-1 derepresses the cytoskeleton regulatory protein twinfilin-1 to provoke cardiac hypertrophy. J Cell Sci. 2010;123(14):2444–52.
Li M, et al. Emergence of fibroblasts with a proinflammatory epigenetically altered phenotype in severe hypoxic pulmonary hypertension. J Immunol. 2011;187(5):2711–22.
Liang G, et al. Analysis of gene induction in human fibroblasts and bladder cancer cells exposed to the methylation inhibitor 5-aza-2′-deoxycytidine. Cancer Res. 2002;62(4):961–6.
Lin RJ, et al. Role of the histone deacetylase complex in acute promyelocytic leukaemia. Nature. 1998;391(6669):811–4.
Lin H, et al. Suppression of intestinal neoplasia by deletion of Dnmt3b. Mol Cell Biol. 2006;26(8):2976–83.
Lopez-Serra L, et al. A profile of methyl-CpG binding domain protein occupancy of hypermethylated promoter CpG islands of tumor suppressor genes in human cancer. Cancer Res. 2006;66(17):8342–6.
Loscalzo J, Handy DE. Epigenetic modifications: basic mechanisms and role in cardiovascular disease (2013 Grover conference series). Pulm Circ. 2014;4(2):169–74.
Lubbert M. DNA methylation inhibitors in the treatment of leukemias, myelodysplastic syndromes and hemoglobinopathies: clinical results and possible mechanisms of action. Curr Top Microbiol Immunol. 2000;249:135.
Lund G, et al. DNA methylation polymorphisms precede any histological sign of atherosclerosis in mice lacking apolipoprotein E. J Biol Chem. 2004;279(28):29147–54.
Maegawa S, et al. Caloric restriction delays age-related methylation drift. Nat Commun. 2017;8(1):539.
Majumdar G, et al. Pan-histone deacetylase inhibitors regulate signaling pathways involved in proliferative and pro-inflammatory mechanisms in H9c2 cells. BMC Genomics. 2012;13:709.
Malkin D. p53 and the Li-Fraumeni syndrome. Cancer Genet Cytogenet. 1993;66(2):83–92.
Manabe I, Owens GK. Recruitment of serum response factor and hyperacetylation of histones at smooth muscle-specific regulatory regions during differentiation of a novel P19-derived in vitro smooth muscle differentiation system. Circ Res. 2001;88(11):1127–34.
Marks PA, et al. Histone deacetylases. Curr Opin Pharmacol. 2003;3(4):344–51.
Martin L, et al. Systematic reconstruction of RNA functional motifs with high-throughput microfluidics. Nat Methods. 2012;9(12):1192–4.
McDonald OG, et al. Control of SRF binding to CArG box chromatin regulates smooth muscle gene expression in vivo. J Clin Invest. 2006;116(1):36–48.
McMurray JJ, Pfeffer MA. Heart failure. Lancet. 2005;365(9474):1877–89.
Mestdagh P, et al. Evaluation of quantitative miRNA expression platforms in the microRNA quality control (miRQC) study. Nat Methods. 2014;11(8):809–15.
Mikkelsen TS, et al. Genome-wide maps of chromatin state in pluripotent and lineage-committed cells. Nature. 2007;448(7153):553–60.

Milne TA, et al. Chromatin immunoprecipitation (ChIP) for analysis of histone modifications and chromatin-associated proteins. Methods Mol Biol (Clifton, N.J.). 2009;538:409–23.

Mitchell PS, et al. Circulating microRNAs as stable blood-based markers for cancer detection. Proc Natl Acad Sci. 2008;105(30):10513–8.

Miyamoto K, Ushijima T. Diagnostic and therapeutic applications of epigenetics. Jpn J Clin Oncol. 2005;35(6):293–301.

Moazed D. Small RNAs in transcriptional gene silencing and genome defence. Nature. 2009;457(7228):413–20.

Moncada S, Higgs A. The L-arginine-nitric oxide pathway. N Engl J Med. 1993;329(27):2002–12.

Montgomery RL, et al. Histone deacetylases 1 and 2 redundantly regulate cardiac morphogenesis, growth, and contractility. Genes Dev. 2007;21(14):1790–802.

Monzon J, et al. CDKN2A mutations in multiple primary melanomas. N Engl J Med. 1998;338(13):879–87.

Movassagh M, et al. Differential DNA methylation correlates with differential expression of angiogenic factors in human heart failure. PLoS One. 2010;5(1):e8564.

Mu S, et al. Epigenetic modulation of the renal beta-adrenergic-WNK4 pathway in salt-sensitive hypertension. Nat Med. 2011;17(5):573–80.

Nagy Z, Tora L. Distinct GCN5/PCAF-containing complexes function as co-activators and are involved in transcription factor and global histone acetylation. Oncogene. 2007;26(37):5341–57.

Napoli C, Ignarro LJ. Nitric oxide and pathogenic mechanisms involved in the development of vascular diseases. Arch Pharm Res. 2009;32(8):1103–8.

Nemerovski CW, et al. Vitamin D and cardiovascular disease. Pharmacother: J Hum Pharmacol Drug Ther. 2009;29(6):691–708.

Ng H-H, et al. MBD2 is a transcriptional repressor belonging to the MeCP1 histone deacetylase complex. Nat Genet. 1999;23:58–61.

Niwa Y, et al. Methylation silencing of SOCS-3 promotes cell growth and migration by enhancing JAK/STAT and FAK signalings in human hepatocellular carcinoma. Oncogene. 2005;24(42):6406–17.

O'Geen H, et al. Using ChIP-seq technology to generate high-resolution profiles of histone modifications. Epigenetics Protoc. 2011;791:265–86.

Oishi K. Plasminogen activator inhibitor-1 and the circadian clock in metabolic disorders. Clin Exp Hypertens. 2009;31(3):208–19.

Oka D, et al. The presence of aberrant DNA methylation in noncancerous esophageal mucosae in association with smoking history. Cancer. 2009;115(15):3412–26.

Okano M, et al. DNA methyltransferases Dnmt3a and Dnmt3b are essential for de novo methylation and mammalian development. Cell. 1999;99(3):247–57.

Olsen EA, et al. Phase IIb multicenter trial of vorinostat in patients with persistent, progressive, or treatment refractory cutaneous T-cell lymphoma. J Clin Oncol. 2007;25(21):3109–15.

Omodei D, Fontana L. Calorie restriction and prevention of age-associated chronic disease. FEBS Lett. 2011;585(11):1537–42.

Ovchinnikova ES, et al. Signature of circulating microRNAs in patients with acute heart failure. Eur J Heart Fail. 2016;18(4):414–23.

Park PJ. ChIP–seq: advantages and challenges of a maturing technology. Nat Rev Genet. 2009;10(10):669–80.

Paul CL, Clark SJ. Cytosine methylation: quantitation by automated genomic sequencing and GENESCAN analysis. BioTechniques. 1996;21(1):126–33.

Perez-Campo FM, et al. The histone acetyl transferase activity of monocytic leukemia zinc finger is critical for the proliferation of hematopoietic precursors. Blood. 2009;113(20):4866–74.

Petretto E, et al. Integrated genomic approaches implicate osteoglycin (Ogn) in the regulation of left ventricular mass. Nat Genet. 2008;40(5):546–52.

Piekarz RL, et al. Phase II multi-institutional trial of the histone deacetylase inhibitor romidepsin as monotherapy for patients with cutaneous T-cell lymphoma. J Clin Oncol. 2009;27(32):5410–7.

Plongthongkum N, et al. Advances in the profiling of DNA modifications: cytosine methylation and beyond. Nat Rev Genet. 2014;15(10):647–61.
Plumb JA, et al. Reversal of drug resistance in human tumor xenografts by 2′-deoxy-5-azacytidine-induced demethylation of the hMLH1 gene promoter. Cancer Res. 2000;60(21):6039–44.
Portela A, Esteller M. Epigenetic modifications and human disease. Nat Biotechnol. 2010;28(10):1057–68.
Post WS, et al. Methylation of the estrogen receptor gene is associated with aging and atherosclerosis in the cardiovascular system. Cardiovasc Res. 1999;43(4):985–91.
Qiu T, et al. Effects of treatment with histone deacetylase inhibitors in solid tumors: a review based on 30 clinical trials. Future Oncol. 2013;9(2):255–69.
Rao PK, et al. Loss of cardiac microRNA-mediated regulation leads to dilated cardiomyopathy and heart failure. Circ Res. 2009;105(6):585–94.
Raposo G, Stoorvogel W. Extracellular vesicles: exosomes, microvesicles, and friends. J Cell Biol. 2013;200(4):373–83.
Ravasi T, et al. Experimental validation of the regulated expression of large numbers of non-coding RNAs from the mouse genome. Genome Res. 2006;16(1):11–9.
Razin A, Riggs AD. DNA methylation and gene function. Science. 1980;210(4470):604–10.
Reddy MA, Natarajan R. Epigenetic mechanisms in diabetic vascular complications. Cardiovasc Res. 2011;90(3):421–9.
Reik W, Lewis A. Co-evolution of X-chromosome inactivation and imprinting in mammals. Nat Rev Genet. 2005;6(5):403–10.
Reik W, Walter J. Genomic imprinting: parental influence on the genome. Nat Rev Genet. 2001;2(1):21–32.
Ren B, et al. Genome-wide location and function of DNA binding proteins. Science. 2000;290(5500):2306–9.
Riggs AD. X inactivation, differentiation, and DNA methylation. Cytogenet Genome Res. 1975;14(1):9–25.
Robert M-F, et al. DNMT1 is required to maintain CpG methylation and aberrant gene silencing in human cancer cells. Nat Genet. 2003;33(1):61–5.
Roberts RJ, et al. A nomenclature for restriction enzymes, DNA methyltransferases, homing endonucleases and their genes. Nucleic Acids Res. 2003;31(7):1805–12.
Robertson KD. DNA methylation and human disease. Nat Rev Genet. 2005;6(8):597–610.
Rochette L, et al. Nitric oxide synthase inhibition and oxidative stress in cardiovascular diseases: possible therapeutic targets? Pharmacol Ther. 2013;140(3):239–57.
Roger VL. Epidemiology of heart failure. Circ Res. 2013;113(6):646–59.
Rountree MR, et al. DNMT1 binds HDAC2 and a new co-repressor, DMAP1, to form a complex at replication foci. Nat Genet. 2000;25(3):269–77.
Rousseau E, et al. CDKN2A, CDKN2B and p14 ARF are frequently and differentially methylated in ependymal tumours. Neuropathol Appl Neurobiol. 2003;29(6):574–83.
Roy S, et al. MicroRNA expression in response to murine myocardial infarction: miR-21 regulates fibroblast metalloprotease-2 via phosphatase and tensin homologue. Cardiovasc Res. 2009;82(1):21–9.
Rush LJ, Plass C. Restriction landmark genomic scanning for DNA methylation in cancer: past, present, and future applications. Anal Biochem. 2002;307(2):191–201.
Sadri R, Hornsby PJ. Rapid analysis of DNA methylation using new restriction enzyme sites created by bisulfite modification. Nucleic Acids Res. 1996;24(24):5058–9.
Sato N, et al. Frequent hypomethylation of multiple genes overexpressed in pancreatic ductal adenocarcinoma. Cancer Res. 2003;63(14):4158–66.
Saxonov S, et al. A genome-wide analysis of CpG dinucleotides in the human genome distinguishes two distinct classes of promoters. Proc Natl Acad Sci. 2006;103(5):1412–7.
Sayed D, et al. MicroRNAs play an essential role in the development of cardiac hypertrophy. Circ Res. 2007;100(3):416–24.
Segura-Pacheco B, et al. Global DNA hypermethylation-associated cancer chemotherapy resistance and its reversion with the demethylating agent hydralazine. J Transl Med. 2006;4(1):32.

Selker EU, et al. The methylated component of the Neurospora crassa genome. Nature. 2003;422(6934):893–7.
Shapiro R, et al. Reactions of uracil and cytosine derivatives with sodium bisulfite. J Am Chem Soc. 1970;92(2):422–4.
Shingara J, et al. An optimized isolation and labeling platform for accurate microRNA expression profiling. RNA. 2005;11(9):1461–70.
Shirodkar AV, Marsden PA. Epigenetics in cardiovascular disease. Curr Opin Cardiol. 2011;26(3):209.
Shu J, et al. Dynamic and modularized microRNA regulation and its implication in human cancers. Sci Rep. 2017;7(1):13356.
Silverman LR, et al. Randomized controlled trial of azacitidine in patients with the myelodysplastic syndrome: a study of the cancer and leukemia group B. J Clin Oncol. 2002;20(10):2429–40.
Singer-Sam J, et al. A quantitative HpaII-PCR assay to measure methylation of DNA from a small number of cells. Nucleic Acids Res. 1990;18(3):687.
Singh N, et al. Molecular modeling and molecular dynamics studies of hydralazine with human DNA methyltransferase 1. Chem Med Chem. 2009;4(5):792–9.
Sobulo OM, et al. MLL is fused to CBP, a histone acetyltransferase, in therapy-related acute myeloid leukemia with a t (11; 16)(q23; p13. 3). Proc Natl Acad Sci. 1997;94(16):8732–7.
Song F, et al. Association of tissue-specific differentially methylated regions (TDMs) with differential gene expression. Proc Natl Acad Sci U S A. 2005;102(9):3336–41.
Šorm F, et al. 5-Azacytidine, a new, highly effective cancerostatic. Cell Mol Life Sci. 1964;20(4):202–3.
Spencer VA, et al. Chromatin immunoprecipitation: a tool for studying histone acetylation and transcription factor binding. Methods. 2003;31(1):67–75.
Stenvinkel P, et al. Impact of inflammation on epigenetic DNA methylation–a novel risk factor for cardiovascular disease? J Intern Med. 2007;261(5):488–99.
Stewart D, et al. A phase I pharmacokinetic and pharmacodynamic study of the DNA methyltransferase 1 inhibitor MG98 administered twice weekly. Ann Oncol. 2003;14(5):766–74.
Suzuki H, et al. Epigenetic inactivation of SFRP genes allows constitutive WNT signaling in colorectal cancer. Nat Genet. 2004;36(4):417–22.
Suzuki K, et al. Global DNA demethylation in gastrointestinal cancer is age dependent and precedes genomic damage. Cancer Cell. 2006;9(3):199–207.
Swedberg K, et al. Guidelines for the diagnosis and treatment of chronic heart failure: executive summary (update 2005). Eur Heart J. 2005;26(11):1115–40.
Thiagalingam S, et al. Loss of heterozygosity as a predictor to map tumor suppressor genes in cancer: molecular basis of its occurrence. Curr Opin Oncol. 2002;14(1):65–72.
Thiagarajan D, et al. Mechanisms of transcription factor acetylation and consequences in hearts. Biochim Biophys Acta. 2016;1862(12):2221–31.
Thum T, et al. MicroRNA-21 contributes to myocardial disease by stimulating MAP kinase signalling in fibroblasts. Nature. 2008;456(7224):980–4.
Tijsen AJ, et al. Circulating microRNAs as diagnostic biomarkers for cardiovascular diseases. Am J Phys Heart Circ Phys. 2012;303(9):H1085–95.
Tobi EW, et al. DNA methylation differences after exposure to prenatal famine are common and timing- and sex-specific. Hum Mol Genet. 2009;18(21):4046–53.
Toyota M, Issa J-PJ. CpG island methylator phenotypes in aging and cancer. Semin Cancer Biol. 1999;9(5):349–57.
Trivedi MS, et al. Short-term sleep deprivation leads to decreased systemic redox metabolites and altered epigenetic status. PLoS One. 2017;12(7):e0181978.
Tsai M-C, et al. Long noncoding RNA as modular scaffold of histone modification complexes. Science. 2010;329(5992):689–93.
Turunen MP, et al. Epigenetics and atherosclerosis. Biochim Biophys Acta (BBA)-Gen Subj. 2009;1790(9):886–91.
Usui M, et al. Increased endogenous nitric oxide synthase inhibitor in patients with congestive heart failure. Life Sci. 1998;62(26):2425–30.

Usui T, et al. HDAC4 mediates development of hypertension via vascular inflammation in spontaneous hypertensive rats. Am J Physiol Heart Circ Physiol. 2012;302(9):H1894–904.
Vaissière T, et al. Epigenetic interplay between histone modifications and DNA methylation in gene silencing. Mutat Res/Rev Mutat Res. 2008;659(1):40–8.
Valencia A, et al. Wnt signaling pathway is epigenetically regulated by methylation of Wnt antagonists in acute myeloid leukemia. Leukemia. 2009;23(9):1658–66.
Van Aelst LN, et al. Osteoglycin prevents cardiac dilatation and dysfunction after myocardial infarction through infarct collagen strengthening. Circ Res. 2014;116:425–36.
Varrone F, et al. The circulating level of FABP3 is an indirect biomarker of microRNA-1. J Am Coll Cardiol. 2013;61(1):88–95.
Villar AV, et al. Myocardial and circulating levels of microRNA-21 reflect left ventricular fibrosis in aortic stenosis patients. Int J Cardiol. 2013;167(6):2875–81.
Voellenkle C, et al. MicroRNA signatures in peripheral blood mononuclear cells of chronic heart failure patients. Physiol Genomics. 2010;42(3):420–6.
Waddington C. The epigenetics of birds. New York: Cambridge University Press; 1952.
Wadhwa PD, et al. Developmental origins of health and disease: brief history of the approach and current focus on epigenetic mechanisms. Semin Reprod Med. 2009;27(5):358–68.
Wagner JM, et al. Histone deacetylase (HDAC) inhibitors in recent clinical trials for cancer therapy. Clin Epigenetics. 2010;1(3):117.
Wang LQ, Chim CS. DNA methylation of tumor-suppressor miRNA genes in chronic lymphocytic leukemia. Epigenomics. 2015;7(3):461–73.
Wang Y, et al. Dysregulation of histone acetyltransferases and deacetylases in cardiovascular diseases. Oxidative Med Cell Longev. 2014;2014:641979.
Waterland RA, Jirtle RL. Transposable elements: targets for early nutritional effects on epigenetic gene regulation. Mol Cell Biol. 2003;23(15):5293–300.
Waterland RA, Michels KB. Epigenetic epidemiology of the developmental origins hypothesis. Annu Rev Nutr. 2007;27:363–88.
Weber J, et al. Expression of the MAGE-1 tumor antigen is up-regulated by the demethylating agent 5-aza-2′-deoxycytidine. Cancer Res. 1994;54(7):1766–71.
Weber M, et al. Chromosome-wide and promoter-specific analyses identify sites of differential DNA methylation in normal and transformed human cells. Nat Genet. 2005;37(8):853–62.
Weber M, et al. Distribution, silencing potential and evolutionary impact of promoter DNA methylation in the human genome. Nat Genet. 2007;39(4):457–66.
West AC, Johnstone RW. New and emerging HDAC inhibitors for cancer treatment. J Clin Invest. 2014;124(1):30–9.
Whittaker SJ, et al. Final results from a multicenter, international, pivotal study of romidepsin in refractory cutaneous T-cell lymphoma. J Clin Oncol. 2010;28(29):4485–91.
Widschwendter M, et al. DNA hypomethylation and ovarian cancer biology. Cancer Res. 2004;64(13):4472–80.
Willenbrock H, et al. Quantitative miRNA expression analysis: comparing microarrays with next-generation sequencing. RNA. 2009;15(11):2028–34.
Wilson Tang WH, et al. Differential effects of arginine methylation on diastolic dysfunction and disease progression in patients with chronic systolic heart failure. Eur Heart J. 2008;29(20):2506–13.
Wimalawansa SJ. Vitamin D and cardiovascular diseases: causality. J Steroid Biochem Mol Biol. 2016;175:29–43.
Xiao H, et al. Both Sp1 and Sp3 are responsible for p21waf1 promoter activity induced by histone deacetylase inhibitor in NIH3T3 cells. J Cell Biochem. 1999;73(3):291–302.
Xie M, Hill JA. HDAC-dependent ventricular remodeling. Trends Cardiovasc Med. 2013;23(6):229–35.
Yamada Y, et al. Opposing effects of DNA hypomethylation on intestinal and liver carcinogenesis. Proc Natl Acad Sci U S A. 2005;102(38):13580–5.
Yamashita K, et al. Pharmacologic unmasking of epigenetically silenced tumor suppressor genes in esophageal squamous cell carcinoma. Cancer Cell. 2002;2(6):485–95.

Yanazume T, et al. Cardiac p300 is involved in myocyte growth with decompensated heart failure. Mol Cell Biol. 2003;23(10):3593–606.

Yang XJ, Seto E. HATs and HDACs: from structure, function and regulation to novel strategies for therapy and prevention. Oncogene. 2007;26(37):5310–8.

Yang XJ, Seto E. Lysine acetylation: codified crosstalk with other posttranslational modifications. Mol Cell. 2008;31(4):449–61.

Yang X, et al. Targeting DNA methylation for epigenetic therapy. Trends Pharmacol Sci. 2010;31(11):536–46.

Yang Q, et al. Pulmonary artery smooth muscle cell proliferation and migration in fetal lambs acclimatized to high-altitude long-term hypoxia: role of histone acetylation. Am J Physiol Lung Cell Mol Physiol. 2012;303(11):L1001–10.

Yoder JA, et al. Cytosine methylation and the ecology of intragenomic parasites. Trends Genet. 1997;13(8):335–40.

Yoo CB, et al. Long-term epigenetic therapy with oral zebularine has minimal side effects and prevents intestinal tumors in mice. Cancer Prev Res. 2008;1(4):233–40.

Yoshida M, et al. Potent and specific inhibition of mammalian histone deacetylase both in vivo and in vitro by trichostatin A. J Biol Chem. 1990;265(28):17174–9.

Yoshikawa H, et al. SOCS-1, a negative regulator of the JAK/STAT pathway, is silenced by methylation in human hepatocellular carcinoma and shows growth-suppression activity. Nat Genet. 2001;28(1):29–35.

Zaina S, et al. Nutrition and aberrant DNA methylation patterns in atherosclerosis: more than just hyperhomocysteinemia? J Nutr. 2005;135(1):5–8.

Zhang Y, et al. Model-based analysis of ChIP-Seq (MACS). Genome Biol. 2008;9(9):R137.

Zhao W, et al. Abnormal activation of the synuclein-gamma gene in hepatocellular carcinomas by epigenetic alteration. Int J Oncol. 2006;28(5):1081–8.

Zhao Y, et al. Dysregulation of cardiogenesis, cardiac conduction, and cell cycle in mice lacking miRNA-1-2. Cell. 2007;129(2):303–17.

Zhong J, et al. Cardiac autonomic dysfunction: particulate air pollution effects are modulated by epigenetic immunoregulation of *toll-like receptor 2* and dietary flavonoid intake. J Am Heart Assoc. 2015;4(1):e001423.

Zhu S, et al. Inactivation of monocarboxylate transporter MCT3 by DNA methylation in atherosclerosis. Circulation. 2005;112(9):1353–61.

Zilberman D, Henikoff S. Genome-wide analysis of DNA methylation patterns. Development. 2007;134(22):3959–65.

Chapter 4
Mitochondria Genome Mutations and Cardiovascular Diseases

Hui Shi, Ying Yu, Minghui Li, and Ruizhen Chen

Abstract The cardiovascular diseases (CVDs) are one of the leading causes of morbidity and mortality across the world. CVDs have multifactorial etiologies including environmental, lifestyle, and genetic factors. Mitochondria (mt) are indispensable organelles harboring a 16,569 base-pair, circular DNA genome that encodes 13 proteins involved in electron transport and oxidative phosphorylation system (OXPHOS). Critical components of OXPHOS are encoded by the mitochondrial DNAs (mtDNAs). Any mutation in this genome might impair the mitochondrial function leading to increased oxidative stress, inflammation, and cell death, which are deleterious factors to cardiovascular system. There is a growing body of evidence in support of the roles of mitochondrial genome mutations in the pathogenesis of CVDs, including cardiomyopathies, hypertension, atherosclerosis, inflammation, etc. Consequently, the mtDNA genome mutations might represent promising molecular biomarkers or therapeutic targets for CVDs. In this review, we focused on the state-of-the-art studies about mitochondrial genome mutations associated with CVDs, seeking for their potential diagnosis and treatment significances in clinical practices.

Keywords Cardiovascular diseases · Mitochondrial genome · mtDNA mutation · Mitochondria dysfunction

H. Shi · M. Li · R. Chen (✉)
Department of Cardiology, Shanghai Institute of Cardiovascular Diseases, Zhongshan Hospital, Shanghai Medical College of Fudan University, Shanghai, China
e-mail: chen.ruizhen@zs-hospital.sh.cn

Y. Yu
Department of General Practice, Zhongshan Hospital, Shanghai Medical College of Fudan University, Shanghai, China

H. Jiang, M. Liu (eds.), *Heart Genomics*, Translational Bioinformatics 16,
https://doi.org/10.1007/978-981-13-1429-2_4

1 Introduction

Cardiovascular diseases (CVDs), responsible for more than 17 million death annually (Du et al. 2016), are one of the leading causes of mortality across the world. Accordingly, the annual death rate due to CVDs is 236.1 per 100,000 persons (Go et al. 2013). The etiology of CVDs is multifractional, such as the plasma lipid and lipoprotein levels, blood pressure and blood glucose regulation, and lifestyle. Correspondingly, various pathophysiological procedures are involved at the cellular level like energy starvation, oxidative stress abnormality, and apoptosis.

Besides the conventional factors, there is a growing number of evidence indicating the roles of genetic mutation in the etiology of CVDs. Previously, the nuclear genome abnormality has drawn the main attention of the scientists (McNally et al. 2013) until some researchers found that the inheritance patterns of some CVDs did not obey the classic Mendelian genetics but maternal inheritance pattern, such as hypertension (Lu et al. 2011) and cardiomyopathy (Mohammed et al. 2012), which indicated that mitochondrial genome might participate in the development and progression of CVDs.

For all the nucleated cells, mitochondria are indispensable, particularly for those with high-energy demands such as skeletal muscle cells, neurons, and cardiomyocytes. Among them, the heart is the most energy-consuming organ which consumes 100 times its weight of ATP to support more than 100,000 myocardial contractions each day. The mitochondria, accounting for almost 35% of the heart volume, must constantly generate new ATP to maintain the basic heart function. Consequently, the heart is highly sensitive to mitochondria dysfunction which always means an energy depletion condition. Mitochondria dysfunction could lead to a decline in the energy production, loss of myocytes contractility, and increased cardiomyocyte apoptosis during the CVDs development (Marin-Garcia and Goldenthal 2008).

Various factors might attribute to the mitochondria dysfunction, in which mitochondrial genome damages are one of the most common. Mitochondria damages could reduce mitochondrial respiration chain activity, decrease ATP production, and disrupt protein synthesis. Mitochondrial defects, once considered to lead to only rare severe metabolic and neurological diseases, are now believed to have close relationship with a wide range of disorders (Wallace 1999). Although various mitochondria diseases such as mitochondrial myopathy, encephalopathy, lactic acidosis, and stroke-like episodes syndrome (MELAS), Leigh's encephalopathy, and myoclonic epilepsy with ragged-red fibers (MERF) display a variety of clinical manifestation, cardiac symptoms are often involved in the clinical manifestation of these genetic mitochondrial diseases, for example, cardiomyopathy occurs in 18–30% of individuals with MELAS syndrome (Finsterer and Kothari 2014). And, biological aging in human is accompanied by an accumulation of mtDNA damage caused by oxidative activity. In addition, advanced age is closely associated with prevalence of CVDs. Over 80% of coronary artery diseases (CADs) and more than 75% of congestive heart failure (HF) are observed in elderly patients (Marzetti et al. 2013). Thus, it is rational to assume that mitochondria genome damages might be associated with CVDs. And recently, there have been growing number of evidence indicating the roles of mtDNA in the pathogenesis of CVDs including exacerbating inflammation, causing oxidative stress and triggering cell apoptosis. Therefore, in this review, we focused on the state-of-the-art studies about mitochondrial genome mutations associated with CVDs, seeking for the potential diagnosis and treatment significances in clinical practices.

2 Mitochondria Structure, Genome, and Physiological Function

Mitochondria are 2–8-μm-long, membrane-bound, rod-shaped organelles found in all nucleated cells, each of which contains numerous mitochondria in the cytoplasm. The mitochondrial outer membrane is smooth, while the inner membrane is highly folded forming cristae. The respiratory chain or electron transport chain (ETC) containing energy-generating multipolypeptide enzyme complexes is located at the inner mitochondrial membrane.

Uniquely among organelles, mitochondria are the only organelle containing extranuclear source of DNA. Approximately 1500 proteins are involved in maintaining mitochondrial structure and function; however, only <1% are encoded by mtDNA, while the rest are under nuclear DNA's control.

The human mitochondrial genome is a double strands loop with 16,569 bps which encodes 37 genes. The DNA strands are heavy strands, rich in guanines, and light strands, rich in cytosines. There are no introns in mtDNA, and the majority of the genome is composed of coding regions. The displacement loop (D-loop) is the only major noncoding portion (1.1 kb) that is crucial for the mtDNA replication and transcription since it contains the origin of H-strand replication and promoters of the two strands, from which transcripts are produced and then processed to yield the individual message RNA (mRNA), encoding 13 proteins and 24 different RNAs, including 2 ribosomal RNAs (rRNAs) and 22 transfer RNAs (tRNAs). The 13 proteins are the components of the key subunits of the respiratory chain complexes, including 7 subunits of rotenone-sensitive NADH-ubiquinone oxidoreductase (complex I), 1 subunit (cytochrome b) of ubiquinol-cytochrome-c oxidoreductase (complex III), 3 subunits (COI, COII, and COIII) of cytochrome-c oxidase (complex IV), and 2 subunits (ATPase 6 and 8) of complex V. While, all the components of complex II are encoded by the nuclear DNA.

Mitochondria conduct many basic functions; the most important one is to generate energy required for cells in the form of ATP through oxidative phosphorylation (OXPHOS) carried out by the ETC complexes. Mitochondria also regulate cell apoptosis. Under the stimulation of several intracellular stress, the mitochondria permeability transition pores would open, leading to the rupture of mitochondrial membranes, release of proapoptotic factors, and ultimately cell death. What is more, mitochondria could deal with cellular reactive oxygen species (ROS), which are the main by-products of mitochondrial respiration that promote cell death. Mitochondria are involved in the homeostatic regulation of cellular cations such as Ca^{2+}, Na^{+}, and K^{+}. In the heart, mitochondria participate in calcium homeostasis through sarcoplasmic reticulum-mitochondrial calcium cross talk, energized mitochondria being able to accumulate calcium. Therefore, mitochondria participate in many aspects of cellular biological systems and play an important role in maintaining cellular homeostasis.

3 Mitochondrial Genome Mutation and Mitochondria Dysfunction

The replication rate of mtDNA is several folds higher than nDNA's, which increases the possibility of enzymes' errors during genome edition, causing increased risk of genome mutations. In addition, mitochondrial genome has nonhistone proteins, which means lacking of protective barrier against random attacks. Besides, DNA repair systems absence and reactive oxygen species product proximity make mitochondria more vulnerable than nuclear genome when exposing in damaging environments. It is estimated that mtDNA mutation rate is approximately 10 to 20-fold higher than that of nuclear DNA (DiMauro and Schon 2001).

Higher mtDNA mutation rate means higher variability level, as well as the emergence of somatic mtDNA mutation with aging. Multiple copies of mtDNAs exist in the plasm of most cells, and the levels differ in cells depending on the energy demands. If the cell contains identical genome, all wild type, or all mutation, this condition is called homoplasmic, while the mtDNA mutations coexist with wild-type mtDNA in a single cell known as heteroplasmy. It is thought that the degree of heteroplasmy determines the clinical phenotype. In most cases, these mutations do not cause abnormal manifestation until the ratio of mutant mtDNA to wild-type mtDNA exceeds a certain level. This phenomenon is known as "threshold effect," and it can vary depending on the mutation, the cell type, and the tissue. Most of the mitochondria in fertilized eggs come from oocytes since mtDNA bottleneck and sperm contains 1000 times less mitochondrial than unfertilized egg. This inheritance pattern is called maternal inheritance. Maternal inheritance pattern is a very important evidence to diagnose a suspected mitochondrial disease since this transmission occurs only through the mother.

As mentioned above, the high replication rate and absence of repair systems result in higher mutation rate of mtDNA than nDNA. It has been reported that at least 1 in 200 healthy human carries pathogenic mtDNA mutations (Elliott et al. 2008). Unlike nDNA, which replicates with each cell division and equally divides the DNA into the daughter cells, mitochondria are randomly distributed during cell division, and daughter cells can harbor different degrees of mutant mtDNAs. Consequently, the percentage of mutant mtDNA can vary in different cells and tissues, resulting in the variety of mitochondrial diseases.

mtDNA mutation could be categorized into the following three types: point mutation, rearrangement, and reduction in number. Point mutation can result in insertion, deletion, or substitution on identical sites, which could affect the genes encoding protein of tRNAs or rRNAs. Currently, more than half of the point mutation associated with diseases are found in the tRNA genes. Most of the mtDNA rearrangements are large-scale depletion, which vary in size from 1.3 to 8.0 kb and encompass several genes.

The amount of mtDNA deletion and its distribution in tissues are the major factors in the occurrence of clinical symptoms. The reduction in mtDNA copy number (or content) means the mtDNA copy number is less than normal, which plays a role in a range of cardiac manifestations as well as aging. Mutations can be detected in genes that encode subunits of ETC complexes, in genes that encode ancillary proteins needed for the assembly and transport or in genes that regulate activities of the mitochondria. All these mutations are maternally inherited and typically associated with variable phenotypes. However, not all the mutations could result in pathogenic manifestation. Mutation load of the mitochondrial genome, which is determined by the percentage of heteroplasmic mutations, may vary in different tissues or within the same tissue. If the percentage of mutant mtDNA increases to a certain threshold, the mitochondria function and the subsequent cellular function would be impaired, and the disease phenotypes would appear.

mtDNA mutations affect mitochondrial functions by influencing the genes encoding ETC subunits as well as by a decline in the mtRNA transcripts and protein synthesis and subsequently cause energy shortage.

Mutations in these genes result in mitochondrial dNTP pool depletion, leading to significant reduction in mtDNA content. Inadequate amount of mtDNA results in impaired synthesis of key subunits of ETC complexes. These subunits are crucial to sustain OXPHOS and to generate sufficient energy to meet the needs for various organs, especially those with high-energy demands. Consequently, ATP deficiency occurs. What is more, the consequences of mitochondrial dysfunction include excessive reactive oxygen species (ROS) production, nitric oxide deficiency, and apoptosis.

Mitochondria are a source of ROS and are also regarded as a major target for ROS-mediated damage. By regulating redox signaling in the cell, mitochondria have been considered as a signal center in large part. A small part of oxygen turns into ROS during OXPHOS. Under physiological condition, trace amount of ROS is useful for the gene transcription and enzymatic activity and could be eliminated by many enzymes. However, too much ROS production would damage mitochondrial macromolecules including mtDNA injury like single- and double-strand breaks, basic sites, and base mutations. Those mitochondria which isolated from failing heart show that increased ROS production in association with oxidative stress-related damage contributed to mtDNA mutations (Ide et al. 2001), indicating the relationship between mtDNA mutation and ROS in CVDs. In addition, mtDNA damage can introduce increased ROS production by disruption of OXPHOS, ATP synthesis, and calcium homeostasis, thus creating a self-perpetuation, positive-feedback loop. ROS production also evokes the activation of inflammation and induces the expression and secretion of some CVD-related proinflammatory factors. Additionally, mitochondrial dysfunction caused by mtDNA mutation induces cellular apoptosis. mtDNA damage results in ATP deficiency and lower mitochondrial membrane potential, both of which could enhance mitochondrial permeability transition pore (mPTP) opening and promote apoptosis. MPTP opening increases mitochondria inner membrane permeability and gives rise to the equilibration of the ions between the mitochondrial matrix and cytosol, leading to mitochondria swelling and ruptures releasing proapoptotic proteins such as cytochrome-c, which trigger latent forms of caspases and finally lead to cell death due to apoptosis or necrosis. And accumulation of mtDNA mutations is associated with the induction of apoptotic makers in transgenic mice (Kujoth et al. 2005).

Mitochondria also serve as a dynamic buffer for sarcolemmal Ca^{2+} by uptaking and extruding Ca^{2+} to modulate cardiomyocyte function. In addition, enough mitochondrial ATP production is prerequisite for calcium pumps in the plasma membrane and endoplasmic reticulum. Therefore, mitochondria dysfunctions could affect Ca^{2+} concentration in cells and then influence mitochondria energy production. This might explain the frequent involvement of muscle and nerve tissues in mitochondria diseases, in which normal cell function relies on ATP production and intracellular calcium fluctuation.

Finally, there is growing body of evidence indicating that NO deficiency is involved in mitochondria diseases like MELAS syndrome via affecting microvasculature blood perfusion in various organs and always manifested with stroke-like episodes (El-Hattab et al. 2014). NO deficiency in mitochondria disorders might attribute to multiple factors, such as the impaired NO production and postproduction sequestration. Lacking the NO precursors citrulline can lead to the NO deficiency in mitochondria diseases, because mitochondria dysfunction caused by mtDNA mutations impair ATP production required for citrulline synthesis (El-Hattab et al. 2012). NO decreasing results in endothelial cells apoptosis and dysfunction, which play a major role in CVD pathology. Therefore, NO deficiency plays a vital role in the pathogenesis of several complications in mitochondria diseases.

4 Myocardia Mitochondrial Genome Mutation and Cardiovascular Diseases

Myocardium is an energy-consuming tissue that heavily relies on normal function of mitochondria. Many researches have been devoted to focusing on the relationship between mtDNA mutation and CVDs pathogenesis, seeking to find new ways to treat CVDs. We summarized here in the following manuscript.

4.1 mtDNA Mutation and Hypertension

Hypertension is a worldwide health problem, affecting approximately one billion people around the world. Hypertension is also a risk factor for many other diseases, like coronary heart disease, stroke, heart failure, and renal diseases. Over 90% cases of hypertension have undetermined causes, which is called essential hypertension (EH).

Until now, there are many studies on the etiology of EH, but it has had no definite conclusion yet because of its multifactorial causes. Now, it is generally believed that EH could be caused by single-gene or multifactorial conditions resulting from interaction between environmental and inherited risk factors. Familial aggregation phenomenon can be seen in high blood pressure, which gives evidence that genetic factors play a role in the etiology of hypertension (Zinner et al. 1971). In the past, efforts to identify genetic factors of hypertension have focused on the nuclear genome. However, Fuentes et al. reported that the association of blood pressure between genetically related subjects (parents-offspring) was larger than the association of BP between nongenetically related subjects (Fuentes et al. 2000), indicating the close relationship between mother and offspring in blood pressure levels. This phenomenon suggests that the genetic model of hypertension may be matrilineal inheritance, which is consistent with mitochondria. From then on, many studies have bore out that some mtDNA mutations are involved in the development and progression of EH.

Mitochondria heritability for systolic blood pressure was about 5%, and mitochondrial effects may account for about 35% of hypertensive pedigrees (Lu et al. 2014). Schwartz et al. analyzed the sequence of the entire mitochondrial genome in patients from 20 pedigrees, consisting of 10 African American and 10 white families. In this study, 297 variants were detected, including 24 in the rRNA genes, 15 in the tRNA genes, and 46 in amino acid substitutions, the remainder involving the noncoding regions or synonymous changes, indicating that mitochondrial genome mutations played an important role in hypertension (Schwartz et al. 2004). Watson et al. reported the results of a systematic study of human mitochondrial genome in black American with hypertension. It showed that six variants were identified with significant differences between the hypertensive and normotensive groups. One of them occurred in the gene of OXPHOS complex I, A10086G mutation. This mutation may stimulate free radical production, which in turn lead to further damage of mtDNA and inner membrane damage (Watson et al. 2001). Liu et al. systematically analyzed the whole mitochondrial DNA sequence in nine maternally inherited hypertension probands from Chinese Han families. The sequence analysis revealed 172 base changes. They also identified seven amino acid changes, including four mutations in APT synthase F0 subunit 6 and three in cytochrome b. Among these mutations, ATPase6 C8794T is associated with coronary atherosclerosis risk, and Cytb G15662A mutation would lead to mitochondriopathy (Liu et al. 2014). Wang et al. performed genetic characterization of a Han Chinese family with maternally transmitted hypertension. The mutation analysis of the mitochondrial genome revealed a novel A4263G mutation in this family. This mutation caused an A to G transition at the 5′ end of the tRNA gene, disrupting the processing of the tRNA 5′-end precursor and finally inducing the impairment of mitochondrial translation and deficient respiration (Wang et al. 2011). Teng et al. showed that the T12338C mutation and A12330G mutation existed in a Chinese family with essential hypertension. A12330G variant decreased the stability level of tRNA, giving rise to the failure of tRNA metabolism. T12338C variant not only resulted in the replacement of the first amino acid in ND5 gene but also affected the processing of RNA precursors. And then this variant led to a reduction in tRNA level. Taking together, the combination of T12338C and A12330G variant may contribute to the high penetrance of hypertension (Teng et al. 2012). Xu et al. reported that A4401G and T5655C mutations were related with essential hypertension. A4401G mutations were localized near the $tRNA^{Met}$, and T5655C mutation occurred at the 5′ end of the $tRAN^{Ala}$ (Xu et al. 2017).

In conclusion, numerous studies have indicated that the frequency and quantity of mitochondria DNA mutations is higher in hypertensive patients than in normal individuals. Most of the mutations are located at the noncoding D-loop region, especially in RNA gene. For example, the A4435G mutation located at the anticodon could bring to 40% reduction in the level of $tRNA^{Met}$ stability (Liu et al. 2009). And the A4263G mutation at the processing site reduces 46% reduction in the steady-state level of $tRNA^{Lle}$ (Wang et al. 2011). Similarly, A4401G mutation at the junction site of $tRNA^{Met}$ and $tRNA^{Glu}$ significantly reduces the steady state of $tRNA^{Met}$ and $tRNA^{Glu}$ and decreases the oxygen consumption rate in cell lines carrying the mutation. Failing to maintain the balance of oxygen consumption causes a series of cardiovascular disorders (Zhu et al. 2009). Moreover, mutations also influence the amount of RNA. For example, T5655C mutation produces an approximate 50% reduction in the $tRNA^{Ala.}$ level, which plays a crucial role in the recognition of aminoacyl-tRNA synthetase (Xu et al. 2017). Consequently, these pathogenic mitochondrial genome mutations alter the structure and function of mtRNA through destabilizing the tRNA tertiary structure, altering processing of RNA precursors, losing nucleotide modification, lacking aminoacylation, and subsequently leading to the failure in mtRNA metabolism which will decrease the tRNA steady-state level. And then these dysfunctions caused by the mutations may evoke the impairment mitochondrial translation and respiration, which causes increase in the production of reactive oxygen species and decrease in the levels of ATP production. Subsequently, energy supply cannot satisfy the demand of endothelial cell function, thereby contributing to the elevated blood pressure (Jiang et al. 2016). In clinical practice, the mean age of hypertension onset was significantly lower in patients carried with tRNA mutations than those without. The level of ventricular septal thickness in patients with tRNA mutations was higher than those without mutations (Lu et al. 2014). In addition, deletion mutation also contributed to the development of EH. Elango et al. observed a 50 bp deletion spanning the position from 298 to 347 in the hypervariable region II (HVSII) after sequencing the complete mtDNA of a family with hypertension. Some transcription factors such as MT-TFY (276-303), MT-CBS2 (299-315), MT-HPR (317-321), and MT-CBS3 (346-367) are located at the deletion region of mtDNA, which are crucial to the regulation of mtDNA replication. Therefore, it is reasonable to predict that 50 bp deletion may cause the reduction in the copy number of mtDNA and decline in the level of cellular energy, leading to the mitochondrial dysfunction and development of hypertension (Elango et al. 2011). Systematic study of the relationship between the hypertension and mtDNA mutations not only helps to better understand the underlying mechanisms but also improves the diagnosis, precaution, and treatment of the disease. Despite numerous studies that have revealed the relationship between mtDNA mutations and hypertension, these mutations alone are insufficient to manifest clinical phenotypes. There are many other factors that simultaneously contribute to the development of hypertension like nuclear genes, environmental factors, and so on. However, it is worth mentioning that screening for mtDNAs with hypertension phenotype may provide a new promising approach to understand and treat hypertension.

4.2 mtDNA Mutation and Cardiomyopathies

Because myocardium is one of the high-energy demanding tissues, the heart tissue per se is involved in many mitochondrial diseases. Cardiomyopathies have been one of the most common cardiac manifestation of mitochondrial diseases.

Mitochondrial cardiomyopathy (MCM) is a series of myocardial conditions characterized by abnormal heart structure and/or function secondary to the genetic defects involving mitochondrial respiratory chain, in absence of concomitant coronary artery disease, hypertension, valvular disease, and congenial heart disease. The MCM clinical phenotype is diverse and complicated, which can be expressed as hypertrophic cardiomyopathy, dilated cardiomyopathy, left ventricular myocardial noncompaction, and other cardiomyopathy corresponding clinical features. The cardiac manifestation could worsen acutely during a metabolic crisis. Most MCMs are caused by mtDNA mutations including point mutation and deletion mutation. In addition, most respiratory chain complex subunits are encoded by nDNA, and many enzymes required for mtDNA replication and expression are encoded by nDNA. Consequently, nDNA mutations can also lead to MCM.

MCM is usually manifested as a part of a multisystem involvement and may rarely be isolated. It can be associated with many other mitochondrial disorders such as MELAS, Leigh syndrome (subacute necrotizing encephalomyelopathy), and MERRF. The state may be stable for many years and suddenly deteriorates and leads to heart failure (Lev et al. 2004). It has been reported that 17% of patients with mitochondrial diseases are suffering from cardiomyopathy, and a mortality rate was higher in patients with cardiomyopathy (71%) than in those without cardiac involvements (26%) (Holmgren et al. 2003). Currently, hypertrophic cardiomyopathy (HCM) and dilated cardiomyopathy (DCM) are the two major cardiac manifestations in MCM caused by mtDNA mutations. Hypertrophic form is the most common type of mitochondrial cardiomyopathies. Previously, HCM was considered to be an autosomal dominant inherited disease. In 1999, it was reported that T9997C mutation and A4295G mutation were the causes of HCM (Raha et al. 1999). After that, a lot of attentions have been paid to mtDNAs. Taylor et al. described two families with HCM associated with a homoplasmic A4300G mutation in the mt-tRNA^{Lle} gene which lead to severely decreased respiratory chain activity and very low steady-state levels of mt-tRNA^{Lle} gene (Taylor et al. 2003). Jonckheere et al. identified a molecular genetic defect in a 16-year-old patient presenting with hypertrophic cardiomyopathy and revealed that a homoplasmic mutation G8529A was found in the mitochondrial ATP8 gene, resulting in an improper assembly and reduced activity of complex V holoenzyme (Jonckheere et al. 2009). Allia-fersi et al. reported the presence of the A8527G and C8932T mutations in ATP6 gene of a mitochondrial cardiomyopathy patient. The imbalance in hydrophobicity in ATPase 6 protein caused by these mutations gives rise to a possible alteration of its structure flexibility and a deficiency in the ATP production (Alila-Fersi et al. 2017). Consequently, it can be presumed that mtRNA mutations may affect the respiratory chain protein synthesis and then result in the short energy supply for myocardial cells. And damage to mtDNA in the mitochondrial mutator mouse could cause cardiac hypertrophy, fibrosis, and impairment of systolic and diastolic function, which could be attenuated by catalase. The results above indicate that accumulation of mtDNA mutations in the heart is associated with increased oxidative stress and then results in the development of cardiomyopathy (Dai et al. 2010).

Mitochondrial cardiomyopathy can also manifest dilated cardiomyopathy (DCM) phenotype. DCM is a multi-pathogenesis disease, 20–35% of which is related to genetic factor, including mitochondrial genome. The detected common mitochondrial tRNA gene mutation sites about DCM include A3243G, A3260G, and C3303T. It has been reported that in genes encoding for NADH dehydrogenase subunits, the number of mtDNA mutations detected in patients with DCM is significantly increased. The G5973A and T7042G mutations are in highly conserved domains of gene encoding for cytochrome-c oxidase subunit. In addition, the T16189C mutation in the D-LOOP region that is associated with susceptibility to DCM could be detected in 15.6% of patients (Ruppert et al. 2004). Besides, mtDNA deletion mutation also involved in the development of DCM. mtDNA molecules with 7436 bp deletion were found in patients with cardiomyopathy (Bobba et al. 1995). Ahuja et al. examined the mitochondrial in people with DCM and revealed that DCM was associated with enhanced mtDNA deletion mutations which can cause mitochondrial dysfunction, resulting in an inability of the heart to maintain adequate energy production and continuing to cause oxidative stress (Ahuja et al. 2013).

Some mitochondrial cardiomyopathy patients also suffer from left ventricular noncompaction cardiomyopathy (LVNC). Tang et al. determined complete nucleotide sequences of mitochondrial genomes from 20 LVNC patients. They revealed that mutations in Met31 in ND1 caused by A3397G and T3398C and resultant defects in complex I can be associated with LVNC (Tang et al. 2010). And the mtDNA copy number is lower in myocardium of LVNC patients than in normal controls, indicating that lower mtDNA copy number in mitochondrial genome may be associated with etiology of LVNC (Liu et al. 2013).

There are various other factors influencing the mitochondrial cardiomyopathy. Skou et al. reported a 19-month-old girl who suffered from cardiomyopathy carrying A1555G mutation, who did not have any cardiac symptoms before. During the treatment for acute myelogenous leukemia, she was treated with high-dose chemotherapy and repeated aminoglycoside exposure. Before she underwent stem cell transplantation, cardiomyopathy with a dilated left ventricle and subnormal systolic function was diagnosed (Skou et al. 2014). According to another case report, a 62-year-old woman developed mitochondrial cardiomyopathy after chemotherapy. Although mutation of the mitochondrial A3243G gene was detected by genetic testing, the patient presented no symptoms of heart failure until the treatment with anticancer drug (Yamasaki et al. 2017). Some drugs like anthracyclines and cisplatin could accumulate in the mitochondria resulting in mitochondrial dysfunction by increasing oxidative stress and damaging mitochondrial DNA. Since the cardiomyopathy can be seen in both drug-induced cardiomyopathy and mitochondrial disorders, we speculate that the mitochondrial mutations may increase the susceptibility to drug-induced cardiotoxicity, especially in patients who are already diagnosed with mitochondrial diseases. Consequently, active genetic testing for the mitochondrial mutation is recommended in patients frequently treated with these drugs to be careful of cardiovascular complications caused by mitochondrial cardiomyopathy.

In conclusion, the mechanism of the cardiomyopathy caused by mtDNA mutations may be contributed to energy deficiency and ROS impairment. mtDNA mutations not only impair the oxidative phosphorylation, resulting in ATP production deficiency, but also affect the enzyme subunits of the respiratory chain and increase the level of ROS, which could subsequently lead to the loss of myocyte contractility, increase in cell death, and eventually cause or deteriorate cardiomyopathy.

4.3 *mtDNA Mutation and Atherosclerosis*

Atherosclerosis is the fundamental pathology of many diseases, representing a chronic inflammatory disease. Clinical manifestation of atherosclerosis, coronary artery disease, is the leading cause of death in many countries. Although many factors such as inflammation, lipid metabolism, endothelial dysfunction, and other environmental factors are responsible for the development of atherosclerosis, a role of mitochondrial factors in the pathogenesis of atherosclerosis should not be overlooked.

Somatic mutations of the mitochondrial genome may be attributed to atherosclerosis development in humans. Some mitochondrial genome mutations associated with atherosclerotic lesions have been identified. Sazonova et al. reported that 10 mutations in mitochondrial genes that were concerned with atherosclerosis plaques. They encode rRNA 12S, $tRNA^{Leu}$ (UUR recognition codon), $tRNA^{Leu}$ (CUN recognition codon), subunits of 1, 2, 5, and 6 NADH-dehydrogenase, and cytochrome B. Among these mutations, G13513A was the highest frequency in the lipofibrosis plaques. What is more, compared to healthy vessel tissue, the mutations percentage for G13513A and G15059A is 1.3 to 2-fold higher in the atherosclerotic plaques (Sazonova et al. 2009). In clinical studies, Qin et al. performed a clinical and mutational screening of 22 mitochondrial genes in Han Chinese with coronary heart disease and identified 16 variants through mutational analysis. Of these variants, only $tRNA^{Gln}$ A5592G and $tRNA^{Thr}$ G15927A mutations may be associated with coronary heart disease, thereby altering the structure and function of their tRNAs and causing long-standing increase of ROS in cells. Although these two are by themselves insufficient to generate a clinical phenotype, they can promote the development of coronary heart disease combined with other factors such as nuclear modifier, environmental and epigenetic factors, as well as personal lifestyle (Qin et al. 2014). Mitrofanov et al. presented that four mitochondrial mutations (G12315A, C3256T, G15059A, and G13513A) explained at least 60% variability of the clinical manifestations of atherosclerosis (Mitrofanov et al. 2016). Thus, these four mutations accounted a panel of genetic makers. Apart from point mutations, other mtDNA damages including deletions, single-strand breaks, double strands, base modification, and mis-pairing have also been implicated in the pathogenesis of atherosclerosis (Tian et al. 2016).

However, whether mtDNA damage directly facilitates atherosclerosis or is just a by-product of the risk factors that promote atherosclerosis is unclear. In ApoE(-/-) mice, a model of atherosclerosis, mtDNA damage is an early event in atherogenesis and can be detected even before obvious atherosclerotic plaque has formed. The early onset of mtDNA damage raises the possibility that mtDNA damage initially promotes the atherosclerosis rather than a late consequence of ROS and tissue damage. What's more, mtDNA damage induces AS through increasing the release of interleukin-1β and tumor necrosis factor-α, two proinflammatory cytokines that play significant roles in the development of AS (Yu et al. 2013). Further, mtDNA deletion could downregulate the expression of COX III protein resulting in impairment of oxidative phosphorylation process. Accumulated mtDNA deletion in the aortic root of ApoE(-/-) mice occurred prior to the formation of atherosclerotic plaques, implying that mtDNA deletion may be a representative of a major casual factor in both the initiation and progression of atherosclerosis in this animal (Tian et al. 2016). In conclusion, mtDNA damages are essential for the development of atherosclerosis and related diseases like coronary heart disease. But further researches are needed to explore the underlying mechanisms.

4.4 mtDNA Mutation and Arrhythmia

Arrhythmia is another common cardiovascular disease in clinical practice. One-third of the cardiac ATP produced by mitochondria is necessary for ion channels, providing electrical activity of the cardiac cells. Therefore, mitochondrial dysfunction could disrupt the cardiac rhythm through depleting energy that supply to these channels (Yang et al. 2014).

Atrial fibrillation (AF) is one of the most common types of tachyarrhythmia in population. Due to the rapid atrial rate in AF with 350–450 beats per minute, cardiac tissue requires higher oxygen consumption and thus accelerates the accumulation of mtDNA mutations. Tsuboi et al. demonstrated that the incidence of AF was higher in patients with mtDNA deletion than those without a deletion. Additionally, the mtDNA deletion might result in a decrease concentration of adenine nucleotides, disrupting ATP synthesis (Tsuboi et al. 2001). mtDNA 4977 deletion was also detected in AF patients, and its amount of mtDNA 4977 deletion increased significantly in the atrial tissue progressively with aging in adults, which would in turn cause the mitochondrial dysfunction and aggravate the cardiac function in AF patients (Lai et al. 2003).

Cardiac arrhythmias, especially long QT syndrome (LQTS), have been a major cause of sudden cardiac death in young individuals. In addition to nuclear genes, mitochondrial DNA may be involved in LQTS. After analysis of mtDNA from 39 LQTS patients, Khatami et al. found higher substitution frequency and more potential pathogenic mutations in mtDNA in LQTS patients. Besides, patients harboring these mutations were more likely to have familial history of syncope than normal controls, indicating that accumulation of these mutations may be risk factors of syncope in patients with LQTS (Khatami et al. 2010). These mtDNA mutations have been contributed to the onset of the LQTS but had secondary effects such as disruption of ATPase genes, which lead to the defects in the mitochondrial respiration chain and energy production (Khatami et al. 2014).

Brugada syndrome (BrS), one of cardiac disorders, is characterized by typical ECG alteration and a high risk of sudden death in individuals with structurally normal hearts, accounting for 20% of sudden deaths in the absence of structural heart diseases. The 16 BrS patients with mtDNA mutations tended to have higher substitution rate in the mtDNA d-loop compared to control groups. Although mtDNA alterations may not be the cause of Brs, BrS patients with more mtDNA variants are more susceptive to develop a severe clinical phenotype (Stocchi et al. 2016). Fallah et al. revealed that four mutations located in $tRNA^{Glu}$ (T4377A), $tRNA^{Met}$ (G4407A and C4456T), and $tRNA^{Trp}$ (T5580C) were found in BrS patients. Consequently, the translational process of important proteins of the respirator was absent and then potentially results in BrS (Fallah Tafti et al. 2017).

What's more, conduction disorders are the main cardiac symptoms in Kearns-Sayre syndrome, varying from PR-interval prolongation to infranodal high-degree AV block (Limongelli et al. 2010). Cardiac conduction abnormalities including Wolff-Parkinson-White syndrome has been reported in 13–27% of individuals with MELAS syndrome (El-Hattab et al. 2015). Progression in intraventricular conduction disorders is unpredictable and responsible for sudden cardiac death.

In summary, some mitochondria mutations can cause mitochondrial dysfunction, so arrhythmias are developed. Due to the mitochondrial dysfunction, the decrease of ATP/ADP ratio leads to the interruption of ion exchange, which could slow electrical propagation and shorten refractory periods, both promoting arrhythmias to happen. In addition, increased ROS production could directly alter cardiomyocyte excitability and cell-to-cell coupling and disrupt the intracellular ionic homeostasis in cardiomyocytes. Therefore, mtDNA mutations might be a promising target to treat arrhythmias, especially meaningful for the prediction and treatment of sudden cardiac death caused by lethal arrhythmias.

4.5 mtDNA Mutation and Heart Failure

Heart failure (HF) is a complex disturbance at the end stage of many heart diseases. Despite the large medical costs associated with HF and many improvements in therapy, the mortality for HF remains nearly 50% within 5 years after diagnosis. Therefore, it is important to explore new risk factors and biomarkers to prevent, predict, or treat HF.

Mitochondrial dysfunction has been observed in failing heart failure with impaired myocardial energetics and increased cell death (Marin-Garcia and Goldenthal 2008). The mtDNA content was decreased more than 40% accompanied by reductions in mtDNA-encoded proteins in the myocardial tissue from 23 end-stage heart failure patients (Karamanlidis et al. 2010). Conversely, overexpression of mitochondrial transcription factor maintains high levels of mtDNA and significantly delays the development of heart failure (Ikeuchi et al. 2005). In transgenic mice with a decrease in mtDNA copy number, it was shown that instability of mtDNA posed a threat on the balanced cycle of fusion and fission of mitochondrial, leading to the progressive breakdown of mitochondrial dynamics (Lauritzen et al. 2015). Disruption of mitochondrial dynamics could result in mitochondrial senescence, increased reactive oxygen species production, and overall reduction in OXPHOS capacity. Gradually, subsequent severe heart failure and death are taken place.

Taken together, we could conclude that quantitative and/or qualitative defects of mtDNAs lead to heart failure. Therefore, mtDNA may be a promising therapeutic target for heart failure treatment.

5 Blood Cell mtDNA Mutation and Cardiovascular Diseases

Although invasive coronary angiography is considered to be the gold standard for diagnosis of CHD, it has limitations like its invasiveness and some risks of complications. Moreover, while traditional HF is associated with reduced left ventricular ejection fraction, it has been recognized that HF can occur even when ejection fraction is preserved, which is called heart failure with preserved ejection fraction. And up to 25% of patients have asymptomatic left ventricular systolic and/or diastolic dysfunction. Consequently, it is necessary to find some early or noninvasive biomarkers for heart diseases prediction or diagnosis. Analysis of genetic predisposition to the development of chronic diseases in human may be innovative approaches to evaluation of predisposition to different pathologies.

Recently, mitochondrial genome mutations represent a novel type of genetic makers to predict the diseases. Although the emergence of pathology requires high levels of mutant mtDNA that exceed the threshold, there is evidence that lower levels of heteroplasmy of certain mtDNA mutations could increase the risk of some chronic diseases. Recently, the mitochondrial DNA in the peripheral blood cells (PBC) attracted a lot of attentions. mtDNA with 4977 bp deletion mutation is detected in many diseases. This deletion is generated between a pair of 13-bp direct repeats, located at 13447–13459 of mtDNA. Botto et al. reported that a significantly higher mtDNA 4977 deletion mutation accumulated in the blood cells of patients with CHD compared to the healthy subjects (Botto et al. 2005). In addition, mtDNA mutations in blood monocytes also have relationship with atherosclerosis. The mutation C3256T of mtDNA observed in human white blood cells is associated with atherosclerosis. The C3256T heteroplasmy is not only related to the size of carotid atherosclerotic plaques but also to the clinical manifestations of atherosclerosis such as coronary heart disease and myocardial infarction (Sobenin et al. 2012). Besides, mtDNA point and depletion mutations in PBC are associated with various CVDs. Some researches are focused on the relationship between mtDNA content in PBC and CVDs as well. mtDNA copy number derived from 21,870 participants in 3 cohort studies revealed that there was an inverse association between CVDs and mtDNA copy number (Ashar et al. 2017). In addition, mtDNA copy number may have potential clinical utility in CVD severity quantization (Liu et al. 2017). Interestingly, the increase of mtDNA content is also found in some CVDs. Postoperative atrial fibrillation (POAF) is a major clinical problem after cardiac operations occurring in 30–50% of patients, which will increase the risk of other postoperative complications. It has been reported that mtDNA copy number in peripheral blood cells markedly increased in patients with POAF postoperatively, indicating mtDNA may serve as clinical, valuable biomarkers of impending POAF (Zhang et al. 2017).

The integral mitochondrial function is vital to maintain the stable state of cardiovascular system. mtDNA copy number generally remains within a relatively stable range to maintain the energy of the cell to preserve normal physiological functions, while the mitochondria carrying mutant mtDNA could be removed by mitophagy, leading to the reduction in mitochondrial and then energy supply shortage. The number of mtDNA copies correlates with the size and number of mitochondria. In mitochondrial transcription factor A (Tfam) knockout mice, the decline of respiratory chain function caused by depletion of mtDNA resulted in cardiomyopathy (Wang et al. 1999). On the contrary, overexpression of Tfam in mice could alleviate the myocytes hypertrophy and apoptosis. These beneficial effects of Tfam were associated with attenuation of mtDNA decline (Ikeuchi et al. 2005). The mtDNA copy number decline may lead to the impairment of oxidative phosphorylation, enhanced ROS production or mitochondrial dysfunction, which play an important role in the development of heart diseases (May-Panloup et al. 2005). An intimate link among ROS, mtDNA, and defects in the electron transport function may directly trigger the hypertrophy, fibrosis, and apoptosis in cardiomyocytes. The cross talk of peripheral blood and myocardium forms a catastrophic feedback cycle, leading to greater myocardial injury. However, in some arrhythmia patients, they have higher level of mtDNA content than control groups. This can be explained that increase in the mtDNA content may be the complementary mechanism against oxidative damage to mtDNA and respiratory chain components.

In conclusion, mtDNA mutations and copy number levels in peripheral blood cells could be biomarkers of risk factors for CVDs. However, the mechanism that mtDNA measured in peripheral blood cells reflects in some degree, the mtDNA content in cardiomyocytes is unclear and should be further studied in clinical studies.

6 Peripheral mtDNA: Indication of Systemic Inflammation and a Promising Biomarker

It is well acknowledged that inflammation also participates in the development and progression of CVDs. Recent studies have revealed that dysfunctional mitochondria are important sources of danger-associated molecular patterns (DAMPs). And mtDNA could be released into the cytosol as initiators and masters of the innate immune response. MtDNA, containing inflammatogenic unmethylated CpG motifs, plays a growing role in inflammatory diseases. Recently, it has been reported that damaged mtDNA could induce potent inflammatory responses and accelerate the development of AS in LDLR knockout mice (Ding et al. 2013). Takafumi et al. have found that mtDNA escaping from autophagy leaded to toll-like reporter (TLR)-9 mediated inflammation responses in cardiomyocytes and dilated cardiomyopathy, indicating that mtDNA may play an important role in inducing and maintaining inflammation (Oka et al. 2012). Circulating mtDNA is also elevated and activate the innate immune system via TLR9, which contribute to elevated blood pressure and vascular dysfunction in spontaneously hypertensive rats (McCarthy et al. 2015). In acute myocardial infarction (AMI), focal myocardial necrosis leads to release of mtDNA and mtDNA plasma levels parallel with the degree of myocardial damage through impairing mitochondrial function and inducing cardiomyocyte death (Bliksoen et al. 2016). Taken together, mtDNA in circulation has the potential to activate the immune responses directly via TLR-9 which subsequently leads to the unregulation of proinflammatory transcription factors such as nuclear factor-kappa B (NF-κB) followed by the increase expression of proinflammatory factors like TNF-α, IL-6, and IL-1β. Mitochondrial damage including membrane permeability transition, unsuccessful mitophagy, and ROS production results in mtDNA secretion, thereby strongly activating inflammation responses. The increased peripheral mtDNA indicates the degree of tissue injury, causing positive feedback that inflammation promotes mtDNA release, and then elevated circulating mtDNA in turn exaggerates inflammation responses and tissue injury. Many studies focused on mtDNA excretion achieve positive results in reversing mtDNA-induced harmful effects. Pretreatment with DNAs I could alleviate the cytosolic mtDNA and reduce IL-1β and IL-18 secretion in macrophages stimulated by LPS. Pharmacological or genetical inhibition of TLR9 could significantly allay the mtDNA-mediated inflammation responses and tissue damage in spontaneously hypertensive rats (McCarthy et al. 2015). In experimental autoimmune myocarditis, direct inhibition of inflammation responses or alleviation of mitochondrial ROS stress could reduce mtDNA secretion, which could significantly make improvement to the cardiac inflammation and injury in EAM (Wu et al. 2017). In conclusion, the peripheral mtDNA might be severed as an indicator of systemic inflammation and a therapeutic target for CVDs.

7 Significance of mtDNA in Cardiovascular Diseases, Diagnosis, and Treatment

If a patient has the following presentations, mitochondrial genome mutation-related cardiovascular diseases should be suspected: familial history with maternal inheritance pattern and echocardiographic feature such as hypertrophy, abnormal contraction ability, reduction in ejection fraction, and involvement of many other organs, especially in patients with classic mitochondrial syndrome.

Both genetic testing and histochemical analysis of biopsied tissue are helpful for diagnosis. Each mtDNA mutation has its own heteroplasmy threshold that needs to be measured. Therefore, quantitative evaluation of a mutant allele of mitochondria genome may be a foundation of diagnostics of individual genetic predisposition to disease. At present, high technology methods of molecular genetic diagnostics can significantly simplify early presymptomatic recognition of patients with CVDs to detect risk groups (Sazonova et al. 2016). Polymerase chain reaction-restricted fragment length polymorphisms (PCR-RFLP) and direct sequencing like Sanger sequencing are the routine methods to evaluate a mutant allele of mitochondrial genome, but they target only one mutation. It is unable to detect the common mutations with different parts of mitochondrial genome that will limit their application. Currently, the most promising method for detection of mutations with different pathologies is the next-generation sequencing (NGS), which gives advantages of a high productivity, high probability of new mutations, and high sensibility. NGS could analyze the full genome of each DNA, including nuclear DNA and mitochondrial DNA. The biggest difference from Sanger sequencing to NGS is a multiple, simultaneous reading of short DNA fragment. Moreover, NGS could detect a greater number of mutations related to the investigated pathologies. NGS has already been used to sequence the mtDNA in maternally inherited cardiomyopathy, implying that NGS could discover novel mtDNA variants with heteroplasmy. Recently, Russian scientists used NGS method to investigate the mitochondrial genome mutations with carotid atherosclerosis and reported that 59 mitochondrial genome mutations were associated with atherosclerotic lesions (Sazonova et al. 2015). Consequently, with the use of the NGS technology, it proved to be possible to detect more pathological mutations, which could make people have deeper understanding of these diseases and pave the way to treat with these diseases.

However, the relationship between the genotype and phenotype is weak, and it is difficult to distinguish pathogenic mutations from benign variants. The effect of genetic analysis on clinical management is inadequate. Therefore, analysis of muscle biopsy specimen remains the gold standard for mitochondrial disease diagnosis, even if the specificity and sensitivity is not 100%. Previous researches on the relationship between mtDNA mutations and CVDs have been focused on the measurement of mtDNA in tissues, which is not convenient and safe. Instead, screening the mtDNA in the peripheral blood cells is much easier and to some extent could also reflect the genetic background of the patients and predict the risk of CVDs. Therefore, peripheral blood cells may be used as biomarkers to assess the relationship between mtDNA and CVDs risks, predicting and stratifying CVDs.

Currently, there is no specific approach for treating individuals with mitochondrial diseases. Treatment is largely empiric and is tailored for individual patients. For cardiac manifestations, standard medical therapy is used with successful patients. AV block and supraventricular and ventricular arrhythmias are the common cardiac presentations in patients with mitochondrial cardiomyopathies. Besides conventional drugs, complex device therapies like pacing and implantable cardioverter defibrillator are also supportive considerations. Although being controversial, heart transplantation might be successful in patients with mtDNA disease, when clinical expression is limited to myocardium or other manifestations outside the heart are mild and appear nonprogressive (El-Hattab and Scaglia 2016). Coenzyme Q10 (CoQ10) are being used in mitochondrial disease. Coenzyme Q10, an electron carrier of the mitochondrial electron transport chain, is the most commonly used compound for mitochondrial disease therapy because of its antioxidant properties. Mitochondrial cardiomyopathies caused by CoQ10 deficiency may benefit from CoQ10 supplementation, which could prevent the opening of mPTP and protect against apoptosis and then dramatically improve the clinical manifestations (Di Giovanni et al. 2001). Other supplementations like L-arginine, L-carnitine, and creatine are also recommended for treating mitochondrial diseases (Enns 2014).

Non-pharmacological treatments are being increasingly investigated in the treatment of diseases related to mitochondrial genome mutation. Epidemiologically, exercise is associated with greater life expectancy and reduced risk of chronic diseases. It has been reported that endurance exercise was highly effective in increasing the mtDNA copy number in chronic kidney disease (Balakrishnan et al. 2010). Chen et al. mentioned that chronic exercise training increased complex I and mtDNA content and promoted vasodilatation as well (Chen et al. 2016). Moreover, endurance training reduces mtDNA mutation burden and increases lifespan of the mutator mice through promoting mtDNA mutation repair system and maintaining mitochondrial biogenesis (Safdar et al. 2016).

In addition to exercise, diet control may be beneficial for treatment of mitochondrial disorders. A ketogenic diet (KD) is a diet with high-fat, adequate protein and very low carbohydrates. It mimics the state of ketosis induced by starvation. KD protects the mitochondrial genome from oxidative stress, improves mitochondrial redox status, stimulates de novo biosynthesis of mitochondrial glutathione, and enhances mitochondrial bioenergetic capacity. Mitochondrial cardiomyopathy mice fed with KD have longer survival time. This showed that lethal cardiomyopathy could be alleviated by dietary intervention (Krebs et al. 2011). KD also mitigates the epileptic seizure in patients with mitochondrial diseases through remedying mitochondrial respiratory chain complex defects (Kang et al. 2007). Therefore, since there are no proven treatments available for mitochondrial genome mutation, KD and exercise can supplement a pharmaceutical approach, particularly in cases where drug delivery may pose problems.

From the perspective of genetic counseling, improving genetic technologies are used to prevent the transmission of mtDNA to next generation. Preimplantation genetic diagnosis (PGD), used to detect nuclear DNA mutations, is currently applied to determine the mutation load of embryos produced by women carrying mtDNA mutations, reducing the risk of transmitting mtDNA disease from mother to child. For women with homoplasmic mtDNA mutations or with heteroplasmic mtDNA mutation loads close to the disease threshold, transplanting the nuclear genome between oocytes before or after fertilization aiming at uncoupling the inheritance of mtDNA would reduce the risk of transmitting mtDNA mutations as well (Richardson et al. 2015). Other technology like mitochondrial replacements in oocytes including spindle transfer and pronuclear transfer is a very important aspect of disease management (Kang et al. 2016). Overall, current therapeutic approaches have not been successful. Thus, therapeutic regimens for mitochondrial genome mutation in CVD require more extensive research based on multiple aspects.

Table 4.1 Pathogenic mtDNA mutations related to CVDs from a search of online database

Locus	Disease phenotype	Allele	RNA/amino acid change	Homoplasmy/ heteroplasmy
MT-TI	Hypertension	A4263G	$tRNA^{Ile}$	Homoplasmy
MT-NC2	Hypertension	A4401G	NC2 Gln-Met spacer	Homoplasmy
MT-TA	Hypertension	T5655C	$tRNA^{Met}$	Homoplasmy
MT-TA	Hypertension	A5655G	$tRNA^{Ala}$	Homoplasmy
MT-TM	Hypertension	A4435G	$tRNA^{Met}$	Homoplasmy
MT-TI	Cardiomyopathy	A4300G	tRNAIle	Homoplasmy
MT-ATP8/6	Cardiomyopathy	G8529A	W-X(ATP8);M-M(ATP6)	Homoplasmy
MT-TL1	Cardiomyopathy	A3243G	$tRNA^{Leu(UUR)}$	Heteroplasmy
MT-TL1	Cardiomyopathy	A3260G	$tRNA^{Leu(UUR)}$	Heteroplasmy
MT-TL1	Cardiomyopathy	C3303T	$tRNA^{Leu(UUR)}$	Heteroplasmy
MT-CR	Cardiomyopathy	T16189C	Non-coding	Homoplasmy
MT-RNR1	Cardiomyopathy	A1555G	12 s rRNA	Homoplasmy
MT-ND1	LVNC	A3397G	M-V	Homoplasmy
MT-ND1	LVNC	T3398C	M-T	Homoplasmy
MT-TT	CHD	G15927	$tRNA^{Thr}$	Homoplasmy
MT-TA	CHD	A5592G	$tRNA^{Gln}$	Homoplasmy

MITOMAP, http://mitomap.org

8 Conclusions

Given that mitochondrial genome mutations are widespread in CVDs, we focused on the relationship between CVDs and mitochondrial genome mutations and summarized some mtDNA mutations associated with CVDs (Table 4.1). As detailed in this manuscript, numerous studies point to mitochondrial genome mutations as a common occurrence in CVDs and suggest its major role in the pathophysiology. In CVD patients and animal models, the general biochemical abnormalities that are caused by mtDNA mutations include increased ROS production, $Ca2^{+}$ overload, and reduced ATP synthesis, which subsequently promote the development of CVDs. On the other hand, under pahtological conditions, mtDNA escaping from damaged mitochondria could stimulate inflammation through TLR-9 and the following activation of NF-κB signaling to promote the expression of other inflammation cytokines, which play a central role in the development of CVDs (Fig. 4.1).

In spite of progressing understanding about mitochondria function, genetic mutations, and related manifestations over the past decades, it should be emphasized that correlation between a certain mtDNA mutations and CVDs development is still to be studied. Whether mtDNAs alternation itself promotes CVDs directly or is simply a by-product of the risk factors, triggering CVDs is still unknown. Moreover, although there are a lot of challenges in diagnosis and treatment of mitochondrial disease, the improvement and reduced cost in sequencing technology might help us to better understand the genetic contribution to the CVDs. Therefore, mtDNAs abnormality might be an important pathogenic factor, prognostic predictor, or treatment target for CVDs in clinical practices. Further efforts need to be devoted to this promising academic realm.

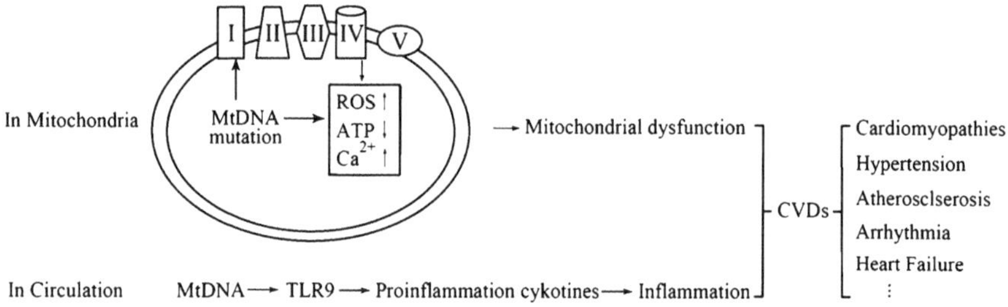

Fig. 4.1 MtDNAs and cardiovascular diseases. mtDNA mutations could lead to mitochondrial dysfunction via decreasing ATP synthesis, enhancing ROS production, and disrupting Ca^{2+} homeostasis, which subsequently promote the development and progression of CVDs. On the other hand, mtDNAs released into circulation could trigger inflammation responses and further contribute to CVDs progression

References

Ahuja P, Wanagat J, Wang Z, et al. Divergent mitochondrial biogenesis responses in human cardiomyopathy. Circulation. 2013;127:1957–67.

Alila-Fersi O, Chamkha I, Majdoub I, et al. Co segregation of the m.1555A>G mutation in the MT-RNR1 gene and mutations in MT-ATP6 gene in a family with dilated mitochondrial cardiomyopathy and hearing loss: a whole mitochondrial genome screening. Biochem Biophys Res Commun. 2017;484:71–8.

Ashar FN, Zhang Y, Longchamps RJ, et al. Association of mitochondrial DNA copy number with cardiovascular disease. JAMA Cardiol. 2017;2:1247–55.

Balakrishnan VS, Rao M, Menon V, et al. Resistance training increases muscle mitochondrial biogenesis in patients with chronic kidney disease. Clin J Am Soc Nephrol. 2010;5:996–1002.

Bliksoen M, Mariero LH, Torp MK, et al. Extracellular mtDNA activates NF-kappaB via toll-like receptor 9 and induces cell death in cardiomyocytes. Basic Res Cardiol. 2016;111:42.

Bobba A, Giannattasio S, Pucci A, Lippolis R, Camaschella C, Marra E. Characterization of mitochondrial DNA in primary cardiomyopathies. Clin Chim Acta. 1995;243:181–9.

Botto N, Berti S, Manfredi S, et al. Detection of mtDNA with 4977 bp deletion in blood cells and atherosclerotic lesions of patients with coronary artery disease. Mutat Res. 2005;570:81–8.

Chen X, An X, Chen D, et al. Chronic exercise training improved aortic endothelial and mitochondrial function via an AMPKalpha2-dependent manner. Front Physiol. 2016;7:631.

Dai DF, Chen T, Wanagat J, et al. Age-dependent cardiomyopathy in mitochondrial mutator mice is attenuated by overexpression of catalase targeted to mitochondria. Aging Cell. 2010;9:536–44.

Di Giovanni S, Mirabella M, Spinazzola A, et al. Coenzyme Q10 reverses pathological phenotype and reduces apoptosis in familial CoQ10 deficiency. Neurology. 2001;57:515–8.

DiMauro S, Schon EA. Mitochondrial DNA mutations in human disease. Am J Med Genet. 2001;106:18–26.

Ding Z, Liu S, Wang X, Khaidakov M, Dai Y, Mehta JL. Oxidant stress in mitochondrial DNA damage, autophagy and inflammation in atherosclerosis. Sci Rep. 2013;3:1077.

Du H, Li L, Bennett D, et al. Fresh fruit consumption and major cardiovascular disease in China. N Engl J Med. 2016;374:1332–43.

Elango S, Govindaraj P, Vishwanadha VP, et al. Analysis of mitochondrial genome revealed a rare 50 bp deletion and substitutions in a family with hypertension. Mitochondrion. 2011;11:878–85.

El-Hattab AW, Scaglia F. Mitochondrial cardiomyopathies. Front Cardiovasc Med. 2016;3:25.

El-Hattab AW, Emrick LT, Craigen WJ, Scaglia F. Citrulline and arginine utility in treating nitric oxide deficiency in mitochondrial disorders. Mol Genet Metab. 2012;107:247–52.

El-Hattab AW, Emrick LT, Chanprasert S, Craigen WJ, Scaglia F. Mitochondria: role of citrulline and arginine supplementation in MELAS syndrome. Int J Biochem Cell B. 2014;48:85–91.

El-Hattab AW, Adesina AM, Jones J, Scaglia F. MELAS syndrome: clinical manifestations, pathogenesis, and treatment options. Mol Genet Metab. 2015;116:4–12.

Elliott HR, Samuels DC, Eden JA, Relton CL, Chinnery PF. Pathogenic mitochondrial DNA mutations are common in the general population. Am J Hum Genet. 2008;83:254–60.

Enns GM. Treatment of mitochondrial disorders: antioxidants and beyond. J Child Neurol. 2014;29:1235–40.

Fallah Tafti M, Khatami M, Rezaei S, Heidari MM, Hadadzadeh M. Novel and heteroplasmic mutations in mitochondrial tRNA genes in Brugada syndrome. Cardiol J. 2018;25:113–9.

Finsterer J, Kothari S. Cardiac manifestations of primary mitochondrial disorders. Int J Cardiol. 2014;177:754–63.

Fuentes RM, Notkola IL, Shemeikka S, Tuomilehto J, Nissinen A. Familial aggregation of blood pressure: a population-based family study in eastern Finland. J Hum Hypertens. 2000;14:441–5.

Go AS, Mozaffarian D, Roger VL, et al. Heart disease and stroke statistics – 2013 update: a report from the American Heart Association. Circulation. 2013;127:e6–e245.

Holmgren D, Wahlander H, Eriksson BO, Oldfors A, Holme E, Tulinius M. Cardiomyopathy in children with mitochondrial disease; clinical course and cardiological findings. Eur Heart J. 2003;24:280–8.

Ide T, Tsutsui H, Hayashidani S, et al. Mitochondrial DNA damage and dysfunction associated with oxidative stress in failing hearts after myocardial infarction. Circ Res. 2001;88:529–35.

Ikeuchi M, Matsusaka H, Kang D, et al. Overexpression of mitochondrial transcription factor a ameliorates mitochondrial deficiencies and cardiac failure after myocardial infarction. Circulation. 2005;112:683–90.

Jiang P, Wang M, Xue L, et al. A hypertension-associated tRNAAla mutation alters tRNA metabolism and mitochondrial function. Mol Cell Biol. 2016;36:1920–30.

Jonckheere AI, Hogeveen M, Nijtmans L, et al. A novel mitochondrial ATP8 gene mutation in a patient with apical hypertrophic cardiomyopathy and neuropathy. BMJ Case Rep. 2009;2009:bcr07.2008.0504.

Kang HC, Lee YM, Kim HD, Lee JS, Slama A. Safe and effective use of the ketogenic diet in children with epilepsy and mitochondrial respiratory chain complex defects. Epilepsia. 2007;48:82–8.

Kang E, Wu J, Gutierrez NM, et al. Mitochondrial replacement in human oocytes carrying pathogenic mitochondrial DNA mutations. Nature. 2016;540:270–5.

Karamanlidis G, Nascimben L, Couper GS, Shekar PS, del Monte F, Tian R. Defective DNA replication impairs mitochondrial biogenesis in human failing hearts. Circ Res. 2010;106:1541–8.

Khatami M, Houshmand M, Sadeghizadeh M, et al. Accumulation of mitochondrial genome variations in Persian LQTS patients: a possible risk factor? Cardiovasc Pathol. 2010;19:e21–7.

Khatami F, Mehdi Heidari M, Houshmand M. The mitochondrial DNA mutations associated with cardiac arrhythmia investigated in an LQTS family. Iran J Basic Med Sci. 2014;17:656–61.

Krebs P, Fan W, Chen YH, et al. Lethal mitochondrial cardiomyopathy in a hypomorphic Med30 mouse mutant is ameliorated by ketogenic diet. Proc Natl Acad Sci U S A. 2011;108:19678–82.

Kujoth GC, Hiona A, Pugh TD, et al. Mitochondrial DNA mutations, oxidative stress, and apoptosis in mammalian aging. Science. 2005;309:481–4.

Lai LP, Tsai CC, Su MJ, et al. Atrial fibrillation is associated with accumulation of aging-related common type mitochondrial DNA deletion mutation in human atrial tissue. Chest. 2003;123:539–44.

Lauritzen KH, Kleppa L, Aronsen JM, et al. Impaired dynamics and function of mitochondria caused by mtDNA toxicity leads to heart failure. Am J Physiol Heart Circ Physiol. 2015;309:H434–49.

Lev D, Nissenkorn A, Leshinsky-Silver E, et al. Clinical presentations of mitochondrial cardiomyopathies. Pediatr Cardiol. 2004;25:443–50.

Limongelli G, Tome-Esteban M, Dejthevaporn C, Rahman S, Hanna MG, Elliott PM. Prevalence and natural history of heart disease in adults with primary mitochondrial respiratory chain disease. Eur J Heart Fail. 2010;12:114–21.

Liu Y, Li R, Li Z, et al. Mitochondrial transfer RNAMet 4435A>G mutation is associated with maternally inherited hypertension in a Chinese pedigree. Hypertension. 2009;53:1083–90.

Liu S, Bai Y, Huang J, et al. Do mitochondria contribute to left ventricular non-compaction cardiomyopathy? New findings from myocardium of patients with left ventricular non-compaction cardiomyopathy. Mol Genet Metab. 2013;109:100–6.

Liu Y, Zhu Q, Zhu C, et al. Systematic analysis of the clinical and biochemical characteristics of maternally inherited hypertension in Chinese Han families associated with mitochondrial. BMC Med Genet. 2014;7:73.

Liu LP, Cheng K, Ning MA, et al. Association between peripheral blood cells mitochondrial DNA content and severity of coronary heart disease. Atherosclerosis. 2017;261:105–10.

Lu Z, Chen H, Meng Y, et al. The tRNAMet 4435A>G mutation in the mitochondrial haplogroup G2a1 is responsible for maternally inherited hypertension in a Chinese pedigree. Eur J Hum Genet. 2011;19:1181–6.

Lu Y, Xiao T, Zhang F, et al. Effect of mitochondrial tRNA(Lys) mutation on the clinical and biochemical characteristics of Chinese essential hypertensive subjects. Biochem Biophys Res Commun. 2014;454:500–4.

Marin-Garcia J, Goldenthal MJ. Mitochondrial centrality in heart failure. Heart Fail Rev. 2008;13:137–50.

Marzetti E, Csiszar A, Dutta D, Balagopal G, Calvani R, Leeuwenburgh C. Role of mitochondrial dysfunction and altered autophagy in cardiovascular aging and disease: from mechanisms to therapeutics. Am J Physiol Heart Circ Physiol. 2013;305:H459–76.

May-Panloup P, Vignon X, Chretien MF, et al. Increase of mitochondrial DNA content and transcripts in early bovine embryogenesis associated with upregulation of mtTFA and NRF1 transcription factors. Reprod Biol Endocrinol. 2005;3:65.

McCarthy CG, Wenceslau CF, Goulopoulou S, et al. Circulating mitochondrial DNA and Toll-like receptor 9 are associated with vascular dysfunction in spontaneously hypertensive rats. Cardiovasc Res. 2015;107:119–30.

McNally EM, Golbus JR, Puckelwartz MJ. Genetic mutations and mechanisms in dilated cardiomyopathy. J Clin Invest. 2013;123:19–26.

Mitrofanov KY, Zhelankin AV, Shiganova GM, et al. Analysis of mitochondrial DNA heteroplasmic mutations A1555G, C3256T, T3336C, capital ES, Cyrillic5178capital A, cyrillic, G12315A, G13513A, G14459A, G14846capital A, cyrillic and G15059A in CHD patients with the history of myocardial infarction. Exp Mol Pathol. 2016;100:87–91.

Mohammed S, Bahitham W, Chan A, Chiu B, Bamforth F, Sergi C. Mitochondrial DNA related cardiomyopathies. Front Biosci (Elite Ed). 2012;4:1706–16.

Oka T, Hikoso S, Yamaguchi O, et al. Mitochondrial DNA that escapes from autophagy causes inflammation and heart failure. Nature. 2012;485:251–5.

Qin Y, Xue L, Jiang P, et al. Mitochondrial tRNA variants in Chinese subjects with coronary heart disease. J Am Heart Assoc. 2014;3:e000437.

Raha S, Merante F, Shoubridge E, et al. Repopulation of rho0 cells with mitochondria from a patient with a mitochondrial DNA point mutation in tRNA(Gly) results in respiratory chain dysfunction. Hum Mutat. 1999;13:245–54.

Richardson J, Irving L, Hyslop LA, et al. Concise reviews: assisted reproductive technologies to prevent transmission of mitochondrial DNA disease. Stem Cells. 2015;33:639–45.

Ruppert V, Nolte D, Aschenbrenner T, Pankuweit S, Funck R, Maisch B. Novel point mutations in the mitochondrial DNA detected in patients with dilated cardiomyopathy by screening the whole mitochondrial genome. Biochem Biophys Res Commun. 2004;318:535–43.

Safdar A, Khrapko K, Flynn JM, et al. Exercise-induced mitochondrial p53 repairs mtDNA mutations in mutator mice. Skelet Muscle. 2016;6:7.

Sazonova M, Budnikov E, Khasanova Z, Sobenin I, Postnov A, Orekhov A. Studies of the human aortic intima by a direct quantitative assay of mutant alleles in the mitochondrial genome. Atherosclerosis. 2009;204:184–90.

Sazonova MA, Zhelankin AV, Barinova VA, et al. Mutations of mitochondrial genome in carotid atherosclerosis. Front Genet. 2015;6:111.

Sazonova MA, Shkurat TP, Demakova NA, et al. Mitochondrial genome sequencing in atherosclerosis: what's next? Curr Pharm Des. 2016;22:390–6.

Schwartz F, Duka A, Sun F, Cui J, Manolis A, Gavras H. Mitochondrial genome mutations in hypertensive individuals. Am J Hypertens. 2004;17:629–35.

Skou AS, Tranebjaerg L, Jensen T, Hasle H. Mitochondrial 12S ribosomal RNA A1555G mutation associated with cardiomyopathy and hearing loss following high-dose chemotherapy and repeated aminoglycoside exposure. J Pediatr. 2014;164:413–5.

Sobenin IA, Sazonova MA, Ivanova MM, et al. Mutation C3256T of mitochondrial genome in white blood cells: novel genetic marker of atherosclerosis and coronary heart disease. PLoS One. 2012;7:e46573.

Stocchi L, Polidori E, Potenza L, et al. Mutational analysis of mitochondrial DNA in Brugada syndrome. Cardiovasc Pathol. 2016;25:47–54.

Tang S, Batra A, Zhang Y, Ebenroth ES, Huang T. Left ventricular noncompaction is associated with mutations in the mitochondrial genome. Mitochondrion. 2010;10:350–7.

Taylor RW, Giordano C, Davidson MM, et al. A homoplasmic mitochondrial transfer ribonucleic acid mutation as a cause of maternally inherited hypertrophic cardiomyopathy. J Am Coll Cardiol. 2003;41:1786–96.

Teng L, Zheng J, Leng J, Ding Y. Clinical and molecular characterization of a Han Chinese family with high penetrance of essential hypertension. Mitochondrial DNA. 2012;23:461–5.

Tian F, Li J, Liu XW, Tong TJ, Zhang ZY. Age-dependent accumulation of mitochondrial DNA deletions in the aortic root of atherosclerosis-prone apolipoprotein E-knockout mice. Arch Gerontol Geriatr. 2016;63:72–7.

Tsuboi M, Hisatome I, Morisaki T, et al. Mitochondrial DNA deletion associated with the reduction of adenine nucleotides in human atrium and atrial fibrillation. Eur J Clin Investig. 2001;31:489–96.

Wallace DC. Mitochondrial diseases in man and mouse. Science. 1999;283:1482–8.

Wang J, Wilhelmsson H, Graff C, et al. Dilated cardiomyopathy and atrioventricular conduction blocks induced by heart-specific inactivation of mitochondrial DNA gene expression. Nat Genet. 1999;21:133–7.

Wang S, Li R, Fettermann A, et al. Maternally inherited essential hypertension is associated with the novel 4263A>G mutation in the mitochondrial tRNAIle gene in a large Han Chinese family. Circ Res. 2011;108:862–70.

Watson B Jr, Khan MA, Desmond RA, Bergman S. Mitochondrial DNA mutations in black Americans with hypertension-associated end-stage renal disease. Am J Kidney Dis. 2001;38:529–36.

Wu B, Ni H, Li J, et al. The impact of circulating mitochondrial DNA on cardiomyocyte apoptosis and myocardial injury after TLR4 activation in experimental autoimmune myocarditis. Cell Physiol Biochem. 2017;42:713–28.

Xu Y, Chen X, Huang H, Liu W. The mitochondrial tRNAAla T5655C mutation may modulate the phenotypic expression of tRNAMet and tRNAGln A4401G mutation in a Han Chinese family with essential hypertension. Int Heart J. 2017;58:95–9.

Yamasaki T, Yanishi K, Tateishi S, et al. Late-onset mitochondrial cardiomyopathy triggered by anticancer treatment. Intern Med. 2017;56:1357–61.

Yang KC, Bonini MG, Dudley SC Jr. Mitochondria and arrhythmias. Free Radic Biol Med. 2014;71:351–61.

Yu E, Calvert PA, Mercer JR, et al. Mitochondrial DNA damage can promote atherosclerosis independently of reactive oxygen species through effects on smooth muscle cells and monocytes and correlates with higher-risk plaques in humans. Circulation. 2013;128:702–12.

Zhang J, Xu S, Xu Y, et al. Relation of mitochondrial DNA copy number in peripheral blood to postoperative atrial fibrillation after isolated off-pump coronary artery bypass grafting. Am J Cardiol. 2017;119:473–7.

Zhu HY, Wang SW, Liu L, et al. A mitochondrial mutation A4401G is involved in the pathogenesis of left ventricular hypertrophy in Chinese hypertensives. Eur J Hum Genet. 2009;17:172–8.

Zinner SH, Levy PS, Kass EH. Familial aggregation of blood pressure in childhood. N Engl J Med. 1971;284:401–4.

Chapter 5
GATA Transcription Factors and Cardiovascular Disease

Bohao Chen

Abstract Development of the mammalian heart is a complex process that involves the specification and differentiation of cardiovascular progenitor cells, their migration to the organ-forming region, interactions and signaling within and between tissues, and later growth and functional maturation of the organ. Normal heart structure and function rely on the precise expression and regulation of the transcription factor network during organogenesis. Over the last three decades, intensive studies from a number of invertebrate and vertebrate species have identified multiple genes encoding cardiac transcription factors, including GATA4/5/6, Nkx2-5, Tbx5, SRF, Mef2c, FOG, and HAND2, which are required for cardiogenesis. Defective expression or interactions between cardiac transcription factors underlie a significant portion of congenital heart disease. In this chapter, we summarize the current understanding of the functions of GATA transcription factors, especially GATA4, GATA5, and GATA6, in cardiovascular development and the molecular interactions of GATA factors with other transcription factors. The association of GATA transcription factors with cardiovascular disease is also discussed.

Keywords GATA4 · GATA5 · GATA6 · Heart · Cardiovascular disease · Mutation differentiation

1 Introduction

GATA proteins belong to a highly conserved family of transcriptional factors with two zinc finger domains that mediate binding to the sequence regulatory elements (A/T) GATA (A/G) and protein-protein interaction with other transcriptional partners and/or cofactors. In vertebrates, the six GATA proteins have been identified and divided into two subfamilies, GATA1/2/3 and GATA4/5/6, based on their expression patterns and amino acid sequence homologies. GATA1/2/3 are prominently expressed in hematopoietic cell lineages and are essential for erythroid and

B. Chen (✉)
Department of Medicine, University of Chicago, Chicago, IL, USA
e-mail: bchen@medicine.bsd.uchicago.edu

H. Jiang, M. Liu (eds.), *Heart Genomics*, Translational Bioinformatics 16,
https://doi.org/10.1007/978-981-13-1429-2_5

megakaryocyte differentiation, the proliferation of hematopoietic stem cells, and the development of T lymphocytes. GATA1/2/3 are also expressed in the brain, spinal cord, and inner ear where they play important roles in their structural development. The second subfamily members, GATA4/5/6, are predominantly expressed in the mesodermal- and endodermal-derived organs and are important for the regulation of cardiac development and terminal differentiation of the heart, liver, gut, gonads, and pancreas. We will focus more on GATA4/5/6 proteins and their roles in heart development and cardiovascular disease.

2 GATA Transcription Factors

2.1 Molecular Features of GATA Transcription Proteins

Six GATA family members of transcription factors have been characterized from vertebrate species (Evans and Felsenfeld 1989; Laverriere et al. 1994; Evans 1997). Each protein shares a highly conserved region that contains two C-X2-C-X17-C-X2-C zinc fingers (referred to as N-terminal zinc finger and C-terminal zinc finger) and adjacent residues. The C-terminal zinc finger and adjacent basic residues constitute a minimal DNA-binding domain sufficient for sequence-specific recognition of a "GATA" cis-element, usually WGATAR (W=A/T; R=A/G) or a related DNA sequence, present in promoters and/or enhancers of target genes (Evans and Felsenfeld 1989; Visvader et al. 1995; Pedone et al. 1997; Newton et al. 2001). The N-terminal zinc finger can stabilize DNA-protein interactions. The N-terminal zinc finger, especially which of GATA2 and GATA3, can also bind DNA independently but with a slightly different sequence preference (GATC). On certain sequences with two proximal GATA sites, including the palindromic GATA motif (ATC(A/T)GATAAG) found in the promoter of GATA1, the N-terminal zinc finger can also participate in specific protein-protein interactions together with the C-finger, resulting in a GATA/DNA complex with markedly increased affinity (Tsai et al. 1991; Trainor et al. 1996, 2000).

Multidimensional heteronuclear magnetic resonance spectroscopy analysis revealed the three-dimensional solution structure of a complex between the DNA-binding domain of the chicken erythroid transcription factor GATA1 and its cognate DNA site. The core of DNA-binding domain is composed of two irregular antiparallel beta sheets and an alpha helix, followed by a long loop that leads into the carboxyl-terminal tail. The DNA target site in contact with the protein spans eight base pairs. The helix and the loop connecting the two antiparallel beta sheets interact with the major groove of the DNA. The carboxyl-terminal tail wraps around into the minor groove of the target sequence (Omichinski et al. 1993). The crystal structures and accompanying biochemical analyses reveal two distinct modes of DNA binding by GATA to closely arranged sites. One mode involves cooperative binding by two GATA factors through protein-protein interactions. Another involves simultaneous binding of the N-terminal zinc finger (N-finger) and the C-terminal zinc finger by the same GATA factor (Bates et al. 2008). Although a number of reports have shown that the C-terminal zinc finger and the adjacent basic domain of the GATA protein are sufficient for specific DNA binding, in some circumstances, GATA1 has been shown to require both zinc finger domains for high-affinity binding (Martin and Orkin 1990; Trainor et al. 1996, 2000). The zinc finger domain of GATA1 mediates the interaction of GATA1 with itself and with other GATA family members, GATA2 and GATA3, and forms homo-oligomers or hetero-oligomers in vivo and in vitro. In addition, the zinc finger domain of GATA1 also interacts with the C2H2 zinc fingers of other transcription factors, such as Sp1, erythroid Krüppel-like factor (EKLF), and forms dimers and large multimers. These self-association and heterotypic interactions are thought to play important roles in GATA-dependent transcriptional control and gene regulation (Crossley et al. 1995; Mackay et al. 1998). Further study demonstrated that the interaction of both zinc finger domains of GATA4 at a GATA motif is necessary for efficient binding and is essential for target gene promoter activity (Murakami et al. 2002). The studies on the structure-function relationships for GATA4 using systematic mutagenesis and luciferase reporter assays reveal that two N-terminal transcriptional activation domains are required for transcriptional activation and are highly conserved within the GATA4/5/6 subfamily but not in the GATA1/2/3 subfamily of transcription factors. The C-terminal zinc finger and adjacent basic domain of GATA4 are bifunctional, modulating both DNA-binding and nuclear localization activities (Morrisey et al. 1997a, b). In vitro transactivation experiments indicated that an alternative GATA5 isoform, which retains only one zinc finger and the COOH-terminal activation domain, still transactivates the atrial natriuretic factor (ANF) promoter, although to a lesser extent than full-length GATA5 or GATA4 (Chen et al. 2009).

2.2 Expression of GATA Transcription Factors in the Cardiovascular System

The mammalian genome encodes six GATA factors (Table 5.1). GATA1/2/3 are predominantly expressed in the hematopoietic system, while GATA4/5/6 are predominantly expressed in the cardiovascular system, the gonads, and endodermal derivatives (Charron and Nemer 1999). This chapter will focus on the regulation and function of members of the second subclass in cardiovascular cells. During embryonic and fetal development, GATA6 is first present in the visceral endoderm at the advanced egg cylinder stage (E6.5). At E7.5, GATA4 and GATA6 are expressed in the primitive streak mesoderm, the mesoderm subjacent to the headfold region, and the extraembryonic visceral and parietal endoderm. In contrast, GATA5 is present within the precardiac mesoderm that is limited to the cardiogenic plate between E7.0 and E8.0 (Heikinheimo et al. 1994). At E9.5, GATA4/5/6 are expressed in both the endocardium and myocardium of the primitive atria, ventricle, and truncus arteriosus of the

Table 5.1 Expression of GATA4/5/6 transcription factors in the embryonic and postnatal heart

GATA factor	Genomic location	Canonical protein size	Firdt zinc finger	Second zinc finger	Cardiac progenitors	Embryonic heart		Postnatal heart	
						Myocardium	Endocardium	Myocardium	Endocardium
GATA4	8p23.1	44.6 KD	217–241	271–295	+	+	+	++	+
GATA5	20q13.33	41.3 KD	189–213	243–267	+	+/–	++	–	++
GATA6	18q11.2	60 KD	390–414	444–468	+	+	+/–	+	–

two-chambered heart, the septum transversum, and the cells lining the midgut and hindgut regions. GATA4 and GATA6 continued to be expressed throughout the myocardium and endocardium throughout pre- and postnatal development. Interestingly, within the myocardium, almost all GATA6-positive myocytes contained GATA4, but a significant percent of GATA4-positive cells did not express GATA6, suggesting the existence of distinct myocyte subpopulations (Nemer and Nemer 2003). In contrast, GATA5 is exclusively expressed within the atria and becomes restricted primarily to endocardial and endocardial cushion cells in both the atrioventricular cushion and the truncus arteriosis by E12, and it disappears completely by E16.5 (Morrisey et al. 1996, 1997a, b). In the endocardial cushions, GATA5 expression transiently overlaps with GATA4, but only GATA4 is still present in both endocardial and myocardial cells. The deletion of Gata5 in mice leads to bicuspid aortic valve, but Gata5$^{-/-}$ mice do not display detectable defects in other endocardial-derived structures, suggesting that Gata4 may partially compensate for Gata5 in earlier stages of endocardial cell expansion and differentiation (Laforest et al. 2011).

In adults, GATA4 and GATA6 continue to be abundantly expressed throughout the postnatal heart development. GATA4 is additionally found in the ovary, pancreas, testis, placenta, and pancreatic islet during adulthood. GATA6 is detected in the myocardium, aorta, gastrointestinal tract, ovary, placenta, pancreas, testis, and, to a lesser extent, in the liver and lungs. In contrast, GATA5 expression becomes markedly upregulated in the intestine, stomach, lungs, bladder, and endocardium but not in myocardium (Chen et al. 2009).

3 Function of GATA Transcription Factors in Cardiac Specification and Development

The temporal and spatial patterns of GATA4/5/6 expression are consistent with the roles of these factors in the regulation of cardiac myocyte differentiation and heart development. The importance of GATA factors is first evidenced in the finding that GATA4 transactivates the cardiac-specific murine cardiac troponin C promoter (Ip et al. 1994, 1995). Over the past decades, studies of loss-of-function and overexpression in mice and other species have confirmed the critical roles of GATA/4/5/6 in heart development.

3.1 GATA4

The mouse Gata4 cDNA is first identified in the 6.5-day embryo as one of the earliest developmental markers of cardiogenesis (Arceci et al. 1993). By analysis of genomic DNAs from an interspecific backcross, the mouse Gata4 gene was mapped to chromosome 14. By screening a human heart cDNA library, Huang et al. isolated a full-length cDNA clone for GATA4.The human GATA4 gene was first mapped to the proximal region of 8p23.1-p22 using fluorescence in situ hybridization (Huang et al. 1995). Its role in normal heart development is best evidenced by the targeted deletion of Gata4 gene in mice (Kuo et al. 1997; Molkentin et al. 1997). Gata4-null mice died between 8.5 and 10.5 days post coitum (dpc) due to severe defects in cardiac morphogenesis and ventral closure of foregut, suggesting a critical role for Gata4 in regulating the migration or morphogenesis of the rostral-to-caudal and lateral-to-ventral folding of the embryo. However, these defects could be rescued using tetraploid embryo complementation, and clonal embryonic day 9.5 Gata4$^{-/-}$ embryos can be generated directly from embryonic stem cells. The rescued Gata4-null embryos displayed heart defects characterized by disrupted looping morphogenesis, septation, and a hypoplastic ventricular myocardium. These findings demonstrated that Gata4 is essential for formation of the proepicardium and regulates cardiogenesis (Watt et al. 2004).

To investigate the role of Gata4 in endocardial cushion (EC) development, Rivera-Feliciano et al. specifically inactivated Gata4 using a Tie2-Cre in the endothelium and endothelium-derived cushion mesenchyme. Mutant endothelium failed to undergo an epithelial to mesenchymal transition (EMT) to form the mesenchymal progenitors of the AV valves, resulting in hypocellular cushions. These results indicate that Gata4 is required to promote EMT of endocardial cells to generate AV cushion mesenchyme. Gata4 activity in endocardial-derived cells is also required later during AV valve maturation for the growth and remodeling of the AV cushions to septate the ventricular inlet (Rivera-Feliciano et al. 2006).

In cardiac myocytes, GATA4 is thought to function as a key transcriptional regulator of numerous endogenous cardiac genes including numerous embryonic and adult genes that encode transcription factors, contractile proteins, membrane receptors and channels, growth factors, as well as adhesion molecules, such as Nkx2-5, ANF, Tbx5, SRF, Smad1/4, B-type natriuretic peptide (BNP), α-myosin heavy chain (α-MHC), β-myosin heavy chain (β-MHC), cTnI, and platelet-derived growth factor receptor β (PDGFRβ). Inhibition of GATA4 and GATA6 protein production using adenovirus-mediated antisense strategy in neonatal cardiomyocytes leads to significant downregulation of these endogenous cardiac genes (Charron et al. 1999; Brody et al. 2013).

GATA4 also mediates inducible gene expression in response to hypertrophic stimuli, including pressure overload, isoproterenol, phenylephrine, and endothelin-1 (Molkentin 2000). In culture postnatal cardiomyocytes, overexpression of GATA4, as well as GATA6, is sufficient to induce cytoskeletal reorganization and myocyte hypertrophy (Charron et al. 2001). Transgenic mice with 2.5-fold overexpression of GATA4 within the adult heart using recombinant adenovirus demonstrated a slowly progressing increase in heart to body weight ratio, histological features of cardiomyopathy, and activation of hypertrophy-associated genes, suggesting that GATA factors are sufficient regulators of cardiomyocyte hypertrophy in vitro and in vivo (Liang et al. 2001). To address the importance of Gata4 in regulating the maintenance and compensatory responsiveness of the adult heart, Gata4-loxP-targeted mice (Gata4$^{fl/fl}$) were crossed with mice containing a α-MHC or β-MHC promoter-driven Cre transgene, which produced viable mice that survived into adulthood despite a 95% and 70% loss of GATA4 protein, respectively. The cardiac-specific deletion of Gata4 from the adult heart severely compromised basal gene expression, survival of cardiac myocytes, and the ability of the myocardium to hypertrophy and compensate to pressure overload or hypertrophy following exercise stimulation (Oka et al. 2006). Gata4 heterozygous mice exhibited systolic and diastolic dysfunction associated with reduced heart weight and decreased cardiomyocyte number. After transverse aortic constriction, these mice developed overt heart failure and eccentric cardiac hypertrophy, associated with significantly increased fibrosis and cardiomyocyte apoptosis (Bisping et al. 2006). These findings furthermore demonstrated that GATA4 is a necessary regulator of cardiac gene expression, hypertrophy, stress compensation, and myocyte viability.

3.2 GATA5

The GATA5 gene was identified first in *Xenopus* and highly transcribed in the heart and throughout the gut, with lower levels detected in the lung, gonads, liver, and gall bladder.

During heart development, GATA5 is initially expressed in the cardiac crescent prior to the formation of the primordial heart tube and appears to be more restricted to endocardial cells and endocardial cushions of the outflow tract (OFT) and atrioventricular canal. The dynamic expression of GATA5 suggests a specific function for this transcription factor in endocardial development (Kelley et al. 1993). Analysis of faust mutants (which encode Gata5) indicates that Gata5 is required for the production of normal numbers of developing myocardial precursors and the expression of normal levels of several myocardial genes including Nkx2-5 in embryogenesis. Gata5 is also necessary for the elaboration of ventricular tissue, the migration of the cardiac primordia to the embryonic midline, and endodermal morphogenesis (Reiter et al. 1999). Inhibition of GATA5 expression in an in vitro model blocks endocardial terminal differentiation at a pre-endocardial stage expression of endocardial differentiation markers (Nemer and Nemer 2002). Gata5 knockout mice are viable and fertile, suggesting that some of the functions performed by zebrafish Gata5 may be fulfilled by Gata4 or Gata6 in the mouse (Laforest and Nemer 2011; Chen et al. 2014). However, loss of Gata5 in mice leads to a significant reduction in the mRNA levels of other transcription factors (Tbx20, Mef2c, and Bmp4), the Notch targets (Nrg1, Hey1), and endothelial and endocardial cell markers (VEcad, Tie2, and Ephb4). Defective endocardial cell differentiation and disruption of migration or cell-cell interactions in the absence of Gata5 suggest that GATA5 plays a crucial role in AV formation and may be a disease-causing gene (Laforest et al. 2011).

In addition, GATA5 is also expressed in microvascular endothelial cells, and its genetic inactivation in mice (Gata5-null) leads to vascular endothelial dysfunction and hypertension (Messaoudi et al. 2015).

3.3 *GATA6*

Similar to GATA4 expression pattern, GATA6 is present in the precardiac mesoderm, the embryonic heart tube, and the developing endoderm during the early murine heart development. GATA6 is also expressed in arterial smooth muscle cells, the developing bronchi, and the urogenital ridge and bladder (Morrisey et al. 1996; Nemer and Nemer 2003). Forced expression of Gata6 in embryonic stem (ES) cells induces endoderm differentiation, whereas gene disruption of Gata6 in mice blocks endoderm differentiation and leads to embryonic death between E5.5 and E7.5. These studies established an essential role for GATA6 in differentiation of extraembryonic endoderm in early embryogenesis (Morrisey et al. 1998; Koutsourakis et al. 1999; Fujikura et al. 2002; Cai et al. 2008). In addition, GATA6 promotes cell survival by regulating endoderm expression of BMP-2 and basement membrane proteins during embryonic epithelial morphogenesis (Rong et al. 2012). Conditional inactivation of GATA6 specifically in the cardiac neural crest cells that give rise to the septum of OFT causes persistent truncus arteriosus, suggesting an important role in the patterning of OFT and the aortic arch (Lepore et al. 2006) and in endodermal differentiation (Kodo and Yamagishi 2010). In cultured vascular smooth muscle cells (VSMCs), GATA6 expression is rapidly and transiently downregulated following mitogen stimulation, suggesting that GATA6 might have a role in maintaining VSMCs in a quiescent state (Suzuki et al. 1996). GATA6 overexpression resulted in the induction of the Cdk inhibitor p21 and cell cycle inhibition (Perlman et al. 1998). Microarray analysis identified multiple genes activated or repressed by GATA6 in VSMCs, including genes encoding the angiotensin type1a (AT1a) receptor, endothelin-1, and vascular cell adhesion molecule-1 (VCAM-1). It suggested that GATA6 regulates a set of genes involved in cell-cell and cell-matrix interactions associated with synthetic SMC functions such as the response of VSMCs to arterial injuries (Lepore et al. 2005). GATA6 may function in VSMCs to coordinate cell cycle activity with the state of cellular differentiation, and disruption of GATA6 expression may predispose or lead to vascular disease.

In conclusion, GATA transcription factors are essential for embryonic heart formation and required for postnatal development and maintenance of normal cardiovascular function.

4 Interactions of GATA Transcription Factors with Other Transcriptional Factors

Normal heart structure and function rely on the precise expression and regulation of a transcription factor network during organogenesis. GATA transcriptional regulation involves a complex process through interactions with other transcription factors, in which GATA factors form protein-protein complexes or bind to regulatory sequences of other transcriptional factors' promoters to control cell-specific gene expression and cardiac cell fates. Disruption of these interactions has been reported to lead to cardiac defects in experimental models and in human. We will discuss the main collaborators in this section.

4.1 Nkx2-5

Nkx2-5 is a cardiac-restricted transcription factor essential for proper cardiac development (Tanaka et al. 1999; Warren et al. 2011) and conduction system functions (Moskowitz et al. 2007; Briggs et al. 2008). Mutations of Nkx2-5 result in congenital heart disease, electrophysiological abnormalities, and sudden death in animal models (Terada et al. 2011; Brody et al. 2013) and humans (Schott et al. 1998). The expression patterns of the Nkx2 and GATA factors are largely overlapping during early cardiogenesis. It is feasible that they might interact and regulate the expression of each other. The first example of collaboration of GATA protein with other transcription factors is that the binding of C-terminal zinc finger of GATA4 to the C-terminus autorepressive domain of Nkx2-5 induces a conformational change and synergistically activates ANF (Durocher et al. 1997; Lee et al. 1998). In vivo, Nkx2-5 and GATA4 endogenously occupy the proximal and distal cardiac regulatory elements of leucine-rich repeat containing 10 (Lrrc10), a cardiac-specific factor, in the heart, and cooperatively regulate cardiac-specific expression of Lrrc10 (Brody et al. 2013). Moreover, Nkx2-5 and GATA4 physically interact to cooperatively regulate the expression of essential cardiac target genes, including T- and L-type Ca2+ channels (Wang et al. 2007), connexin 40 (Linhares et al. 2004), α-actin (Sepulveda et al. 1998, 2002; Belaguli et al. 2000), and Id2 (Lim et al. 2008). Furthermore, mutations of GATA4 have been shown to cause cardiac septal defects in human (Garg et al. 2003). Cardiac-specific deletion of Gata4 (Oka et al. 2006) or perinatal knockout of Nkx2-5 (Briggs et al. 2008) in mice results in compromised cardiac function and dilated cardiomyopathy, suggesting a prominent role for Nkx2-5- and GATA4-mediated transcription in adult cardiac function and disease. Nkx2-5 was able cooperatively to activate transcription of the ANF reporter only with GATA4 and GATA5. No synergy was observed with either GATA1 or GATA6, suggesting that transcriptional cooperativity between Nkx2-5 and GATA proteins requires specific molecular/structural determinants on the GATA4 and GATA5 proteins (Durocher et al. 1997).

4.2 T-Box Proteins

The T-box family of transcription factors is a large family of proteins which play fundamental roles in determining early cell fate decisions and controlling differentiation and organogenesis. At the earliest stages of heart development, two members of the T-box family, Tbx5 and Tbx20, are co-expressed in the heart-forming region at the same time, where the heart markers Nkx2-5 and Gata4 are also present. TBX5 and TBX20 can physically interact but function in a non-redundant fashion and also have independent roles in heart development (Brown et al. 2005). GATA4 cooperates with TBX5 to activate the ANF promoter (Garg et al. 2003; Maitra et al. 2009). Similar to GATA4, GATA6 and Tbx5 are co-expressed in the embryonic heart and synergistically activate the ANF promoter (Maitra et al. 2009). In contrast, Tbx20 represses ANF promoter activity and inhibits the activation mediated by Tbx5 (Plageman and Yutzey 2004). Double haploinsufficiency of either Gata4 and Tbx5 or Gata6 and Tbx5 in mice results in embryonic or early neonatal lethality and causes severe cardiovascular defects compared to wild-type and single heterozygote littermates. The cardiovascular defects of the Gata4$^{+/-}$ Tbx5$^{+/-}$ heterozygotes included complete atrioventricular (AV) septal defects and thin myocardium. The cardiac phenotype of the Gata6$^{+/-}$ Tbx5$^{+/-}$ compound heterozygotes was less severe and involved only variable thinning of the myocardium, highlighting the unique and functional importance of Gata4 at these stages. Cardiac gene expression analysis revealed that Gata4$^{+/-}$ Tbx5$^{+/-}$ heterozygotes displayed a significantly decreased mRNA level of α-myosin heavy chain, a direct target of Gata4 and Tbx5 that has been implicated as a cause of human atrial septal defects. Furthermore, GATA4 and Tbx5 cooperatively interact to regulate the Nppa gene in the myocardial cells (Georges et al. 2008) and BclXL and Nos3 genes in endocardial cells (Nadeau et al. 2010). Tbx20 is able to activate transcription of cardiac genes through direct and synergistic interactions with cardiac transcription factors Nkx2-5, GATA4, and GATA5. Among cardiac GATA factors, Tbx20 is co-expressed with GATA5 in the myocardium and endocardium and preferentially synergizes with GATA5, implicated in endocardial differentiation (Stennard et al. 2003).

4.3 Friend of GATA Family of Proteins

The Friend of GATA (FOG) family of proteins is an evolutionarily conserved class of large multiple zinc finger cofactors that bind to the amino zinc finger of GATA transcription factors and modulate their activity in multiple tissues. In mammals, there are two FOG proteins, FOG1 and FOG2, which both strongly interact with the N-terminal zinc finger of GATA proteins. FOG1 is expressed highly in hematopoietic tissues and is also expressed in the liver and testis of adult mice, whereas FOG2 is expressed predominantly in the heart and brain. Gene targeting experiments demonstrate FOG1 as an activator of GATA1 function, whereas reporter assays in transfected cells indicate that FOG1 and FOG2 can act as repressors (Deconinck et al. 2000). GATA4-FOG2 interaction has first been evidenced in mice harboring a single amino acid replacement in Gata4 that impairs its physical interaction with its presumptive cardiac cofactor FOG2 or other FOG factors. Such Gata4 point mutants die just after embryonic day E12.5, exhibiting features in common with Fog2$^{-/-}$ embryos as well as additional semilunar cardiac valve defects and a double outlet right ventricle, suggesting that GATA4 function is dependent on interaction with

FOG2 (Crispino et al. 2001). In in vitro promoter-luciferase reporter assays, the use of GATA mutants that are incapable of forming complexes with FOG2 reveals that the formation of GATA-FOG complexes is required for the FOG2-induced suppression in myocytes, but not for the FOG2-mediated enhancement in PC12 cells (Jia and Takimoto 2003). Furthermore, mutation of the FOG2 SUMOylation sites, or de-SUMOylation, with SENP-1 or SENP-8 resulted in stronger transcriptional repression activity of GATA4 in both heterologous cells and cardiomyocytes. Conversely, increased FOG2 SUMOylation by overexpression of SUMO-1 or expression of a SUMO-1-FOG2 fusion protein rendered FOG2 incapable of repressing GATA4-mediated activation of BNP promoter. This study demonstrated the physical interaction between FOG2 and GATA4 and that the function of protein-protein interactions can be altered by posttranslational modifications such as SUMOylation (Perdomo et al. 2012).

During heart development, FOG2 interacts with GATA4 in multiple cell types and regulates GATA4 transcriptional activity on cardiac-specific genes such as atrial natriuretic peptide (ANP), BNP, and α-MHC (Svensson et al. 1999; Tevosian et al. 1999; Cantor and Orkin 2005; Carter et al. 2014). In embryos in which Gata4 was selectively inactivated using a Tie2-Cre transgene, the expression of a Gata4 mutant protein defective in interaction with Friend of GATA (FOG) cofactors resulted in a decreased proliferation of mesenchyme and hypoplastic cushions that failed to septate the ventricular inlet, suggesting that Gata4 interaction with Fog cofactors is required for the growth and remodeling of the atrioventricular cushions (Rivera-Feliciano et al. 2006). GATA4/FOG2 transcription complex also represses expression of Lhx9 gene, a family member of the LIM-HD (LIM-homeodomain) transcription factors, directly through binding to the evolutionary conserved GATA sites in the Lhx9 regulatory regions. The loss of GATA4/FOG2 interaction leads to derepression of the LHX9α/β expression in epicardial cells (Smagulova et al. 2008).

4.4 *Serum Response Factor*

Serum response factor (SRF), a member of an ancient DNA-binding transcription factor family, shares a highly conserved DNA-binding/dimerization domain of 90 amino acids termed the MADS (MCM1, Agamous and Deficiens, SRF) box (Sommer et al. 1990). This protein binds to the serum response element in the c-fos promoter and thereby participates in cell cycle regulation, cell apoptosis, cell growth, and cell differentiation (Sun et al. 2006). The central $CC[AT]_6GG$ motif of serum response element forms the core binding site for SRF and is found in the promoters of many muscle- and growth-factor-associated genes, including cardiac and smooth muscle α-actins. SRF and GATA4 coactivate the cardiac α-actin expression through SRF binding to the most proximal CArG box of the cardiac α-actin promoter. Deletion and specific mutations of the most proximal CArG box of cardiac α-actin promoter will reduce the basal activity of the cardiac α-actin promoter and eliminate the synergistic activation. There are no GATA-binding sites detected in the minimal cardiac α-actin promoter (−100 bp), suggesting the recruitment of GATA4 to the cardiac α-actin promoter by SRF is independent on GATA4 DNA binding to DNA. It was found that the N-terminal and C-terminal activation domains of GATA4 play different roles in the SRF-GATA4 combination. Deletion of N-terminal activation domains of GATA4 did not affect the ability of GATA4 to coactivate with SRF. Interestingly, deletion of the second N-terminal activation domains and the first zinc finger of GATA4 increased the ability of GATA4 to synergize with SRF, suggesting that these domains interfere with the interaction of SRF and GATA4. However, deletion of the C-terminal activation domain of SRF or GATA4 abolished coactivation, indicating that the C-terminal activation domains of both SRF and GATA4 were required for the coactivation of the cardiac α-actin promoter (Belaguli et al. 2000). Other cardiovascular tissue-restricted GATA factors, such as GATA5 and GATA6, were equivalent to GATA4 in coactivating SRF-dependent targets. In vitro transient-transfection study demonstrated GATA proteins and SRF form a ternary complex and synergistically activate ANF and other ET-1-inducible promoters that contain both GATA and SRF binding sites (Morin et al. 2001).

4.5 MEF2

The myocyte enhancer factor-2 (MEF2) proteins are a family of transcription factors and a critical regulator in heart development and cardiac gene expression. Four genes (MEF2A, MEF2B, MEF2C, and MEF2D) in the MEF2 transcription factor family are identified in vertebrates and expressed at specific times during development. MEF2 proteins contain MADS box, MEF2 DNA-binding domain, and transactivation domains. Targeted disruption of MEF2C leads to improper looping, outflow tract abnormalities, and complete lack of the right ventricle during embryogenesis, indicating an essential role of MEF2 in myocardial cell differentiation (Dodou et al. 2004). Through physical contact between the C-terminal zinc finger DNA-binding domain and the adjacent basic region of GATA4 and the MADS domain of MEF2, GATA4 and MEF2 synergistically activate the transcription of several cardiac genes such as ANP, BNP, α-MHC, and cardiac α-actin. This transcriptional synergy between MEF2 and GATA4 may be implicated in the generation of cardiac hypertrophy (Akazawa and Komuro 2003). Morin et al. demonstrated that all GATA factors are not equally competent to synergize with MEF2. GATA4 and GATA6 transcription factors, but not GATA5, synergize with MEF2 to activate ANF and several other MEF2C target promoters (Morin et al. 2000).

4.6 HAND2

The heart and neural crest derivatives expressed 2 (HAND2) belongs to the basic helix-loop-helix family of transcription factors that binds DNA on E-box consensus sequence 5′-CANNTG-3′ and plays an essential role in cardiac morphogenesis, particularly in the formation of the right ventricle and of the aortic arch arteries. GATA4 physically interacts with HAND2 through the bHLH domain of HAND2 and the C-terminal zinc finger domain of GATA4 and forms a higher-order complex to synergistically activate expression of cardiac-specific promoters from the ANF gene, the β-type natriuretic peptide gene, and the α-MHC gene. As shown using artificial reporter constructs, this functional synergy was mediated through GATA, but not E-box DNA-binding sites, suggesting a dHAND-binding site-independent mechanism of regulation (Dai et al. 2002). Direct physical interactions between the N- and C-zinc finger domains of GATA4 and the cysteine/histidine-rich region 3 (C/H3) of p300 were identified in immunoprecipitation and glutathione S-transferase pull-down experiments (Dai and Markham 2001). Through GATA4, the transcriptional coactivator p300 is recruited to form a higher-order complex between GATA4, dHAND, and p300 for synergistic activation of cardiac-specific promoters (Dai et al. 2002). Like GATA4 gene-targeted mice, disruption of the gene of dHAND in mouse embryos resulted in embryonic lethality at embryonic day 10.5 from heart failure, suggesting non-redundant roles for each of these factors in specifying developmental gene expression in vivo (Srivastava et al. 1997).

In addition, GATA transcription proteins have also been reported to interact with other complexes through phosphorylation, acetylation, methylation, and SUMOylation, resulting in increased or reduced DNA binding and transcriptional activity (Table 5.2).

5 The Association of GATA Transcription Factors with Cardiovascular Disease

GATA transcription factors play an important role in differentiation, proliferation, and survival of different cell types in early and postnatal heart development. Disruption of GATA transcription factor expression and their interactions with other cardiac transcription factors is associated with human cardiovascular disease.

GATA4 is highly expressed in myocytes, endocardial cells, and epicardial cells of the heart and is required for the rostral-to-caudal and lateral-to-ventral folding of the embryo in cardiac morphogenesis. Gata4 homozygous mutant mice die between E8.5 and E10.5 because of defects in ventral migration causing cardia bifida (Kuo

Table 5.2 Major GATA collaborators in the human heart

GATA factor	GATA collaborators	Representative references
GATA4/5	Nkx2-5	Durocher et al. (1997)
GATA4/5/6	Tbx2	Laforest and Nemer (2011)
GATA4/5/6	Txb5	Garg et al. (2003)
		Lou et al. (2011)
		Maitra et al. (2009)
GATA4/5/6	Tbx20	Stennard et al. (2003)
		Laforest and Nemer (2011)
GATA4/5/6	FOG1	Daoud et al. (2014)
GATA4/6	FOG2	Crispino et al. (2001)
		Rawnsley et al. (2013)
		Jia and Takimoto (2003)
GATA4/5/6	SRF	Belaguli et al. (2000)
		Morin et al. (2001)
GATA4/6	MEF2	Schlange et al. (2000) and Laforest and Nemer (2011)
GATA4/5/6	HAND2	Dai et al. (2002)
		Trinh et al. (2005)
		Laforest and Nemer (2011)
GATA4/6	NFATc1	van Rooij et al. (2002)
		Wada et al. (2002)
GATA4	NFATc4	Oka et al. (2005)
GATA4/6	Fos	McBride et al. (2003)
		Nishida et al. (2002)
GATA4/6	Hey2	Fischer et al. (2005)
		Shirvani et al. (2006)
GATA4/6	CBP	Silverman et al. (2006)
		Robert et al. (2006)
GATA4/5/6	Smad	Lee et al. (2004)
GATA4/5/6	P300	Miyamoto et al. (2006)
		Kakita et al. (1999)
		Takaya et al. (2008)

et al. 1997; Molkentin et al. 1997). Analysis of rescued Gata4 mutant embryos revealed that Gata4 deletion results in disruption of heart looping, absence of endocardial cushion formation, lack of a proepicardial organ, and hypoplastic ventricular myocardium (Watt et al. 2004). Further studies indicated early myocardial specific deletion of Gata4 results in striking myocardial thinning and hypoplastic endocardial cushions. Zeisberg et al. demonstrated that myocardial expression of Gata4 was required in a stage-dependent manner for proliferation of cardiomyocytes, formation of the endocardial cushions, development of the RV, and septation of the OFT (Zeisberg et al. 2005). Moreover, haploinsufficiency of Gata4 has been associated with cardiac defects including common atrioventricular canal, double outlet right ventricle (DORV), and hypoplastic ventricular myocardium (Pu et al. 2004).

Inactivation of Gata4 in endothelial cells causes aberrant development of the AV valves and embryonic lethality by E12.5 because of failure to promote endocardial cushion formation and remodeling (Rivera-Feliciano et al. 2006).

Mutations of GATA4 in humans were identified to cause congenital heart defect (CHD) characteristic of atrial and/or ventricular septal defects. Pehlivan et al. performed FISH analysis using a GATA4 probe on five patients with interstitial deletions of 8p23.1. and demonstrated hemizygosity for GATA4 in the four patients with congenital heart disease but not in the patient without known cardiac anomalies, suggesting haploinsufficiency of GATA4 may contribute to the congenital heart disease observed in some patients with del(8)(p23.1) (Pehlivan et al. 1999).

Garg et al. identified a heterozygous G296S missense mutation of GATA4 affecting all 16 individuals in 5 generations of a family with an autosomal dominant pattern of inheritance. The mutant allele was not detected in unaffected family members or in 3000 unrelated individuals of diverse ethnicity. All affected individuals had atrial septal defects, and eight had additional forms of congenital heart defects, including ventricular septal defects (VSD), atrioventricular septal defects (AVSD), pulmonary valve thickening, or insufficiency of the cardiac valves. Another mutation of GATA4 (hE359del) was found in all available affected members spanning four generations with autosomal dominant transmission of atrial septal defects. The hE359del mutation in this pedigree resulted in a frame shift and truncation of the last 40 amino acids. The G296S mutation abrogated a physical interaction between Gata4 and TBX5 and displayed less transcriptional activation of the α-MHC and ANF enhancers compared to the wild type. In contrast, the truncated E360del protein was unable to activate transcription of either reporter. These results suggest that GATA4 mutations diminished DNA-binding affinity and transcriptional activity of GATA4 that causes cardiac defects (Garg et al. 2003).

In two brothers and their male cousin from a French family with testicular anomalies and congenital heart disease, Lourenco et al. identified heterozygosity for a missense mutation in the conserved N-terminal zinc finger of GATA4 (p. Gly221Arg). This mutation was not observed in 450 ancestry-matched control individuals. Functional analysis showed that the p.Gly221Arg mutant failed to bind to FOG2 to form FOG2-GATA4 complexes that regulate GATA4 transcriptional activity in multiple cell types, suggesting that the p.Gly221Arg mutation in GATA4 is associated with 46,XY disorder of sex development and congenital heart disease (Lourenco et al. 2011).

More recent studies have reported that many GATA4 mutations in humans are associated with atrial septal defect, ventricular septal defect, pulmonary stenosis, sporadic dilated cardiomyopathy, and TOF (Table 5.3).

In contrast to GATA4, GATA5 expression is more restricted to endocardial cells and endocardial cushions of the OFT and atrioventricular canal during heart development (Morrisey et al. 1997a, b). In zebrafish, faust (which encodes Gata5) mutants lack endocardial cells and have a reduced number of myocytes (Reiter et al. 1999). In vitro studies revealed a requirement for Gata5 for differentiation of committed cardiogenic precursors into endothelial endocardial cells (Nemer and Nemer 2002). An important role for GATA5 in early cardiogenesis was further demonstrated in

Table 5.3 Summary of reported GATA factor mutations in humans

GATA factor	Mutation variant	Disease association	Reference
GATA4	p.Ala6Val p.Glu359Lys p. Ser429Thr p.Ala442Val	VSD	Zhang et al. (2008)
	p.Ala9Pro p.Leu51Val p. Asn285Ser	TOF	Yang et al. (2013)
	p.Val39Leu p.Pro226Gln p. Thr279Ser	Sporadic dilated cardiomyopathy	Li et al. (2014)
	p.Arg43Trp	VSD	Yang et al. (2012a)
	p.Ser52Phe p.Gly296Ser	ASD	Hirayama-Yamada et al. (2005)
	p.Gly93Ala p.Gln316Glu p. Ala411Val p.Asp425Asn	ASD	Tomita-Mitchell et al. (2007)
	p.Pro163Ser	VSD, AVSD	Rajagopal et al. (2007)
	p.Gly221Arg	TACHD	Lourenco et al. (2011)
	p.Thr280Met	ASD	Chen et al. (2010b)
	p.Gly296Cys; p.Leu403Met	ASD	Rajagopal et al. (2007)
	p.Met310Val	ASD	Chen et al. (2010a)
	p.Met310Val	VSD	Wang et al. (2011)
	p.Ala346Val	AVSD	Rajagopal et al. (2007)
	p.Pro407Gln	VSD, TOF	Zhang et al. (2008) and Peng et al. (2010)
GATA5	p.Gln3Arg p.Ser19Trp p. Gly166Ser	BAV	Padang et al. (2012)
	p.Tyr16Asp p.Thr252Pro	BAV	Shi et al. (2014)
	p.Tyr138Phe p.Cys210Gly	Atrial fibrillation	Gu et al. (2012)
	p.Tyr142His	DORV	Kassab et al. (2016)
	p.Gly184Val p.Lys218Thr p. Ala266Pro	Atrial fibrillation	Yang et al. (2012b)
	p.Arg187Gly p.His207Arg	TOF	Wei et al. (2013a, b)
	p.Leu199Val	VSD	Wei et al. (2013a, b)
	p.Trp200Gly	Atrial fibrillation	Wang et al. (2013)
	p.Asp203Glu	TOF	Huang et al. (2013)
	p.Gly240Asp	Familial dilated cardiomyopathy	Zhang et al. (2015)

(continued)

Table 5.3 (continued)

GATA factor	Mutation variant	Disease association	Reference
GATA6	p.Asn466His	Non-syndromic persistent truncus arteriosus	Kodo et al. (2009)
	p.Ala178Val	AVSD	Maitra et al. (2010)
	p.Leu198Val	TOF	Maitra et al. (2010)
	p.Ser184Asn	ASD	Lin et al. (2010)
	p.Thr452Ala p.Arg456Cys p. Arg456His p.Asn466Asp p. Lys473Gln p.Ala467Thr	Pancreatic agenesis and congenital heart defects (PACHD)	Allen et al. (2011)

double heterozygous mice for both GATA4/GATA5 and GATA5/GATA6, which developed severe congenital heart defects resulting in embryonal or perinatal lethality. It suggested Gata5 as an important regulator of mammalian heart development and a candidate CHD-causing gene (Laforest et al. 2011). In humans, diseases associated with heterozygous genetic alterations in GATA5 have been reported, including VSD, ASD, double outlet right ventricle, pulmonary stenosis, TOF, aortic stenosis, and bicuspid aortic valve (BAV). In adult patients, heterozygous alterations in GATA5 have been also linked to atrial fibrillation and dilated cardiomyopathy (Table 5.3).

Similar to GATA4, Gata6 is expressed in myocytes and also in the neural crest as well as endocardial and vascular smooth muscle cells (VSMC) (Nemer and Nemer 2003). Systemic ablation of Gata6 in mice results in embryonic lethality (Morrisey et al. 1998), whereas conditional inactivation of Gata6 specifically in neural crest cells (CNC) is sufficient to cause persistent truncus arteriosus (PTA) and lethality by E18.5-P2, revealing a role for GATA6 in the patterning of the OFT and the aortic arch (Lepore et al. 2006). These observations suggest that defects in GATA6 and its downstream target genes that regulate CNC development could be responsible for human CHD involving the OFT. Consistent with this role in OFT development, two loss-of-function GATA6 mutations were reported to cause human cardiac OFT defects by disrupting semaphorin-plexin signaling. In vitro analysis demonstrated that the mutants were unable to activate the GATA6-dependent cardiac promoters NPPA, WNT2, SEMA3C, and PLXNA2, and there was reduced binding efficiency to the latter two promoters, suggesting that each of the GATA6 mutations disturbs semaphorin-plexin signaling to varying degrees, resulting in abnormal development of the cardiac OFT (Kodo et al. 2009). Multiple GATA6 heterozygous missense mutations have been identified in the patients with TOF, atrioventricular septal defects, and atrial septal defects (Lin et al. 2010; Maitra et al. 2010; Allen et al. 2011).

In summary, multiple heterozygosities of variants of GATA4/5/6 factors could contribute, by additive effects, to the development of individual cases of cardiac malformation (Table 5.3). However, it is uncertain whether the GATA variants identified in current studies are segregated with the cardiac disease and the causal relationship to the above cardiac defects needs to be further explored.

6 Conclusion

During the last three decades, much more has been disclosed about the role of GATA4/5/6 transcription factors in cardiac development and terminal differentiation. GATA factors specifically bind to the transcriptional regulatory regions of promoters and enhancers to activate or repress the downstream genes in the heart, such as Nkx2-5, Tbx5, SRF, Mef2c, FOG, HAND2, MHC, cTnC, VCAM, and BNP. However, the mechanisms regulating GATA factor interactions with their collaborators and the relationship of these interactions to the functions in heart remain incompletely understood. Despite the apparent importance of GATA factors in cell differentiation and development, very little is known about how the transcriptional regulation of GATA gene expression is regulated normally (Chen et al. 2012). Targeting deletion of GATA factors suggests that GATA4 and GATA6 are required for cardiac morphogenesis and early embryogenesis. In contrast, GATA5 does not appear to be required for embryonic viability, but it is required for differentiation of committed cardiogenic precursors into endothelial endocardial cells. Although heterozygous alterations in GATA5 have been reported to be associated with human cardiovascular disease, the role of GATA5 in mammalian cardiac development still needs to be elucidated. More studies need to clarify the causal relationship of heterozygosity of variants of GATA factors to the reported congenital heart defects.

Acknowledgments The author thanks Dr. Julian Solway for helpful discussion and comments. The author thanks all investigators who have contributed to our understanding of the role of GATA transcriptional factors in the heart development and cardiovascular disease.

References

Akazawa H, Komuro I. Roles of cardiac transcription factors in cardiac hypertrophy. Circ Res. 2003;92(10):1079–88.

Allen HL, Flanagan SE, Shaw-Smith C, De Franco E, Akerman I, Caswell R, C. International Pancreatic Agenesis, Ferrer J, Hattersley AT, Ellard S. GATA6 haploinsufficiency causes pancreatic agenesis in humans. Nat Genet. 2011;44(1):20–2.

Arceci RJ, King AA, Simon MC, Orkin SH, Wilson DB. Mouse GATA-4: a retinoic acid-inducible GATA-binding transcription factor expressed in endodermally derived tissues and heart. Mol Cell Biol. 1993;13(4):2235–46.

Bates DL, Chen Y, Kim G, Guo L, Chen L. Crystal structures of multiple GATA zinc fingers bound to DNA reveal new insights into DNA recognition and self-association by GATA. J Mol Biol. 2008;381(5):1292–306.

Belaguli NS, Sepulveda JL, Nigam V, Charron F, Nemer M, Schwartz RJ. Cardiac tissue enriched factors serum response factor and GATA-4 are mutual coregulators. Mol Cell Biol. 2000;20(20):7550–8.

Bisping E, Ikeda S, Kong SW, Tarnavski O, Bodyak N, McMullen JR, Rajagopal S, Son JK, Ma Q, Springer Z, Kang PM, Izumo S, Pu WT. Gata4 is required for maintenance of postnatal cardiac function and protection from pressure overload-induced heart failure. Proc Natl Acad Sci U S A. 2006;103(39):14471–6.

Briggs LE, Takeda M, Cuadra AE, Wakimoto H, Marks MH, Walker AJ, Seki T, Oh SP, Lu JT, Sumners C, Raizada MK, Horikoshi N, Weinberg EO, Yasui K, Ikeda Y, Chien KR, Kasahara H. Perinatal loss of Nkx2-5 results in rapid conduction and contraction defects. Circ Res. 2008;103(6):580–90.

Brody MJ, Cho E, Mysliwiec MR, Kim TG, Carlson CD, Lee KH, Lee Y. Lrrc10 is a novel cardiac-specific target gene of Nkx2-5 and GATA4. J Mol Cell Cardiol. 2013;62:237–46.

Brown DD, Martz SN, Binder O, Goetz SC, Price BM, Smith JC, Conlon FL. Tbx5 and Tbx20 act synergistically to control vertebrate heart morphogenesis. Development. 2005;132(3):553–63.

Cai KQ, Capo-Chichi CD, Rula ME, Yang DH, Xu XX. Dynamic GATA6 expression in primitive endoderm formation and maturation in early mouse embryogenesis. Dev Dyn. 2008;237(10):2820–9.

Cantor AB, Orkin SH. Coregulation of GATA factors by the Friend of GATA (FOG) family of multitype zinc finger proteins. Semin Cell Dev Biol. 2005;16(1):117–28.

Carter DR, Buckle AD, Tanaka K, Perdomo J, Chong BH. Art27 interacts with GATA4, FOG2 and NKX2.5 and is a novel co-repressor of cardiac genes. PLoS One. 2014;9(4):e95253.

Charron F, Nemer M. GATA transcription factors and cardiac development. Semin Cell Dev Biol. 1999;10(1):85–91.

Charron F, Paradis P, Bronchain O, Nemer G, Nemer M. Cooperative interaction between GATA-4 and GATA-6 regulates myocardial gene expression. Mol Cell Biol. 1999;19(6):4355–65.

Charron F, Tsimiklis G, Arcand M, Robitaille L, Liang Q, Molkentin JD, Meloche S, Nemer M. Tissue-specific GATA factors are transcriptional effectors of the small GTPase RhoA. Genes Dev. 2001;15(20):2702–19.

Chen B, Yates E, Huang Y, Kogut P, Ma L, Turner JR, Tao Y, Camoretti-Mercado B, Lang D, Svensson EC, Garcia JG, Gruber PJ, Morrisey EE, Solway J. Alternative promoter and GATA5 transcripts in mouse. Am J Physiol Gastrointest Liver Physiol. 2009;297(6):G1214–22.

Chen Y, Han ZQ, Yan WD, Tang CZ, Xie JY, Chen H, Hu DY. A novel mutation in GATA4 gene associated with dominant inherited familial atrial septal defect. J Thorac Cardiovasc Surg. 2010a;140(3):684–7.

Chen Y, Mao J, Sun Y, Zhang Q, Cheng HB, Yan WH, Choy KW, Li H. A novel mutation of GATA4 in a familial atrial septal defect. Clin Chim Acta. 2010b;411(21–22):1741–5.

Chen B, Hsu R, Li Z, Kogut PC, Du Q, Rouser K, Camoretti-Mercado B, Solway J. Upstream stimulatory factor 1 activates GATA5 expression through an E-box motif. Biochem J. 2012;446(1):89–98.

Chen B, Moore TV, Li Z, Sperling AI, Zhang C, Andrade J, Rodriguez A, Bahroos N, Huang Y, Morrisey EE, Gruber PJ, Solway J. Gata5 deficiency causes airway constrictor hyperresponsiveness in mice. Am J Respir Cell Mol Biol. 2014;50(4):787–95.

Crispino JD, Lodish MB, Thurberg BL, Litovsky SH, Collins T, Molkentin JD, Orkin SH. Proper coronary vascular development and heart morphogenesis depend on interaction of GATA-4 with FOG cofactors. Genes Dev. 2001;15(7):839–44.

Crossley M, Merika M, Orkin SH. Self-association of the erythroid transcription factor GATA-1 mediated by its zinc finger domains. Mol Cell Biol. 1995;15(5):2448–56.

Dai YS, Markham BE. p300 functions as a coactivator of transcription factor GATA-4. J Biol Chem. 2001;276(40):37178–85.

Dai YS, Cserjesi P, Markham BE, Molkentin JD. The transcription factors GATA4 and dHAND physically interact to synergistically activate cardiac gene expression through a p300-dependent mechanism. J Biol Chem. 2002;277(27):24390–8.

Daoud G, Kempf H, Kumar D, Kozhemyakina E, Holowacz T, Kim DW, Ionescu A, Lassar AB. BMP-mediated induction of GATA4/5/6 blocks somitic responsiveness to SHH. Development. 2014;141(20):3978–87.

Deconinck AE, Mead PE, Tevosian SG, Crispino JD, Katz SG, Zon LI, Orkin SH. FOG acts as a repressor of red blood cell development in Xenopus. Development. 2000;127(10):2031–40.

Dodou E, Verzi MP, Anderson JP, Xu SM, Black BL. Mef2c is a direct transcriptional target of ISL1 and GATA factors in the anterior heart field during mouse embryonic development. Development. 2004;131(16):3931–42.

Durocher D, Charron F, Warren R, Schwartz RJ, Nemer M. The cardiac transcription factors Nkx2-5 and GATA-4 are mutual cofactors. EMBO J. 1997;16(18):5687–96.

Evans T. Regulation of cardiac gene expression by GATA-4/5/6. Trends Cardiovasc Med. 1997;7(3):75–83.

Evans T, Felsenfeld G. The erythroid-specific transcription factor Eryf1: a new finger protein. Cell. 1989;58(5):877–85.

Fischer A, Klattig J, Kneitz B, Diez H, Maier M, Holtmann B, Englert C, Gessler M. Hey basic helix-loop-helix transcription factors are repressors of GATA4 and GATA6 and restrict expression of the GATA target gene ANF in fetal hearts. Mol Cell Biol. 2005;25(20):8960–70.

Fujikura J, Yamato E, Yonemura S, Hosoda K, Masui S, Nakao K, Miyazaki Ji J, Niwa H. Differentiation of embryonic stem cells is induced by GATA factors. Genes Dev. 2002;16(7):784–9.

Garg V, Kathiriya IS, Barnes R, Schluterman MK, King IN, Butler CA, Rothrock CR, Eapen RS, Hirayama-Yamada K, Joo K, Matsuoka R, Cohen JC, Srivastava D. GATA4 mutations cause human congenital heart defects and reveal an interaction with TBX5. Nature. 2003;424(6947):443–7.

Georges R, Nemer G, Morin M, Lefebvre C, Nemer M. Distinct expression and function of alternatively spliced Tbx5 isoforms in cell growth and differentiation. Mol Cell Biol. 2008;28(12):4052–67.

Gu JY, Xu JH, Yu H, Yang YQ. Novel GATA5 loss-of-function mutations underlie familial atrial fibrillation. Clinics (Sao Paulo). 2012;67(12):1393–9.

Heikinheimo M, Scandrett JM, Wilson DB. Localization of transcription factor GATA-4 to regions of the mouse embryo involved in cardiac development. Dev Biol. 1994;164(2):361–73.

Hirayama-Yamada K, Kamisago M, Akimoto K, Aotsuka H, Nakamura Y, Tomita H, Furutani M, Imamura S, Takao A, Nakazawa M, Matsuoka R. Phenotypes with GATA4 or NKX2.5 mutations in familial atrial septal defect. Am J Med Genet A. 2005;135(1):47–52.

Huang WY, Cukerman E, Liew CC. Identification of a GATA motif in the cardiac alpha-myosin heavy-chain-encoding gene and isolation of a human GATA-4 cDNA. Gene. 1995;155(2):219–23.

Huang RT, Xue S, Xu YJ, Yang YQ. Somatic mutations in the GATA6 gene underlie sporadic tetralogy of Fallot. Int J Mol Med. 2013;31(1):51–8.

Ip HS, Wilson DB, Heikinheimo M, Tang Z, Ting CN, Simon MC, Leiden JM, Parmacek MS. The GATA-4 transcription factor transactivates the cardiac muscle-specific troponin C promoter-enhancer in nonmuscle cells. Mol Cell Biol. 1994;14(11):7517–26.

Ip HS, Wilson DB, Heikinheimo M, Leiden JM, Parmacek MS. The GATA-4 transcription factor transactivates the cardiac-specific troponin C promoter-enhancer in non-muscle cells. Adv Exp Med Biol. 1995;382:117–24.

Jia Y, Takimoto K. GATA and FOG2 transcription factors differentially regulate the promoter for Kv4.2 K(+) channel gene in cardiac myocytes and PC12 cells. Cardiovasc Res. 2003;60(2):278–87.

Kakita T, Hasegawa K, Morimoto T, Kaburagi S, Wada H, Sasayama S. p300 protein as a coactivator of GATA-5 in the transcription of cardiac-restricted atrial natriuretic factor gene. J Biol Chem. 1999;274(48):34096–102.

Kassab K, Hariri H, Gharibeh L, Fahed AC, Zein M, El-Rassy I, Nemer M, El-Rassi I, Bitar F, Nemer G. GATA5 mutation homozygosity linked to a double outlet right ventricle phenotype in a Lebanese patient. Mol Genet Genomic Med. 2016;4(2):160–71.

Kelley C, Blumberg H, Zon LI, Evans T. GATA-4 is a novel transcription factor expressed in endocardium of the developing heart. Development. 1993;118(3):817–27.

Kodo K, Yamagishi H. GATA transcription factors in congenital heart defects: a commentary on a novel GATA6 mutation in patients with tetralogy of Fallot or atrial septal defect. J Hum Genet. 2010;55(10):637–8.

Kodo K, Nishizawa T, Furutani M, Arai S, Yamamura E, Joo K, Takahashi T, Matsuoka R, Yamagishi H. GATA6 mutations cause human cardiac outflow tract defects by disrupting semaphorin-plexin signaling. Proc Natl Acad Sci U S A. 2009;106(33):13933–8.

Koutsourakis M, Langeveld A, Patient R, Beddington R, Grosveld F. The transcription factor GATA6 is essential for early extraembryonic development. Development. 1999;126(9):723–32.

Kuo CT, Morrisey EE, Anandappa R, Sigrist K, Lu MM, Parmacek MS, Soudais C, Leiden JM. GATA4 transcription factor is required for ventral morphogenesis and heart tube formation. Genes Dev. 1997;11(8):1048–60.

Laforest B, Nemer M. GATA5 interacts with GATA4 and GATA6 in outflow tract development. Dev Biol. 2011;358(2):368–78.

Laforest B, Andelfinger G, Nemer M. Loss of Gata5 in mice leads to bicuspid aortic valve. J Clin Invest. 2011;121(7):2876–87.

Laverriere AC, MacNeill C, Mueller C, Poelmann RE, Burch JB, Evans T. GATA-4/5/6, a subfamily of three transcription factors transcribed in developing heart and gut. J Biol Chem. 1994;269(37):23177–84.

Lee Y, Shioi T, Kasahara H, Jobe SM, Wiese RJ, Markham BE, Izumo S. The cardiac tissue-restricted homeobox protein Csx/Nkx2.5 physically associates with the zinc finger protein GATA4 and cooperatively activates atrial natriuretic factor gene expression. Mol Cell Biol. 1998;18(6):3120–9.

Lee KH, Evans S, Ruan TY, Lassar AB. SMAD-mediated modulation of YY1 activity regulates the BMP response and cardiac-specific expression of a GATA4/5/6-dependent chick Nkx2.5 enhancer. Development. 2004;131(19):4709–23.

Lepore JJ, Cappola TP, Mericko PA, Morrisey EE, Parmacek MS. GATA-6 regulates genes promoting synthetic functions in vascular smooth muscle cells. Arterioscler Thromb Vasc Biol. 2005;25(2):309–14.

Lepore JJ, Mericko PA, Cheng L, Lu MM, Morrisey EE, Parmacek MS. GATA-6 regulates semaphorin 3C and is required in cardiac neural crest for cardiovascular morphogenesis. J Clin Invest. 2006;116(4):929–39.

Li J, Liu WD, Yang ZL, Yuan F, Xu L, Li RG, Yang YQ. Prevalence and spectrum of GATA4 mutations associated with sporadic dilated cardiomyopathy. Gene. 2014;548(2):174–81.

Liang Q, De Windt LJ, Witt SA, Kimball TR, Markham BE, Molkentin JD. The transcription factors GATA4 and GATA6 regulate cardiomyocyte hypertrophy in vitro and in vivo. J Biol Chem. 2001;276(32):30245–53.

Lim JY, Kim WH, Kim J, Park SI. Induction of Id2 expression by cardiac transcription factors GATA4 and Nkx2.5. J Cell Biochem. 2008;103(1):182–94.

Lin X, Huo Z, Liu X, Zhang Y, Li L, Zhao H, Yan B, Liu Y, Yang Y, Chen YH. A novel GATA6 mutation in patients with tetralogy of Fallot or atrial septal defect. J Hum Genet. 2010;55(10):662–7.

Linhares VL, Almeida NA, Menezes DC, Elliott DA, Lai D, Beyer EC, Campos de Carvalho AC, Costa MW. Transcriptional regulation of the murine Connexin40 promoter by cardiac factors Nkx2-5, GATA4 and Tbx5. Cardiovasc Res. 2004;64(3):402–11.

Lou X, Deshwar AR, Crump JG, Scott IC. Smarcd3b and Gata5 promote a cardiac progenitor fate in the zebrafish embryo. Development. 2011;138(15):3113–23.

Lourenco D, Brauner R, Rybczynska M, Nihoul-Fekete C, McElreavey K, Bashamboo A. Loss-of-function mutation in GATA4 causes anomalies of human testicular development. Proc Natl Acad Sci U S A. 2011;108(4):1597–602.

Mackay JP, Kowalski K, Fox AH, Czolij R, King GF, Crossley M. Involvement of the N-finger in the self-association of GATA-1. J Biol Chem. 1998;273(46):30560–7.

Maitra M, Schluterman MK, Nichols HA, Richardson JA, Lo CW, Srivastava D, Garg V. Interaction of Gata4 and Gata6 with Tbx5 is critical for normal cardiac development. Dev Biol. 2009;326(2):368–77.

Maitra M, Koenig SN, Srivastava D, Garg V. Identification of GATA6 sequence variants in patients with congenital heart defects. Pediatr Res. 2010;68(4):281–5.
Martin DI, Orkin SH. Transcriptional activation and DNA binding by the erythroid factor GF-1/NF-E1/Eryf 1. Genes Dev. 1990;4(11):1886–98.
McBride K, Charron F, Lefebvre C, Nemer M. Interaction with GATA transcription factors provides a mechanism for cell-specific effects of c-Fos. Oncogene. 2003;22(52):8403–12.
Messaoudi S, He Y, Gutsol A, Wight A, Hebert RL, Vilmundarson RO, Makrigiannis AP, Chalmers J, Hamet P, Tremblay J, McPherson R, Stewart AF, Touyz RM, Nemer M. Endothelial Gata5 transcription factor regulates blood pressure. Nat Commun. 2015;6:8835.
Miyamoto S, Kawamura T, Morimoto T, Ono K, Wada H, Kawase Y, Matsumori A, Nishio R, Kita T, Hasegawa K. Histone acetyltransferase activity of p300 is required for the promotion of left ventricular remodeling after myocardial infarction in adult mice in vivo. Circulation. 2006;113(5):679–90.
Molkentin JD. The zinc finger-containing transcription factors GATA-4, −5, and −6. Ubiquitously expressed regulators of tissue-specific gene expression. J Biol Chem. 2000;275(50):38949–52.
Molkentin JD, Lin Q, Duncan SA, Olson EN. Requirement of the transcription factor GATA4 for heart tube formation and ventral morphogenesis. Genes Dev. 1997;11(8):1061–72.
Morin S, Charron F, Robitaille L, Nemer M. GATA-dependent recruitment of MEF2 proteins to target promoters. EMBO J. 2000;19(9):2046–55.
Morin S, Paradis P, Aries A, Nemer M. Serum response factor-GATA ternary complex required for nuclear signaling by a G-protein-coupled receptor. Mol Cell Biol. 2001;21(4):1036–44.
Morrisey EE, Ip HS, Lu MM, Parmacek MS. GATA-6: a zinc finger transcription factor that is expressed in multiple cell lineages derived from lateral mesoderm. Dev Biol. 1996;177(1):309–22.
Morrisey EE, Ip HS, Tang Z, Lu MM, Parmacek MS. GATA-5: a transcriptional activator expressed in a novel temporally and spatially-restricted pattern during embryonic development. Dev Biol. 1997a;183(1):21–36.
Morrisey EE, Ip HS, Tang Z, Parmacek MS. GATA-4 activates transcription via two novel domains that are conserved within the GATA-4/5/6 subfamily. J Biol Chem. 1997b;272(13):8515–24.
Morrisey EE, Tang Z, Sigrist K, Lu MM, Jiang F, Ip HS, Parmacek MS. GATA6 regulates HNF4 and is required for differentiation of visceral endoderm in the mouse embryo. Genes Dev. 1998;12(22):3579–90.
Moskowitz IP, Kim JB, Moore ML, Wolf CM, Peterson MA, Shendure J, Nobrega MA, Yokota Y, Berul C, Izumo S, Seidman JG, Seidman CE. A molecular pathway including Id2, Tbx5, and Nkx2-5 required for cardiac conduction system development. Cell. 2007;129(7):1365–76.
Murakami A, Ishida S, Dickson C. GATA-4 interacts distinctively with negative and positive regulatory elements in the Fgf-3 promoter. Nucleic Acids Res. 2002;30(4):1056–64.
Nadeau M, Georges RO, Laforest B, Yamak A, Lefebvre C, Beauregard J, Paradis P, Bruneau BG, Andelfinger G, Nemer M. An endocardial pathway involving Tbx5, Gata4, and Nos3 required for atrial septum formation. Proc Natl Acad Sci U S A. 2010;107(45):19356–61.
Nemer G, Nemer M. Cooperative interaction between GATA5 and NF-ATc regulates endothelial-endocardial differentiation of cardiogenic cells. Development. 2002;129(17):4045–55.
Nemer G, Nemer M. Transcriptional activation of BMP-4 and regulation of mammalian organogenesis by GATA-4 and -6. Dev Biol. 2003;254(1):131–48.
Newton A, Mackay J, Crossley M. The N-terminal zinc finger of the erythroid transcription factor GATA-1 binds GATC motifs in DNA. J Biol Chem. 2001;276(38):35794–801.
Nishida W, Nakamura M, Mori S, Takahashi M, Ohkawa Y, Tadokoro S, Yoshida K, Hiwada K, Hayashi K, Sobue K. A triad of serum response factor and the GATA and NK families governs the transcription of smooth and cardiac muscle genes. J Biol Chem. 2002;277(9):7308–17.
Oka T, Dai YS, Molkentin JD. Regulation of calcineurin through transcriptional induction of the calcineurin A beta promoter in vitro and in vivo. Mol Cell Biol. 2005;25(15):6649–59.

Oka T, Maillet M, Watt AJ, Schwartz RJ, Aronow BJ, Duncan SA, Molkentin JD. Cardiac-specific deletion of Gata4 reveals its requirement for hypertrophy, compensation, and myocyte viability. Circ Res. 2006;98(6):837–45.

Omichinski JG, Clore GM, Schaad O, Felsenfeld G, Trainor C, Appella E, Stahl SJ, Gronenborn AM. NMR structure of a specific DNA complex of Zn-containing DNA binding domain of GATA-1. Science. 1993;261(5120):438–46.

Padang R, Bagnall RD, Richmond DR, Bannon PG, Semsarian C. Rare non-synonymous variations in the transcriptional activation domains of GATA5 in bicuspid aortic valve disease. J Mol Cell Cardiol. 2012;53(2):277–81.

Pedone PV, Omichinski JG, Nony P, Trainor C, Gronenborn AM, Clore GM, Felsenfeld G. The N-terminal fingers of chicken GATA-2 and GATA-3 are independent sequence-specific DNA binding domains. EMBO J. 1997;16(10):2874–82.

Pehlivan T, Pober BR, Brueckner M, Garrett S, Slaugh R, Van Rheeden R, Wilson DB, Watson MS, Hing AV. GATA4 haploinsufficiency in patients with interstitial deletion of chromosome region 8p23.1 and congenital heart disease. Am J Med Genet. 1999;83(3):201–6.

Peng T, Wang L, Zhou SF, Li X. Mutations of the GATA4 and NKX2.5 genes in Chinese pediatric patients with non-familial congenital heart disease. Genetica. 2010;138(11–12):1231–40.

Perdomo J, Jiang XM, Carter DR, Khachigian LM, Chong BH. SUMOylation regulates the transcriptional repression activity of FOG-2 and its association with GATA-4. PLoS One. 2012;7(11):e50637.

Perlman H, Suzuki E, Simonson M, Smith RC, Walsh K. GATA-6 induces p21(Cip1) expression and G1 cell cycle arrest. J Biol Chem. 1998;273(22):13713–8.

Plageman TF Jr, Yutzey KE. Differential expression and function of Tbx5 and Tbx20 in cardiac development. J Biol Chem. 2004;279(18):19026–34.

Pu WT, Ishiwata T, Juraszek AL, Ma Q, Izumo S. GATA4 is a dosage-sensitive regulator of cardiac morphogenesis. Dev Biol. 2004;275(1):235–44.

Rajagopal SK, Ma Q, Obler D, Shen J, Manichaikul A, Tomita-Mitchell A, Boardman K, Briggs C, Garg V, Srivastava D, Goldmuntz E, Broman KW, Benson DW, Smoot LB, Pu WT. Spectrum of heart disease associated with murine and human GATA4 mutation. J Mol Cell Cardiol. 2007;43(6):677–85.

Rawnsley DR, Xiao J, Lee JS, Liu X, Mericko-Ishizuka P, Kumar V, He J, Basu A, Lu M, Lynn FC, Pack M, Gasa R, Kahn ML. The transcription factor Atonal homolog 8 regulates Gata4 and Friend of Gata-2 during vertebrate development. J Biol Chem. 2013;288(34):24429–40.

Reiter JF, Alexander J, Rodaway A, Yelon D, Patient R, Holder N, Stainier DY. Gata5 is required for the development of the heart and endoderm in zebrafish. Genes Dev. 1999;13(22):2983–95.

Rivera-Feliciano J, Lee KH, Kong SW, Rajagopal S, Ma Q, Springer Z, Izumo S, Tabin CJ, Pu WT. Development of heart valves requires Gata4 expression in endothelial-derived cells. Development. 2006;133(18):3607–18.

Robert NM, Miyamoto Y, Taniguchi H, Viger RS. LRH-1/NR5A2 cooperates with GATA factors to regulate inhibin alpha-subunit promoter activity. Mol Cell Endocrinol. 2006;257–258:65–74.

Rong L, Liu J, Qi Y, Graham AM, Parmacek MS, Li S. GATA-6 promotes cell survival by up-regulating BMP-2 expression during embryonic stem cell differentiation. Mol Biol Cell. 2012;23(18):3754–63.

van Rooij E, Doevendans PA, de Theije CC, Babiker FA, Molkentin JD, de Windt LJ. Requirement of nuclear factor of activated T-cells in calcineurin-mediated cardiomyocyte hypertrophy. J Biol Chem. 2002;277(50):48617–26.

Schlange T, Andree B, Arnold HH, Brand T. BMP2 is required for early heart development during a distinct time period. Mech Dev. 2000;91(1–2):259–70.

Schott JJ, Benson DW, Basson CT, Pease W, Silberbach GM, Moak JP, Maron BJ, Seidman CE, Seidman JG. Congenital heart disease caused by mutations in the transcription factor NKX2-5. Science. 1998;281(5373):108–11.

Sepulveda JL, Belaguli N, Nigam V, Chen CY, Nemer M, Schwartz RJ. GATA-4 and Nkx-2.5 coactivate Nkx-2 DNA binding targets: role for regulating early cardiac gene expression. Mol Cell Biol. 1998;18(6):3405–15.

Sepulveda JL, Vlahopoulos S, Iyer D, Belaguli N, Schwartz RJ. Combinatorial expression of GATA4, Nkx2-5, and serum response factor directs early cardiac gene activity. J Biol Chem. 2002;277(28):25775–82.

Shi LM, Tao JW, Qiu XB, Wang J, Yuan F, Xu L, Liu H, Li RG, Xu YJ, Wang Q, Zheng HZ, Li X, Wang XZ, Zhang M, Qu XK, Yang YQ. GATA5 loss-of-function mutations associated with congenital bicuspid aortic valve. Int J Mol Med. 2014;33(5):1219–26.

Shirvani S, Xiang F, Koibuchi N, Chin MT. CHF1/Hey2 suppresses SM-MHC promoter activity through an interaction with GATA-6. Biochem Biophys Res Commun. 2006;339(1):151–6.

Silverman E, Yivgi-Ohana N, Sher N, Bell M, Eimerl S, Orly J. Transcriptional activation of the steroidogenic acute regulatory protein (StAR) gene: GATA-4 and CCAAT/enhancer-binding protein beta confer synergistic responsiveness in hormone-treated rat granulosa and HEK293 cell models. Mol Cell Endocrinol. 2006;252(1–2):92–101.

Smagulova FO, Manuylov NL, Leach LL, Tevosian SG. GATA4/FOG2 transcriptional complex regulates Lhx9 gene expression in murine heart development. BMC Dev Biol. 2008;8:67.

Sommer H, Beltran JP, Huijser P, Pape H, Lonnig WE, Saedler H, Schwarz-Sommer Z. Deficiens, a homeotic gene involved in the control of flower morphogenesis in *Antirrhinum majus*: the protein shows homology to transcription factors. EMBO J. 1990;9(3):605–13.

Srivastava D, Thomas T, Lin Q, Kirby ML, Brown D, Olson EN. Regulation of cardiac mesodermal and neural crest development by the bHLH transcription factor, dHAND. Nat Genet. 1997;16(2):154–60.

Stennard FA, Costa MW, Elliott DA, Rankin S, Haast SJ, Lai D, McDonald LP, Niederreither K, Dolle P, Bruneau BG, Zorn AM, Harvey RP. Cardiac T-box factor Tbx20 directly interacts with Nkx2-5, GATA4, and GATA5 in regulation of gene expression in the developing heart. Dev Biol. 2003;262(2):206–24.

Sun Q, Chen G, Streb JW, Long X, Yang Y, Stoeckert CJ Jr, Miano JM. Defining the mammalian CArGome. Genome Res. 2006;16(2):197–207.

Suzuki E, Evans T, Lowry J, Truong L, Bell DW, Testa JR, Walsh K. The human GATA-6 gene: structure, chromosomal location, and regulation of expression by tissue-specific and mitogen-responsive signals. Genomics. 1996;38(3):283–90.

Svensson EC, Tufts RL, Polk CE, Leiden JM. Molecular cloning of FOG-2: a modulator of transcription factor GATA-4 in cardiomyocytes. Proc Natl Acad Sci U S A. 1999;96(3):956–61.

Takaya T, Kawamura T, Morimoto T, Ono K, Kita T, Shimatsu A, Hasegawa K. Identification of p300-targeted acetylated residues in GATA4 during hypertrophic responses in cardiac myocytes. J Biol Chem. 2008;283(15):9828–35.

Tanaka M, Chen Z, Bartunkova S, Yamasaki N, Izumo S. The cardiac homeobox gene Csx/Nkx2.5 lies genetically upstream of multiple genes essential for heart development. Development. 1999;126(6):1269–80.

Terada R, Warren S, Lu JT, Chien KR, Wessels A, Kasahara H. Ablation of Nkx2-5 at mid-embryonic stage results in premature lethality and cardiac malformation. Cardiovasc Res. 2011;91(2):289–99.

Tevosian SG, Deconinck AE, Cantor AB, Rieff HI, Fujiwara Y, Corfas G, Orkin SH. FOG-2: a novel GATA-family cofactor related to multitype zinc-finger proteins friend of GATA-1 and U-shaped. Proc Natl Acad Sci U S A. 1999;96(3):950–5.

Tomita-Mitchell A, Maslen CL, Morris CD, Garg V, Goldmuntz E. GATA4 sequence variants in patients with congenital heart disease. J Med Genet. 2007;44(12):779–83.

Trainor CD, Omichinski JG, Vandergon TL, Gronenborn AM, Clore GM, Felsenfeld G. A palindromic regulatory site within vertebrate GATA-1 promoters requires both zinc fingers of the GATA-1 DNA-binding domain for high-affinity interaction. Mol Cell Biol. 1996;16(5):2238–47.

Trainor CD, Ghirlando R, Simpson MA. GATA zinc finger interactions modulate DNA binding and transactivation. J Biol Chem. 2000;275(36):28157–66.

Trinh LA, Yelon D, Stainier DY. Hand2 regulates epithelial formation during myocardial diferentiation. Curr Biol. 2005;15(5):441–6.

Tsai SF, Strauss E, Orkin SH. Functional analysis and in vivo footprinting implicate the erythroid transcription factor GATA-1 as a positive regulator of its own promoter. Genes Dev. 1991;5(6):919–31.

Visvader JE, Crossley M, Hill J, Orkin SH, Adams JM. The C-terminal zinc finger of GATA-1 or GATA-2 is sufficient to induce megakaryocytic differentiation of an early myeloid cell line. Mol Cell Biol. 1995;15(2):634–41.

Wada H, Hasegawa K, Morimoto T, Kakita T, Yanazume T, Abe M, Sasayama S. Calcineurin-GATA-6 pathway is involved in smooth muscle-specific transcription. J Cell Biol. 2002;156(6):983–91.

Wang Y, Morishima M, Zheng M, Uchino T, Mannen K, Takahashi A, Nakaya Y, Komuro I, Ono K. Transcription factors Csx/Nkx2.5 and GATA4 distinctly regulate expression of Ca2+ channels in neonatal rat heart. J Mol Cell Cardiol. 2007;42(6):1045–53.

Wang J, Fang M, Liu XY, Xin YF, Liu ZM, Chen XZ, Wang XZ, Fang WY, Liu X, Yang YQ. A novel GATA4 mutation responsible for congenital ventricular septal defects. Int J Mol Med. 2011;28(4):557–64.

Wang XH, Huang CX, Wang Q, Li RG, Xu YJ, Liu X, Fang WY, Yang YQ. A novel GATA5 loss-of-function mutation underlies lone atrial fibrillation. Int J Mol Med. 2013;31(1):43–50.

Warren CM, Tanaka JW, Holroyd CB. What can topology changes in the oddball N2 reveal about underlying processes? Neuroreport. 2011;22(17):870–4.

Watt AJ, Battle MA, Li J, Duncan SA. GATA4 is essential for formation of the proepicardium and regulates cardiogenesis. Proc Natl Acad Sci U S A. 2004;101(34):12573–8.

Wei D, Bao H, Liu XY, Zhou N, Wang Q, Li RG, Xu YJ, Yang YQ. GATA5 loss-of-function mutations underlie tetralogy of fallot. Int J Med Sci. 2013a;10(1):34–42.

Wei D, Bao H, Zhou N, Zheng GF, Liu XY, Yang YQ. GATA5 loss-of-function mutation responsible for the congenital ventriculoseptal defect. Pediatr Cardiol. 2013b;34(3):504–11.

Yang YQ, Li L, Wang J, Liu XY, Chen XZ, Zhang W, Wang XZ, Jiang JQ, Liu X, Fang WY. A novel GATA4 loss-of-function mutation associated with congenital ventricular septal defect. Pediatr Cardiol. 2012a;33(4):539–46.

Yang YQ, Wang J, Wang XH, Wang Q, Tan HW, Zhang M, Shen FF, Jiang JQ, Fang WY, Liu X. Mutational spectrum of the GATA5 gene associated with familial atrial fibrillation. Int J Cardiol. 2012b;157(2):305–7.

Yang YQ, Gharibeh L, Li RG, Xin YF, Wang J, Liu ZM, Qiu XB, Xu YJ, Xu L, Qu XK, Liu X, Fang WY, Huang RT, Xue S, Nemer G. GATA4 loss-of-function mutations underlie familial tetralogy of fallot. Hum Mutat. 2013;34(12):1662–71.

Zeisberg EM, Ma Q, Juraszek AL, Moses K, Schwartz RJ, Izumo S, Pu WT. Morphogenesis of the right ventricle requires myocardial expression of Gata4. J Clin Invest. 2005;115(6):1522–31.

Zhang W, Li X, Shen A, Jiao W, Guan X, Li Z. GATA4 mutations in 486 Chinese patients with congenital heart disease. Eur J Med Genet. 2008;51(6):527–35.

Zhang XL, Dai N, Tang K, Chen YQ, Chen W, Wang J, Zhao CM, Yuan F, Qiu XB, Qu XK, Yang YQ, Xu YW. GATA5 loss-of-function mutation in familial dilated cardiomyopathy. Int J Mol Med. 2015;35(3):763–70.

Chapter 6
The Genetic Paradigm of WT1 Gene in Heart Development and Cardiac Repair

Xueqing Liu, Min Chen, Xiaoming Deng, and Zhaoping Ding

Abstract Wilms' tumor gene (WT1) is known as a tumor repressor gene causatively involved in not only eponymous nephroblastoma as initially described but also in many other incidences of pediatric malignancies. In recent decades, WT1 is also recognized as a master switch that tips the balance of epithelial-to-mesenchymal transition (EMT) that imparts the formation of multiple mesodermal organs in the embryonic development and tissue integrity in adulthood. In this chapter, we review the genetic paradigm and the developmental significance of WT1 and, particularly, recent findings demonstrating how the WT1-expressing cells transcriptionally contribute to the formation of the second heart field and constitute partially the cellular composition of the compact heart, including cardiomyocytes, cardiac fibroblasts, and coronary complex. Drawing on these findings, we highlight the epicardial redeployment of WT1 expression in response to ischemic injury and the reparative role of the fetal reprogramming that confers the formation of multipotent cardiac progenitors that may potentially be utilized for cell-based therapy for ischemic heart disease and heart failure.

X. Liu
Department of Cardiology, Danyang People's Hospital, Danyang, Jiangsu, China

M. Chen
Department of Physiology, School of Medicine, Jiangsu University, Zhenjiang, Jiangsu, China

X. Deng
Department of Anesthesiology, Changhai Hospital, Second Military Medical University, Shanghai, China

Z. Ding (✉)
Department of Anesthesiology, Changhai Hospital, Second Military Medical University, Shanghai, China

Institute of Molecular Cardiology, Heinrich-Heine University Düsseldorf, Düsseldorf, Germany
e-mail: ding@uni-duesseldorf.de

H. Jiang, M. Liu (eds.), *Heart Genomics*, Translational Bioinformatics 16,
https://doi.org/10.1007/978-981-13-1429-2_6

Keywords Wilms' tumor gene factor 1 · Epithelial-to-mesenchymal transition (EMT) · Epicardial derive cells (EPDC) · Cardiac stem cells · Heart development · Myocardial infarction · Cardiac regeneration

1 Introduction

The nomenclature of Wilms' tumor factor 1 gene, WT1, comes from the eponymous nephroblastoma that was first described by a germen surgeon, Dr. Max Wilms, in a young patient with malignant neoplasm of the kidney in 1899 (Haber and Buckler 1992). Wilms' tumor occurs in the subset of designated kidney cancers when the suppressive activity of WT1 is aberrant (Little and Wells 1997), leading to the pluripotent mesenchymal cells in the developing kidney fail to differentiate into glomeruli and tubules but, instead, persist to proliferate in a uncontrolled manner (Wagner et al. 2003a). The incidence of Wilms' tumor is found to affect around 1 in 10,000 children, which represents about 8% incidence of entire pediatric malignancies (Chau and Hastie 2012). However, despite small mobility and the fact that most of patients nowadays can be surgically treated with optimistic prognosis, the fascinating biology of WT1 was drastically extended far beyond its nephropathy (Wagner et al. 2003a) and perhaps represents the most striking example of occurrence of oncogenic transformation when normal developmental control goes awry (Hohenstein and Hastie 2006; Hastie 2017).

The complexity of our life is indeed much more intricate than we presently know. Mutations in WT1, as primarily found in nephroblastoma, are detected in a variety of syndromes, including Denys-Drash syndrome (Patek et al. 1999), Frasier syndrome (Klamt et al. 1998), and WAGR syndrome (Little and Wells 1997), and also linked to many other malignancies including leukemia (King-Underwood and Pritchard-Jones 1998), desmoplastic small round cell tumors (Little and Wells 1997), breast cancer (Silberstein et al. 1997), lung cancer (Cano et al. 2013), and retinoblastoma (Wagner et al. 2002). Surprisingly, genetical deletion of WT1 gene in mice leads to severe defects of several mesodermally derived organs including the kidney, heart, gonads, and nervous system (Kreidberg et al. 1993; Wilhelm and Englert 2002; Scholz et al. 2009) in the mid-gestation, indicating that WT1 protein plays pivotal roles in the embryonic organ formation. Furthermore, WT1 is found to sit atop of the molecular hierarchy operating in the sex determination (Hammes et al. 2001; Hastie 2001) and maintaining the tissue homeostasis in adulthood (Hastie 2017). Therefore, WT1 may in nature have much broader biological activities than just a tumor repressor as we initially presumed (Wagner et al. 2003a).

Human genome sequence operates the complexity of cellular function in the body by generating a myriad of distinct proteomes from relatively small number of genes that are regulated (Hastie 2001). This situation is best epitomized by WT1 gene that confers the distinct functional isoforms at both transcriptional and post-transcriptional levels by means of alternative splicing (Larsson et al. 1995), RNA editing (Caricasole et al. 1996), and alternative translation initiation sites (Haber et al. 1993; Morrison et al. 2008), among others. In recent years, significant evidence has been gleaned, and our understanding of the complexity of WT1 biology has been substantially advanced (Hastie 2017). In this chapter, we review recent findings showing the WT1 gene as a key regulator of mesenchyme maintenance and the mesenchyme to epithelial balance in the development of certain mesodermal organs (Hastie 2017). We also summarize recent experiments demonstrating how WT1 regulates the process of epithelial-to-mesenchymal transition (EMT) (von Gise et al. 2011; Takeichi et al. 2013) particularly in the embryonic formation of the second heart field (Zhou et al. 2008). Furthermore, we highlight the reactivation of WT1 gene in epicardial cells in response to ischemic injury in adulthood and the role of the fetal reprogramming in the formation of multipotent cardiac progenitors that may potentially be utilized for cell-based therapy for ischemic heart disease. Owing to a limited length of this chapter, we will not patronize WT1 function in tumorigenesis, which has been discussed in details by more specialized reviewers (Chau and Hastie 2012; Hastie 2017; Wagner et al. 2003a).

2 The Biology of Wilms' Tumor Factor 1

2.1 WT1 Gene and Protein Isoforms

WT1 was cloned first in 1990 and is so far only a strong and relevant predisposition gene for Wilms' tumor (Call et al. 1990). The mammalian WT1 gene is found to locate on chromosome 11p13 and consists of ten exons spanning 50 kb of the genomic sequence (Scharnhorst et al. 2001). Transcriptional and posttranscriptional regulation of WT1 gene may involve in the combination of alternative mRNA splicing (Hewitt et al. 1996), initiation of translation at variable start codons, and RNA editing (Bharathavikru et al. 2017), which generates more than 36 different isoforms from the same genomic sequence with distinct functional properties in the context of activation, co-activation, and repress transcription of target genes (Hewitt et al. 1996). Exon 5 is a cassette exon that adds an additional 17 amino acids in the middle of the protein, while the exon 9 is evolutionarily conserved and can be found in all WT1 genes from zebrafish to human (Hohenstein and Hastie 2006). WT1 genome encodes proteins with 52–65 kD molecular mass that share a high degree of structural homology with the early growth response (EGR) family of transcriptional factors (Rauscher et al. 1990). WT1 functions as a typical transcription factor acting on GC-rich sequences (5′-GCGGGGGCG-3′) with three base pairs recognized by each zinc finger (Toska and Roberts 2014). The WT1 protein structurally constitutes two functional domains: a proline-glutamine-rich domain at the NH_2-terminus and a zinc finger domain composed of four Cys_2-His_2 zinc fingers at the COOH-terminus (Stoll et al. 2007). The most notable posttranscriptional regulation is the second splicing event, which involves the use of two alternative splice donor sites at the end of exon 9, leading to the insertion/omission of three amino acids (lysine, threonine, and serine; KTS) between zinc fingers three and four (Wagner et al. 2003a). The inclusion/exclusion of three amino acids in zinc fingers 3 and 4 creates two important isoforms: WT1(−KTS) and WT1(+KTS), respectively, which differ in their selectivity of DNA-binding site (Wagner et al. 2003a). The isoform of WT1(+KTS) exhibits a high affinity to RNA and co-localizes with and binds to nuclear splicing factors, suggesting that WT1(+KTS) protein plays a posttranscriptional role in mRNA splicing rather at transcriptional level (Morrison et al. 2008). In the absent form of KTS, i.e., WT1(−KTS), the linker sequence promotes a C-terminal helix-capping interaction, suggesting that this isoform works best as a transcription factor (Hamilton et al. 1995). In the early development of organogenesis, the ratio of the two types of isoforms, typically 60:40%, is tightly regulated by an unknown mechanism which likely depends on the precise structure of the splice junction (Laity et al. 2000). Misbalance of the ratio may result in severe urogenital abnormalities (Patek et al. 1999) and characteristic of Frasier syndrome (Klamt et al. 1998). Furthermore, active form WT1 can be also modified through other means of modification such as phosphorylation and sumoylation, which adds further complexity of WT1 functionality (Ye et al. 1996).

2.2 WT1 Function: Oncogenesis vs. Kidney Development

WT1 is primarily recognized as a tumor suppressor, as genetic analysis revealed that germline mutations of WT1 gene may lead to not only the eponymous kidney tumor but also to many other types of malignancies (Little and Wells 1997). Functional analysis by transient co-transfection experiments also demonstrates that WT1 protein is able to endogenously repress several cellular promoters, most of which include growth factor and cognate receptor genes (Haber et al. 1993). Loss of the repressive activity of WT1 is considered to cause oncogenic transformation in the undifferentiated metanephric mesenchyme (Park et al. 1993).

Over past decades, the Janus face of WT1 has been revealed (Scharnhorst et al. 2001). While WT1 is widely accepted as a tumor repressor gene, the growing body of evidence has shown that the expression of WT1 in adult tumors from different origins, including colorectal, breast, desmoid, and brain tumors, is detected in tissues that normally do not express WT1 (Hastie 2017). In fact, 90% of sporadic Wilms' tumors continue to express this protein either because cells have acquired secondary mutations in other genes which affect WT1 tumor suppressor function and/or because WT1 provides an important survival function in these tumors (Mayo et al. 1999). This finding suggests that, besides acting as a potent transcriptional repressor, WT1 may also function as a transcription factor with strong trans-activating potential (Wagner et al. 2003a). In vitro evidence using antisense oligonucleotides has shown that WT1 is required for proliferation via inhibiting apoptosis of tumor cells in culture (Mayo et al. 1999). Furthermore, WT1 is not spatially restricted to the nucleus: it shuttles between nucleus and cytoplasm (Niksic et al. 2004) and is present in functional polyribosomes, suggesting WT1 may also exert cytoplasmic roles, for example, in the regulation of mRNA export, translation, turnover, or localization (Laity et al. 2000). Recently, the +KTS isoform is demonstrated to mainly bind mRNA targets via its zinc fingers and interacts with several splice factors (Caricasole et al. 1996; Morrison et al. 2008). Therefore, beyond its suppressive activity, WT1 may function as important transcriptional activator or posttranscriptional modulator and, by acting on both DNA and mRNA targets, causatively involved in various tumorigeneses in both embryonic stage and adulthood. The apparent contrasting roles of WT1 in suppression versus activation and the context-dependent bi-directional control of mesenchymal–epithelial fate might partially explain how WT1 can act as a tumor suppressor gene in some tissues (Haber and Buckler 1992) and as a potential oncogene in others (Hohenstein and Hastie 2006). The diversity of WT1 function is seemingly dependent on its protein isoforms generated by distinct genetical modifications (Hohenstein and Hastie 2006) (Fig. 6.1a).

In addition to its oncogenic role, WT1 is found to play its crucial role in the normal development of several mesodermal organs (Scholz and Kirschner 2005). Within the developing kidney, WT1 expression levels and patterns of WT1 vary throughout the process of kidney development, which may indicate that it may play different roles at the intermediate stage of nephrogenesis (Hastie 2017). At the early time of gestation, WT1 is found to express at low level in the metanephric blastema where the ureteric bud divides to form branches and the blastema starts to condense at the tip region (Kreidberg et al. 1993). This condensed mesenchyme then undergoes a burst of proliferation and differentiates into the renal vesicle, a hollow ball of epithelial cells in which a persisting expression of WT1 is observed. Thereafter, the renal vesicle undergoes a series of morphological changes: it first forms a comma-

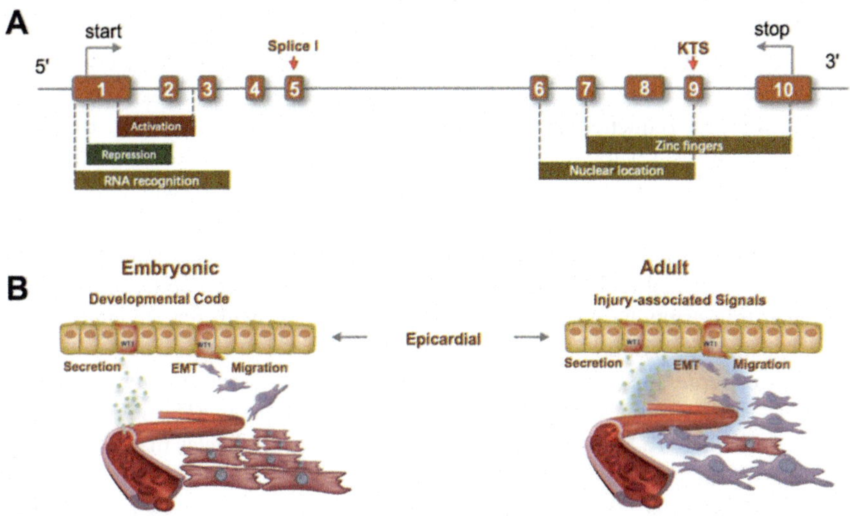

Fig. 6.1 (**a**) The WT1 locus encodes ten exons which by alternative start codons, exons, and splices and RNA editing generate distinct isoforms; (**b**) expression regulates the heart development and, after injury, is reinstated in epicardial cells that regain the cellular plasticity and secretion profile similar to their embryonic counterparts

shaped then an S-shaped body where WT1 expression is localized to the cells at the proximal part. These WT1-expressing cells developmentally form the glomerular podocytes in the kidney (Kreidberg et al. 1993). After birth, expression of WT1 persists in these cells, which implies that WT1 gene also has a role in the maintenance of the differentiated state in adult (Kann et al. 2015). Therefore, the expression evidence from the various stages of kidney development pinpoints the imperative role of WT1 in the nephrogenesis in mammalian (Kreidberg et al. 1993).

The most striking findings perhaps are the results derived from genetically modified animals: homozygous deletion of the WT1 gene in the mouse germline results in embryonic lethality at day 11 of gestation due to a failure of urogenital development (Moore et al. 1999). In the mutant embryos, a predominant feature is the lack of kidneys and gonads, together with defects in mesothelial tissues (i.e., diaphragm, peritoneum, epicardium), heart, adrenal glands, and spleen, and maybe neuronal tissues and vertebrate retina (Chau et al. 2011), strongly suggesting an obligatory role of WT1 in the formation of multiple mesodermal organs. Taken together, a wealth of circumstantial evidence supports the notion that WT1, in addition to its tumorigenic role, also functions as a physiologically relevant regulator that participates crucially the entire process of kidney development.

2.3 *Molecular Basis of WT1 Function*

WT1 may act as a transcription factor that binds to DNA targets through its four zinc fingers (Hastie 2017). The first zinc finger mainly enables WT1 anchoring to the RNA to endow a series of steps of WT1-RNA interaction, while the zinc fingers 2–4 insert deeply into the major groove, where they make base-specific contacts to target consensuses (Stoll et al. 2007). The WT1 target consensuses are mainly related to EGR1 sequence that has three zinc fingers identical to WT1. By using ChIP-seq analysis, several thousand potential target genes have been identified, but only a small subset of these have been experimentally validated (Kim et al. 2009). The potential cardiac-related target genes include p53 (TRP53), Snail1/2, E-cadherin (Cdh2), Radh2, Cxcl10, and β-cadherin; most of them are pivotal initiators of epicardial EMT in the heart development (Chau and Hastie 2012). So far, most of the validated transcriptional targets appear to KTS specific, with few genuine +KTS targets identified. Within those gene targets, WT1 can act as either a transcriptional activator or repressor, mainly depending on its binding partners in the regulatory domains at the residues 71–101 and 185–250, respectively (Hastie 2017). For instance, BASP1 is a novel transcriptional partner that coordinates the downregulation of WT1 targets (Essafi et al. 2011).

Recent studies have revealed that WT1 appears to be involved in global changes in chromatin access (Essafi et al. 2011). Genetical deletion of WT1 in the epicardium switches the chromatin between these two CCCTC-binding factor (CTCF) sites from a repressed state to an active conformation. This switching of chromatin states is epigenetically referred to as "chromatin flip-flop" and differentially regulates the activation of Wnt4 (Essafi et al. 2011). In the kidney mesenchyme, WT1 activated Wnt4 expression, which allows the access of other co-activators, including β -catenin and SIX2, to upregulate the downstream targets, whereas in the epicardium WT1 keeps the locus in a silent state, which prevents the access of Wnt4 activators (Essafi et al. 2011). This finding suggests that WT1 may modulate the state of chromatin domains beyond its binding sites, which confers distinct regulatory function upon the chromatin accessibility in various organs.

Additionally, WT1 may modulate its targets at the posttranscriptional level by interaction with RNA sequences (Hohenstein and Hastie 2006). The WT1 interacts through the first zinc finger to establish RNA-WT1 contact, which mainly favored by the isoform without KTS (Hastie 2001). Three high-affinity RNA-binding motifs have been identified for the -KTS isoform using systematic evolution of ligands by exponential enrichment (SELEX), and both the sequence and secondary structure of RNA determine binding specificity and affinity for WT1 (Bickmore et al. 1992). Furthermore, WT1 shuttles between the nucleus and the cytoplasmic compartment, locating on actively translating polysomes, suggesting a posttranscriptional role of WT1. Additional circumstantial evidence for posttranscriptional functions comes from reports that WT1 actively interacts with RNA-binding proteins including the splice factors (U2AF65, RBM4, HNRPU, HNRNPU) and the Wilms' tumor-associated protein WTAP in an isoform-dependent manner (Hastie 2017). Therefore, these data support the notion that WT1 has posttranscriptional checkpoint that closely regulates the expression of its target genes.

Therefore, upon its isoforms, WT1 protein physically interacts with downstream targets and, by recruiting distinct molecular partners, spatiotemporally regulate cellular functions. In this regard, WT1 sits atop the molecular hierarchy that controls the complex network of its target genes at either the transcriptional or posttranscriptional level at the different embryonic stages of organogenesis.

2.4 *Role of WT1 in Heart Development*

2.4.1 Embryonic Formation of Mammalian Heart

The heart is the first functional organ formed during embryonic development, with cardiac progenitors specified in early gastrulation (Brade et al. 2013). The earliest heart formation may initiate with about 50 founder cells that are located on both sides of the midline in the epiblast of early-gastrula stage embryos (Abu-Issa et al. 2004). Three spatially and temporally distinct cardiac precursors, namely, cardiac mesodermal cells, proepicardial cells, and cardiac neural crest cell, essentially contribute the formation of the contractile, vascular, and interstitial compartments of the mature heart. During developmental stage, the specification of individual lineages is restrictively regulated by the intrinsic signals generated by developmental codes. The first heart field (FHF) progenitors in the cardiac crescent move into the heart relatively earlier and are exposed to local cytokines and growth factors, which induce differentiation and upregulation of essential cardiac transcriptional factors such as Nkx2.5, T-box transcription factor (Tbx)-5, and GATA4, among others. The induction of cardiac transcriptomes efficiently drives the multipotent progenitors into cardiac commitment by the expression of sarcomeric proteins, and those cells spatially contribute to the major portion of the left ventricular and some cells of the both atria (Keith and Bolli 2015). Subsequently, the second heart field (SHF) precursors derived from a cluster of cells locating in the region of pharyngeal mesoderm begin to proliferate. The proepicardial progenitors are a relatively heterogeneous population marked by expression of islet-1 transcription factor (Isl1) (Laugwitz et al. 2005), WT1 (Zhou et al. 2008), and KDR (kinase insert domain receptor) (Brade et al. 2013), sometimes overlapping with SHF markers such as Nkx2.5 (Zhou et al. 2008). At this stage, the tubular myocardium possesses only a thin endocardial lining within the contorting primitive heart tube and is essentially naked, lacking adventitia, perforating vasculature, or surrounding epicardium. These progenitors migrate over the surface of the exposed myocardium and form the epicardium (Abu-Issa et al. 2004). Once the formation of epicardium being complete, directed proliferation parallel or perpendicular to the cellular basement membrane leads to the expansion of epicardium-derived cells (EPDCs) which, under the influence of local fibroblast growth factors (FGFs), bone morphogenic proteins (BMPs), and Wnt-signaling, complementarily contribute to the heart mass, including adventitial and smooth muscle lineages and some vascular endothelia (Brade et al. 2013). Ultimately, the mesenchymally transformed EPDCs locating in the subepicardial region migrate inward to form the coronary plexus (which later becomes the coronary vasculature, with contributions of endocardium-derived endothelial cells) and cardiac adventitial fibroblasts (Keith and Bolli 2015).

In addition to cellular contribution, the proepicardium-derived cells are known to support the myocardium by secreting a variety of beneficial growth factors (Zhou et al. 2011). The specific paracrine mediators responsible for these beneficial effects likely involve a host of pathways including microparticles and microRNA-mediated effects as well as release of growth factors and cytokines such as stromal cell-derived factor 1, vascular endothelial growth factor, and many others, which may stimulate further myocardial growth and myocyte division, specifically to aid formation of compact myocardium (Brade et al. 2013; Abu-Issa 2014).

2.4.2 Role of WT1-Expressing Cells

In mammalian, expression of WT1 is first observed as early as E8.5 in the intermediate mesoderm and adjoining lateral plate mesoderm (Wu et al. 2008; Parenti et al. 2015). The WT1-expressing mesothelial layer, arising from the coelomic epithelium, lines the body cavity and the visceral organs (Brade et al. 2013). At E9.5, WT1 expression is mainly restricted in the mesothelial surface lining of the heart, epicardium, derived from the proepicardial organ, migrates to form a layer around the developing heart (Duim et al. 2016). In vitro study demonstrated that the isolated proepicardial explants are able to spontaneously differentiate into cardiomyocytes, and this is modulated by BMP, fibroblast growth factor (FGF), and Notch signaling, suggesting that the proepicardium and its derivatives constitute the myogenic potential (Kruithof et al. 2006). In normal developing heart, a small subpopulation of WT1-expressing EPDCs may proceed through a further round of EMT to generate coronary vasculature progenitors that constitute the coronary complex formation (Martínez-Estrada et al. 2010; Scholz et al. 2009) and cardiomyocytes (Zhou et al. 2008). Therefore, the roadmap of WT1 expression alludes that WT1 may act as a key regulator to direct a subset of mesenchymal progenitors to commit into cardiac lineages that partially make cellular contribution to heart formation (Velecela et al. 2013; Zhou et al. 2008).

The detailed engagement of WT1 in heart development was recently disclosed in mouse embryos with a homozygous knockout of the WT1 gene (WT1 KO) that presumably died from heart failure at mid-gestation (Chau et al. 2011). The WT1 KO heart was phenotypically reduced in size and displayed extreme thinning of the muscular walls with lethal transmural bleeding into the pericardial cavity (Moore et al. 1999), suggesting that WT1 is essentially required for the embryonic formation of mature heart. The direct proof that WT1 expression virtually contributes to the heart development in mammals is derived from genetic Cre-loxP-based lineage tracing strategies which consistently label the derivate of the WT1 expression cells throughout the early developmental stage of organogenesis (Zhou and Pu 2012a). In this setting, a reporter gene like green fluorescent protein (GFP) or β-galactosidase (LacZ) was placed under the control of a WT1 promoter ($WT1^{GFPCre}$), whereby the presence of the reporter signals indicates an active WT1 expression at the time of analysis, or by Cre/LoxP technique that constitutively and irreversibly labels the WT1 expression cells and their progeny (Zhou et al. 2008; Smart et al. 2011). Lineage tracing in $WT1^{GFPCre}$ mice reveals that WT1 cells readily locate in the heart at E10.5–E11.5 and, notably, are overlapped with some important cardiac genes (Nkx 2.5, Tbx 18) that govern cardiac commitment in the formation of tubular heart (Wu et al. 2006). Notably, $WT1^{GFP}$ cells are able to differentiate into functional cardiomyocytes that anatomically locate in the myocardium of all four cardiac chambers and in the interventricular septum, constituting 7–10% of cardiomyocytes in ventricles and 18% in atria (Zhou et al. 2008). This experiment provides definitive significance of WT1-expressing cells that partially populate the compact mammalian heart (Hastie 2017).

Moreover, WT1 cells may indirectly support the robust growth of the tubular heart by secreting multiple pleiotropic factors that stimulate cardiac mitosis. Studies performed on chick heart slices testified that trophic signals from the epicardium are required for continued cardiac myocyte proliferation and survival (Weeke-Klimp et al. 2010). Considering these findings in the light of the cardiac phenotype of the WT1 deficient embryos, one is tempted to speculate that WT1 may activate yet-unknown mitogenic factor(s) in the epicardium to promote the proliferation of cardiac myocytes (Scholz and Kirschner 2005). Thus, WT1 cells may help to build the mature heart in two ways: they directly give rise to myocytic lineages and in part contribute to the formation of functioning heart and, by secreting pleiotropic factors, foster the second proliferative wave of cardiomyocytes in an indirect manner to establish the mature form of heart mass (Hastie 2017).

A functioning heart requires essentially contractile components and adequate blood supply through coronary vasculature (Brade et al. 2013). Although the origin of coronary vascular endothelial cells is still controversial (Tian et al. 2015), there are growing evidence that the WT1 cells are specified into vascular lineages (endothelial and smooth muscle cells) that essentially make up the coronary complex (Martínez-Estrada et al. 2010; Zhou and Pu 2012a). This idea comes also from the characteristic phenotype of WT1 KO mutants which show severally defect vasculogenesis in the developing heart (Kreidberg et al. 1993). By utilizing the aforementioned strategy of genetically labeling the WT1-positive cells in the embryonic heart, fate-mapping experiments provide direct evidence of the spatiotemporal distribution of the endogenous WT1 in the developing heart, showing that, around embryonic day 12.5, WT1-positive cells migrate into the myocardium where they give rise to coronary vascular smooth muscle cells, perivascular, and intermyocardial fibroblasts and presumably also vascular endothelial cells (Rudat and Kispert 2012; Zhou and Pu 2012a). Although Cre-based lineage tracing approach bears the technical caveats that WT1 labeling is potentially contaminated by the endogenous expression of WT1 (Zhou and Pu 2012a), the general consensus is now that WT1 epicardial cells contribute in small proportion to the formation of coronary complex during heart development (Tian et al. 2015).

In addition to cardiac lineages, WT1 epicardial cells abundantly give rise to cardiac fibroblasts through EMT to build up the fibrous architecture of the heart (see below) (Moore et al. 1999). Epicardial derivatives are especially around the AV groove, and in this region EPDCs differentiate into cells of the annulus fibrosis (Zhou et al. 2010), the fibrous insulation that separates atrial and ventricular myocardium to permit sequential atrial and then ventricular contraction (Groot et al. 1998). The epicardium-derived cardiac fibroblasts may postnatally retain some molecular signatures similar to their embryonic counterparts (Furtado et al. 2014), suggesting those cells are able to potentially give rise to cardiomyocytes under specified condition (Liu et al. 2017).

Taken together, WT1 seems to sit atop of a hierarchy of embryonic epicardial gene loci and directly controls the transcription of several cardiac genes that play essential role in heart development and disease through two distinct mechanisms. First, it undergoes epithelial-to-mesenchymal transition (EMT) to form EPDCs, which subsequently give rise to both cardiac and most of the non-cardiomyocyte cell types of the myocardial wall. Second, the myocardial-epicardial interface forms a critical signaling center that is necessary for both the myocardial growth and the coronary vessel development (Fig. 6.1b).

2.4.3 WT1-Mediated Epithelial-to-Mesenchymal Transition

A significant proportion of the vertebrate body comprises two types of cells: polarized epithelial and more motile mesenchymal cells. Typical epithelia are organized as sheets of cells polarized along an apical-basal axis, perpendicular to the epithelial plane. An apical belt-like arrangement of tight and adherent junctions interconnects the cells within the epithelial sheet, providing mechanical stability and partitioning the cells into apical and basolateral membrane domains, while mesenchymal cells feature rather an amoeboid or stellate morphology that facilitates their movement as individual cells through the extracellular matrix (Lim and Thiery 2012). The interchangeable transition of the two cell types by EMT is an evolutionarily conserved developmental process that enables the formation of more complex body plans featuring new types of internal organs (Thiery et al. 2009). In developing heart, all of cells that ultimately populate the mature heart come from a few funder cells that undergo several steps of EMT processes (Abu-Issa 2014). In the early gastrulation, epiblast cells are primarily converted into mesenchymal cells that form the lateral plate mesoderm, and, by secondary EMT, these cells further become heart precursors that ultimately give rise to cardiomyocytes in the FHF (Abu-Issa 2014). Additional EMT event occurs in the cardiac cushion (SHF) where the valve progenitors are generated to build up the valves and in the epicardial layer where the epicardial cells delaminate and become EPDCs that contribute to vascular cell fibroblasts and possibly a subset of cardiomyocytes in the mature heart (von Gise and Pu 2012; Zhou et al. 2008, 2010; Moore-Morris et al. 2016; Martínez-Estrada et al. 2010). The EMT event is carefully regulated and efficiently executed through multiple mechanisms in different contexts but typically shares general cell biological features and bears similarities in molecular regulation (Lim and Thiery 2012). At a molecular level, a number of signaling pathways have been implicated in EMT process and intensively reviewed elsewhere (Lim and Thiery 2012; Thiery et al. 2009; von Gise and Pu 2012). In this chapter, we only focus on the role of WT1-mediated signaling in epicardial cells on heart development.

The most direct evidences rely on the findings in the loss-of-function experiments which show impaired formation of an intact epicardium in WT1 lacking mice (Moore et al. 1999) and a reduced number of EPDCs in WT1 GATA5-Cre conditional knockout mice (Martínez-Estrada et al. 2010). WT1 expression appears initially in the proepicardial population locating in the outflow tract and continues in the mesothelial lining of the heart surface. As a key transcriptional activator, WT1 directly binds to the promoters of downstream targets and thus induces expression of Snail1, an important EMT regulator, and, in meantime, represses the expression of epithelial component E-cadherin, which ensures the epicardial cells effectively undergoing EMT process (Martínez-Estrada et al. 2010). Likewise, WT1 expression also promotes canonical and non-canonical Wnt pathways (von Gise and Pu 2012). Interestingly, WT1 also functions by ambiguously targeting Wnt4 gene which is found to be activated in kidney mesenchyme undergoing MET (Kreidberg et al. 1993), but repressed in epicardial EMT, although both processes directly mediated by WT1 binding to the same transcriptional regulatory site of Wnt4 (Essafi et al. 2011). This reflects the functional complex of WT1 in different context of co-activators in developmental organogenesis (Chau and Hastie 2012). In the kidney mesenchyme, the CBP co-activator is recruited to the WT1-bound Wnt4 locus, whereas, in the epicardium, the brain-abundant membrane-attached signal protein 1 (Basp1) co-repressor is recruited (Carpenter et al. 2004). The boundary of the Wnt4 locus is marked by CCCTC-binding factor (CTCF) and by cohesion, and loss of WT1 switches the Wnt4 locus chromatin domain into either a repressed or an activated state (Essafi et al. 2011).

Not merely morphological changes, one of the important features of EMT is that a series of molecular engagement endows epicardial cells stem cell-like properties and becoming epicardial progenitors or epicardium-derived cells (EPDCs) (Moore et al. 1999). The multipotent EPDCs are also developmentally pliable, often capable of differentiating into a variety of distinct types of cardiac cells that populate the mature heart (von Gise and Pu 2012). This molecular feature that EPDCs acquire stemness after EMT process is of particular interest in understanding the transcriptional hierarchy governing cardiovascular lineage commitment and generating somatical stem cells that are feasible to cardiac regeneration (vide infra).

3 Epicardial Reactivation of WT1 in Response to Injury Signals

WT1 expression is postnatally switched off soon after birth when organogenesis is complete although a rare population in the cortex of the adult adrenal gland and in the podocytes in the kidney retains WT1 expression (Hastie 2017). In the heart, the cells in adult epicardium, unlike in their embryonic counterparts, are typically quiescent, and the functional role of the outermost epithelial layer of the heart has been neglected for many years as it was considered as an external cover devoid of any functional impact on heart performance (Dueñas et al. 2017). The discovery that WT1 expression can be rapidly reactivated in response to cardiac injury has branded the epicardium as a source of cardiac stem cells (Zhou and Pu 2008). The recapitulation of embryonic gene program may represent a hallmark of attempts by the adult heart to compensate for cardiovascular stress and disease (Smits and Riley 2014).

3.1 Cellular Contribution to Cardiac Repair

Myocardial infarction (MI) stimulates a robust epicardial response, marked by re-expression of a panel of fetal epicardial markers of not only WT1 but also Raldh2 and Tbx18, leading to the proliferation and migration of EPDCs inwards the damaged myocardium as an innate attempt to reconstitute the injured part. The most exciting evidence comes from the adult zebrafish heart which completely regenerated the 20% of amputated part of the heart apex by reactivation of epicardium. In response to injury signal, upregulation of several fetal epicardial genes WT1 as well as RaldH2 and Tbx18 occurs within the 1st day after injury, a scenario found in adult zebrafish in which WT1 express redeploys (Zhou et al. 2011). As a result, the single cell epithelial layer of the heart surface undergoes a series steps of EMT and delamination, which happen similarly in the developing heart and become mitotically active together with upregulation of EMT-related genes including Snail Slag Twist and Smad1 (Zhou et al. 2011). The activated epicardial cells expand substantially in members as early as 3 days and the thickness of the epicardial layer peaks at approximately 5 days post-MI. In order to follow the fate of the activated epicardium, GFP-producing lentiviral vectors were injected directly into the pericardial sac to label the underlying epicardium. After 7 and 21 days, GFP-positive cells were found within the left ventricular infarcted wall and histologically confirmed by virtue of WT1 co-expression as seen in epicardial origin (Limana et al. 2007). Further analysis in $WT1^{GFPCre}$ mice demonstrates that WT1 cells invade inward the subepicardial compartment where they gave rise to fibroblast population, myofibroblasts, and coronary endothelium and in a small proportion to heart muscle cells (van Wijk et al. 2012). Notably, the WT1-derived myogenic cells still remain round sharp, form clusters, and fail to exhibit the classical rod-shaped morphology, although they do co-express several other markers specific for cardiomyocytes including SERCA2a as an indicative of an early cardiac commitment (van Wijk et al. 2012). Therefore, it seems that the differentiation capacity of WT1 cells in adulthood is less potent than that present in fetal development. In default, the activated EPDCs gain the expression of mesenchymal markers (Ding et al. 2016), but inefficiently invade the subjacent myocardium, rather remaining in a thickened epicardial layer on the surface of the heart (Zhou et al. 2011). Nevertheless, the general consensus is that ischemic injury alone represents a strong activator of WT1 expression in the epicardium and that the mammalian heart seems to still retain the evolutionarily conserved reparative mechanism, like in the fish heart, resulting in the formation of the majority of fibrotic scar tissue together only with a limited number of newly formed cardiomyocytes (Asli et al. 2014).

EPDCs primarily differentiated into myofibroblast/fibroblast and smooth muscle lineages and mainly adopted a perivascular location in the thickened epicardium (Zhou and Pu 2008). The limited range of cell types derived from post-MI EPDCs may have been attributable to "partial" EMT of epicardial precursors or lack of EPDC mobilization and migration into a suitable environment within the myocardium. However, the paradigm was shifted into optimistic situation in which full mature cardiomyocytes formed from the progeny of WT1 cells when mice were primed by a G-actin monomer binding protein thymosin beta 4 (Tβ4) (Smart et al. 2012). Exogenous Tβ4 increased epicardial progenitor cell plasticity and mobilization and increased vascular cell outgrowths from fetal epicardial explants. Tβ4 "priming" in the context of myocardial injury stimulated de novo formation of cardiomyocytes from WT1-expressing epicardial precursors, suggesting that epicardial cells might represent a resident cardiac progenitor that could be recruited for therapeutic myocardial regeneration (Smart et al. 2011). Interestingly, myogenic potential of the WT1 cells is detected when Tβ4 is given concurrent with induction of MI, suggesting that some important signals are necessarily established prior to the activation of WT1 epicardial cells (Masters and Riley 2014). Although the number of epicardium-derived myocytes remained low, this study provides proof-of-concept evidence showing that adult epicardial cells or at least a subpopulation thereof could differentiate into fully functional cardiomyocytes in vivo (Smart et al. 2011). In future work, it will be important to determine the signaling mechanisms that synergistically impart WT1-expressing epicardial cells into myocytic commitment under Tβ4 priming.

The formation of new blood vessels is critical not only for normal heart development but also for myocardial tissue repair after MI. Recapitulation of WT1 expression in the injured heart cues to instigate an angiogenic program in the epicardial cells that similarly happened in the embryonic state (Tian et al. 2015; Dubé et al. 2017). In the study of WT1 lineage tracing, von Wijk et al. demonstrated that WT1 reporter co-expressed with the endothelial marker PECAM and a smooth muscle actin at the wall of coronary vessels in the infarcted myocardium, suggesting that WT1 represents a novel family member of proteins that control the genetic program for the formation of new blood vessels not only in the developing fetal heart but also in the ischemic heart (van Wijk et al. 2012). However, this result is not experimentally confirmed by another cohort of groups using inducible WT1 trace who fail to detect the cellular contribution of WT1 cell directly to the vasculature network (Rudat and Kispert 2012; Zhou et al. 2011, 2012; Dubé et al. 2017). More recently, Dubé et al. reported that WT1 epicardial cells exclusively contributed to pericytes in the adult heart after MI, but without progression to mature endothelial and smooth muscle cells (Dubé et al. 2017). Thus, it remains to be determined whether the activated WT1 cells offer an analogous potential to contribute toward neovasculature after ischemic injury in a similar manner in developing heart (Martínez-Estrada et al. 2010). In vivo tracking studies are essentially limited to functional imaging (MRI), which has yet to reach cellular resolution for tracking migration events during ischemia. Recent study by Ding et al. using 19F image (Flögel et al. 2011) illustrated that injury-associated epicardial cells actively take up nanoparticles containing 19F and their migration events can be followed by 19F image (Ding et al. 2016). They also demonstrated that 19F-labeled epicardial cells contribute to about 10% of conduit vessels (Ding et al. 2016), suggesting an angiogenic contribution of epicardial cell in the ischemic heart.

3.2 *Paracrine Activity*

Despite the lack of direct contribution, WT1 cells were found to closely associate with arteriolar smooth muscle cells, perhaps contributing adventitial fibroblasts or supporting the coalescence and differentiation of new vessels indirectly, via modulation of extracellular matrix (ECM) formation or paracrine secretion (Zhou et al. 2011). Redeployment of WT1 expression in epicardial cells is associated with pro-angiogenic factors that robustly stimulate blood vessel formation in angiogenesis assays (Zhou et al. 2011). The reactivated WT1 cells upregulated a number of genes encoding potent secreted factors, including Vegfa, Fgf2, Tgfb2, Sdf1, and Mcp1. In zebrafish, a subpopulation of these epicardial cells undergoes EMT, invades the wound, and provides new vasculature to regenerating muscle via activating fibroblast growth factor (FGF) pathways, suggesting injury responses by myocardial and epicardial tissues that collaborate in a FGF-dependent manner to achieve cardiac regeneration (Lepilina et al. 2006). Moreover, injection of this conditioned media of WT1 cells into the ischemic area and border zone at the time of coronary artery ligation improved capillary density and ventricular function and reduced adverse left ventricular remodeling, indicating that enhanced vessel density contributes to the beneficial effects of WT1 cell injection. Furthermore, the proangiogenic evidence of WT1 cells relies on the experiments by injecting human epicardial WT1 cells into murine myocardium. Exogenous WT1 cells are able to improve ventricular function of the infarct heart and to augment the density of coronary vasculature (Winter et al. 2007). Additionally, WT1 cells clearly have a distinct role in orchestrating vessel formation in the infarcted heart by providing stability to the newly formed vessels (via ECM formation) as well as providing factors to induce de novo formation of coronary vasculature. Epicardial cells also are able to provide a suitable ECM to promote expansion of the neovessel network toward the infarcted myocardium (Dubé et al. 2017).

Therefore, the WT1 cells in the epicardium are developmentally malleable and uniquely situate to influence both myocardial and coronary vessel function through cellular differentiation and paracrine mechanisms.

3.3 *Molecular Basis of WT Reactivation*

Although downstream target genes of WT1 have been well illustrated, surprisingly, little is known about the pathways and factors that act upstream to regulate WT1 expression. In general, activation of WT1 is likely a complicated molecular process that may temporally and spatially involve multiple factors and co-activators at transcription level and also at post-transcription level. It was reported that WT1 is induced by hypoxia (8%% O_2), with compelling evidence showing a binding site of the hypoxia-inducible factor HIF-1 in WT1 promoters which confers transactivation of WT1 directly in response to hypoxic stimuli (Wagner et al. 2003b). Recently, Huang et al. identified the 14 conserved regions of WT1 sequence that contain putative CCAAT-/enhancer-binding protein (C/EBP) binding sites. C/EBP proteins recognize a consensus sequence of (A/C) TTNCNN (A/C) A, and epicardial activation of these enhancers depends on a combinatorial transcriptional code centered on C/EBP transcription factors (Huang et al. 2012). These results support the conclusion that C/EBP proteins mediate injury-induced activation of Raldh2 and WT1 epicardial enhancers and gene expression in the adult epicardium. Furthermore, BRG1, the essential ATPase subunit of the SWI/SNF chromatin remodeling complex, is also described to be essential for the activation of WT1 transcription. BRG1 physically interacts with Tβ4 and is recruited by C/EBP type B to discrete regulatory elements in the WT1 locus (Vieira et al. 2017). Therefore, reactivation of WT1 expression in epicardial cells involves multiple factors that converge to act at either WT1 locus or at the upstream regulatory elements that coordinate transcriptional response to injury signals. Nevertheless, the translational regulation of WT1 isoforms that act at the distinct downstream targets remains unknown and needs to be further explored.

4 WT1-Activated Endogenous Source of Progenitor Cells for Cell-Based Therapy

The reactivation of fetal genes after injury is a hallmark of attempts by the adult heart to compensate the loss of functional cardiomyocytes after ischemic insult (Laflamme and Murry 2011). However, the endogenous reparative mechanisms, including activation of WT1 expression in epicardial cells, are unable to instigate a regenerative response and frustratingly insufficient to counteract the pathology and suppress adverse remodeling. Following MI in adult mammals, billions of cardiomyocytes are lost and the damaged myocardium is subsequently replaced by the proliferation and activation of fibroblasts which deposit ECM (most notably collagen), resulting in scar formation that is eventually lack of contractile activity. The consequence of losing over a billion functional cardiomyocytes in mammals is replaced by a non-contractile scar, pathological remodeling, and progression to heart failure. Since the mammalian heart is unable to adequately regenerate beyond early postnatal stages, transplantation is the only possible cure, although confounded by host immune rejection and a limited supply of donor hearts. Consequently, there has been an intensive effort to develop stem cell-based strategies for cardiac repair to both regenerate heart muscle and promote coronary vasculogenesis after MI, culminating in a number of clinical trials. However, patient benefit has been modest at best, and the majority of such trials have relied on the use of autologous cell types, including bone marrow stem cells (Orlic et al. 2001). So far, there is no consensus that these cells can differentiate to contribute new muscle efficiently (Eschenhagen et al. 2017). Perhaps for this reason, the clinical improvement has been generally disappointing (Kennedy-Lydon and Rosenthal 2017). Hence, there continues to be an urgent need to identify the most promising cardiovascular stem cells, either for transplantation and engraftment or activation of a resident population for induction toward a cardiomyocyte fate (Cai and Molkentin 2017).

Several types of cardiac progenitors have been broadly described in the literature and mainly characterized according to the expression of specific stem cell markers, such as stem cell antigen-1 (Sca-1) and c-kit (Keith and Bolli 2015), or based on their in vitro culture properties (Bollini et al. 2011). These different cardiac stem cells have been described as independent populations, although it is not clear whether they are derived from different cell sources during development or whether they represent cells derived from a common precursor subsequently isolated at different stages of differentiation and maturation (Aguirre et al. 2013). Among all the different stem cell sources defined so far, there are two distinct populations, the origin of which can be specifically defined and traced back to embryonic development: notably Isl-1-positive cells (Laugwitz et al. 2005) and epicardium-derived cells (Bollini et al. 2014). In the adult heart, the epicardial cells are typically dormant but recapitulate WT1 expression in response to injury signaling, as an exemplary paradigm of reinstating embryonic potential in putative therapeutic progenitor cell populations in the adult heart (Smits and Riley 2014). Therefore, epicardial WT1 cells appear to represent a font of progenitors with potential for cardiac regeneration and heart repair (Smart and Riley 2011).

The definition of a stem cell is referred to the relatively undifferentiated, proliferative cells that maintain their number while at the same time producing a range of differentiated progeny that may continue to divide (Wessels and Pérez-Pomares 2004; Zhang et al. 2015). After EMT, epicardial WT1 cells gain cellular "stemness" characterized by expression of undifferentiated genes such as slug (Takeichi et al. 2013; Zhou et al. 2010) and, in general, of no markers for advanced stage of lineage commitment (Wessels and Pérez-Pomares 2004). Notably, recent data support the notion that the differentiation capacities of the epicardial WT1 cells, found within the normal embryonic environment in vivo, seem to be retained in adult life and disease states (Vieira and Riley 2011). WT1 expression may upregulate multiple targets that are transcriptionally involved in lineage commitment such as GATA4, Nkx2.5, and Mef2c (unpublished data) and secrete myriad pleiotropic factors, including some growth factors (IGF-1, PDGF-α, TGF-β, and FGF), that support neoangiogenesis and cardiac proliferation, suggesting reactivated epicardial cells may function in a similar manner in heart development (Limana et al. 2011). Moreover, activated epicardial cells express several surface markers identical to mesenchymal stem cells (Limana et al. 2010; Ding et al. 2016). In this regard, epicardial WT1 cells phenotypically meet these criteria for cardiac stem cells (Wessels and Pérez-Pomares 2004) and thus represent an attractive cell source for stem cell-based therapy.

The initial attempt to isolate adult epicardial cells was derived from atrial surface and cultured in vitro (Wessels and Pérez-Pomares 2004). Thereafter, rapid and robust methods for generating proepicardial (epicardial progenitor)-like cells in both aggregates and monolayer cultures with intrinsic cardiac potential (Zhou and Pu 2012b; Smart and Riley 2007; Vieira and Riley 2011) and even in reprogrammed cells (Witty et al. 2014) have been intensively developed. In vitro cultivation, WT1 cells show an adherrent cobblestone morphology and a highly mitotic activity that enables maintainance in culture (Zhou and Pu 2012b) with exponentially increase of cell number (Temme et al. 2017). Most interestingly, their cardiac commitment is shown by expression of GATA4 and cTnT, while specific makers for cardiomyocyte (Winter et al. 2007) and for smooth muscle cell phenotype (van Tuyn et al. 2007) are absent. However, recent study revealed that, despite the globe reactivation of developmental program via re-expression of the embryonic gene WT1, the epicardial cells adopt a different phenotype in response to cardiac injury relative to embryonic contribution to the developing heart (Bollini et al. 2014). For instance, the Sca-1 subpopulation of WT1 cells is not only the most active in terms of initiating the developmental epicardial gene program, but also pre-empted cardiovascular lineage commitment, via significantly higher expression of both Isl1 and Flk1, the transcriptional signature of common primordial/multipotent cardiac progenitors during heart development (Bollini et al. 2014). Therefore, reactivated WT1 cells are relatively heterogeneous population that when activated can restore an embryonic gene program, but not revert entirely to adopt an embryonic phenotype (Bollini et al. 2014). Further studies are needed to create functionally homogenous population suitable for cardiomyocyte replacement. This can be achieved by either chemical genetic approach using small molecules (Zhao et al. 2016) or drug-like compounds such as Tβ4 (Smart et al. 2007) and, a combination of RA and BMP4 in pluripotent stem cells (Guadix et al. 2017). Understanding the subpopulation in the reactivated epicardial cells alongside the cardiomyogenic or cardiovascular potential of the resident subpopulations may provide important insight into the deployment of resident cardiac stem cells for cardiac repair.

Remarkably, transplantation of human epicardial cells into the infarct murine heart preserves cardiac function and attenuates ventricular remodeling, and the beneficial effects is mediated by the action of augmented cellular survival and paracrine effects, a scenario occurred in the embryonic counterpart (Mayo et al. 1999; Hastie 2017). Indeed, we find epicardial cells constitute a myriad of secretomes that may overtly account for epicardial cell-based therapy (Temme et al. 2017). Therefore, the adult epicardium can be considered a source of dormant cardiac progenitor cells, which, upon appropriate stimulation, can become reactivated, such that adult WT1 cells have the potential to contribute to cardiac homeostasis and repair. Potential biases in cardiovascular cell fate suggest that discrete subpopulation of WT1 cells might be clinically relevant for regenerative therapy.

5 Concluding Remarks

The evolutionary enigma why our body employs oncogenic gene WT1 as a pivotal transcriptional factor to regulate the embryonic formation of certain mesodermal organs and to maintain tissue integrity in adulthood remains elusive. The functional complexity of WT1 is apparently due at least in part to the variety of proteins that are generated from a single gene. To fully understand the physiology of WT1, an effort must be made to obtain a more complete picture of the target genes that are regulated by each of the different WT1 molecules. We believe that the experimental avenues suggested here have the potential to reveal key cells, molecules, and pathways regulating tissue homeostasis and repair. Finally, it will be important to determine with rigor that WT1 expression not only plays a significant role in adult cancer but also can be translated into clinical benefit and therapeutic applications.

In this regard, the epicardial express of WT1 is centered not only in the embryonic stage of heart development but also in the intrinsic response to ischemic injury.

A contemporary paradigm in regenerative medicine is that tissue repair in the adult is frequently underpinned by a reactivation of the embryonic program that created the tissue in the first instance. As such, there is much to gain from understanding the embryonic mechanisms of vasculogenesis and cardiogenesis induced by WT1 protein. Due to their fundamental role in heart development, the epicardial WT1 cells have emerged as a population that fulfills this remit and have come under intense scrutiny as a new source for myocardial regeneration. In this setting, activated epicardial WT1 cells play a critical role via paracrine secretion of key signaling factors in myocardial compaction. Epigenetic induction, with co-factor amplification, of fetal genes to activate adult progenitor populations may represent a general paradigm in cell-based tissue repair. More specifically, targeting dynamic chromatin signatures at key embryonic genes that underlie epicardial cell activation could represent a therapeutic strategy to regenerate the injured-adult mammalian heart. Improved understanding and modulation of epicardial signals to restrict scarring/fibrosis and to promote restoration of lost muscle and coronary vasculature hold genuine potential for translating the regenerative capacities of lower vertebrates and immature mammals to human patients with heart failure.

Acknowledgment This study was funded by NSFC 81570244.

References

Abu-Issa R. Heart fields: spatial polarity and temporal dynamics. Anat Rec. 2014;297:175–82. https://doi.org/10.1002/ar.22831.

Abu-Issa R, Waldo K, Kirby ML. Heart fields: one, two or more? Dev Biol. 2004;272:281–5. https://doi.org/10.1016/j.ydbio.2004.05.016.

Aguirre A, Sancho-Martinez I, Izpisua Belmonte JC. Reprogramming toward heart regeneration: stem cells and beyond. Cell Stem Cell. 2013;12:275–84. https://doi.org/10.1016/j.stem.2013.02.008.

Asli NS, Xaymardan M, Harvey RP. Epicardial origin of resident mesenchymal stem cells in the adult mammalian heart. J Dev Biol. 2014;2:117–37. https://doi.org/10.3390/jdb2020117.

Bharathavikru R, Dudnakova T, Aitken S, Slight J, Artibani M, Hohenstein P, Tollervey D, Hastie N. Transcription factor Wilms' tumor 1 regulates developmental RNAs through 3′ UTR interaction. Genes Dev. 2017;31:347–52. https://doi.org/10.1101/gad.291500.116.

Bickmore WA, Oghene K, Little MH, Seawright A, van Heyningen V, Hastie ND. Modulation of DNA binding specificity by alternative splicing of the Wilms tumor wt1 gene transcript. Science. 1992;257:235–7. https://doi.org/10.1126/science.1321494.

Bollini S, Smart N, Riley PR. Resident cardiac progenitor cells: at the heart of regeneration. J Mol Cell Cardiol. 2011;50:296–303. Special Issue: Cardiovascular Stem Cells Revisited. https://doi.org/10.1016/j.yjmcc.2010.07.006.

Bollini S, Vieira JMN, Howard S, Dubè KN, Balmer GM, Smart N, Riley PR. Re-activated adult epicardial progenitor cells are a heterogeneous population molecularly distinct from their embryonic counterparts. Stem Cells Dev. 2014;23:1719–30. https://doi.org/10.1089/scd.2014.0019.

Brade T, Pane LS, Moretti A, Chien KR, Laugwitz K-L. Embryonic heart progenitors and cardiogenesis. Cold Spring Harb Perspect Med. 2013;3:a013847. https://doi.org/10.1101/cshperspect.a013847.

Cai C-L, Molkentin JD. The elusive progenitor cell in cardiac regeneration: slip slidin' away. Circ Res. 2017;120:400–6. https://doi.org/10.1161/CIRCRESAHA.116.309710.

Call KM, Glaser T, Ito CY, Buckler AJ, Pelletier J, Haber DA, Rose EA, Kral A, Yeger H, Lewis WH, Jones C, Housman DE. Isolation and characterization of a zinc finger polypeptide gene at the human chromosome 11 Wilms' tumor locus. Cell. 1990;60:509–20. https://doi.org/10.1016/0092-8674(90)90601-A.

Cano E, Carmona R, Muñoz-Chápuli R. Wt1-expressing progenitors contribute to multiple tissues in the developing lung. Am J Physiol Lung Cell Mol Physiol. 2013;305:L322–32. https://doi.org/10.1152/ajplung.00424.2012.

Caricasole A, Duarte A, Larsson SH, Hastie ND, Little M, Holmes G, Todorov I, Ward A. RNA binding by the Wilms tumor suppressor zinc finger proteins. Proc Natl Acad Sci. 1996;93:7562–6.

Carpenter B, Hill KJ, Charalambous M, Wagner KJ, Lahiri D, James DI, Andersen JS, Schumacher V, Royer-Pokora B, Mann M, Ward A, Roberts SGE. BASP1 is a transcriptional cosuppressor for the Wilms' tumor suppressor protein WT1. Mol Cell Biol. 2004;24:537–49. https://doi.org/10.1128/MCB.24.2.537-549.2004.

Chau Y-Y, Hastie ND. The role of Wt1 in regulating mesenchyme in cancer, development, and tissue homeostasis. Trends Genet. 2012;28:515–24. https://doi.org/10.1016/j.tig.2012.04.004.

Chau Y-Y, Brownstein D, Mjoseng H, Lee W-C, Buza-Vidas N, Nerlov C, Jacobsen SE, Perry P, Berry R, Thornburn A, Sexton D, Morton N, Hohenstein P, Freyer E, Samuel K, van't Hof R, Hastie N. Acute multiple organ failure in adult mice deleted for the developmental regulator Wt1. PLoS Genet. 2011;7:e1002404. https://doi.org/10.1371/journal.pgen.1002404.

Ding Z, Temme S, Quast C, Friebe D, Jacoby C, Zanger K, Bidmon H-J, Grapentin C, Schubert R, Flögel U, Schrader J. Epicardium-derived cells formed after myocardial injury display phagocytic activity permitting in vivo labeling and tracking. Stem Cells Transl Med. 2016;5:639–50. https://doi.org/10.5966/sctm.2015-0159.

Dubé KN, Thomas TM, Munshaw S, Rohling M, Riley PR, Smart N. Recapitulation of developmental mechanisms to revascularize the ischemic heart. JCI Insight. 2017;2. https://doi.org/10.1172/jci.insight.96800.

Dueñas A, Aranega AE, Franco D. More than just a simple cardiac envelope; cellular contributions of the epicardium. Front Cell Dev Biol. 2017;5:44. https://doi.org/10.3389/fcell.2017.00044.

Duim SN, Smits AM, Kruithof BPT, Goumans M-J. The roadmap of WT1 protein expression in the human fetal heart. J Mol Cell Cardiol. 2016;90:139–45. https://doi.org/10.1016/j.yjmcc.2015.12.008.

Eschenhagen T, Bolli R, Braun T, Field LJ, Fleischmann BK, Frisén J, Giacca M, Hare JM, Houser S, Lee RT, Marbán E, Martin JF, Molkentin JD, Murry CE, Riley PR, Ruiz-Lozano P, Sadek HA, Sussman MA, Hill JA. Cardiomyocyte regeneration: a consensus statement. Circulation. 2017;136:680–6. https://doi.org/10.1161/CIRCULATIONAHA.117.029343.

Essafi A, Webb A, Berry RL, Slight J, Burn SF, Spraggon L, Velecela V, Martinez-Estrada OM, Wiltshire JH, Roberts SGE, Brownstein D, Davies JA, Hastie ND, Hohenstein P. A Wt1-controlled chromatin switching mechanism underpins tissue-specific Wnt4 activation and repression. Dev Cell. 2011;21:559–74. https://doi.org/10.1016/j.devcel.2011.07.014.

Flögel U, Su S, Kreideweiß I, Ding Z, Galbarz L, Fu J, Jacoby C, Witzke O, Schrader J. Noninvasive detection of graft rejection by in vivo 19F MRI in the early stage. Am J Transplant. 2011;11:235–44. https://doi.org/10.1111/j.1600-6143.2010.03372.x.

Furtado MB, Costa MW, Pranoto EA, Salimova E, Pinto AR, Lam NT, Park A, Snider P, Chandran A, Harvey RP, Boyd R, Conway SJ, Pearson J, Kaye DM, Rosenthal NA. Cardiogenic genes expressed in cardiac fibroblasts contribute to heart development and repairnovelty and significance. Circ Res. 2014;114:1422–34. https://doi.org/10.1161/CIRCRESAHA.114.302530.

Groot ACG, Peeters M-PFMV, Mentink MMT, Gourdie RG, Poelmann RE. Epicardium-derived cells contribute a novel population to the myocardial wall and the atrioventricular cushions. Circ Res. 1998;82:1043–52. https://doi.org/10.1161/01.RES.82.10.1043.

Guadix JA, Orlova VV, Giacomelli E, Bellin M, Ribeiro MC, Mummery CL, Pérez-Pomares JM, Passier R. Human pluripotent stem cell differentiation into functional epicardial progenitor cells. Stem Cell Rep. 2017;9:1754–64. https://doi.org/10.1016/j.stemcr.2017.10.023.
Haber DA, Buckler AJ. WT1: a novel tumor suppressor gene inactivated in Wilms' tumor. New Biol. 1992;4:97–106.
Haber DA, Park S, Maheswaran S, Englert C, Re GG, Hazen-Martin DJ, Sens DA, Garvin AJ. WT1-mediated growth suppression of Wilms tumor cells expressing a WT1 splicing variant. Science. 1993;262:2057–9. https://doi.org/10.1126/science.8266105.
Hamilton TB, Barilla KC, Romaniuk PJ. High affinity binding sites for the Wilms' tumour suppressor protein WT1. Nucleic Acids Res. 1995;23:277–84. https://doi.org/10.1093/nar/23.2.277.
Hammes A, Guo J-K, Lutsch G, Leheste J-R, Landrock D, Ziegler U, Gubler M-C, Schedl A. Two splice variants of the Wilms' tumor 1 gene have distinct functions during sex determination and nephron formation. Cell. 2001;106:319–29. https://doi.org/10.1016/S0092-8674(01)00453-6.
Hastie ND. Life, sex, and WT1 isoforms—three amino acids can make all the difference. Cell. 2001;106:391–4. https://doi.org/10.1016/S0092-8674(01)00469-X.
Hastie ND. Wilms' tumour 1 (WT1) in development, homeostasis and disease. Development. 2017;144:2862–72. https://doi.org/10.1242/dev.153163.
Hewitt SM, Fraizer GC, Wu Y-J, Rauscher FJ, Saunders GF. Differential function of Wilms tumor gene WT1 splice isoforms in transcriptional regulation. J Biol Chem. 1996;271:8588–92. https://doi.org/10.1074/jbc.271.15.8588.
Hohenstein P, Hastie ND. The many facets of the Wilms' tumour gene, WT1. Hum Mol Genet. 2006;15:R196–201. https://doi.org/10.1093/hmg/ddl196.
Huang GN, Thatcher JE, McAnally J, Kong Y, Qi X, Tan W, DiMaio JM, Amatruda JF, Gerard RD, Hill JA, Bassel-Duby R, Olson EN. C/EBP transcription factors mediate epicardial activation during heart development and injury. Science. 2012;338:1599–603. https://doi.org/10.1126/science.1229765.
Kann M, Ettou S, Jung YL, Lenz MO, Taglienti ME, Park PJ, Schermer B, Benzing T, Kreidberg JA. Genome-wide analysis of Wilms' tumor 1-controlled gene expression in podocytes reveals key regulatory mechanisms. J Am Soc Nephrol. 2015;26:2097–104. https://doi.org/10.1681/ASN.2014090940.
Keith MCL, Bolli R. "String Theory" of c-kitpos cardiac cells: a new paradigm regarding the nature of these cells that may reconcile apparently discrepant results. Circ Res. 2015;116:1216–30. https://doi.org/10.1161/CIRCRESAHA.116.305557.
Kennedy-Lydon T, Rosenthal N. Cardiac regeneration: all work and no repair? Sci Transl Med. 2017;9:eaad9019. https://doi.org/10.1126/scitranslmed.aad9019.
Kim MK-H, McGarry TJ, Broin PÓ, Flatow JM, Golden AA-J, Licht JD. An integrated genome screen identifies the Wnt signaling pathway as a major target of WT1. Proc Natl Acad Sci. 2009;106:11154–9. https://doi.org/10.1073/pnas.0901591106.
King-Underwood L, Pritchard-Jones K. Wilms' tumor (WT1) gene mutations occur mainly in acute myeloid leukemia and may confer drug resistance. Blood. 1998;91:2961–8.
Klamt B, Koziell A, Poulat F, Wieacker P, Scambler P, Berta P, Gessler M. Frasier syndrome is caused by defective alternative splicing of WT1 leading to an altered ratio of WT1 +/−KTS splice isoforms. Hum Mol Genet. 1998;7:709–14. https://doi.org/10.1093/hmg/7.4.709.
Kreidberg JA, Sariola H, Loring JM, Maeda M, Pelletier J, Housman D, Jaenisch R. WT-1 is required for early kidney development. Cell. 1993;74:679–91. https://doi.org/10.1016/0092-8674(93)90515-R.
Kruithof BPT, van Wijk B, Somi S, Kruithof-de Julio M, Pérez Pomares JM, Weesie F, Wessels A, Moorman AFM, van den Hoff MJB. BMP and FGF regulate the differentiation of multipotential pericardial mesoderm into the myocardial or epicardial lineage. Dev Biol. 2006;295:507–22. https://doi.org/10.1016/j.ydbio.2006.03.033.
Laflamme MA, Murry CE. Heart regeneration. Nature. 2011;473:326–35., nature10147. https://doi.org/10.1038/nature10147.

Laity JH, Dyson HJ, Wright PE. Molecular basis for modulation of biological function by alternate splicing of the Wilms' tumor suppressor protein. Proc Natl Acad Sci. 2000;97:11932–5. https://doi.org/10.1073/pnas.97.22.11932.

Larsson SH, Charlieu J-P, Miyagawa K, Engelkamp D, Rassoulzadegan M, Ross A, Cuzin F, van Heyningen V, Hastie ND. Subnuclear localization of WT1 in splicing or transcription factor domains is regulated by alternative splicing. Cell. 1995;81:391–401. https://doi.org/10.1016/0092-8674(95)90392-5.

Laugwitz K-L, Moretti A, Lam J, Gruber P, Chen Y, Woodard S, Lin L-Z, Cai C-L, Lu MM, Reth M, Platoshyn O, Yuan JX-J, Evans S, Chien KR. Postnatal isl1+ cardioblasts enter fully differentiated cardiomyocyte lineages. Nature. 2005;433:647–53. https://doi.org/10.1038/nature03215.

Lepilina A, Coon AN, Kikuchi K, Holdway JE, Roberts RW, Burns CG, Poss KD. A dynamic epicardial injury response supports progenitor cell activity during zebrafish heart regeneration. Cell. 2006;127:607–19. https://doi.org/10.1016/j.cell.2006.08.052.

Lim J, Thiery JP. Epithelial-mesenchymal transitions: insights from development. Development. 2012;139:3471–86. https://doi.org/10.1242/dev.071209.

Limana F, Zacheo A, Mocini D, Mangoni A, Borsellino G, Diamantini A, Mori RD, Battistini L, Vigna E, Santini M, Loiaconi V, Pompilio G, Germani A, Capogrossi MC. Identification of myocardial and vascular precursor cells in human and mouse epicardium. Circ Res. 2007;101:1255–65. https://doi.org/10.1161/CIRCRESAHA.107.150755.

Limana F, Bertolami C, Mangoni A, Di Carlo A, Avitabile D, Mocini D, Iannelli P, De Mori R, Marchetti C, Pozzoli O, Gentili C, Zacheo A, Germani A, Capogrossi MC. Myocardial infarction induces embryonic reprogramming of epicardial c-kit+ cells: role of the pericardial fluid. J Mol Cell Cardiol. 2010;48:609–18. https://doi.org/10.1016/j.yjmcc.2009.11.008.

Limana F, Capogrossi MC, Germani A. The epicardium in cardiac repair: from the stem cell view. Pharmacol Ther Stem Cells. 2011;129:82–96. https://doi.org/10.1016/j.pharmthera.2010.09.002.

Little M, Wells C. A clinical overview of WT1 gene mutations. Hum Mutat. 1997;9:209–25. https://doi.org/10.1002/(SICI)1098-1004(1997)9:3<209::AID-HUMU2>3.0.CO;2-2.

Liu Z, Wang L, Welch JD, Ma H, Zhou Y, Vaseghi HR, Yu S, Wall JB, Alimohamadi S, Zheng M, Yin C, Shen W, Prins JF, Liu J, Qian L. Single-cell transcriptomics reconstructs fate conversion from fibroblast to cardiomyocyte. Nature. 2017;551:100–4., nature24454. https://doi.org/10.1038/nature24454.

Martínez-Estrada OM, Lettice LA, Essafi A, Guadix JA, Slight J, Velecela V, Hall E, Reichmann J, Devenney PS, Hohenstein P, Hosen N, Hill RE, Muñoz-Chapuli R, Hastie ND. Wt1 is required for cardiovascular progenitor cell formation through transcriptional control of Snail and E-cadherin. Nat Genet. 2010;42:89. https://doi.org/10.1038/ng.494.

Masters M, Riley PR. The epicardium signals the way towards heart regeneration. Stem Cell Res., Heart Regeneration and Rejuvenation. 2014;13:683–92. https://doi.org/10.1016/j.scr.2014.04.007.

Mayo MW, Wang C-Y, Drouin SS, Madrid LV, Marshall AF, Reed JC, Weissman BE, Baldwin AS. WT1 modulates apoptosis by transcriptionally upregulating the bcl-2 proto-oncogene. EMBO J. 1999;18:3990–4003. https://doi.org/10.1093/emboj/18.14.3990.

Moore AW, McInnes L, Kreidberg J, Hastie ND, Schedl A. YAC complementation shows a requirement for Wt1 in the development of epicardium, adrenal gland and throughout nephrogenesis. Development. 1999;126:1845–57.

Moore-Morris T, Cattaneo P, Puceat M, Evans SM. Origins of cardiac fibroblasts. J Mol Cell Cardiol. 2016;91:1–5. https://doi.org/10.1016/j.yjmcc.2015.12.031.

Morrison AA, Viney RL, Ladomery MR. The post-transcriptional roles of WT1, a multifunctional zinc-finger protein. Biochim Biophys Acta BBA Rev Cancer. 2008;1785:55–62. https://doi.org/10.1016/j.bbcan.2007.10.002.

Niksic M, Slight J, Sanford JR, Caceres JF, Hastie ND. The Wilms' tumour protein (WT1) shuttles between nucleus and cytoplasm and is present in functional polysomes. Hum Mol Genet. 2004;13:463–71. https://doi.org/10.1093/hmg/ddh040.

Orlic D, Kajstura J, Chimenti S, Jakoniuk I, Anderson SM, Li B, Pickel J, McKay R, Nadal-Ginard B, Bodine DM, Leri A, Anversa P. Bone marrow cells regenerate infarcted myocardium. Nature. 2001;410:701. https://doi.org/10.1038/35070587.

Parenti R, Salvatorelli L, Musumeci G, Parenti C, Giorlandino A, Motta F, Magro G. Wilms' tumor 1 (WT1) protein expression in human developing tissues. Acta Histochem., Immunomarkers in human developing and pediatric neoplastic tissues. 2015;117:386–96. https://doi.org/10.1016/j.acthis.2015.03.009.

Park S, Schalling M, Bernard A, Maheswaran S, Shipley GC, Roberts D, Fletcher J, Shipman R, Rheinwald J, Demetri G, Griffin J, Minden M, Housman DE, Haber DA. The Wilms tumour gene WT1 is expressed in murine mesoderm–derived tissues and mutated in a human mesothelioma. Nat Genet. 1993;4:415. https://doi.org/10.1038/ng0893-415.

Patek CE, Little MH, Fleming S, Miles C, Charlieu J-P, Clarke AR, Miyagawa K, Christie S, Doig J, Harrison DJ, Porteous DJ, Brookes AJ, Hooper ML, Hastie ND. A zinc finger truncation of murine WT1 results in the characteristic urogenital abnormalities of Denys– Drash syndrome. Proc Natl Acad Sci. 1999;96:2931–6. https://doi.org/10.1073/pnas.96.6.2931.

Rauscher FJ, Morris JF, Tournay OE, Cook DM, Curran T. Binding of the Wilms' tumor locus zinc finger protein to the EGR-1 consensus sequence. Science. 1990;250:1259–62. https://doi.org/10.1126/science.2244209.

Rudat C, Kispert A. Wt1 and epicardial fate mapping novelty and significance. Circ Res. 2012;111:165–9. https://doi.org/10.1161/CIRCRESAHA.112.273946.

Scharnhorst V, van der Eb AJ, Jochemsen AG. WT1 proteins: functions in growth and differentiation. Gene. 2001;273:141–61. https://doi.org/10.1016/S0378-1119(01)00593-5.

Scholz H, Kirschner KM. A role for the Wilms' tumor protein WT1 in organ development. Physiology. 2005;20:54–9. https://doi.org/10.1152/physiol.00048.2004.

Scholz H, Wagner K-D, Wagner N. Role of the Wilms' tumour transcription factor, Wt1, in blood vessel formation. Pflüg Arch Eur J Physiol. 2009;458:315–23. https://doi.org/10.1007/s00424-008-0621-3.

Silberstein GB, Horn KV, Strickland P, Roberts CT, Daniel CW. Altered expression of the WT1 Wilms tumor suppressor gene in human breast cancer. Proc Natl Acad Sci. 1997;94:8132–7.

Smart N, Riley PR. Derivation of epicardium-derived progenitor cells (EPDCs) from adult epicardium. In: Current protocols in stem cell biology. Wiley; 2007, https://doi.org/10.1002/9780470151808.sc02c02s8.

Smart N, Riley PR. The epicardium as a candidate for heart regeneration. Futur Cardiol. 2011;8:53–69. https://doi.org/10.2217/fca.11.87.

Smart N, Risebro CA, Melville AAD, Moses K, Schwartz RJ, Chien KR, Riley PR. Thymosin β4 induces adult epicardial progenitor mobilization and neovascularization. Nature. 2007;445:177. https://doi.org/10.1038/nature05383.

Smart N, Bollini S, Dubé KN, Vieira JM, Zhou B, Davidson S, Yellon D, Riegler J, Price AN, Lythgoe MF, Pu WT, Riley PR. De novo cardiomyocytes from within the activated adult heart after injury. Nature. 2011;474:640–4. https://doi.org/10.1038/nature10188.

Smart N, Bollini S, Dubé KN, Vieira JM, Zhou B, Riegler J, Price AN, Lythgoe MF, Davidson S, Yellon D, Pu WT, Riley PR. Myocardial regeneration: expanding the repertoire of thymosin β4 in the ischemic heart. Ann N Y Acad Sci. 2012;1269:92–101. https://doi.org/10.1111/j.1749-6632.2012.06708.x.

Smits AM, Riley PR. Epicardium-derived heart repair. J Dev Biol. 2014;2:84–100. https://doi.org/10.3390/jdb2020084.

Stoll R, Lee BM, Debler EW, Laity JH, Wilson IA, Dyson HJ, Wright PE. Structure of the Wilms tumor suppressor protein zinc finger domain bound to DNA. J Mol Biol. 2007;372:1227–45. https://doi.org/10.1016/j.jmb.2007.07.017.

Takeichi M, Nimura K, Mori M, Nakagami H, Kaneda Y. The transcription factors Tbx18 and Wt1 control the epicardial epithelial-mesenchymal transition through bi-directional regulation of slug in murine primary epicardial cells. PLoS One. 2013;8:e57829. https://doi.org/10.1371/journal.pone.0057829.

Temme S, Friebe D, Schmidt T, Poschmann G, Hesse J, Steckel B, Stühler K, Kunz M, Dandekar T, Ding Z, Akhyari P, Lichtenberg A, Schrader J. Genetic profiling and surface proteome analysis of human atrial stromal cells and rat ventricular epicardium-derived cells reveals novel insights into their cardiogenic potential. Stem Cell Res. 2017;25:183–90. https://doi.org/10.1016/j.scr.2017.11.006.

Thiery JP, Acloque H, Huang RYJ, Nieto MA. Epithelial-mesenchymal transitions in development and disease. Cell. 2009;139:871–90. https://doi.org/10.1016/j.cell.2009.11.007.

Tian X, Pu WT, Zhou B. Cellular origin and developmental program of coronary angiogenesis. Circ Res. 2015;116:515–30. https://doi.org/10.1161/CIRCRESAHA.116.305097.

Toska E, Roberts SGE. Mechanisms of transcriptional regulation by WT1 (Wilms' tumour 1). Biochem J. 2014;461:15–32. https://doi.org/10.1042/BJ20131587.

van Tuyn J, Atsma DE, Winter EM, van der Velde-van Dijke I, Pijnappels DA, Bax NAM, Knaän-Shanzer S, Gittenberger-de Groot AC, Poelmann RE, van der Laarse A, van der Wall EE, Schalij MJ, de Vries AAF. Epicardial cells of human adults can undergo an epithelial-to-mesenchymal transition and obtain characteristics of smooth muscle cells in vitro. Stem Cells. 2007;25:271–8. https://doi.org/10.1634/stemcells.2006-0366.

van Wijk B, Gunst QD, Moorman AFM, van den Hoff MJB. Cardiac regeneration from activated epicardium. PLoS One. 2012;7:e44692. https://doi.org/10.1371/journal.pone.0044692.

Velecela V, Lettice LA, Chau Y-Y, Slight J, Berry RL, Thornburn A, Gunst QD, van den Hoff M, Reina M, Martínez FO, Hastie ND, Martínez-Estrada OM. WT1 regulates the expression of inhibitory chemokines during heart development. Hum Mol Genet. 2013;22:5083–95. https://doi.org/10.1093/hmg/ddt358.

Vieira JM, Riley PR. Epicardium-derived cells: a new source of regenerative capacity. Heart. 2011;97:15–9. https://doi.org/10.1136/hrt.2010.193292.

Vieira JM, Howard S, del Campo CV, Bollini S, Dubé KN, Masters M, Barnette DN, Rohling M, Sun X, Hankins LE, Gavriouchkina D, Williams R, Metzger D, Chambon P, Sauka-Spengler T, Davies B, Riley PR. BRG1-SWI/SNF-dependent regulation of the *Wt1* transcriptional landscape mediates epicardial activity during heart development and disease. Nat Commun. 2017;8:16034. https://doi.org/10.1038/ncomms16034.

von Gise A, Pu WT. Endocardial and epicardial epithelial to mesenchymal transitions in heart development and disease. Circ Res. 2012;110:1628–45. https://doi.org/10.1161/CIRCRESAHA.111.259960.

von Gise A, Zhou B, Honor LB, Ma Q, Petryk A, Pu WT. WT1 regulates epicardial epithelial to mesenchymal transition through β-catenin and retinoic acid signaling pathways. Dev Biol. 2011;356:421–31. https://doi.org/10.1016/j.ydbio.2011.05.668.

Wagner N, Wagner KD, Schley G, Coupland SE, Heimann H, Grantyn R, Scholz H. The Wilms' tumor suppressor Wt1 is associated with the differentiation of retinoblastoma cells. Cell Growth Differ Mol Biol J Am Assoc Cancer Res. 2002;13:297–305.

Wagner K-D, Wagner N, Schedl A. The complex life of WT1. J Cell Sci. 2003a;116:1653–8. https://doi.org/10.1242/jcs.00405.

Wagner K-D, Wagner N, Wellmann S, Schley G, Bondke A, Theres H, Scholz H. Oxygen-regulated expression of the Wilms' tumor suppressor Wt1 involves hypoxia-inducible factor-1 (HIF-1). FASEB J. 2003b;17:1364–6. https://doi.org/10.1096/fj.02-1065fje.

Weeke-Klimp A, Bax NAM, Bellu AR, Winter EM, Vrolijk J, Plantinga J, Maas S, Brinker M, Mahtab EAF, Gittenberger-de Groot AC, van Luyn MJA, Harmsen MC, Lie-Venema H. Epicardium-derived cells enhance proliferation, cellular maturation and alignment of cardiomyocytes. J Mol Cell Cardiol. 2010;49:606–16. https://doi.org/10.1016/j.yjmcc.2010.07.007.

Wessels A, Pérez-Pomares JM. The epicardium and epicardially derived cells (EPDCs) as cardiac stem cells. Anat Rec A Discov Mol Cell Evol Biol. 2004;276A:43–57. https://doi.org/10.1002/ar.a.10129.

Wilhelm D, Englert C. The Wilms tumor suppressor WT1 regulates early gonad development by activation of Sf1. Genes Dev. 2002;16:1839–51. https://doi.org/10.1101/gad.220102.

Winter EM, Grauss RW, Hogers B, van Tuyn J, van der Geest R, Lie-Venema H, Steijn RV, Maas S, DeRuiter MC, deVries A a F, Steendijk P, Doevendans PA, van der Laarse A, Poelmann RE, Schalij MJ, Atsma DE, Groot ACG. Preservation of left ventricular function and attenuation of remodeling after transplantation of human epicardium-derived cells into the infarcted mouse heart. Circulation. 2007;116:917–27. https://doi.org/10.1161/CIRCULATIONAHA.106.668178.

Witty AD, Mihic A, Tam RY, Fisher SA, Mikryukov A, Shoichet MS, Li R-K, Kattman SJ, Keller G. Generation of the epicardial lineage from human pluripotent stem cells. Nat Biotechnol. 2014;32:1026. https://doi.org/10.1038/nbt.3002.

Wu SM, Fujiwara Y, Cibulsky SM, Clapham DE, Lien C, Schultheiss TM, Orkin SH. Developmental origin of a bipotential myocardial and smooth muscle cell precursor in the mammalian heart. Cell. 2006;127:1137–50. https://doi.org/10.1016/j.cell.2006.10.028.

Wu SM, Chien KR, Mummery C. Origins and fates of cardiovascular progenitor cells. Cell. 2008;132:537–43. https://doi.org/10.1016/j.cell.2008.02.002.

Ye Y, Raychaudhuri B, Gurney A, Campbell CE, Williams BR. Regulation of WT1 by phosphorylation: inhibition of DNA binding, alteration of transcriptional activity and cellular translocation. Regulation of WT1 by phosphorylation: inhibition of DNA binding, alteration of transcriptional activity and cellular translocation. EMBO J. 1996;15:5606–15.

Zhang Y, Mignone J, MacLellan WR. Cardiac regeneration and stem cells. Physiol Rev. 2015;95:1189–204. https://doi.org/10.1152/physrev.00021.2014.

Zhao J, Cao H, Tian L, Huo W, Zhai K, Wang P, Ji G, Ma Y. Efficient differentiation of TBX18+/WT1+ epicardial-like cells from human pluripotent stem cells using small molecular compounds. Stem Cells Dev. 2016;26:528–40. https://doi.org/10.1089/scd.2016.0208.

Zhou B, Pu WT. More than a cover: epicardium as a novel source of cardiac progenitor cells. Regen Med. 2008;3:633–5. https://doi.org/10.2217/17460751.3.5.633.

Zhou B, Pu WT. Genetic Cre-loxP assessment of epicardial cell fate using Wt1-driven Cre alleles. Circ Res. 2012a;111:e276–80. https://doi.org/10.1161/CIRCRESAHA.112.275784.

Zhou B, Pu WT. Isolation and characterization of embryonic and adult epicardium and epicardium-derived cells. In: Cardiovascular development, methods in molecular biology. Totowa: Humana Press; 2012b. pp. 155–168. https://doi.org/10.1007/978-1-61779-523-7_15.

Zhou B, Ma Q, Rajagopal S, Wu SM, Domian I, Rivera-Feliciano J, Jiang D, von Gise A, Ikeda S, Chien KR, Pu WT. Epicardial progenitors contribute to the cardiomyocyte lineage in the developing heart. Nature. 2008;454:109. https://doi.org/10.1038/nature07060.

Zhou B, von Gise A, Ma Q, Hu YW, Pu WT. Genetic fate mapping demonstrates contribution of epicardium-derived cells to the annulus fibrosis of the mammalian heart. Dev Biol. 2010;338:251–61. https://doi.org/10.1016/j.ydbio.2009.12.007.

Zhou B, Honor LB, He H, Ma Q, Oh J-H, Butterfield C, Lin R-Z, Melero-Martin JM, Dolmatova E, Duffy HS, von Gise A, Zhou P, Hu YW, Wang G, Zhang B, Wang L, Hall JL, Moses MA, McGowan FX, Pu WT. Adult mouse epicardium modulates myocardial injury by secreting paracrine factors. J Clin Invest. 2011;121:1894–904. https://doi.org/10.1172/JCI45529.

Zhou B, Honor LB, Ma Q, Oh J-H, Lin R-Z, Melero-Martin JM, von Gise A, Zhou P, Hu T, He L, Wu KH, Zhang H, Zhang Y, Pu WT. Thymosin beta 4 treatment after myocardial infarction does not reprogram epicardial cells into cardiomyocytes. J Mol Cell Cardiol. 2012;52:43–7. https://doi.org/10.1016/j.yjmcc.2011.08.020.

Chapter 7
Progress of Genomics in Hypertension–Cardiac Hypertrophy

Xiaoyan Wang and Yunzeng Zou

Abstract Hypertension is a common risk factor for cardiac hypertrophy, and the pathological process of hypertension to cardiac hypertrophy has been widely studied. Except for the classic proteins and its related signaling pathways involved in the process from hypertension to cardiac hypertrophy, the genetic basis and epigenetic mechanism have been hotpots in recent years. Herein, we provide a comprehensive overview of currently known genetic or epigenetic pathways involved in the progression from hypertension to cardiac hypertrophy.

Keywords Hypertension · Cardiac hypertrophy · Genetic · Epigenetic

1 Introduction

Hypertension is the important determinant of LV hypertrophy, which is considered a complex genetic disease, likely representing an interaction of several genes with the environment. Genes encoding proteins involved in LV structure, as well as genes encoding cell signal transduction, hormones, growth factors, calcium homeostasis, substrate metabolism, and blood pressure, are likely candidates for the development of common forms of LV hypertrophy. An overview of the pathophysiology of LV hypertrophy and dysfunction is provided, in addition to evidence of the genetic basis for LV hypertrophy in humans and animal models.

X. Wang · Y. Zou (✉)
Shanghai Institute of Cardiovascular Diseases, Zhongshan Hospital and Institutes of Biomedical Science, Fudan University, Shanghai, China
e-mail: zou.yunzeng@zs-hospital.sh.cn

H. Jiang, M. Liu (eds.), *Heart Genomics*, Translational Bioinformatics 16,
https://doi.org/10.1007/978-981-13-1429-2_7

Table 7.1 Categories of BP in adults

BP category	SBP		DBP
Normal	<120 mmHg	and	<80 mmHg
Elevated	120–129 mmHg	and	<80 mmHg
Hypertension			
Stage 1	130–139 mmHg	or	80–89 mmHg
Stage 2	≥140 mm Hg	or	≥90 mmHg

Individuals with SBP and DBP in two categories should be designated to the higher BP category. BP indicates blood pressure (based on an average of ≥2 careful readings obtained on ≥2 occasions, as detailed in Sect. 4)
DBP diastolic blood pressure and *SBP* systolic blood pressure

1.1 Epidemic Characteristics of Hypertension

Hypertension is an important global risk factor for cardiovascular diseases (Global Burden of Metabolic Risk Factors for Chronic Diseases Collaboration 2014), and the number of individuals with raised blood pressure has increased from 594 million in 1975 to 1.13 billion in 2015 (Zhou et al. 2017). Moreover, according to the new guidelines for hypertension published in the conference of AHA/ACC (Whelton et al. 2018), the definition of hypertension have decreased from 140 to 130 mmHg (systolic pressure) and from 90 to 80 mmHg (diastolic pressure) (Table 7.1), and the incidence of patients with hypertension or had self-reported antihypertensive medication has increased from 32% to 46%.

Hypertension can be categorized into primary hypertension and secondary hypertension. The pathogeneses of primary hypertension have been described by Lrivin Page and Arthur Guyton. The basis of the so-called Page's Mosaic developed by Lrivin Page described the association between biological and environmental factors (Harrison 2013; Page 1949); the biological factors here include genetic, anatomic, metabolic, nervous, and hemodynamic factors. This theoretical basis of the pathogeneses of primary hypertension indicates that the process from high BP to hypertension is influenced by multifactor, and although many new mechanisms involved in hypertension have been discovered, this theory is still effective (Harrison 2013).

Another theory developed by Arthur Guyton is based on the relationship between volume and peripheral vascular resistance. It explains the main mechanisms in regulating peripheral blood flow, along with adaptations of other neural, hormonal, and autacoid systems concerning the changes in resistance–volume relations. Moreover, it describes not only physiological regulations of BP and peripheral flow but also explains the pathogenesis of most of the secondary forms of arterial hypertension. Recent findings from experimental and clinical studies have suggested that metabolic abnormalities, sympathetic nervous system (SNS) alterations, immune activation, early arterial changes, and their interrelationships may in fact represent inciting events in the development of sustained primary hypertension.

The common causes for secondary hypertension include renovascular disease, primary aldosteronism, obstructive sleep apnea, and drug or alcohol induced. And the uncommon causes included pheochromocytoma/paraganglioma, Cushing's syndrome, hypothyroidism, aortic coarctation (undiagnosed or repaired), primary hyperparathyroidism, congenital adrenal hyperplasia, mineralocorticoid excess syndromes other than primary aldosteronism, and acromegaly. Treatment of the causes for secondary hypertension can often correct the level of blood pressure.

1.2 *Pathological Process of Hypertension to Cardiac Hypertrophy*

Cardiac hypertrophy is a cardinal manifestation of preclinical cardiovascular disease (Devereux and Alderman 1993) that strongly predicts myocardial infarction (MI), stroke, and cardiovascular death in patients with hypertension (Levy et al. 1990; Bikkina et al. 1994). In the beginning stages, cardiac hypertrophy is an adaptive response that reduces wall stress in an attempt to augment or preserve cardiac function (Frey et al. 2004). Sustained cardiac hypertrophy in response to pathologic insults is maladaptive and can eventually lead to systolic insufficiency and heart failure characterized by increased rates of myocyte apoptosis, interstitial fibrosis, and ventricular dilation (Frey et al. 2004).

With the development of molecular biology, more studies have focused on the signaling pathways of cell size expansion and apoptosis; some of which are related to cardiac hypertrophy, such as the mitogen-activated protein kinase (MAPK) (Ravingerová et al. 2003; Sopontammarak et al. 2005), phosphatidylinositol 3-kinase/AKT (Sopontammarak et al. 2005), and nuclear factor-κB (Young et al. 2008) pathways. There is evidence that cardiac hypertrophy had a close relationship with cardiomyocyte metabolism. For instance, Ca^{2+} plays a crucial role in the strictly regulated supply of ATP to meet the energy requirements of the cardiac myofibrils (Balaban 2009). This indicates that cardiac hypertrophy is closely related to body metabolism and signaling pathway changes.

Hypertension being inherited in families provides an important clue for the presence of susceptibility factors on our genomes that predispose some of us to develop high blood pressure. During the last 25 years, there has been enormous growth in the availability and quality of tools to permit genetic analysis, such tools emerged rapidly for human, mouse, and rat. These tools include a representative genome sequence (Gibbs et al. 2004; Worley et al. 2008) that provides an assembly against which next-generation, short-read sequence of additional rat strains (Atanur et al. 2013), including spontaneously hypertensive rat (SHR) and so on. The progress in those gene tool have moved the genetic study about hypertensive from the early development of polymorphic simple sequence length polymorphisms (SSLP) useful for genetic mapping (Gigante et al. 2001) to a large collection of genome-wide, stable single nucleotide polymorphisms (Consortium et al. 2008) and onward to include a comprehensive inventory of rat protein-coding gene sequences (Shimoyama et al. 2015), expressed proteins (Low et al. 2013), and rat noncoding

RNAs (Gopalakrishnan et al. 2015), as well as high-coverage genome sequences for several SHR and numerous other inbred rat lines. In addition, large-scale human population genetic studies have continued to seek progress in identifying loci, genetic variation, and functional pathways associated with blood pressure levels in humans. The progress includes the discovery of a large number of replicable blood pressure loci, some associated with genes that function in known pathways by which cardiovascular function is pharmacologically influenced.

Moreover, a growing number of studies indicate that epigenetic mechanisms also involved in the process from hypertension to cardiac hypertrophy. Epigenetics refers to mitotically and meiotically stable (heritable) and functionally relevant DNA and chromatin modifications that are not caused by alterations or mutations in the primary DNA sequence itself (Waddington 2012). They can be either inherited or accumulated throughout a lifetime. Furthermore, the epigenetic state of a cell is malleable, it evolves in an ordered manner during the cellular differentiation and development of an organism, and epigenetic changes are responsible for cellular plasticity that enables cellular reprogramming and response to the environment. More specifically, by altering the chromatin architecture and the accessibility of DNA coding and regulatory regions, they orchestrate the spatiotemporal gene expression in a cell-type and even allele-specific (maternally or paternally imprinted loci) manner (Bonasio et al. 2010).

At the molecular level, epigenetic regulation involves four major types of epigenetic mechanisms: DNA methylation and hydroxymethylation, covalent histone modification and incorporation of histone variants, ATP-dependent chromatin remodeling, and chromatin and gene regulation by noncoding RNAs (ncRNAs) including small microRNAs (microRNAs/miRs), long noncoding RNAs (lncRNAs), and circular RNAs (circRNAs) (Kim et al. 2009; Esteller 2011; Zink and Hake 2016; Rizki and Boyer 2015; Narlikar et al. 2013; Arrowsmith et al. 2012). These mutually interdependent epigenetic alterations, collectively named the epigenome, have profound effects on cellular repertoire of active genes.

Herein, we provide a comprehensive overview of currently known genetic or epigenetic pathways involved in the progression from hypertension to cardiac hypertrophy.

1.2.1 Genetic Basis

Hypertension being inherited in families serves as an important clue for the presence of susceptibility factors on our genomes that predispose some of us but not others to develop high BP. After the year of 1953, in which Watson and Crick made the landmark discovery of the molecular structure of DNA (Watson and Crick 1953a, b), a great number of correlative observations were made, paving the fundamental insights in to key concepts that sealed our understanding of RNA as the genetic material that is inherited. Physiologists interested in the etiology of hypertension began to look into genes related to hypertension since the definitions of what constitutes a "gene" were being worked on in the 1950s to the 1980s by molecular biologists.

With the development in gene technology, transgenic experiments in mice and manipulations of physiological systems are very informative on a physiological level because they provide information on how those candidate genes influence BP in the context of the whole animal with all its regulatory and compensatory systems intact. What's more, more precise evidence for genetic factors to influence BP came from correlative observation of a history of hypertension in parents of hypertensive subjects (Mo et al. 1995; Shear et al. 1986, 1987; Hunt et al. 1991). Despite the questions about additional shared nongenetic, environment factors such as food habits within a family that could influence blood pressure, evidence was obtained through population studies wherein a greater concordance of blood pressure was observed within families than between families (Longini et al. 1984), indicating that gene plays an important role in hypertension.

At the molecular level, the onset of cardiac hypertrophy and heart failure is associated with reprogramming of gene expression, including downregulation of the alpha-myosin heavy chain (α-MHC) gene and sarcoplasmic reticulum Ca^{2+} ATPase genes and reactivation of specific fetal cardiac genes such as atrial natriuretic factor (ANF) and brain natriuretic peptide (BNP). These deviations in gene expression result in structural and electrophysiological changes, which eventually progress to HF.

Here, we reviewed the theoretical basis for the genetic basis and recent progress in understanding genetic hypertension and cardiac hypertrophy.

1.3 Genes Involved in Hypertension

The evidence for single gene inheritance originally came from Weitz (1923) and was based on the assumption that a line can be drawn between normal and abnormal pressures at 160 mmHg. This conception was supported by Platt who concluded from his studies of hypertension families that primary hypertension was a heritable disease in 1947 (Platt 1947). However, other researchers, most notably Pickering, had a long controversy over the nature of inheritance and the extent to which it contributes to the pressures found in essential hypertension. Pickering provided evidence that hypertension is a quantitative trait, supporting multifactorial genetic inheritance of blood pressure; he also suggested a role of environmental factors. Blood pressure is a moderately hereditary trait (30–50%, although such estimates may be inflated) with a multifactorial pattern of inheritance in the general population (Padmanabhan et al. 2015). Today, blood pressure is recognized to exhibit a near normal distribution within the general population, which hints at the fact that blood pressure is determined by cumulation of several small-scale factors (Arora and Newton-Cheh 2010).

Early insights in the genetic investigation are derived from the study of twins, stating in the early twentieth century. Environmental cofounders in "Twins study" are minimized because twin children are usually exposed to similar environments, and blood pressure was first shown to be heritable in a twin study. The advent of molecular genetics has made twin studies more useful than ever because of the power of quantitative trait loci analyses (Luft 2001). By thorough physiological examination of twins, researchers have demonstrated that the pathological processes of hypertension have correlation with genetic information in different populations (Rao et al. 1993).

The rapid rise of genetics and genomics has made it possible to make further genetic analysis on hypertension. Early studies relied on linkage analysis to map Mendelian disease genes, particularly in large families with high penetrance, a concept that had already been described for drosophila in 1913. Over the last decade, the advent of next-generation sequencing has enabled the identification of additional high frequency of de novo variants, a small number of families, or locus heterogeneity (Warren et al. 2017). Since 2009, numerous genome-wide association studies (GWSA) on hypertension have been published.

The CHARGE Consortium conducted by Levy identified four loci attained genome-wide significance for SBP (ATP2B1, CYP17A1, PLEKAH7, SH2B3), six for DBP (ATP2B1, CACNB2, CSK-ULK3, SH2B3, TBX3-TBX5, ULK4), and one for hypertension (ATP2B1) in a meta-analysis of the CHARGE and Global BPgen cohorts ($n = 63{,}569$) (Levy et al. 2009). The effect size of the identified variants was small; the proportion of BP variation explained by the top 10 single nucleotide polymorphisms (SNPs) associated with systolic and diastolic BP was estimated to be 1%. ATP2B1 encodes a plasma membrane calcium ATPase isoform, which is expressed in vascular endothelium and removes calcium from the cell, CACNB2 encodes a voltage-gated calcium channel beta subunit, and CYP17A1 encodes an enzyme involved in adrenal mineralocorticoid and glucocorticoid biosynthesis. Christopher Newton-Cheh et al. (2009) tested 2.5 million genotyped and imputed SNPs for association with systolic and diastolic blood pressure in 34,433 subjects of European ancestry from the Global BPgen consortium and followed up findings with direct genotyping and in silico comparison. And finally they identified association between SBP or DBP and common variants in eight regions near the CYP17A1, CYP1A2, FGF5, SH2B3, MTHFR, c10orf107, ZNF652, and PLCD3 genes. Wain et al. (2011) conducted a genome-wide association meta-analysis of pulse pressure (PP) and mean arterial pressure (MAP) in 74,064 individuals of European ancestry from 35 studies. In this study, they identified four new PP loci (at 4q12 near CHIC2, 7q22.3 near PIK3CG, 8q24.12 in NOV, and 11q24.3 near ADAMTS8), two new MAP loci (3p21.31 in MAP4 and 10q25.3 near ADRB1), and one locus associated with both of these traits (2q24.3 near FIGN) that has also recently been associated with SBP in east Asians. A genome-wide association study in 200,000 individuals of European descent identified 16 novel loci: 6 of these loci contain genes previously known or suspected to regulate blood pressure (GUCY1A3-GUCY1B3, NPR3-C5orf23, ADM, FURIN-FES, GOSR2, GNAS-EDN3); the other 10 provide new clues to blood pressure physiology. Moreover, a variant near ALDH2 (alcohol dehydrogenase) was associated with BP (East Asian specific), and this association was presumably mediated by alcohol tolerance and intake.

Despite of the studies conducted in populations of European descent, Norihiro Kato et al. carried out a study in east Asian ancestry (Kato et al. 2011), providing more information about genetic variants on blood pressure in specific ethnic groups.

This study identified genome-wide significant associations with SBP and DBP, which included variants at four new loci (ST7L-CAPZA1, FIGN-GRB14, ENPEP, and NPR3) and a newly discovered variant near TBX3; all except NPR3 replicated in independent samples. They also confirmed seven loci previously identified in population of European descent. Moreover, a variant near ALDH2 (alcohol dehydrogenase) was associated with BP (East Asian specific), and this association was presumably mediated by alcohol tolerance and intake. In addition, to identify novel susceptibility loci for blood pressure in people of African ancestry, Jingjing Liang et al. (2017) performed both single- and multiple-trait genome-wide association analyses. They analyzed 21 genome-wide association studies comprised of 31,968 individuals of African ancestry and validated their results with additional 54,395 individuals from multiethnic studies. The results demonstrated that four loci (TARID/TCF21, LLPH/TMBIM4, FRMD3, GRP20/CDH17) associated variants had allels common only in African ancestry population. For patients with Hispanic/Latino background, investigators carried out a GWAS of blood pressure in 12,278 participants from Hispanics Community Health Study/Study of Latinos (HCHS/SOL) (Sofer et al. 2017), and they identified two potential novel loci for SBP and DBP and identified three potential loci in the Caribbean subgroup. The results of this study showed heterogeneity of effects across different subgroups and validated several established loci for BP traits identified in GWAS of different races.

Several additional GWAS were published in this year. A meta-analysis of 150,134 individuals from 54 GWAS of European ancestry with imputation based on the 1000 Genomes Project Phase 1 was carried out to discover additional novel blood pressure loci (Wain et al. 2017). The researchers reported six new signals of association in or near HSPB7, TNXB, LRP12, LOC283335, SEPT9, and AKT2 and provide new replication evidence for a further two signals in EBF2 and NFKBIA. They further combined large whole-blood gene expression resources totaling 12,607 individuals and identified 48 genes with evidence for involvement in blood pressure regulation that are significant in multiple resources. They also reported three signals that were unique to the kidney and previously reported (C4orf34, HIP2, and ASIC1) and confirmed a previously reported kidney eQTL signal for an antisense RNA for PSMD5. Louise V wain et al. (Warren et al. 2017) reported genetic association of blood pressure (systolic, diastolic, pulse pressure) among UK Biobank participants of European ancestry with independent replication in other cohorts, and robust validation of 107 independent loci. They also identify new independent variants at previously reported blood pressure loci. In combination with results from a range of in silico functional analyses and wet bench experiments, our findings highlight new biological pathways for blood pressure regulation enriched for genes expressed in vascular tissues and identify potential therapeutic targets for hypertension. Results from genetic risk score models raise the possibility of a precision medicine approach through early lifestyle intervention to offset the impact of blood pressure–raising genetic variants on future cardiovascular disease risk. Overall, those enhanced discovery of BP loci increases the potential targets for therapeutic advances.

1.4 Genes Involved in Cardiac Hypertrophy

1.4.1 Receptor Interacting Protein-2 (RIP2)

Receptor interacting protein-2 (RIP2) plays an essential role in host defense (Pellegrini et al. 2017; Tigno-Aranjuez and Abbott 1823; Rahman et al. 2014). It is capable of inducing NF-kappa B activation and cell death (Jeong et al. 2014). It is reported that RIP2 played an essential role in regulating the pathologic cardiac hypertrophy, inflammation, and fibrosis in animals and cells (Zhao et al. 2017). In this study, mice were subjected to aortic banding (AB) surgery to explore the pathological, echocardiographic, and molecular mechanisms. RIP2 expressed highly in cardiomyocytes after AB operation in wild-type (WT) mice. RIP2 knockout (KO) attenuated cardiac hypertrophy, inflammation, and fibrosis in mice 4 weeks after AB surgery. First, RIP2 knockout downregulated hypertrophic markers of atrial natriuretic peptide (ANP), brain natriuretic peptide (BNP), and b-myosin heavy chain (b-MHC) in the heart of AB-operated mice. In addition, RIP2-deficiency reduced toll-like receptor 4/myeloid differentiation factor 88/nuclear factor-kappa B (TLR4/MyD88/NF-kB) activation, mitogen-activated protein kinases (MAPKs) phosphorylation, and transforming growth factor-b1 (TGF-b1)/SMADs expressions, contributing to the suppression of inflammatory response and fibrosis, as further evidenced by downregulated pro-inflammatory cytokines, including tumor necrosis factor-a (TNF-a), interleukin (IL)-1b, IL-6, and IL-18, as well as fibrosis markers of collagen I, collagen III, and a-smooth muscle actin (a-SMA). Taken together, these data indicated that RIP2-deficience ameliorated cardiac hypertrophy, inflammation, and fibrosis through modulating multiple signaling pathways.

1.5 TRPV1

Transient receptor potential vanilloid type 1 (TRPV 1) is a 6-transmembrane tetrameric nonselective cation channel found mainly in the nociceptive neurons of the peripheral nervous system. TRPV 1 is activated by a wide variety of physical and chemical stimuli, such as capsaicin (CAP) and temperatures >43 °C, leading to a painful sensation. In addition to the peripheral sensory neurons, TRPV1 is found in the cardiovascular system, for example, cardiomyocytes, cardiac blood vessels, and coronary endothelial cells.

Previous results showed that increased TRPV1 expression was associated with cardiac hypertrophy (Zhong and Wang 2009; Buckley and Stokes 2011). Furthermore, functional knockout of TRPV1 and administration of a TRPV1 antagonist can improve heart function and reduce hypertrophic, fibrotic, and apoptotic markers in cardiac hypertrophy models (Buckley and Stokes 2011; Zhong and Wang 2007). It is suggested that multiple signals and inputs, such as pH, endogenous anandamide, protein kinase C phosphorylation, mechanosensitivity, and baroreflex, which activate TRPV1, are also associated with cardiohypertrophic conditions

(Vaughan-Jones et al. 2009; Malinowska et al. 2012; Studer and McNaughton 2010; Yamazaki and Yazaki 2000; Inoue et al. 2009). In addition, Ca2+ CaMKII, extracellular signal-regulated kinases, and p38 and Serca2a were shown to involved in TRPV1 activation-induced cardiac hypertrophy (Chen et al. 2016).

1.6 SERCA2

Diastolic dysfunction is commonly observed in humans with long-standing pressure overload as a result of chronic aortic stenosis or systemic hypertension. Patients with this form of heart failure manifest slowed isovolumic relaxation, delayed rapid filling, and increased myocardial stiffness. The mechanisms responsible for diastolic dysfunction in pressure overload may include structural and/or functional abnormalities in cardiac myocyte intracellular Ca^{2+} transport, alterations in intracellular Ca^{2+} transport, alterations in extracellular matrix collagen content, or simply altered mechanical properties of the hypertrophied ventricular muscle itself (Buttrick et al. 1994).

Reduced Ca^{2+} transport by the sarcoplasmic reticulum (SR) has been suggested as a cause for altered myocardial relaxation in the pressure-overloaded rodent heart. It has been accepted that reduced SR Ca^{2+} pump activity in severe left ventricular hypertrophy (LVH) resulted from a reduction in the density of functioning SR Ca^{2+}-adenosine triphosphatase (ATPase) (SERCA2) molecules, which in turn resulted from decreased expression of the SERCA2 gene (Buttrick et al. 1994; de la Bastie et al. 1990; Feldman et al. 1993).

In man, three genes (ATP2A1-3) generate multiple isoforms of SERCA through alternative splicing (SERCA1a and b, SERCA2a-c, and SERCA3a-f). SERCA2a is the predominant isoform in cardiomyocytes. In both failing human myocardium and various animal models of heart failure, mRNA levels (Qi et al. 1997; Zarain-Herzberg et al. 1996; Cadre et al. 1998) and activity of SERCA2a are reduced (Hasenfuss et al. 1994; Schwinger et al. 1995; Flesch et al. 1996). Further support for the importance of SERCA2a activity in heart failure comes from the following observations: (1) a SERCA2a knockout mouse develops systolic and diastolic dysfunction in vivo (Andersson et al. 2009), (2) inhibition of SERCA2a in isolated human myocytes creates a heart failure phenotype (Davia et al. 1997), and (3) increasing the activity of SERCA2a with gene therapy improves important cardiac parameters in heart failure (see below).

Recognition of the pivotal role that Ca2+ and SERCA2a play in normal physiology and in the development of heart failure has led to SERCA2a gene therapy being the vanguard for clinical trials of gene therapy in heart failure. The convergence of the acquired phenotype from different etiologies of heart failure suggested that targeting this deficit in calcium handling could correct a final common pathway of disease.

Despite those genes, previous reports have indicated that under hypertrophic stimuli such as pressure overload, a so-called fetal gene program is induced

(Parker et al. 1990) that includes the downregulation of adult isoforms of several cardiac proteins and the upregulation of genes such as beta-myosin heavy chain (MHC), skeletal alpha-actin, atrial natriuretic factor (ANF), the GLUT1 glucose transporter, and aMLC-1 (Parker et al. 1990; Izumo et al. 1987; Izumo and Nadal-Ginard 1988; Morano et al. 1997; Razeghi et al. 2001; Schwartz et al. 1986).

1.7 Genes Involved in Hypertrophic Cardiomyopathy

Adult-onset hypertrophic cardiomyopathy is a prevalent genetic condition caused by inherited or new mutations in genes that encode sarcomere proteins, including cardiac β-myosin heavy chain (MYH7), cardiac myosin-binding protein C (MYBPC3), cardiac troponin T (TNNT2), cardiac troponin I (TNNI3), essential myosin light chain (MYL3), regulatory myosin light chain (MYL2), α-tropomyosin (TPM1), cardiac actin (ACTC), and titin (TTN) (Seidman and Seidman 2001). Less commonly, hypertrophic cardiomyopathy is caused by mutations in other genes (Lind et al. 2006; Osio et al. 2007). The range of ages at clinical diagnosis of hypertrophic cardiomyopathy is broad; however, manifestations before 14 years of age are atypical, even in children with an inherited gene mutation (Maron et al. 2003).

Mutations affecting the γ2 regulatory subunit of AMP-activated protein kinase (PRKAG2) (which regulates substrate use for energy production) cause early-onset left ventricular hypertrophy with arrhythmias (Arad et al. 2002) and, more rarely, fatal infantile cardiac glycogenosis (Burwinkel et al. 2005). Mutations in the gene encoding lysosome-associated membrane protein 2 (LAMP2), which is located on the X chromosome, cause massive left ventricular hypertrophy in boys in whom systemic manifestations (the Danon disease) may also develop (Arad et al. 2005). Because LAMP2 mutations are usually clinically silent in female carriers, affected boys appear to have sporadic rather than inherited disease.

1.7.1 Chromatin Remodeling, Modification of DNA, and Histones in Hypertension and Cardiac Hypertrophy

The genetic information is encoded by DNA sequence, and epigenetic alternations of chromatin can regulate the expression of genetic information. Chromatin is evolved to associate DNA with histone proteins, providing a scaffold to package the long 1.7-m. DNA strands into a small 6-μm nucleus. The basic packaging unit of chromatin is the nucleosome, which contains ~147 bp DNA enveloping an octamer core of histones (two molecules of each of the canonical H2A, H2B, H3, and H4 histones). Chemical interactions between histones and DNA, as well as the position of nucleosomes in the genome, can alter the accessibility of DNA sequence to transcription regulation. These changes of chromatin are mediated by three major mechanisms: covalent modifications of histone protein side chains, covalent modifications of DNA bases, and non-covalent changes of the composition and positioning of nucleosomes (Clapier and Cairns 2009; Li 2002; Yang and Seto 2008). Such covalent and non-covalent changes of nucleosomes enable the chromatin scaffold to respond to environmental signals to modulate gene expression and alter the physiological state of cells.

Histones can be modified by various histone-modifying enzymes that catalyze methylation, acetylation, or other chemical processes, whereas the cytosine base of DNA can be methylated by DNA methyltransferases. On the other hand, the position and histone composition (canonical vs noncanonical histones) of nucleosomes are regulated by a group of ATP-dependent nucleosome/chromatin-remodeling complexes. A critical component of the chromatin-remodeling complex is the ATPase subunit, which hydrolyzes ATP to mobilize nucleosomes or change their histone composition. In vertebrates, there are four families of chromatin-remodeling complexes defined by the evolutionarily conserved ATPase subunit: SWI/SNF (Switch defective/sucrose nonfermenting), ISWI (imitation switch), CHD (chromodomain, helicase, DNA binding), and INO80 (inositol requiring 80) (Clapier and Cairns 2009; Li 2002; Yang and Seto 2008; Ho and Crabtree 2010). Each ATPase protein consists of a conserved catalytic domain (for ATP hydrolysis) and various functional domains that recognize modified histones, regulate ATP catalytic activity, or mediate protein–protein interactions. These specific domains characterize the ATPase from each of the four families of chromatin-remodeling complexes.

1.8 Chromatin Remodeling Through ATP-Dependent Enzymes

ATP-dependent factors that participate in chromatin assembly include Chd1 (chromo-ATPase/helicase DNA-binding protein 1), ATRX (alpha thalassemia/mental retardation syndrome X-linked), and several ISWI (imitation switch)-containing complexes, such as ACF (ATP-utilizing chromatin assembly and remodeling factor), CHRAC (chromatin accessibility complex), RSF (remodeling and spacing factor), and ToRC (Toutatis-containing chromatin remodeling complex) (Ito et al. 1997; Varga-Weisz et al. 1997; Loyola et al. 2001; Lusser et al. 2005; Lewis et al. 2010; Emelyanov et al. 2012). These motor proteins exhibit both chromatin assembly and remodeling activities and are members of the SNF2 (sucrose nonfermenting 2) protein family, which comprises the ATPases that are known to be involved in chromatin remodeling (Clapier and Cairns 2009; Flaus and Owen-Hughes 2011; Hargreaves and Crabtree 2011; Ryan and Owen-Hughes 2011).

1.9 DNA Methylation

DNA methylation is the most common epigenetic modification in the mammalian genome. Three DNA methyltransferases (DNMTs) have been identified in mammals (DNMT1, 3A, 3B). The epigenetic profile undergoes alterations in end-stage cardiomyopathy. Increased expression of certain genes in the failing heart is associated with promoter demethylation, but hypermethylation does not show a clear correlation with reduced gene expression (Movassagh et al. 2011). Significant hypomethylation at satellite regions has also been found in failing hearts and is correlated with a 27-fold increase of the corresponding transcripts (Haider et al. 2012). Accordingly, further studies are warranted to determine the role of alterations of DNA methylation in the progression of heart failure.

DNA methylation is, generally, attributed to gene silencing by hampering the accessibility of cis-DNA binding elements present in the promoter regions of genes of the transcriptional machinery (Suzuki and Bird 2008) and plays a crucial role in the regulation of chromatin structure including X chromosome inactivation, genomic imprinting, silencing of repetitive DNA elements, and transposon transcription (Li 2002; Li et al. 1993; Panning and Jaenisch 1996). Moreover, DNA methylation has been linked to biological processes underlying various diseases from cancer (Feinberg and Tycko 2004) to CVD, such as hypertension (Millis 2011), diabetes (Ling and Groop 2009), atherosclerosis, and inflammation (Wierda et al. 2010).

1.10 Histone Modification

The eukaryotic DNA is tightly compact and organized in chromatin. The nucleosome is the central unit of chromatin and is composed of an octomer center of two copies of each histone protein (H2A, H2B, H3, and H4) (Jenuwein and Allis 2001) around which a DNA segment of 14–150 base pairs is looped.

Each histone has an amino-terminal tail that protrudes from the surface of the nucleosome and which can be subjected to various posttranscriptional modifications such as phosphorylation, sumoylation, ubiquitination, methylation, ADP-ribosylation, proline isomerization, deimination, and acetylation (Handy et al. 2011). These modifications lead to conformational changes in the chromatin resulting in altered gene expression (Margueron and Reinberg 2010) depending on whether DNA becomes accessible (euchromatin) or inaccessible (heterochromatin) for transcription.

1.11 Histone Acetylation

Histone acetylation occurs at the lysine residues of the histone tails resulting in decondensation of the chromatin structure and serving as a binding site for bromodomain proteins and transcriptional activators, and eventually leading to transcriptional activation (Elliott and McKenna 2004; Rodriguez et al. 2009). Conversely, histone deacetylation induces chromatin condensation and therefore transcriptional repression (Clayton et al. 2006; Shahbazian and Grunstein 2007). Acetylation of histones is a dynamic process mediated by two counteracting enzyme families, the histone acetyltransferases (HATs) and histone deacetylases (HDACs). The harmony between the activities of these two sets of enzymes is a crucial element during regulation of gene expression, and its deregulation is linked to several pathological conditions varying from cancer to CVD (Ordovás and Smith 2010; Burgess 2012).

Acetylation of histones is regulated by histone acetyltransferases (HATs) and histone deacetylases (HDACs). Over-activation of HATs, such as CREB-binding protein (CBP) and p300, induces hypertrophy and left ventricular remodeling in mice (Gusterson et al. 2003; Yanazume et al. 2003). Hypertrophy of non-necrotic tissue in a myocardial infarction model also requires p300 HAT activity (Miyamoto et al. 2006), and phenylephrine-induced cardiac hypertrophy is associated with high p300 activity. HDACs have an opposing role to HATs by mediating the removal of acetyl groups. There are three main classes of HDACs, which are class I (HDAC1, 2, 3, and 8), class II (HDAC4, 5, 6, 7, 9, and 10), and class III. Class II HDACs have an anti-hypertrophic effect (Lu et al. 2000). In mice, depletion of HDAC5 or HDAC9 was reported to increase sensitivity to stress signals and enhance cardiac hypertrophy, owing to the role of these enzymes in silencing MEF2C (Chang et al. 2004). Recently, an increased load on the heart was shown to induce nuclear export of HDAC4, demethylation of H3K9, and activation of ANP, with these responses not requiring an increase of histone acetylation (Hohl et al. 2013). Unlike class II HDACs, the class I HDACs mediate hypertrophic responses. HDAC2 induces cardiac hypertrophy by repression of INPP5F, the gene coding for phosphatidylinositide phosphatase SAC2, which is a negative regulator of the Akt/GSK-3β pathway (Trivedi et al. 2007). Similar to class II HDACs, class III HDACs are associated with inhibition of cardiac hypertrophy and with improved cardiomyocyte survival. Modulation of HDAC expression, including suppression of specific prohypertrophic HDACs, has the potential to become next-generation therapy for heart failure.

Acetylation of histones on lysine residues promotes the relaxation of chromatin, leading to transcriptional activation, while suppression of histone acetylation leads to condensation of chromatin and inhibition of gene expression. Active promoters are characterized by high levels of monoacetylated lysine 9 and lysine 14 (H3K9ac and H3K14ac) and trimethylated (me3) lysine 4 (H3K4me3) on histone H3. Inactive promoters are characterized by methylated DNA, trimethylation of lysine 9 and lysine 27 on histone H3 (H3K9me3 and H3K27me3), and deacetylation of histones. Active enhancers are enriched in H3K27ac, while lack of H3K27ac and a high level of H3K27me3 define an enhancer as "poised."

2 Histone Methylation

Another key modulator of posttranslational regulation is histone methylation which can occur on all basic amino acid residues of the histone tail: arginines, lysine, and histidines (Cheung and Lau 2005). In addition, different amino acids can be methylated to a different extent, and while lysine can be subjected to mono-, di-, and trimethylation, arginine residues can only become mono- or dimethylated (Cheung and Lau 2005). Methylation of histones is a dynamic process mediated by histone methyltransferases (HMTs) and histone demethylases (HDMs), and, unlike acetylation, histone methylation can induce either activation or repression of gene expression depending on the target sites and degree of methylation (Lachner and Jenuwein 2002). In contrast to histone acetylation, histone methylation governed mainly by HMTs SUV39H1 and G9a (Martin and Zhang 2005; Shi and Whetstine 2007) has long been considered to be a permanent epigenetic mark (Jenuwein and Allis 2001). However, the discovery of new players such as HDMs has shifted the paradigm, and, in fact, several studies showed that histone methylation is tightly regulated in inflammatory and metabolic disorders (Doerks et al. 2002; Villeneuve et al. 2008; Brasacchio et al. 2009).

2.1 *MicroRNAs*

2.1.1 The Biology of MicroRNA

MicroRNAs (miRs) are endogenous, single-stranded, short RNA sequences (22 nucleotides (Ambros 2004)), more than 2500 miRNAs have been annotated in humans (miRBase v.21, September 2016) (Griffiths-Jones et al. 2008), and considerable details have been known about their biogenesis (Lin and Gregory 2015), mechanisms of action, and functions (Jonas and Izaurralde 2015). miRNAs are small noncoding RNAs which belongs to a class of small silencing RNAs that are critically important in the posttranscriptional regulation of genes with which it shares extensive sequence complementarity (Lee et al. 1993), serving as the target recognition element of an RNA-protein complex known as the RNA-induced silencing complex (RISC), which contains an Argonaute (AGO) family protein that binds the miRNA, along with a range of accessory components (Kawamata and Tomari 2010).

The 5′ end of the miRNA (the seed region) forms the major target recognition element (Lai 2002; Lewis et al. 2005) and is of such importance that miRNAs are grouped into families on the basis of shared seed sequences. Structural studies show that the 5′ end of the miRNA is stacked in the AGO protein such that the Watson–Crick edges of bases (Lin and Gregory 2015; Jonas and Izaurralde 2015; Kawamata and Tomari 2010; Lai 2002; Wang et al. 2008) are ideally positioned for interaction with target mRNAs (Wang et al. 2008; Elkayam et al. 2012; Schirle et al. 2014). When bound to a target mRNA, the RISC complex reduces the rate of translation of the mRNA and accelerates the shortening of the poly(A) tail, resulting in faster mRNA degradation (Jonas and Izaurralde 2015).

MicroRNA has been demonstrated to play an important role in the differentiation and development of many cells and tissues, including the heart (Ohtani and Dimmeler 2011). In the cardiovascular field, microRNAs are acknowledged as fundamental in regulating the expression of genes that governs physiological and pathological myocardial adaptation to stress. Here, we review recent progress in the understanding of their role in hypertension and cardiac hypertrophy.

2.1.2 MicroRNAs Involved in Hypertension and Cardiac Hypertrophy

Blood pressure is regulated by an ensemble of cardiac, renal, vascular, and neurohormonal mechanisms (Chopra et al. 2011). The pathogenesis of hypertension is complex, involving ECs and VSMCs dysfunction and altered activation of RAAS and SNS together with increase in oxidative stress and inflammation (Klimczak et al. 2017). Previous animal and human studies indicate that miRNAs contribute to the regulation of these aforementioned mechanisms, representing a new promising potential target for therapies.

Endothelial cells are subjected to shear stress exerted by circulating blood on the vessel wall. They are able to receive mechanical stimuli and transform them into intracellular signals, inducing alteration of cellular structure and function. This adaptation is mediated by miRNA. For instance, miR-19a leads to an antiproliferative effect of shear stress as it inhibits cyclin D1, preventing the process of excessive cellular proliferation which could loosen the intercellular junctions and increase the permeability to lipoproteins (Qin et al. 2010; Chien et al. 1998). miR-34a expression in endothelial cells was downregulated by atheroprotective physiological high shear stress (HSS), whereas it was upregulated by atheroprone oscillatory shear stress (OSS). Blockade of endogenous miR-34a significantly decreased basal vascular cell adhesion molecule-1 (VCAM-1) and intercellular adhesion molecule-1 (ICAM-1) protein expression levels. In-depth investigation demonstrated that miR-34a increased nuclear factor-kappa B (NF-κB) p65 subunit acetylation and silencing NF-κB signaling reduced miR-34a-induced VCAM-1 and ICAM-1 protein expression (Fan et al. 2015a). miR-126 which is expressed by endothelial cells can inhibit VCAM-1 expression, thus protecting endothelial cells from inflammation. Transfection of endothelial cells with an oligonucleotide that decreases miR-126 permits an increase in TNF-stimulated VCAM-1 expression. Conversely, overexpression of the precursor to miR-126 increases miR-126 levels and decreases VCAM-1 expression. Additionally, decreasing endogenous miR-126 levels increases leukocyte adherence to endothelial cells (Harris et al. 2008). The transcription factor Kruppel-like factor 2 (KLF2) is critical integrator for endothelial lineage development and vascular functions. Bioinformatics analysis predicted that miR-92a could bind to the 3′ untranslated region of KLF2 mRNA. Overexpression of miR-92a decreased the expression of KLF2 and KLF2-regulated endothelial nitric oxide synthase and thrombomodulin at mRNA and protein levels. Inhibition of miR-92a could increase the mRNA and protein expression of KLF2, endothelial nitric oxide synthase, and thrombomodulin. Atheroprotective laminar flow downregulated the level of miR-92a precursor to induce KLF2, and the level of this flow-induced KLF2 was reduced by miR-92a precursor. Furthermore, miR-92a level was lower in human umbilical vein endothelial cells exposed to the atheroprotective pulsatile shear flow than under atheroprone oscillatory shear flow. Overall, atheroprotective flow patterns decrease the level of miR-92a, which in turn increases KLF2 expression to maintain endothelial homeostasis (Marin et al. 2013).

The RAAS system has a pivotal role in the pathogenesis of hypertension. Several microRNAs have been shown to interact with the RAAS (Batkai and Thum 2012). MicroRNA-181a, a negative regulator of renin mRNA, was shown to be a potential mediator of the BP elevation in hypertension (Jackson et al. 2013). What's more, Marques et al. (2011) showed that hypertensive individuals have marked under expression of miR-181a in the kidney, accompanied by marked overexpression of renin mRNA. Such a pattern was seen in BPH/2 J mice in our study. Expression of miR-181a in monocytes has since been reported to be negatively correlated with systolic BP in obese subjects (Hulsmans et al. 2012).

Except the association between microRNA and hypertension, SNPs in the miRNA binding sites of several RAAS-related genes are associated with alterations in blood pressure in humans (Nossent et al. 2011; Sethupathy et al. 2007). The AGTR1 target site harbors three SNPs, one of which, rs5186 (1166A/C), is located in the region that interacts with the miR-155 seed. Using reporter silencing assays, Praveen Sethupathy et al. showed that hsa-miR-155 downregulates the expression of only the 1166A, and not the 1166C, allele of rs5186. Remarkably, the 1166C allele has been associated with hypertension in many studies (Sethupathy et al. 2007). Patients with SNPs in the 3′ UTR of the vasopressin receptor (AVPR1A) that leads to a loss of miR-526b and miR-578 binding exhibit increased blood pressure. In addition, several miRNA loss-of-function SNPs in potential miRNA binding sites of transcripts encoding the bradykinin 2 (BDKRB2) and thromboxane A2 (TBXA2R) receptors are associated with reduced blood pressure. Intriguingly, an additional SNP in the 3′UTR of the aldosterone receptor (NR3C2) is not linked with alterations in blood pressure but is associated with an increased risk of myocardial infarction in young males (Nossent et al. 2011). While the effect of these SNPs on the expression of their associated genes remains to be elucidated in vivo, their linkage to blood pressure phenotypes implicates miRNAs as possible modulators of RAAS signaling.

2.1.3 Clinical Value of MicroRNAs

Various studies have shown that miRNAs play an important role in cardiovascular disease, such as hypertension, cardiac hypertrophy, and so on, suggesting their potential to be a therapeutic target, which has been well reviewed elsewhere (Li and Rana 2014). What's more, considering their stable existence in human plasma, semen, saliva, vaginal secretions, and menstrual blood, miRNA is a promising alternative approach for diagnosis of human disease (Hanson et al. 2009).

miRNAs and miRNA-targeting oligonucleotides (RNAi) have several advantages over traditional small-molecule drugs, most notably the ease with which oligonucleotides can be chemically modified to enhance their PK/PD profiles and the ability of miRNAs to target multiple genes simultaneously. Thus, miRNA-targeting therapies are an area of intense interest to pharmaceutical companies. For example, miR-208 is one of the specific miRNAs that is highly expressed in the heart (van Rooij et al. 2007). It is encoded in the intron region of the human and mouse αMHC (α-myosin heavy chain) gene (van Rooij et al. 2007). miR-208 knockout mice experienced gradual loss of cardiac function and failed to initiate cardiac hypertrophic growth in response to pressure-overload stress. miR-208 inhibition by an LNA-modified anti-miR could protect rats from hypertension-induced heart failure (Montgomery et al. 2011), suggesting a highly therapeutic potential in the inhibition of miR-208. The therapy is being developed by miRagen Therapeutics and is currently at the preclinical stage. The miR-15 family miRNA miR-195 was highly induced in cardiac ventricles during the postnatal switch to the terminally differentiated stage, when neonatal cardiomyocytes begin to withdraw from the proliferation stage of the cell cycle (Porrello et al. 2011). Inhibition of miR-195 by LNA-modified anti-miRs in mice and pigs showed strong effects on cardiac regeneration, and treated animals were protected from myocardial infarction, which is the most common antecedent of heart failure in humans (Hullinger et al. 2012; Porrello et al. 2013). Currently, an agent targeting miR-15/195 is being co-developed and is at the preclinical stage.

Current knowledge of miRNA function has largely focused on posttranscriptional gene silencing induced by the binding of miRISC to the 3′ untranslated region of the targeted mRNA. However, this represents only one aspect of miRNA function, and we will undoubtedly discover other miRNA-modulated biological processes that could serve as novel therapeutic targets. In addition, considerable work will be necessary to develop more efficient vehicles for the targeted delivery of oligonucleotides to specific organs, tissues, and cell types. What's more, it will be interesting to determine whether miRNA-targeting therapeutics could be combined other chemical or biological drugs for multidrug therapy. Despite those obstacles, an increasing number of these molecules will progress through clinical development and become approved treatments in the coming years.

2.2 LncRNAs

Long noncoding RNA (lncRNA) is a kind of noncoding RNA, which is considered to be greater than 200 nucleotides and can comprise up to thousands of nucleotides. LncRNAs are pervasively transcribed throughout the genome, and the resulting transcripts display remarkable similarities to classical mRNAs in that they are transcribed by RNA polymerase II (RNAP2) and are generally, but not always, alternatively spliced, 5′-capped, and polyadenylated (Derrien et al. 2012). In contrast to protein-coding genes, lncRNAs have limited coding potential as indicated by the lack of protein domains or significant open reading frames (ORFs). Moreover, lncRNAs display random codon usage and no significant bias toward silent nucleotide substitutions underscoring their low protein-coding potential; and these transcripts are rarely translated despite their engagement with ribosomes in some cases (Cabili et al. 2011; Lin et al. 2011; Ulitsky et al. 2011). Based largely on these criteria, a large number of lncRNA have been identified across eukaryotes, thousands of them in mouse and human. Remarkably, lncRNAs appear to be rapidly evolving and generally display low levels of sequence conservation. About one-third of all human lncRNAs have arisen only within the primate lineage, while only a small subgroup of lncRNAs appear to be maintained throughout a range of species with conservation being generally most evident in their promoter regions (Ulitsky et al. 2011; Chodroff et al. 2010). Together, these data suggest that regulation of lncRNA expression patterns is important for the function of this class of transcripts.

LncRNAs can be classified into four main categories based on the gene-regulating mechanism: (1) signal lncRNA, which can regulate gene expression in a time and space manner; (2) decoy lncRNA, which can titrate transcription factors and other proteins away from chromatin, or titrate miRNAs out from their target; (3) guide lncRNA, which can recruit chromatin modifying enzymes to target genes, either in cis (near the site of lncRNA production) or in trans to distant target genes; and (4) scaffold lncRNA, which can facilitate the assembling of multiple proteins to form ribonucleoprotein (RNP) complexes, which may act on chromatin, affecting histone modifications (Shen et al. 2017).

LncRNAs were shown to modulate gene expression at the epigenetic, transcriptional, and posttranscriptional levels. They are involved in many diverse biological and pathological processes such as RNA progressing, cell differentiation and proliferation, chromatin modification, cellular apoptosis, etc., and thus facilitate or suppress the development of diseases such as Alzheimer disease and cancers. Recent studies also revealed the essential role of lncRNAs in cardiovascular system such as the regulating myocardial hypertrophy (Han et al. 2017).

2.2.1 LncRNAs Involved in Cardiac Hypertrophy (Han et al. 2017; Ma et al. 2015)

Some investigations have clarified that the interaction between lncRNAs and miRNAs plays a significant role in regulating cardiac pathophysiological functions. One of the most exciting and impactful findings indicating that lncRNAs could impede cardiac hypertrophy is the case of the lncRNA Mhrt (Han et al. 2014). Han and colleagues identified a cluster of lncRNA transcripts from *Myh7* loci and a new lncRNA-chromatin mechanism for heart failure. In mice, myosin heavy-chain-associated RNA transcripts (MyHEART or Mhrt) are cardiac-specific and abundant in adult hearts. LncRNA Mhrt binds to the Brg1 chromatin-remodeling protein complex and competitively inhibits its binding to genomic DNA targets such as Myh6 and Myh7, thus blocking aberrant fetal gene reactivation during cardiac hypertrophy. The finding that the lncRNA Mhrt directly regulates chromatin remodeling and targets gene expression best exemplifies the great opportunities for developing lncRNA-based therapeutic interventions to treat cardiovascular diseases.

Chaer (cardiac hypertrophy-associated epigenetic regulator), which is a heart-enriched lncRNA, has been identified to be necessary for the development of cardiac hypertrophy (Wang et al. 2016a). *Chaer* directly interacts with the catalytic subunit of polycomb repressor complex 2 (PRC2). This interaction, which is mediated by a 66-mer motif in *Chaer*, interferes with PRC2 targeting to genomic loci, thereby inhibiting histone H3 lysine 27 methylation at the promoter regions of genes involved in cardiac hypertrophy. The interaction between *Chaer* and PRC2 is transiently induced after hormone or stress stimulation in a process involving mammalian target of rapamycin complex 1, and this interaction is a prerequisite for epigenetic reprogramming and induction of genes involved in hypertrophy. Inhibition of *Chaer* expression in the heart before, but not after, the onset of pressure overload substantially attenuates cardiac hypertrophy and dysfunction.

Wang and colleagues revealed the role of CHRF (cardiac hypertrophy-related factor) lncRNA in regulating cardiac hypertrophy (Wang et al. 2014). CHRF acts as an endogenous sponge of miR-489, which downregulates miR-489 expression levels. MiR-489 exhibited reduced hypertrophic response on angiotensin II treatment, and myeloid differentiation primary response gene 88 (Myd88) is a miR-489 target to mediate the function of miR-489 in cardiac hypertrophy. What's more, CHRF is able to directly bind to miR-489 and regulate Myd88 expression and hypertrophy.

A recent study revealed a novel function of H19 lncRNA-miR-675 axis targeting CaMKII δ as a negative regulator of cardiac hypertrophy (Liu et al. 2016). Overexpression or knockdown of H19 and miR-675 demonstrated that both H19 and miR-675 inhibited cardiomyocyte hypertrophy. What's more, miR-675 acts as a direct target of H19. The H19-miR-675 axis further targets CaMKII δ to retard cardiac hypertrophy. It is worth noting that apart from retarding cardiac hypertrophy, H19 lncRNA also plays a critical role in the regulation of cardiac fibrosis as aforementioned, suggesting that it has multiple functions in managing heart failure.

It has been reported that lncRNA myocardial infarction-associated transcript (MIAT) facilitated the pathological in angiotension II (AngII)-induced cardiac hypertrophy. MIAT was upregulated and miR-93 was downregulated in AngII-treated cardiomyocytes. MIAT functioned as a molecular sponge of miR-93 in cardiomyocytes (Li et al. 2017). Additionally, TLR4 was identified as a target of miR-93, and MIAT promoted TLR4 expression by sponging miR-93. Furthermore, MIAT knockdown or miR-93 overexpression inactivated the PI3K/Akt/mTOR pathway via TLR4 in AngII-induced cardiac hypertrophy. Taken together, these data suggested that MIAT knockdown inhibited AngII-induced cardiac hypertrophy by regulating miR-93/TLR4 axis.

Chast (cardiac hypertrophy-associated transcript) lncRNA was recently recognized as a prohypertrophic lncRNA by Viereck and colleagues (2016). *Chast* is upregulated in cardiomyocytes in transverse aortic constriction (TAC)-operated mice, and CHAST homolog in humans was significantly upregulated in hypertrophic heart tissue. Overexpression of *Chast* was sufficient to induce cardiomyocyte hypertrophy, and silencing of *Chast* can prevent and attenuate TAC-induced pathological cardiac remodeling. Further analysis revealed that *Chast* negatively regulated Pleckstrin homology domain-containing protein family M member 1 (opposite strand of *Chast*), impeding cardiomyocyte autophagy and driving hypertrophy.

These results indicate that *Chast* can be a potential target to prevent cardiac remodeling.

LncRNA-ROR was shown to play important roles in the pathogenesis of cardiac hypertrophy (Jiang et al. 2016). The expression of ROR is dramatically increased in hypertrophic heart and cardiomyocytes, and downregulation of ROR attenuates the hypertrophic response. MiR-133 is the target of ROR; overexpression of miR-133 prevents the elevation of ROR and re-expression of ANP and BNP. Mingjing Shao et al. found that lncRNA-TINCR can attenuate cardiac hypertrophy in TAC mice (Shao et al. 2017). TINCR could directly bind with enhancer of zeste homolog 2 (EZH2) in cardiomyocytes. EZH2 could directly bind to CaMKII promoter region and mediate H3K27me3 modification. Knockdown of TINCR was found to reduce EZH2 occupancy and H3K27me3 binding in the promoter of CaMKII in cardiomyocytes. In addition, enforced expression of TINCR was found to decrease CaMKII expression and attenuate AngII-induced cardiomyocyte hypertrophy. Taken all together, TINCR could attenuate myocardial hypertrophy by epigenetically silencing of CaMKII. LncRNA-HOTAIR was proved to be involved in cardiac hypertrophy (Lai et al. 2017), acting as a competing endogenous RNA (ceRNA) for miR-19, thereby modulating the disinhibition of its endogenous target PTEN and playing an important role in inhibiting cardiac hypertrophy progress.

Apart from the aforementioned and well-defined lncRNAs that are involved in cardiac hypertrophy, computational and bioinformatics approaches have predicted several lncRNAs that are highly related to cardiac hypertrophy. Song and colleagues recently identified three lncRNAs (SLC26A4-AS1, RP11-344E13.3, and MAGI1-IT1) that were closely related to cardiac hypertrophy based on the competitive endogenous RNA (ceRNA) mechanism by which the lncRNA could interact with microRNAs (miRNA) and indirectly interact with mRNAs through competing interactions (Song et al. 2016). However, these results must be carefully verified. Peters and colleagues recently reported that the Malat-1 lncRNA that acts as a ceRNA for microRNA-133 (Peters et al. 2016), which is a powerful inhibitor of cardiac hypertrophy, is absent during TAC-induced cardiac hypertrophy and remodeling in mice (Hunt et al. 1991). This negative result stresses the importance of validating the functions of lncRNAs that are proposed by computational and bioinformatics approaches in clinically relevant disease models.

2.2.2 LncRNAs as Biomarkers and Possible Therapeutic Targets for Heart Failure (Han et al. 2017)

Sustained cardiac hypertrophy in response to pathologic insults can eventually lead to systolic insufficiency and heart failure. To date, few reports regarding lncRNA-targeted therapy to treat heart failure clinically were published. However, the notion that lncRNAs could serve as biomarkers for clinical heart failure has been well established.

A comprehensive study using microarray analysis compared expression alterations of lncRNAs in the heart, whole blood, and plasma in mouse model of acute HF (Li et al. 2013). The results showed that the patterns of expression of lncRNA differed between plasma and other issue during HF. LncRNA expression was associated with lncRNA length in all samples but not in plasma during HF. This study revealed 32 lncRNAs all expressed in the heart, whole blood, and plasma showed changed expression with HF, so they may be biomarkers in HF. Another study conducted with serum samples from HF patients reported an early decrease in the mitochondrial lncRNA LIPCAR after myocardial infarction. However, the levels increased during the later stages. Plasma LIPCAR levels have the potential to predict survival in patients with HF (Kitow et al. 2016). In addition, two cardiac-specific and cardiac-related lncRNAs (NRON and MHRT) in plasma samples from patients with HF have been identified as circulating biomarkers for HF (Xuan et al. 2017); the area under ROC curve was 0.865 for NRON and 0.702 for MHRT.

Furthermore, studies have been done to identify and characterize lncRNAs deregulated in patients affected by ischemic heart failure. Fourteen lncRNAs were significantly modulated in non-end-stage heart failure patients (Greco et al. 2016). Nine of these lncRNAs (CDKN2B-AS1/ANRIL, EGOT, H19, HOTAIR, LOC285194/TUSC7, RMRP, RNY5, SOX2-OT, and SRA1) were also confirmed in end-stage failing hearts. Among the conserved lncRNAs, h19, rmrp, and hotair were also induced in a mouse model of heart hypertrophy. CDKN2B-AS1/ANRIL, HOTAIR, and LOC285194/TUSC7 showed similar modulation in peripheral blood mononuclear cells and heart tissue, suggesting a potential role as disease biomarkers.

Taken together, these results strongly suggest a role for lncRNAs in the molecular mechanisms underlying clinical heart failure. LncRNAs could regulate multiple pathological processes of cardiac remodeling, including myocardial hypertrophy, autophagy, necrosis, apoptosis, fibrosis, as well as vascular cell apoptosis and proliferation. Therefore, to manipulate lncRNA expression levels through either the inhibition of disease-upregulated lncRNAs or increasing disease-downregulated lncRNAs represents novel therapeutic strategies for cardiac remodeling. Although no clinical trials has been performed due to the innate immunity response and unpredictable toxicity of lncRNAs, more meaningful translational studies will be done with the development of interfering technology.

2.3 *CircRNAs*

CircRNA is a recently rediscovered noncoding RNA class, which have been shown to be involved in multiple functions such as early embryonic development (Fan et al. 2015b), the process of aging (Wang et al. 2015), and disease-related pathways. Moreover, circRNAs can not only be found in different tissues (Salzman et al. 2012, 2013) and organs (Rybak-Wolf et al. 2015; Tan et al. 2017) but also enriched in body fluids (Bahn et al. 2015; Li et al. 2015; Dong et al. 2016), such as blood, exosomes, saliva, and seminal plasma, indicating that circRNAs are promising therapy targets and promising biomarkers for human disease.

Specifically, circRNAs are widely expressed in the heart of human (Tan et al. 2017; Werfel et al. 2016), rats (Werfel et al. 2016), and mice (Werfel et al. 2016; Jakobi et al. 2016) and have been shown to be associated with many cardiovascular diseases. In this review, we summarize the biogenesis and expression of circular RNA, their function mechanism, the common software, or online tool for circular RNA research. Specially, we discuss circular RNAs and their potential roles in diagnosis, prognosis, and therapeutic strategies for cardiovascular diseases.

In unicellular organisms, circRNAs mostly stem from self-splicing introns of pre-ribosomal RNA (Zhang et al. 2014; Chen 2016), but can also arise from protein-coding genes in archaea (Zhang et al. 2014). In the validated circRNAs in animals, circRNAs are predominantly derived from exons, whose 5′ and 3′ ends are covalently linked to form a "head-to-tail" splice junction. Although spliceosomal machinery is implicated in the pre-mRNA splicing (Wahl et al. 2009), the mechanisms involved in exon circularization remain elusive.

It is supposed that circRNA generated from the editing process of pre-mRNA. There are two major mechanisms of circRNA formation: (1) direct back-splicing and (2) exon skipping (Barrett et al. 2015; Wang and Wang 2015). Direct back-splicing, where exons are spliced in noncanonical order, connects the downstream 5′ splice site (splice donor) to an upstream 3′ splice site (splice acceptor), thereby generating a circular transcript (Cocquerelle et al. 1993). The second mechanism is lariat splicing, which involves generation of a lariat intermediate containing exons, and then, the introns in the lariat are removed, generating ecircRNAs (Zaphiropoulos 1996). Both mechanisms function in vivo and correlate with the canonical spliceosome, but direct back-splicing may occur more frequently than exon skipping in ecircRNA biogenesis (Jeck and Sharpless 2014). Recent findings also indicate that RNA pairing, which can be formed by either repetitive elements (Jeck et al. 2013) or nonrepetitive but complementary sequences (Zhang et al. 2014), across flanking introns is positively associated with exon circularization. Additional observations have revealed that RNA-binding proteins such as ADAR, MBL, and QKI function in specific ecircRNA biogenesis (Ivanov et al. 2015; Ashwal-Fluss et al. 2014; Conn et al. 2015). Of note, these three proteins are known to be implicated in alternative splicing. In addition, compared with all expressed exons, the exons generating single-exon circRNA are threefold longer in length (Salzman et al. 2012; Jeck et al. 2013). Therefore, the length of a given exon may also correlate with circularization.

Based on the above progress of circRNA formation, circRNAs have five categories: (1) exonic circRNA (ecircRNAs), spliced from the exonic segments; (2) intronic circRNA (icircRNAs), derived wholly from an intron of the linear transcript; (3) exon–intron circRNA (EIicRNA), consisted of both exons and introns; (4) antisense, contained one or more exons of the linear transcript on the opposite strand; (5) bidirectional or intragenic, transcribed from same gene locus of the linear transcript but in close genomic proximity and not classified as "sense" and "intronic"; and (6) intergenic, when it is located in the genomic interval between two genes.

2.3.1 CircRNAs–miRNAs Axes in Cardiovascular Disease

As the miRNA sponge, several studies demonstrate the key role of circRNAs in regulating miRNA in cardiovascular disease.

For example, in myocardial infarction (MI) mice, Cdr1as (or CiRS-7) (Geng et al. 2016) was reported to be upregulated with increased cardiac infarct size or cardiomyocytes under hypoxia treatment. In Geng's study, Cdr1as and miR-7 were both overexpressed in MI mice, but the function of Cdr1as can be downregulated by miR-7. SP1 and poly (ADP-ribose) polymerase (PARP) were the downstream targets of miR-7. Zhang et al. (2015) find that the circRNA hsa-circ-000595 was highly expressed in tissues from aortic aneurysm patients and hypoxic aortic smooth muscle cells. Knockdown of hsa-circ-000595 resulted in decreased apoptotic rate of human aortic smooth muscle cells. It was determined that miR-19a is a target of hsa-circ-000595. Tang et al. (2017) found that circRNA_000203 was observed upregulated in the diabetic mouse myocardium and in AngII-induced mouse cardiac fibroblasts. CircRNA_000203 was containing two potential binding sites for microRNA-26b-5p (miR-26b-5p), which exerted the anti-fibrotic effect by targeting Colla2 and CTGF. Overexpression of circRNA_000203 could eliminate the anti-fibrotic effect of miR-26b-5p in cardiac fibroblasts. Furthermore, blocking the function of miR-26b-5p contributes to the pro-fibrosis effect of circRNA_000203 in cardiac fibroblasts. Holdt et al. (2016) found that circANRIL can protect the tissue from atherosclerosis by controlling ribosomal RNA (rRNA) maturation and modulating pathways of atherogenesis. Pescadillo homolog 1 (PES1), which was an essential 60S-preribosomal assembly factor, was the downstream target of circANRIL. CircANRIL can impair exonuclease-mediated pre-rRNA processing and ribosome biogenesis in vascular smooth muscle cells and macrophages through binding to PES1.

A few studies have also been reported on heart failure and cardiomyopathy. In the study of heart failure, Kun Wang et al. showed that a heart-related circRNA (HRCR) attenuated the development of cardiac hypertrophy and heart failure in mice by targeting miR-223 and ARC (apoptosis repressor with caspase recruitment domain) (Wang et al. 2016b). This study revealed a novel cardiac hypertrophic regulating model that was composed of HRCR, miR-223, and ARC. miR-223 promotes the process of cardiac hypertrophy and heart failure, while ARC is the downstream of miR-223. However, the pathological effects of miR-223 can be inhibited by HRCR, which is regarded as the upstream regulator of miR-223. circ-Foxo3 was highly expressed in heart samples of aged patients and mice and was shown to be correlated with cellular senescence (Du et al. 2017). Doxorubicin-induced cardiomyopathy was aggravated by ectopic expression of circ-Foxo3 but was relieved by silencing endogenous circ-Foxo3. circ-Foxo3 induce senescence by interacting with the anti-senescent protein ID-1 and the transcription factor E2F1, as well as the anti-stress proteins FAK and HIF-1α. Strikingly, in their study, in vivo delivery of circ-Foxo3 targeted siRNA into mouse embryonic fibroblasts and primary cardiac myocytes resulted in lower levels of circ-Foxo3 expression and beta-gal staining, which is a standard marker for senescence, suggesting a potential therapeutic approach for myocardial protection.

Except for those circRNAs that have been shown the interaction with miRNAs or proteins, there are also some circRNAs correlated with cardiovascular diseases which have not been found the function mechanism (Jakobi et al. 2016; Boeckel et al. 2015). cZNF292 is a hypoxic-induced circRNA, which has been shown to be regulated hypoxia and have biological functions (Boeckel et al. 2015); however, no validated microRNA-binding sites for cZNF292 were detected, suggesting that cZNF292 does not act as a microRNA sponge. Jakobi et al. (2016) explored the circRNA landscape of heart tissue using a novel DCC software. Among those circRNAs, several circRNAs can be directly attributed to host genes that have been previously described as associated with cardiovascular disease. Further studies of these candidate circRNAs are needed to reveal disease-relevant properties or functions of specific circRNAs.

2.3.2 Characters of CircRNAs as Biomarker for Cardiovascular Disease

CircRNAs are widely found in blood, saliva, and seminal fluid, and they are stable due to their circular structure, which are preconditions for their usage as biomarkers for various diseases. Furthermore, the expression levels of circRNAs vary under different conditions, making them possible to be biomarkers for diseases, including hypertension, cardiac hypertrophy, and heart failure.

3 Conclusions

With the development in modern technology, great progresses have been made in the diagnosis and management of patients with hypertension, cardiac hypertrophy, and heart failure. A better understanding of genetic and epigenetic information and their variable relation to clinical situation can not only facilitate identification of patients with cardiovascular disease but also provide new therapeutic strategies toward the treatment of those disease and reduction of cardiac complications. What's more, although many promising strategies that have been proposed and developed, a collective and integrated translational effort is needed to find the most effective and safe strategy to reach a satisfactory goal in treating cardiovascular disease, including hypertension, cardiac hypertrophy, and heart failure.

Acknowledgment This chapter was modified from the paper published by our group in *Non-coding RNA investigation* (Xiaoyan Wang, Zhiwen Ding, Jian Wu, Shijun Wang, Yunzeng Zou), 13 September 2017; 1; 5). The related contents are reused with the permission.

References

Ambros V. The functions of animal microRNAs. Nature. 2004;431:350–5.

Andersson KB, Birkeland J, Finsen AV, Louch WE, Sjaastad I, Wang Y, Chen J, Molkentin JD, Chien KR, Sejersted OM, Christensen G. Moderate heart dysfunction in mice with inducible cardiomyocyte-specific excision of the Serca2 gene. J Mol Cell Cardiol. 2009;47:180–7.

Arad M, Benson DW, Perez-Atayde AR, McKenna WJ, Sparks EA, Kanter RJ, McGarry K, Seidman JG, Seidman CE. Constitutively active AMP kinase mutations cause glycogen storage disease mimicking hypertrophic cardiomyopathy. J Clin Investig. 2002;109:357–62.

Arad M, Maron B, Gorham JM, Johnson WH Jr, Saul JP, Perez-Atayde AR, Spirito P, Wright GB, Kanter RJ, Seidman CE, Seidman JG. Glycogen storage diseases presenting as hypertrophic cardiomyopathy. N Engl J Med. 2005;352:362–72.

Arora P, Newton-Cheh C. Blood pressure and human genetic variation in the general population. Curr Opin Cardiol. 2010;25:229–37.

Arrowsmith CH, Bountra C, Fish PV, Lee K, Schapira M. Epigenetic protein families: a new frontier for drug discovery. Nat Rev Drug Discov. 2012;11:384–400.

Ashwal-Fluss R, Meyer M, Pamudurti NR, Ivanov A, Bartok O, Hanan M, Evantal N, Memczak S, Rajewsky N, Kadener S. circRNA biogenesis competes with pre-mRNA splicing. Mol Cell. 2014;56:55–66.

Atanur SS, Diaz AG, Maratou K, Sarkis A, Rotival M, Game L, Tschannen MR, Kaisaki PJ, Otto GW, Ma MC, Keane TM, Hummel O, Saar K, Chen W, Guryev V, Gopalakrishnan K, Garrett MR, Joe B, Citterio L, Bianchi G, McBride M, Dominiczak A, Adams DJ, Serikawa T, Flicek P, Cuppen E, Hubner N, Petretto E, Gauguier D, Kwitek A, Jacob H, Aitman TJ. Genome sequencing reveals loci under artificial selection that underlie disease phenotypes in the laboratory rat. Cell. 2013;154:691–703.

Bahn JH, Zhang Q, Li F, Chan TM, Lin X, Kim Y, Wong DT, Xiao X. The landscape of microRNA, Piwi-interacting RNA, and circular RNA in human saliva. Clin Chem. 2015;61:221–30.

Balaban RS. The role of Ca(2+) signaling in the coordination of mitochondrial ATP production with cardiac work. Biochim Biophys Acta. 2009;1787:1334–41.

Barrett SP, Wang P, Salzman J. Circular RNA biogenesis can proceed through an exon-containing lariat precursor. Elife. 2015;4:e07540.

Batkai S, Thum T. MicroRNAs in hypertension: mechanisms and therapeutic targets. Curr Hypertens Rep. 2012;14:79–87.

Bikkina M, Levy D, Evans JC, Larson MG, Benjamin EJ, Wolf PA, Castelli WP. Left ventricular mass and risk of stroke in an elderly cohort. The Framingham Heart Study. JAMA. 1994;272:33–6.

Boeckel JN, Jae N, Heumuller AW, Chen W, Boon RA, Stellos K, Zeiher AM, John D, Uchida S, Dimmeler S. Identification and characterization of hypoxia-regulated endothelial circular RNA. Circ Res. 2015;117:884–90.

Bonasio R, Tu S, Reinberg D. Molecular signals of epigenetic states. Science. 2010;330:612–6.

Brasacchio D, Okabe J, Tikellis C, Balcerczyk A, George P, Baker EK, Calkin AC, Brownlee M, Cooper ME, El-Osta A. Hyperglycemia induces a dynamic cooperativity of histone methylase and demethylase enzymes associated with gene-activating epigenetic marks that coexist on the lysine tail. Diabetes. 2009;58:1229–36.

Buckley CL, Stokes AJ. Mice lacking functional TRPV1 are protected from pressure overload cardiac hypertrophy. Channels (Austin). 2011;5:367–74.

Burgess DJ. Cancer genomics: histone modification at the gene level. Nat Rev Genet. 2012;13:148–9.

Burwinkel B, Scott J, Bührer C, van Landeghem FK, Cox GF, Wilson CJ, Grahame Hardie D, Kilimann MW. Fatal congenital heart glycogenosis caused by a recurrent activating R531Q mutation in the g2-subunit of AMP-activated protein kinase (PRKAG2), not by phosphorylase kinase deficiency. Am J Hum Genet. 2005;76:1034–49.

Buttrick PM, Kaplan M, Leinwand LA, Scheuer J. Alterations in gene expression in the rat heart after chronic pathological and physiological loads. J Mol Cell Cardiol. 1994;26:61–7.

Cabili MN, Trapnell C, Goff L, Koziol M, Tazon-Vega B, Regev A, Rinn JL. Integrative annotation of human large intergenic noncoding RNAs reveals global properties and specific subclasses. Genes Dev. 2011;25:1915–27.

Cadre BM, Qi M, Eble DM, Shannon TR, Bers DM, Samarel AM. Cyclic stretch down-regulates calcium transporter gene expression in neonatal rat ventricular myocytes. J Mol Cell Cardiol. 1998;30:2247–59.

Chang S, McKinsey TA, Zhang CL, Richardson JA, Hill JA, Olson EN. Histone deacetylases 5 and 9 govern responsiveness of the heart to a subset of stress signals and play redundant roles in heart development. Mol Cell Biol. 2004;24:8467–76.

Chen LL. The biogenesis and emerging roles of circular RNAs. Nat Rev Mol Cell Biol. 2016;17:205–11.

Chen M, Xin J, Liu B, Luo L, Li J, Yin W Li M. Mitogen-activated protein kinase and intracellular polyamine signaling is involved in TRPV1 activation-induced cardiac hypertrophy. J Am Heart Assoc. 2016;5.

Cheung P, Lau P. Epigenetic regulation by histone methylation and histone variants. Mol Endocrinol. 2005;19:563–73.

Chien S, Li S, Shyy YJ. Effects of mechanical forces on signal transduction and gene expression in endothelial cells. Hypertension. 1998;31:162–9.

Chodroff RA, Goodstadt L, Sirey TM, Oliver PL, Davies KE, Green ED, Molnár Z, Ponting CP. Long noncoding RNA genes: conservation of sequence and brain expression among diverse amniotes. Genome Biol. 2010;11:R72.

Chopra S, Baby C, Jacob JJ. Neuro-endocrine regulation of blood pressure. Indian J Endocrinol Metab. 2011;15(Suppl 4):S281–8.

Clapier CR, Cairns B. The biology of chromatin remodeling complexes. Annu Rev Biochem. 2009;78:273–304.

Clayton AL, Hazzalin CA, Mahadevan LC. Enhanced histone acetylation and transcription: a dynamic perspective. Mol Cell. 2006;23:289–96.

Cocquerelle C, Mascrez B, Hétuin D, Bailleul B. Mis-splicing yields circular RNA molecules. FASEB J. 1993;7:155–60.

Conn SJ, Pillman KA, Toubia J, Conn VM, Salmanidis M, Phillips CA, Roslan S, Schreiber AW, Gregory PA, Goodall GJ. The RNA binding protein quaking regulates formation of circRNAs. Cell. 2015;160:1125–34.

Consortium S, Saar K, Beck A, Bihoreau MT, Birney E, Brocklebank D, Chen Y, Cuppen E, Demonchy S, Dopazo J, Flicek P, Foglio M, Fujiyama A, Gut IG, Gauguier D, Guigo R, Guryev V, Heinig M, Hummel O, Jahn N, Klages S, Kren V, Kube M, Kuhl H, Kuramoto T, Kuroki Y, Lechner D, Lee YA, Lopez-Bigas N, Lathrop GM, Mashimo T, Medina I, Mott R, Patone G, Perrier-Cornet JA, Platzer M, Pravenec M, Reinhardt R, Sakaki Y, Schilhabel M, Schulz H, Serikawa T, Shikhagaie M, Tatsumoto S, Taudien S, Toyoda A, Voigt B, Zelenika D, Zimdahl H, Hubner N. SNP and haplotype mapping for genetic analysis in the rat. Nat Genet. 2008;40:560–6.

Davia K, Davies C, Harding SE. Effects of inhibition of sarcoplasmic reticulum calcium uptake on contraction in myocytes isolated from failing human ventricle. Cardiovasc Res. 1997;33:88–97.

de la Bastie D, Levitsky D, Rappaport L, Mercadier JJ, Marotte F, Wisnewsky C, Brovkovich V, Schwartz K, Lompré AM. Function of the sarcoplasmic reticulum and expression of its Ca2(+)-ATPase gene in pressure overload-induced cardiac hypertrophy in the rat. Circ Res. 1990;66:554–64.

Derrien T, Johnson R, Bussotti G, Tanzer A, Djebali S, Tilgner H, Guernec G, Martin D, Merkel A, Knowles DG, Lagarde J, Veeravalli L, Ruan X, Ruan Y, Lassmann T, Carninci P, Brown JB, Lipovich L, Gonzalez JM, Thomas M, Davis CA, Shiekhattar R, Gingeras TR, Hubbard TJ, Notredame C, Harrow J, Guigo R. The GENCODE v7 catalog of human long noncoding RNAs: analysis of their gene structure, evolution, and expression. Genome Res. 2012;22:1775–89.

Devereux RB, Alderman M. Role of preclinical cardiovascular disease in the evolution from risk factor exposure to development of morbid events. Circulation. 1993;88:1444–55.
Doerks T, Copley RR, Schultz J, Ponting CP, Bork P. Systematic identification of novel protein domain families associated with nuclear functions. Genome Res. 2002;12:47–56.
Dong WW, Li HM, Qing XR, Huang DH, Li HG. Identification and characterization of human testis derived circular RNAs and their existence in seminal plasma. Sci Rep. 2016;6:39080.
Du WW, Yang W, Chen Y, Wu ZK, Foster FS, Yang Z, Li X, Yang BB. Foxo3 circular RNA promotes cardiac senescence by modulating multiple factors associated with stress and senescence responses. Eur Heart J. 2017;38(18):1402–12.
Elkayam E, Kuhn CD, Tocilj A, Haase AD, Greene EM, Hannon GJ, Joshua-Tor L. The structure of human argonaute-2 in complex with miR-20a. Cell. 2012;150:100–10.
Elliott P, McKenna WJ. Hypertrophic cardiomyopathy. Lancet. 2004;363:1881–91.
Emelyanov AV, Vershilova E, Ignatyeva MA, Pokrovsky DK, Lu X, Konev AY, Fyodorov DV. Identification and characterization of ToRC, a novel ISWI-containing ATP-dependent chromatin assembly complex. Genes Dev. 2012;26:603–14.
Esteller M. Non-coding RNAs in human disease. Nat Rev Genet. 2011;12:861–74.
Fan W, Fang R, Wu X, Liu J, Feng M, Dai G, Chen G, Wu G. Shear-sensitive microRNA-34a modulates flow-dependent regulation of endothelial inflammation. J Cell Sci. 2015a;128:70–80.
Fan X, Zhang X, Wu X, Guo H, Hu Y, Tang F, Huang Y. Single-cell RNA-seq transcriptome analysis of linear and circular RNAs in mouse preimplantation embryos. Genome Biol. 2015b;16:148.
Feinberg AP, Tycko B. The history of cancer epigenetics. Nat Rev Cancer. 2004;4:143–53.
Feldman AM, Weinberg E, Ray PE, Lorell BH. Selective changes in cardiac gene expression during compensated hypertrophy and the transition to cardiac decompensation in rats with chronic aortic banding. Circ Res. 1993;73:184–92.
Flaus A, Owen-Hughes T. Mechanisms for ATP-dependent chromatin remodelling: the means to the end. FEBS J. 2011;278:3579–95.
Flesch M, Schwinger R, Schnabel P, Schiffer F, van Gelder I, Bavendiek U, Südkamp M, Kuhn-Regnier F, Böhm M. Sarcoplasmic reticulum Ca2+ATPase and phospholamban mRNA and protein levels in end-stage heart failure due to ischemic or dilated cardiomyopathy. J Mol Med (Berl). 1996;74:321–32.
Frey N, Katus H, Olson EN, Hill JA. Hypertrophy of the heart a new therapeutic target? Circulation. 2004;109:1580–9.
Geng HH, Li R, Su YM, Xiao J, Pan M, Cai XX, Ji XP. The circular RNA Cdr1as promotes myocardial infarction by mediating the regulation of miR-7a on its target genes expression. PLoS One. 2016;11:e0151753.
Gibbs RA, Weinstock G, Metzker ML, et al. Genome sequence of the Brown Norway rat yields insights into mammalian evolution. Nature. 2004;428:493–521.
Gigante B, Rubattu S, Zee RY, Volpe M, Lindpaintner K. Non-random chromosomal distribution of SSLPs: systematic assessment using a novel genetic linkage map between two closely related rat strains. Cytogenet Cell Genet. 2001;95:64–72.
Global Burden of Metabolic Risk Factors for Chronic Diseases Collaboration. Cardiovascular disease, chronic kidney disease, and diabetes mortality burden of cardiometabolic risk factors from 1980 to 2010: a comparative risk assessment. Lancet Diabetes Endocrinol. 2014;2:634–47.
Gopalakrishnan K, Kumarasamy S, Mell B, Joe B. Genome-wide identification of long noncoding RNAs in rat models of cardiovascular and renal disease. Hypertension. 2015;65:200–10.
Greco S, Zaccagnini G, Perfetti A, Fuschi P, Valaperta R, Voellenkle C, Castelvecchio S, Gaetano C, Finato N, Beltrami AP, Menicanti L, Martelli F. Long noncoding RNA dysregulation in ischemic heart failure. J Transl Med. 2016;14:183.
Griffiths-Jones S, Saini HK, van Dongen S, Enright AJ. miRBase: tools for microRNA genomics. Nucleic Acids Res. 2008;36:D154–8.
Gusterson RJ, Jazrawi E, Adcock IM, Latchman DS. The transcriptional co-activators CREB-binding protein (CBP) and p300 play a critical role in cardiac hypertrophy that is dependent on their histone acetyltransferase activity. J Biol Chem. 2003;278:6838–47.

Haider S, Cordeddu L, Robinson E, Movassagh M, Siggens L, Vujic A, Choy MK, Goddard M, Lio P, Foo R. The landscape of DNA repeat elements in human heart failure. Genome Biol. 2012;13:R90.

Han P, Li W, Lin CH, Yang J, Shang C, Nuernberg ST, Jin KK, Xu W, Lin CY, Lin CJ, Xiong Y, Chien H, Zhou B, Ashley E, Bernstein D, Chen PS, Chen HV, Quertermous T, Chang CP. A long noncoding RNA protects the heart from pathological hypertrophy. Nature. 2014;514:102–6.

Han D, Gao Q, Cao F. Long noncoding RNAs (LncRNAs) – the dawning of a new treatment for cardiac hypertrophy and heart failure. Biochim Biophys Acta. 2017;1863:2078–84.

Handy DE, Castro R, Loscalzo J. Epigenetic modifications: basic mechanisms and role in cardiovascular disease. Circulation. 2011;123:2145–56.

Hanson EK, Lubenow H, Ballantyne J. Identification of forensically relevant body fluids using a panel of differentially expressed microRNAs. Anal Biochem. 2009;387:303–14.

Hargreaves DC, Crabtree G. ATP-dependent chromatin remodeling: genetics, genomics and mechanisms. Cell Res. 2011;21:396–420.

Harris TA, Yamakuchi M, Ferlito M, Mendell JT, Lowenstein CJ. MicroRNA-126 regulates endothelial expression of vascular cell adhesion molecule 1. Proc Natl Acad Sci U S A. 2008;105:1516–21.

Harrison DG. The mosaic theory revisited: common molecular mechanisms coordinating diverse organ and cellular events in hypertension. J Am Soc Hypertens. 2013;7:68–74.

Hasenfuss G, Reinecke H, Studer R, Meyer M, Pieske B, Holtz J, Holubarsch C, Posival H, Just H, Drexler H. Relation between myocardial function and expression of sarcoplasmic reticulum Ca(2+)-ATPase in failing and nonfailing human myocardium. Circ Res. 1994;75:434–42.

Ho L, Crabtree GR. Chromatin remodelling during development. Nature. 2010;463:474–84.

Hohl M, Wagner M, Reil JC, Muller SA, Tauchnitz M, Zimmer AM, Lehmann LH, Thiel G, Bohm M, Backs J, Maack C. HDAC4 controls histone methylation in response to elevated cardiac load. J Clin Invest. 2013;123:1359–70.

Holdt LM, Stahringer A, Sass K, Pichler G, Kulak NA, Wilfert W, Kohlmaier A, Herbst A, Northoff BH, Nicolaou A, Gabel G, Beutner F, Scholz M, Thiery J, Musunuru K, Krohn K, Mann M, Teupser D. Circular non-coding RNA ANRIL modulates ribosomal RNA maturation and atherosclerosis in humans. Nat Commun. 2016;7:12429.

Hullinger TG, Montgomery RL, Seto AG, Dickinson BA, Semus HM, Lynch JM, Dalby CM, Robinson K, Stack C, Latimer PA, Hare JM, Olson EN, van Rooij E. Inhibition of miR-15 protects against cardiac ischemic injury. Circ Res. 2012;110:71–81.

Hulsmans M, Sinnaeve P, Van der Schueren B, Mathieu C, Janssens S, Holvoet P. Decreased miR-181a expression in monocytes of obese patients is associated with the occurrence of metabolic syndrome and coronary artery disease. J Clin Endocrinol Metab. 2012;97:E1213–8.

Hunt SCSS, Hopkins PN, Williams RR. Predictors of an increased risk of future hypertension in Utah. A screening analysis. Hypertension. 1991;17:969–76.

Inoue R, Jian Z, Kawarabayashi Y. Mechanosensitive TRP channels in cardiovascular pathophysiology. Pharmacol Ther. 2009;123:371–85.

Ito T, Bulger M, Pazin MJ, Kobayashi R, Kadonaga JT. ACF an ISWI-containing and ATP-utilizing chromatin assembly and remodeling factor. Cell. 1997;90:145–55.

Ivanov A, Memczak S, Wyler E, Torti F, Porath HT, Orejuela MR, Piechotta M, Levanon EY, Landthaler M, Dieterich C, Rajewsky N. Analysis of intron sequences reveals hallmarks of circular RNA biogenesis in animals. Cell Rep. 2015;10:170–7.

Izumo S, Lompré A, Matsuoka R, Koren G, Schwartz K, Nadal-Ginard B, Mahdavi V. Myosin heavy chain messenger RNA and protein isoform transitions during cardiac hypertrophy. Interaction between hemodynamic and thyroid hormone-induced signals. J Clin Invest. 1987;79:970–7.

Izumo S, Nadal-Ginard B, Mahdavi V. Protooncogene induction and reprogramming of cardiac gene expression produced by pressure overload. Proc Natl Acad Sci U S A. 1988;85:339–43.

Jackson KL, Marques FZ, Watson AM, Palma-Rigo K, Nguyen-Huu TP, Morris BJ, Charchar FJ, Davern PJ, Head GA. A novel interaction between sympathetic overactivity and aberrant regulation of renin by miR-181a in BPH/2J genetically hypertensive mice. Hypertension. 2013;62:775–81.

Jakobi T, Czaja-Hasse LF, Reinhardt R, Dieterich C. Profiling and validation of the circular RNA repertoire in adult murine hearts. Genomics Proteomics Bioinformatics. 2016;14:216–23.

Jeck WR, Sharpless NE. Detecting and characterizing circular RNAs. Nat Biotechnol. 2014;32:453–61.

Jeck WR, Sorrentino J, Wang K, Slevin MK, Burd CE, Liu J, Marzluff WF, Sharpless NE. Circular RNAs are abundant, conserved, and associated with ALU repeats. RNA. 2013;19:426.

Jenuwein T, Allis C. Translating the histone code. Science. 2001;293:1074–80.

Jeong HJ, Han NR, Kim KY, Choi IS, Kim HM. Gomisin A decreases the LPS-induced expression of iNOS and COX-2 and activation of RIP2/NF-kappaB in mouse peritoneal macrophages. Immunopharmacol Immunotoxicol. 2014;36:195–201.

Jiang F, Zhou X, Huang J. Long non-coding RNA-ROR mediates the reprogramming in cardiac hypertrophy. PLoS One. 2016;11:e0152767.

Jonas S, Izaurralde E. Towards a molecular understanding of microRNA-mediated gene silencing. Nat Rev Genet. 2015;16:421–33.

Kato N, Takeuchi F, Tabara Y, Kelly TN, Go MJ, Sim X, Tay WT, Chen CH, Zhang Y, Yamamoto K, Katsuya T, Yokota M, Kim YJ, Ong RT, Nabika T, Gu D, Chang LC, Kokubo Y, Huang W, Ohnaka K, Yamori Y, Nakashima E, Jaquish CE, Lee JY, Seielstad M, Isono M, Hixson JE, Chen YT, Miki T, Zhou X, Sugiyama T, Jeon JP, Liu JJ, Takayanagi R, Kim SS, Aung T, Sung YJ, Zhang X, Wong TY, Han BG, Kobayashi S, Ogihara T, Zhu D, Iwai N, Wu JY, Teo YY, Tai ES, Cho YS, He J. Meta-analysis of genome-wide association studies identifies common variants associated with blood pressure variation in east Asians. Nat Genet. 2011;43:531–8.

Kawamata T, Tomari Y. Making RISC. Trends Biochem Sci. 2010;35:368–76.

Kim JK, Samaranayake M, Pradhan S. Epigenetic mechanisms in mammals. Cell Mol Life Sci. 2009;66:596–612.

Kitow J, Derda AA, Beermann J, Kumarswarmy R, Pfanne A, Fendrich J, Lorenzen JM, Xiao K, Bavendiek U, Bauersachs J, Thum T. Mitochondrial long noncoding RNAs as blood based biomarkers for cardiac remodeling in patients with hypertrophic cardiomyopathy. Am J Physiol Heart Circ Physiol. 2016;311:H707–12.

Klimczak D, Jazdzewski K, Kuch M. Regulatory mechanisms in arterial hypertension: role of microRNA in pathophysiology and therapy. Blood Press. 2017;26:2–8.

Lachner M, Jenuwein T. The many faces of histone lysine methylation. Curr Opin Cell Biol. 2002;14:286–98.

Lai EC. Micro RNAs are complementary to 3′ UTR sequence motifs that mediate negative post-transcriptional regulation. Nat Genet. 2002;30:363–4.

Lai Y, He S, Ma L, Lin H, Ren B, Ma J, Zhu X, Zhuang S. HOTAIR functions as a competing endogenous RNA to regulate PTEN expression by inhibiting miR-19 in cardiac hypertrophy. Mol Cell Biochem. 2017;432:179–87.

Lee RC, Feinbaum R, Ambros V. The *C. elegans* heterochronic gene lin-4 encodes small RNAs with antisense complementarity to lin-14. Cell. 1993;75:843–54.

Levy D, Garrison R, Savage DD, Kannel WB, Castelli WP. Prognostic implications of echocardiographically determined left ventricular mass in the Framingham Heart Study. N Engl J Med. 1990;322:1561–6.

Levy D, Ehret GB, Rice K, Verwoert GC, Launer LJ, Dehghan A, Glazer NL, Morrison AC, Johnson AD, Aspelund T, Aulchenko Y, Lumley T, Kottgen A, Vasan RS, Rivadeneira F, Eiriksdottir G, Guo X, Arking DE, Mitchell GF, Mattace-Raso FU, Smith AV, Taylor K, Scharpf RB, Hwang SJ, Sijbrands EJ, Bis J, Harris TB, Ganesh SK, O'Donnell CJ, Hofman A, Rotter JI, Coresh J, Benjamin EJ, Uitterlinden AG, Heiss G, Fox CS, Witteman JC, Boerwinkle E, Wang TJ, Gudnason V, Larson MG, Chakravarti A, Psaty BM, van Duijn CM. Genome-wide association study of blood pressure and hypertension. Nat Genet. 2009;41:677–87.

Lewis BP, Burge CB, Bartel DP. Conserved seed pairing, often flanked by adenosines, indicates that thousands of human genes are microRNA targets. Cell. 2005;120:15–20.

Lewis PW, Elsaesser S, Noh KM, Stadler SC, Allis CD. Daxx is an H3.3-specific histone chaperone and cooperates with ATRX in replication-independent chromatin assembly at telomeres. Proc Natl Acad Sci U S A. 2010;107:14075–80.

Li E. Chromatin modification and epigenetic reprogramming in mammalian development. Nat Rev Genet. 2002;3:662–73.

Li Z, Rana TM. Therapeutic targeting of microRNAs: current status and future challenges. Nat Rev Drug Discov. 2014;13:622–38.

Li E, Beard C, Jaenisch R. Role for DNA methylation in genomic imprinting. Nature. 1993;366:362–5.

Li D, Chen G, Yang J, Fan X, Gong Y, Xu G, Cui Q, Geng B. Transcriptome analysis reveals distinct patterns of long noncoding RNAs in heart and plasma of mice with heart failure. PLoS One. 2013;8:e77938.

Li Y, Zheng Q, Bao C, Li S, Guo W, Zhao J, Chen D, Gu J, He X, Huang S. Circular RNA is enriched and stable in exosomes: a promising biomarker for cancer diagnosis. Cell Res. 2015;25:981–4.

Li Y, Wang J, Sun L, Zhu S. LncRNA myocardial infarction-associated transcript (MIAT) contributed to cardiac hypertrophy by regulating TLR4 via miR-93. Eur J Pharmacol. 2017;S0014-2999(17):30764–1.

Liang J, Le TH, Edwards DRV, Tayo BO, Gaulton KJ, Smith JA, Lu Y, Jensen RA, Chen G, Yanek LR, Schwander K, Tajuddin SM, Sofer T, Kim W, Kayima J, McKenzie CA, Fox E, Nalls MA, Young JH, Sun YV, Lane JM, Cechova S, Zhou J, Tang H, Fornage M, Musani SK, Wang H, Lee J, Adeyemo A, Dreisbach AW, Forrester T, Chu PL, Cappola A, Evans MK, Morrison AC, Martin LW, Wiggins KL, Hui Q, Zhao W, Jackson RD, Ware EB, Faul JD, Reiner AP, Bray M, Denny JC, Mosley TH, Palmas W, Guo X, Papanicolaou GJ, Penman AD, Polak JF, Rice K, Taylor KD, Boerwinkle E, Bottinger EP, Liu K, Risch N, Hunt SC, Kooperberg C, Zonderman AB, Laurie CC, Becker DM, Cai J, Loos RJF, Psaty BM, Weir DR, Kardia SLR, Arnett DK, Won S, Edwards TL, Redline S, Cooper RS, Rao DC, Rotter JI, Rotimi C, Levy D, Chakravarti A, Zhu X, Franceschini N. Single-trait and multi-trait genome-wide association analyses identify novel loci for blood pressure in African-ancestry populations. PLoS Genet. 2017;13:e1006728.

Lin S, Gregory RI. MicroRNA biogenesis pathways in cancer. Nat Rev Cancer. 2015;15:321–33.

Lin MF, Jungreis I, Kellis M. PhyloCSF: a comparative genomics method to distinguish protein coding and non-coding regions. Bioinformatics. 2011;27:i275–82.

Lind JM, Chiu C, Semsarian C. Genetic basis of hypertrophic cardiomyopathy. Expert Rev Cardiovasc Ther. 2006;4:927–34.

Ling C, Groop L. Epigenetics: a molecular link between environmental factors and type 2 diabetes. Diabetes. 2009;58:2718–25.

Liu L, An X, Li Z, Song Y, Li L, Zuo S, Liu N, Yang G, Wang H, Cheng X, Zhang Y, Yang X, Wang J. The H19 long noncoding RNA is a novel negative regulator of cardiomyocyte hypertrophy. Cardiovasc Res. 2016;111:56–65.

Longini IM Jr, Higgins M, Hinton PC, Moll PP, Keller JB. Environmental and genetic sources of familial aggregation of blood pressure in Tecumseh, Michigan. Am J Epidemiol. 1984;120:131–44.

Low TY, van Heesch S, van den Toorn H, Giansanti P, Cristobal A, Toonen P, Schafer S, Hubner N, van Breukelen B, Mohammed S, Cuppen E, Heck AJ, Guryev V. Quantitative and qualitative proteome characteristics extracted from in-depth integrated genomics and proteomics analysis. Cell Rep. 2013;5:1469–78.

Loyola A, LeRoy G, Wang YH, Reinberg D. Reconstitution of recombinant chromatin establishes a requirement for histone-tail modifications during chromatin assembly and transcription. Genes Dev. 2001;15:2837–51.

Lu J, McKinse T, Nicol RL, Olson EN. Signal-dependent activation of the MEF2 transcription factor by dissociation from histone deacetylases. Proc Natl Acad Sci U S A. 2000;97:4070–5.

Luft FC. Twins in cardiovascular genetic research. Hypertension. 2001;37:350–6.

Lusser A, Urwin D, Kadonaga JT. Distinct activities of CHD1 and ACF in ATP-dependent chromatin assembly. Nat Struct Mol Biol. 2005;12:160–6.

Ma Y, Ma W, Huang L, Feng D, Cai B. Long non-coding RNAs, a new important regulator of cardiovascular physiology and pathology. Int J Cardiol. 2015;188:105–10.

Malinowska B, Baranowska-Kuczko M, Schlicker E. Triphasic blood pressure responses to cannabinoids: do we understand the mechanism? Br J Pharmacol. 2012;165:2073–88.

Margueron R, Reinberg D. Chromatin structure and the inheritance of epigenetic information. Nat Rev Genet. 2010;11:285–96.

Marin T, Gongol B, Chen Z, Woo B, Subramaniam S, Chien S, Shyy JY. Mechanosensitive microRNAs-role in endothelial responses to shear stress and redox state. Free Radic Biol Med. 2013;64:61–8.

Maron BJ, McKenna WJ, Danielson GK, Kappenberger LJ, Kuhn HJ, Seidman CE, Shah PM, Spencer WH, Spirito P, Ten Cate FJ, Wigle ED, Vogel RA, Abrams J, Bates ER, Brodie BR, Danias PG, Gregoratos G, Hlatky MA, Hochman JS, Kaul S, Lichtenberg RC, Lindner JR, O'rourke RA, Pohost GM, Schofield RS, Tracy CM, Winters WL, Klein WW, Priori SG, Alonso-Garcia A, Blomström-Lundqvist C, De Backer G, Deckers J, Flather M, Hradec J, Oto A, Parkhomenko A, Silber S, Torbicki A. American College of Cardiology/European Society of Cardiology clinical expert consensus document on hypertrophic cardiomyopathy. J Am Coll Cardiol. 2003;42:1687–713.

Marques FZ, Campain AE, Tomaszewski M, Zukowska-Szczechowska E, Yang YH, Charchar FJ, Morris BJ. Gene expression profiling reveals renin mRNA overexpression in human hypertensive kidneys and a role for microRNAs. Hypertension. 2011;58:1093–8.

Martin C, Zhang Y. The diverse functions of histone lysine methylation. Nat Rev Mol Cell Biol. 2005;6:838–49.

Millis RM. Epigenetics and hypertension. Curr Hypertens Rep. 2011;13:21–8.

Miyamoto S, Kawamura T, Morimoto T, Ono K, Wada H, Kawase Y, Matsumori A, Nishio R, Kita T, Hasegawa K. Histone acetyltransferase activity of p300 is required for the promotion of left ventricular remodeling after myocardial infarction in adult mice in vivo. Circulation. 2006;113:679–90.

Mo R, Omvik P, Lund-Johansen P. The Bergen blood pressure study: offspring of two hypertensive parents have significantly higher blood pressures than offspring of one hypertensive and one normotensive parent. J Hypertens. 1995;13:1614–7.

Montgomery RL, Hullinger TG, Semus HM, Dickinson BA, Seto AG, Lynch JM, Stack C, Latimer PA, Olson EN, van Rooij E. Therapeutic inhibition of miR-208a improves cardiac function and survival during heart failure. Circulation. 2011;124:1537–47.

Morano I, Haase K, Haase H, Böhm M, Erdmann E, Schaub MC. Changes in essential myosin light chain isoform expression provide a molecular basis for isometric force regulation in the failing human heart. J Mol Cell Cardiol. 1997;29:1177–87.

Movassagh M, Choy MK, Knowles DA, Cordeddu L, Haider S, Down T, Siggens L, Vujic A, Simeoni I, Penkett C, Goddard M, Lio P, Bennett MR, Foo RS. Distinct epigenomic features in end-stage failing human hearts. Circulation. 2011;124:2411–22.

Narlikar GJ, Sundaramoorthy R, Owen-Hughes T. Mechanisms and functions of ATP-dependent chromatin-remodeling enzymes. Cell. 2013;154:490–503.

Newton-Cheh C, Johnson T, Gateva V, Tobin MD, Bochud M, Coin L, Najjar SS, Zhao JH, Heath SC, Eyheramendy S, Papadakis K, Voight BF, Scott LJ, Zhang F, Farrall M, Tanaka T, Wallace C, Chambers JC, Khaw KT, Nilsson P, van der Harst P, Polidoro S, Grobbee DE, Onland-Moret NC, Bots ML, Wain LV, Elliott KS, Teumer A, Luan J, Lucas G, Kuusisto J, Burton PR, Hadley D, McArdle WL, Wellcome Trust Case Control C, Brown M, Dominiczak A, Newhouse SJ, Samani NJ, Webster J, Zeggini E, Beckmann JS, Bergmann S, Lim N, Song K, Vollenweider P, Waeber G, Waterworth DM, Yuan X, Groop L, Orho-Melander M, Allione A, Di Gregorio A, Guarrera S, Panico S, Ricceri F, Romanazzi V, Sacerdote C, Vineis P, Barroso I, Sandhu MS, Luben RN, Crawford GJ, Jousilahti P, Perola M, Boehnke M, Bonnycastle LL, Collins FS, Jackson AU, Mohlke KL, Stringham HM, Valle TT, Willer CJ, Bergman RN, Morken MA, Doring A, Gieger C, Illig T, Meitinger T, Org E, Pfeufer A, Wichmann HE, Kathiresan S, Marrugat J, O'Donnell CJ, Schwartz SM, Siscovick DS, Subirana I, Freimer

NB, Hartikainen AL, McCarthy MI, O'Reilly PF, Peltonen L, Pouta A, de Jong PE, Snieder H, van Gilst WH, Clarke R, Goel A, Hamsten A, Peden JF, Seedorf U, Syvanen AC, Tognoni G, Lakatta EG, Sanna S, Scheet P, Schlessinger D, Scuteri A, Dorr M, Ernst F, Felix SB, Homuth G, Lorbeer R, Reffelmann T, Rettig R, Volker U, Galan P, Gut IG, Hercberg S, Lathrop GM, Zelenika D, Deloukas P, Soranzo N, Williams FM, Zhai G, Salomaa V, Laakso M, Elosua R, Forouhi NG, Volzke H, Uiterwaal CS, van der Schouw YT, Numans ME, Matullo G, Navis G, Berglund G, Bingham SA, Kooner JS, Connell JM, Bandinelli S, Ferrucci L, Watkins H, Spector TD, Tuomilehto J, Altshuler D, Strachan DP, Laan M, Meneton P, Wareham NJ, Uda M, Jarvelin MR, Mooser V, Melander O, Loos RJ, Elliott P, Abecasis GR, Caulfield M, Munroe PB. Genome-wide association study identifies eight loci associated with blood pressure. Nat Genet. 2009;41:666–76.

Nossent AY, Hansen JL, Doggen C, Quax PH, Sheikh SP, Rosendaal FR. SNPs in microRNA binding sites in 3′-UTRs of RAAS genes influence arterial blood pressure and risk of myocardial infarction. Am J Hypertens. 2011;24:999–1006.

Ohtani K, Dimmeler S. Control of cardiovascular differentiation by microRNAs. Basic Res Cardiol. 2011;106:5–11.

Ordovás JM, Smith C. Epigenetics and cardiovascular disease. Nat Rev Cardiol. 2010;7:510–9.

Osio A, Tan L, Chen SN, Lombardi R, Nagueh SF, Shete S, Roberts R, Willerson JT, Marian AJ. Myozenin 2 is a novel gene for human hypertrophic cardiomyopathy. Circ Res. 2007;100:766–8.

Padmanabhan S, Caulfield M, Dominiczak AF. Genetic and molecular aspects of hypertension. Circ Res. 2015;116:937–59.

Page IH. Pathogenesis of arterial hypertension. J Am Med Assoc. 1949;140:451–8.

Panning B, Jaenisch R. DNA hypomethylation can activate Xist expression and silence X-linked genes. Genes Dev. 1996;10:1991–2002.

Parker TG, Packer S, Schneider MD. Peptide growth factors can provoke "fetal" contractile protein gene expression in rat cardiac myocytes. J Clin Invest. 1990;85:507–14.

Pellegrini E, Signor L, Singh S, Boeri Erba E, Cusack S. Structures of the inactive and active states of RIP2 kinase inform on the mechanism of activation. PLoS One. 2017;12:e0177161.

Peters T, Hermans-Beijnsberger S, Beqqali A, Bitsch N, Nakagawa S, Prasanth KV, de Windt LJ, van Oort RJ, Heymans S, Schroen B. Long non-coding RNA Malat-1 is dispensable during pressure overload-induced cardiac remodeling and failure in mice. PLoS One. 2016;11:e0150236.

Platt R. Heredity in hypertension. Q J Med. 1947;16:311.

Porrello ER, Johnson BA, Aurora AB, Simpson E, Nam YJ, Matkovich SJ, Dorn GW 2nd, van Rooij E, Olson EN. MiR-15 family regulates postnatal mitotic arrest of cardiomyocytes. Circ Res. 2011;109:670–9.

Porrello ER, Mahmoud AI, Simpson E, Johnson BA, Grinsfelder D, Canseco D, Mammen PP, Rothermel BA, Olson EN, Sadek HA. Regulation of neonatal and adult mammalian heart regeneration by the miR-15 family. Proc Natl Acad Sci U S A. 2013;110:187–92.

Qi M, Shannon T, Euler DE, Bers DM, Samarel AM. Downregulation of sarcoplasmic reticulum Ca(2+)-ATPase during progression of left ventricular hypertrophy. Am J Phys. 1997;272:H2416–24.

Qin X, Wang X, Wang Y, Tang Z, Cui Q, Xi J, Li YS, Chien S, Wang N. MicroRNA-19a mediates the suppressive effect of laminar flow on cyclin D1 expression in human umbilical vein endothelial cells. Proc Natl Acad Sci U S A. 2010;107:3240–4.

Rahman MA, Sundaram K, Mitra S, Gavrilin MA, Wewers MD. Receptor interacting protein-2 plays a critical role in human lung epithelial cells survival in response to Fas-induced cell-death. PLoS One. 2014;9:e92731.

Rao RM, Reddy G, Grim CE. Relative role of genes and environment on BP: twin studies in Madras, India. J Hum Hypertens. 1993;7:451–5.

Ravingerová T, Barancik M, Strnisková M. Mitogen-activated protein kinases- a new therapeutic target in cardiac pathology. Mol Cell Biochem. 2003;247:127–38.

Razeghi P, Young M, Alcorn JL, Moravec CS, Frazier OH, Taegtmeyer H. Metabolic gene expression in fetal and failing human heart. Circulation. 2001;104:2923–31.

Rizki G, Boyer LA. Lncing epigenetic control of transcription to cardiovascular development and disease. Circ Res. 2015;117:192–206.

Rodriguez JE, McCudden CR, Willis MS. Familial hypertrophic cardiomyopathy: basic concepts and future molecular diagnostics. Clin Biochem. 2009;42:755–65.

Ryan DP, Owen-Hughes T. Snf2-family proteins: chromatin remodellers for any occasion. Curr Opin Chem Biol. 2011;15:649–56.

Rybak-Wolf A, Stottmeister C, Glazar P, Jens M, Pino N, Giusti S, Hanan M, Behm M, Bartok O, Ashwal-Fluss R, Herzog M, Schreyer L, Papavasileiou P, Ivanov A, Ohman M, Refojo D, Kadener S, Rajewsky N. Circular RNAs in the mammalian brain are highly abundant, conserved, and dynamically expressed. Mol Cell. 2015;58:870–85.

Salzman J, Gawad C, Wang PL, Lacayo N, Brown PO. Circular RNAs are the predominant transcript isoform from hundreds of human genes in diverse cell types. PLoS One. 2012;7:e30733.

Salzman J, Chen RE, Olsen MN, Wang PL, Brown PO. Cell-type specific features of circular RNA expression. PLoS Genet. 2013;9:e1003777.

Schirle NT, Sheu-Gruttadauria J, MacRae IJ. Structural basis for microRNA targeting. Science. 2014;346:608–13.

Schwartz K, de la Bastie D, Bouveret P, Oliviéro P, Alonso S, Buckingham M. A-skeletal muscle actin mRNA's accumulate in hypertrophied adult rat hearts. Circ Res. 1986;59:551–5.

Schwinger RH, Böhm M, Schmidt U, Karczewski P, Bavendiek U, Flesch M, Krause EG, Erdmann E. Unchanged protein levels of SERCA II and phospholamban but reduced Ca2+ uptake and Ca(2+)-ATPase activity of cardiac sarcoplasmic reticulum from dilated cardiomyopathy patients compared with patients with nonfailing hearts. Circulation. 1995;92:3220–8.

Seidman JG, Seidman C. The genetic basis for cardiomyopathy- from mutation identification to mechanistic paradigms. Cell. 2001;104:557–67.

Sethupathy P, Borel C, Gagnebin M, Grant GR, Deutsch S, Elton TS, Hatzigeorgiou AG, Antonarakis SE. Human microRNA-155 on chromosome 21 differentially interacts with its polymorphic target in the AGTR1 3′ untranslated region: a mechanism for functional single-nucleotide polymorphisms related to phenotypes. Am J Hum Genet. 2007;81:405–13.

Shahbazian MD, Grunstein M. Functions of site-specific histone acetylation and deacetylation. Annu Rev Biochem. 2007;76:75–100.

Shao M, Chen G, Lv F, et al. LncRNA TINCR attenuates cardiac hypertrophy by epigenetically silencing CaMKII. Oncotarget. 2017;8:47565–73.

Shear CL, Burke G, Freedman DS, Berenson GS. Value of childhood blood pressure measurements and family history in predicting future blood pressure status: results from 8 years of follow-up in the Bogalusa Heart Study. Pediatrics. 1986;77:862–9.

Shear CL, Freedman D, Burke GL, Harsha DW, Berenson GS. Body fat patterning and blood pressure in children and young adults. The Bogalusa Heart Study. Hypertension. 1987;9:236–44.

Shen S, Jiang H, Bei Y, Xiao J, Li X. Long non-coding RNAs in cardiac remodeling. Cell Physiol Biochem. 2017;41:1830–7.

Shi Y, Whetstine JR. Dynamic regulation of histone lysine methylation by demethylases. Mol Cell. 2007;25:1–14.

Shimoyama M, De Pons J, Hayman GT, Laulederkind SJ, Liu W, Nigam R, Petri V, Smith JR, Tutaj M, Wang SJ, Worthey E, Dwinell M, Jacob H. The Rat Genome Database 2015: genomic, phenotypic and environmental variations and disease. Nucleic Acids Res. 2015;43:D743–50.

Sofer T, Wong Q, Hartwig FP, Taylor K, Warren HR, Evangelou E, Cabrera CP, Levy D, Kramer H, Lange LA, Horta BL, Consortium C-B, Kerr KF, Reiner AP, Franceschini N. Genome-wide association study of blood pressure traits by Hispanic/Latino background: the Hispanic community health study/study of Latinos. Sci Rep. 2017;7:10348.

Song C, Zhang J, Liu Y, et al. Construction and analysis of cardiac hypertrophy-associated lncRNA-mRNA network based on competitive endogenous RNA reveal functional lncRNAs in cardiac hypertrophy. Oncotarget. 2016;7:10827–40.

Sopontammarak S, Aliharoob A, Ocampo C, Arcilla RA, Gupta MP, Gupta M. Mitogen-activated protein kinases (p38 and c-Jun NH2-terminal kinase) are differentially regulated during cardiac volume and pressure overload hypertrophy. Cell Biochem Biophys. 2005;43:61–76.
Studer M, McNaughton PA. Modulation of single-channel properties of TRPV1 by phosphorylation. J Physiol. 2010;588:3743–56.
Suzuki MM, Bird A. DNA methylation landscapes: provocative insights from epigenomics. Nat Rev Genet. 2008;9:465–76.
Tan WL, Lim BT, Anene-Nzelu CG, Ackers-Johnson M, Dashi A, See K, Tiang Z, Lee DP, Chua WW, Luu TD, Li PY, Richards AM, Foo RS. A landscape of circular RNA expression in the human heart. Cardiovasc Res. 2017;113:298–309.
Tang CM, Zhang M, Huang L, Hu ZQ, Zhu JN, Xiao Z, Zhang Z, Lin QX, Zheng XL, Yang M, Wu SL, Cheng JD, Shan ZX. CircRNA_000203 enhances the expression of fibrosis-associated genes by derepressing targets of miR-26b-5p, Col1a2 and CTGF, in cardiac fibroblasts. Sci Rep. 2017;7:40342.
Tigno-Aranjuez JT, Abbott DW. Ubiquitination and phosphorylation in the regulation of NOD2 signaling and NOD2-mediated disease. Biochim Biophys Acta. 2012;1823:2022–8.
Trivedi CM, Luo Y, Yin Z, Zhang M, Zhu W, Wang T, Floss T, Goettlicher M, Noppinger PR, Wurst W, Ferrari VA, Abrams CS, Gruber PJ, Epstein JA. Hdac2 regulates the cardiac hypertrophic response by modulating Gsk3 beta activity. Nat Med. 2007;13:324–31.
Ulitsky I, Shkumatava A, Jan CH, Sive H, Bartel DP. Conserved function of lincRNAs in vertebrate embryonic development despite rapid sequence evolution. Cell. 2011;147:1537–50.
van Rooij E, Sutherland L, Qi X, Richardson JA, Hill J, Olson EN. Control of stress-dependent cardiac growth and gene expression by a microRNA. Science. 2007;316:575–9.
Varga-Weisz PD, Wilm M, Bonte E, Dumas K, Mann M, Becker PB. Chromatin-remodelling factor CHRAC contains the ATPases ISWI and topoisomerase II. Nature. 1997;388:598–602.
Vaughan-Jones RD, Spitzer KW, Swietach P. Intracellular pH regulation in heart. J Mol Cell Cardiol. 2009;46:318–31.
Viereck J, Kumarswamy R, Foinquinos A, et al. Long noncoding RNA Chast promotes cardiac remodeling. Sci Transl Med. 2016;8:326ra22.
Villeneuve LM, Reddy M, Lanting LL, Wang M, Meng L, Natarajan R. Epigenetic histone H3 lysine 9 methylation in metabolic memory and inflammatory phenotype of vascular smooth muscle cells in diabetes. Proc Natl Acad Sci U S A. 2008;105:9047–52.
Waddington CH. The epigenotype. 1942. Int J Epidemiol. 2012;41:10–3.
Wahl MC, Will CL, Luhrmann R. The spliceosome: design principles of a dynamic RNP machine. Cell. 2009;136:701–18.
Wain LV, Verwoert GC, O'Reilly PF, Shi G, Johnson T, Johnson AD, Bochud M, Rice KM, Henneman P, Smith AV, Ehret GB, Amin N, Larson MG, Mooser V, Hadley D, Dorr M, Bis JC, Aspelund T, Esko T, Janssens AC, Zhao JH, Heath S, Laan M, Fu J, Pistis G, Luan J, Arora P, Lucas G, Pirastu N, Pichler I, Jackson AU, Webster RJ, Zhang F, Peden JF, Schmidt H, Tanaka T, Campbell H, Igl W, Milaneschi Y, Hottenga JJ, Vitart V, Chasman DI, Trompet S, Bragg-Gresham JL, Alizadeh BZ, Chambers JC, Guo X, Lehtimaki T, Kuhnel B, Lopez LM, Polasek O, Boban M, Nelson CP, Morrison AC, Pihur V, Ganesh SK, Hofman A, Kundu S, Mattace-Raso FU, Rivadeneira F, Sijbrands EJ, Uitterlinden AG, Hwang SJ, Vasan RS, Wang TJ, Bergmann S, Vollenweider P, Waeber G, Laitinen J, Pouta A, Zitting P, McArdle WL, Kroemer HK, Volker U, Volzke H, Glazer NL, Taylor KD, Harris TB, Alavere H, Haller T, Keis A, Tammesoo ML, Aulchenko Y, Barroso I, Khaw KT, Galan P, Hercberg S, Lathrop M, Eyheramendy S, Org E, Sober S, Lu X, Nolte IM, Penninx BW, Corre T, Masciullo C, Sala C, Groop L, Voight BF, Melander O, O'Donnell CJ, Salomaa V, d'Adamo AP, Fabretto A, Faletra F, Ulivi S, Del Greco F, Facheris M, Collins FS, Bergman RN, Beilby JP, Hung J, Musk AW, Mangino M, Shin SY, Soranzo N, Watkins H, Goel A, Hamsten A, Gider P, Loitfelder M, Zeginigg M, Hernandez D, Najjar SS, Navarro P, Wild SH, Corsi AM, Singleton A, de Geus EJ, Willemsen G, Parker AN, Rose LM, Buckley B, Stott D, Orru M, Uda M, LifeLines Cohort S, van der Klauw MM, Zhang W, Li X, Scott J, Chen YD, Burke GL, Kahonen M, Viikari J, Doring A, Meitinger T,

Davies G, Starr JM, Emilsson V, Plump A, Lindeman JH, Hoen PA, Konig IR, EchoGen C, Felix JF, Clarke R, Hopewell JC, Ongen H, Breteler M, Debette S, Destefano AL, Fornage M, AortaGen C, Mitchell GF, Group CCHFW, Smith NL, KidneyGen C, Holm H, Stefansson K, Thorleifsson G, Thorsteinsdottir U, Consortium CK, Cardiogenics C, CardioGram, Samani NJ, Preuss M, Rudan I, Hayward C, Deary IJ, Wichmann HE, Raitakari OT, Palmas W, Kooner JS, Stolk RP, Jukema JW, Wright AF, Boomsma DI, Bandinelli S, Gyllensten UB, Wilson JF, Ferrucci L, Schmidt R, Farrall M, Spector TD, Palmer LJ, Tuomilehto J, Pfeufer A, Gasparini P, Siscovick D, Altshuler D, Loos RJ, Toniolo D, Snieder H, Gieger C, Meneton P, Wareham NJ, Oostra BA, Metspalu A, Launer L, Rettig R, Strachan DP, Beckmann JS, Witteman JC, Erdmann J, van Dijk KW, Boerwinkle E, Boehnke M, Ridker PM, Jarvelin MR, Chakravarti A, Abecasis GR, Gudnason V, Newton-Cheh C, Levy D, Munroe PB, Psaty BM, Caulfield MJ, Rao DC, Tobin MD, Elliott P, van Duijn CM. Genome-wide association study identifies six new loci influencing pulse pressure and mean arterial pressure. Nat Genet. 2011;43:1005–11.

Wain LV, Vaez A, Jansen R, Joehanes R, van der Most PJ, Erzurumluoglu AM, O'Reilly PF, Cabrera CP, Warren HR, Rose LM, Verwoert GC, Hottenga JJ, Strawbridge RJ, Esko T, Arking DE, Hwang SJ, Guo X, Kutalik Z, Trompet S, Shrine N, Teumer A, Ried JS, Bis JC, Smith AV, Amin N, Nolte IM, Lyytikainen LP, Mahajan A, Wareham NJ, Hofer E, Joshi PK, Kristiansson K, Traglia M, Havulinna AS, Goel A, Nalls MA, Sober S, Vuckovic D, Luan J, Del Greco MF, Ayers KL, Marrugat J, Ruggiero D, Lopez LM, Niiranen T, Enroth S, Jackson AU, Nelson CP, Huffman JE, Zhang W, Marten J, Gandin I, Harris SE, Zemunik T, Lu Y, Evangelou E, Shah N, de Borst MH, Mangino M, Prins BP, Campbell A, Li-Gao R, Chauhan G, Oldmeadow C, Abecasis G, Abedi M, Barbieri CM, Barnes MR, Batini C, Beilby J, Blake T, Boehnke M, Bottinger EP, Braund PS, Brown M, Brumat M, Campbell H, Chambers JC, Cocca M, Collins F, Connell J, Cordell HJ, Damman JJ, Davies G, de Geus EJ, de Mutsert R, Deelen J, Demirkale Y, Doney ASF, Dorr M, Farrall M, Ferreira T, Franberg M, Gao H, Giedraitis V, Gieger C, Giulianini F, Gow AJ, Hamsten A, Harris TB, Hofman A, Holliday EG, Hui J, Jarvelin MR, Johansson A, Johnson AD, Jousilahti P, Jula A, Kahonen M, Kathiresan S, Khaw KT, Kolcic I, Koskinen S, Langenberg C, Larson M, Launer LJ, Lehne B, Liewald DCM, Lin L, Lind L, Mach F, Mamasoula C, Menni C, Mifsud B, Milaneschi Y, Morgan A, Morris AD, Morrison AC, Munson PJ, Nandakumar P, Nguyen QT, Nutile T, Oldehinkel AJ, Oostra BA, Org E, Padmanabhan S, Palotie A, Pare G, Pattie A, Penninx B, Poulter N, Pramstaller PP, Raitakari OT, Ren M, Rice K, Ridker PM, Riese H, Ripatti S, Robino A, Rotter JI, Rudan I, Saba Y, Saint Pierre A, Sala CF, Sarin AP, Schmidt R, Scott R, Seelen MA, Shields DC, Siscovick D, Sorice R, Stanton A, Stott DJ, Sundstrom J, Swertz M, Taylor KD, Thom S, Tzoulaki I, Tzourio C, Uitterlinden AG, Volker U, Vollenweider P, Wild S, Willemsen G, Wright AF, Yao J, Theriault S, Conen D, Attia J, Sever P, Debette S, Mook-Kanamori DO, Zeggini E, Spector TD, van der Harst P, Palmer CNA, Vergnaud AC, Loos RJF, Polasek O, Starr JM, Girotto G, Hayward C, Kooner JS, Lindgren CM, Vitart V, Samani NJ, Tuomilehto J, Gyllensten U, Knekt P, Deary IJ, Ciullo M, Elosua R, Keavney BD, Hicks AA, Scott RA, Gasparini P, Laan M, Liu Y, Watkins H, Hartman CA, Salomaa V, Toniolo D, Perola M, Wilson JF, Schmidt H, Zhao JH, Lehtimaki T, van Duijn CM, Gudnason V, Psaty BM, Peters A, Rettig R, James A, Jukema JW, Strachan DP, Palmas W, Metspalu A, Ingelsson E, Boomsma DI, Franco OH, Bochud M, Newton-Cheh C, Munroe PB, Elliott P, Chasman DI, Chakravarti A, Knight J, Morris AP, Levy D, Tobin MD, Snieder H, Caulfield MJ, Ehret GB. Novel blood pressure locus and gene discovery sing genome-wide association study and expression data sets from blood and the kidney. Hypertension. 2017.

Wang Y, Wang Z. Efficient backsplicing produces translatable circular mRNAs. RNA. 2015;21:172–9.

Wang Y, Juranek S, Li H, Sheng G, Tuschl T, Patel DJ. Structure of an argonaute silencing complex with a seed-containing guide DNA and target RNA duplex. Nature. 2008;456:921–6.

Wang K, Liu F, Zhou LY, Long B, Yuan SM, Wang Y, Liu CY, Sun T, Zhang XJ, Li PF. The long noncoding RNA CHRF regulates cardiac hypertrophy by targeting miR-489. Circ Res. 2014;114:1377–88.

Wang YH, Yu XH, Luo SS, Han H. Comprehensive circular RNA profiling reveals that circular RNA100783 is involved in chronic CD28-associated CD8(+)T cell ageing. Immun Ageing: I & A. 2015;12:17.

Wang Z, Zhang XJ, Ji YX, Zhang P, Deng KQ, Gong J, Ren S, Wang X, Chen I, Wang H, Gao C, Yokota T, Ang YS, Li S, Cass A, Vondriska TM, Li G, Deb A, Srivastava D, Yang HT, Xiao X, Li H, Wang Y. The long noncoding RNA Chaer defines an epigenetic checkpoint in cardiac hypertrophy. Nat Med. 2016a;22:1131–9.

Wang K, Long B, Liu F, Wang JX, Liu CY, Zhao B, Zhou LY, Sun T, Wang M, Yu T, Gong Y, Liu J, Dong YH, Li N, Li PF. A circular RNA protects the heart from pathological hypertrophy and heart failure by targeting miR-223. Eur Heart J. 2016b;37:2602–11.

Warren HR, Evangelou E, Cabrera CP, Gao H, Ren M, Mifsud B, Ntalla I, Surendran P, Liu C, Cook JP, Kraja AT, Drenos F, Loh M, Verweij N, Marten J, Karaman I, Lepe MP, O'Reilly PF, Knight J, Snieder H, Kato N, He J, Tai ES, Said MA, Porteous D, Alver M, Poulter N, Farrall M, Gansevoort RT, Padmanabhan S, Magi R, Stanton A, Connell J, Bakker SJ, Metspalu A, Shields DC, Thom S, Brown M, Sever P, Esko T, Hayward C, van der Harst P, Saleheen D, Chowdhury R, Chambers JC, Chasman DI, Chakravarti A, Newton-Cheh C, Lindgren CM, Levy D, Kooner JS, Keavney B, Tomaszewski M, Samani NJ, Howson JM, Tobin MD, Munroe PB, Ehret GB, Wain LV, International Consortium of Blood Pressure GA, Consortium B, Lifelines Cohort S, Understanding Society Scientific g, Consortium CHDE, Exome BPC, Consortium TDG, Go TDC, Cohorts for H, Ageing Research in Genome Epidemiology BPEC, International Genomics of Blood Pressure C, Group UKBCCBw. Genome-wide association analysis identifies novel blood pressure loci and offers biological insights into cardiovascular risk. Nat Genet. 2017;49:403–15.

Watson JD, Crick F. The structure of DNA. Cold Spring Harb Symp Quant Biol. 1953a;18:123–31.

Watson JD, Crick F. Molecular structure of nucleic acids; a structure for deoxyribose nucleic acid. Nature. 1953b;171:737–8.

Werfel S, Nothjunge S, Schwarzmayr T, Strom TM, Meitinger T, Engelhardt S. Characterization of circular RNAs in human, mouse and rat hearts. J Mol Cell Cardiol. 2016;98:103–7.

Whelton PK, Carey RM, Aronow WS, Casey DE Jr, Collins KJ, Dennison Himmelfarb C, DePalma SM, Gidding S, Jamerson KA, Jones DW, MacLaughlin EJ, Muntner P, Ovbiagele B, Smith SC Jr, Spencer CC, Stafford RS, Taler SJ, Thomas RJ, Williams KA Sr, Williamson JD, Wright JT Jr. 2017 ACC/AHA/AAPA/ABC/ACPM/AGS/APhA/ASH/ASPC/NMA/PCNA guideline for the prevention, detection, evaluation, and management of high blood pressure in adults: executive summary: a report of the American College of Cardiology/American Heart Association Task Force on Clinical Practice Guidelines. J Am Coll Cardiol. 2018;71(19):2199–269.

Wierda RJ, Geutskens S, Jukema JW, Quax PH, van den Elsen PJ. Epigenetics in atherosclerosis and inflammation. J Cell Mol Med. 2010;14:1225–40.

Worley KC, Weinstock GM, Gibbs RA. Rats in the genomic era. Physiol Genomics. 2008;32:273–82.

Xuan L, Sun L, Zhang Y, Huang Y, Hou Y, Li Q, Guo Y, Feng B, Cui L, Wang X, Wang Z, Tian Y, Yu B, Wang S, Xu C, Zhang M, Du Z, Lu Y, Yang BF. Circulating long non-coding RNAs NRON and MHRT as novel predictive biomarkers of heart failure. J Cell Mol Med. 2017;21:1803–14.

Yamazaki T, Yazaki Y. Molecular basis of cardiac hypertrophy. Z Kardiol. 2000;89:1–6.

Yanazume T, Hasegawa K, Morimoto T, Kawamura T, Wada H, Matsumori A, Kawase Y, Hirai M, Kita T. Cardiac p300 is involved in myocyte growth with decompensated heart failure. Mol Cell Biol. 2003;23:3593–606.

Yang XJ, Seto E. The Rpd3/Hda1 family of lysine deacetylases: from bacteria and yeast to mice and men. Nat Rev Mol Cell Biol. 2008;9:206–18.

Young D, Popovic ZB, Jones WK, Gupta S. Blockade of NF-kappaB using IkappaB alpha dominant-negative mice ameliorates cardiac hypertrophy in myotrophin-overexpressed transgenic mice. J Mol Biol. 2008;381:559–68.

Zaphiropoulos PG. Circular RNAs from transcripts of the rat cytochrome P450 2C24 gene: correlation with exon skipping. Proc Natl Acad Sci U S A. 1996;93:6536–41.

Zarain-Herzberg A, Afzal N, Elimban V, Dhalla NS. Decreased expression of cardiac sarcoplasmic reticulum CaZ+-pump ATPase in congestive heart failure due to myocardial infarction. Mol Cell Biochem. 1996;163–164:285–90.

Zhang XO, Wang HB, Zhang Y, Lu X, Chen LL, Yang L. Complementary sequence-mediated exon circularization. Cell. 2014;159:134–47.

Zhao CH, Ma X, Guo HY, Li P, Liu HY. RIP2 deficiency attenuates cardiac hypertrophy, inflammation and fibrosis in pressure overload induced mice. Biochem Biophys Res Commun. 2017;493:1151–8.

Zheng C, Niu H, Li M, Zhang H, Yang Z, Tian L, Wu Z, Li D, Chen X. Cyclic RNA hsacirc000595 regulates apoptosis of aortic smooth muscle cells. Mol Med Rep. 2015;12:6656–62.

Zhong B, Wang D. TRPV1 gene knockout impairs preconditioning protection against myocardial injury in isolated perfused hearts in mice. Am J Physiol Heart Circ Physiol. 2007;293:H1791–8.

Zhong B, Wang D. Protease-activated receptor 2-mediated protection of myocardial ischemia-reperfusion injury: role of transient receptor potential vanilloid receptors. Am J Physiol Regul Integr Comp Physiol. 2009;297:R1681–90.

Zhou B, Bentham J, Di Cesare M, Bixby H, Danaei G, Cowan MJ, Paciorek CJ, Singh G, Hajifathalian K, Bennett JE, Taddei C, Bilano V, Carrillo-Larco RM, Djalalinia S, Khatibzadeh S, Lugero C, Peykari N, Zhang WZ, Lu Y, Stevens GA, Riley LM, Bovet P, Elliott P, Gu D, Ikeda N, Jackson RT, Joffres M, Kengne AP, Laatikainen T, Lam TH, Laxmaiah A, Liu J, Miranda JJ, Mondo CK, Neuhauser HK, Sundström J, Smeeth L, Soric M, Woodward M, Ezzati M, Abarca-Gómez L, Abdeen ZA, Rahim HA, Abu-Rmeileh NM, Acosta-Cazares B, Adams R, Aekplakorn W, Afsana K, Aguilar-Salinas CA, Agyemang C, Ahmadvand A, Ahrens W, Al Raddadi R, Al Woyatan R, Ali MM, Alkerwi AA, Aly E, Amouyel P, Amuzu A, Andersen LB, Anderssen SA, Ängquist L, Anjana RM, Ansong D, Aounallah-Skhiri H, Araújo J, Ariansen I, Aris T, Arlappa N, Aryal K, Arveiler D, Assah FK, Assunção MCF, Avdicová M, Azevedo A, Azizi F, Babu BV, Bahijri S, Balakrishna N, Bandosz P, Banegas JR, Barbagallo CM, Barceló A, Barkat A, Barros AJD, Barros MV, Bata I, Batieha AM, Baur LA, Beaglehole R, Romdhane HB, Benet M, Benson LS, Bernabe-Ortiz A, Bernotiene G, Bettiol H, Bhagyalaxmi A, Bharadwaj S, Bhargava SK, Bi Y, Bikbov M, Bjerregaard P, Bjertness E, Björkelund C, Blokstra A, Bo S, Bobak M, Boeing H, Boggia JG, Boissonnet CP, Bongard V, Braeckman L, Brajkovich I, Branca F, Breckenkamp J, Brenner H, Brewster LM, Bruno G, Bueno-de-Mesquita HB, Bugge A, Burns C, Bursztyn M, de León AC, Cacciottolo J, Cameron C, Can G, Cândido APC, Capuano V, Cardoso VC, Carlsson AC, Carvalho MJ, Casanueva FF, Casas J-P, Caserta CA, Chamukuttan S, Chan AW, Chan Q, Chaturvedi HK, Chaturvedi N, Chen C-J, Chen F, Chen H, Chen S, Chen Z, Cheng C-Y, Dekkaki IC, Chetrit A, Chiolero A, Chiou S-T, Chirita-Emandi A, Cho B, Cho Y, Chudek J, Cifkova R, Claessens F, Clays E, Concin H, Cooper C, Cooper R, Coppinger TC, Costanzo S, Cottel D, Cowell C, Craig CL, Crujeiras AB, Cruz JJ, D'Arrigo G, d'Orsi E, Dallongeville J, Damasceno A, Dankner R, Dantoft TM, Dauchet L, De Backer G, De Bacquer D, de Gaetano G, De Henauw S, De Smedt D, Deepa M, Dehghan A, Delisle H, Deschamps V, Dhana K, Di Castelnuovo AF, Dias-da-Costa JS, Diaz A, Dickerson TT, Do HTP, Dobson AJ, Donfrancesco C, Donoso SP, Döring A, Doua K, Drygas W, Dulskiene V, Džakula A, Dzerve V, Dziankowska-Zaborszczyk E, Eggertsen R, Ekelund U, El Ati J, Ellert U, Elliott P, Elosua R, Erasmus RT, Erem C, Eriksen L, de la Peña JE, Evans A, Faeh D, Fall CH, Farzadfar F, Felix-Redondo FJ, Ferguson TS, Fernández-Bergés D, Ferrante D, Ferrari M, Ferreccio C, Ferrieres J, Finn JD, Fischer K, Föger B, Foo LH, Forslund A-S, Forsner M, Fortmann SP, Fouad HM, Francis DK, Franco MC, Franco OH, Frontera G, Fuchs FD, Fuchs SC, Fujita Y, Furusawa T, Gaciong Z, Gareta D, Garnett SP, Gaspoz J-M, Gasull M, Gates L, Gavrila D, Geleijnse JM, Ghasemian A, Ghimire A, Giampaoli S, Gianfagna F, Giovannelli J, Goldsmith RA, Gonçalves H, Gross MG, Rivas JPG, Gottrand F, Graff-Iversen S, Grafnetter D, Grajda A, Gregor RD, Grodzicki T, Grøntved A, Gruden G, Grujic V, Gu D, Guan OP, Gudnason V, Guerrero R, Guessous I, Guimaraes AL, Gulliford MC, Gunnlaugsdottir J, Gunter M, Gupta PC, Gureje O, Gurzkowska B, Gutierrez L, Gutzwiller F, Hadaegh F, Halkjær J, Hambleton IR, Hardy R, Harikumar R, Hata J, Hayes AJ, He J, Hendriks ME, Henriques A, Cadena LH,

Herrala S, Heshmat R, Hihtaniemi IT, Ho SY, Ho SC, Hobbs M, Hofman A, Dinc GH, Hormiga CM, Horta BL, Houti L, Howitt C, Htay TT, Htet AS, Hu Y, Huerta JM, Husseini AS, Huybrechts I, Hwalla N, Iacoviello L, Iannone AG, Ibrahim MM, Ikram MA, Irazola VE, Islam M, Ivkovic V, Iwasaki M, Jackson RT, Jacobs JM, Jafar T, Jamrozik K, Janszky I, Jasienska G, Jelakovic B, Jiang CQ, Joffres M, Johansson M, Jonas JB, Jørgensen T, Joshi P, Juolevi A, Jurak G, Jureša V, Kaaks R, Kafatos A, Kalter-Leibovici O, Kamaruddin NA, Kasaeian A, Katz J, Kauhanen J, Kaur P, Kavousi M, Kazakbaeva G, Keil U, Boker LK, Keinänen-Kiukaanniemi S, Kelishadi R, Kemper HCG, Kengne AP, Kersting M, Key T, Khader YS, Khalili D, Khang Y-H, Khaw K-T, Kiechl S, Killewo J, Kim J, Klumbiene J, Kolle E, Kolsteren P, Korrovits P, Koskinen S, Kouda K, Koziel S, Kristensen PL, Krokstad S, Kromhout D, Kruger HS, Kubinova R, Kuciene R, Kuh D, Kujala UM, Kula K, Kulaga Z, Kumar RK, Kurjata P, Kusuma YS, Kuulasmaa K, Kyobutungi C, Laatikainen T, Lachat C, Lam TH, Landrove O, Lanska V, Lappas G, Larijani B, Laugsand LE, Laxmaiah A, Bao KLN, Le TD, Leclercq C, Lee J, Lee J, Lehtimäki T, Lekhraj R, León-Muñoz LM, Levitt NS, Li Y, Lilly CL, Lim W-Y, Lima-Costa MF, Lin H-H, Lin X, Linneberg A, Lissner L, Litwin M, Lorbeer R, Lotufo PA, Lozano JE, Luksiene D, Lundqvist A, Lunet N, Lytsy P, Ma G, Ma J, Machado-Coelho GLL, Machi S, Maggi S, Magliano DJ, Majer M, Makdisse M, Malekzadeh R, Malhotra R, Rao KM, Malyutina S, Manios Y, Mann JI, Manzato E, Margozzini P, Marques-Vidal P, Marrugat J, Martorell R, Mathiesen EB, Matijasevich A, Matsha TE, Mbanya JCN, Posso AJMD, McCarlane SR, McGarvey ST, McLachlan S, McLean RM, McNulty BA, Khir ASM, Mediene-Benchekor S, Medzioniene J, Meirhaeghe A, Meisinger C, Menezes AMB, Menon GR, Meshram II, Metspalu A, Mi J, Mikkel K, Miller JC, Miquel JF, Mišigoj-Durakovic M, Mohamed MK, Mohammad K, Mohammadifard N, Mohan V, Yusoff MFM, Møller NC, Molnár D, Momenan A, Mondo CK, Monyeki KDK, Moreira LB, Morejon A, Moreno LA, Morgan K, Moschonis G, Mossakowska M, Mostafa A, Mota J, Motlagh ME, Motta J, Muiesan ML, Müller-Nurasyid M, Murphy N, Mursu J, Musil V, Nagel G, Naidu BM, Nakamura H, Námešná J, Nang EEK, Nangia VB, Narake S, Navarrete-Muñoz EM, Ndiaye NC, Neal WA, Nenko I, Nervi F, Nguyen ND, Nguyen QN, Nieto-Martínez RE, Niiranen TJ, Ning G, Ninomiya T, Nishtar S, Noale M, Noboa OA, Noorbala AA, Noorbala T, Noto D, Al Nsour M, O'Reilly D, Oh K, Olinto MTA, Oliveira IO, Omar MA, Onat A, Ordunez P, Osmond C, Ostojic SM, Otero JA, Overvad K, Owusu-Dabo E, Paccaud FM, Padez C, Pahomova E, Pajak A, Palli D, Palmieri L, Panda-Jonas S, Panza F, Papandreou D, Parnell WR, Parsaeian M, Pecin I, Pednekar MS, Peer N, Peeters PH, Peixoto SV, Pelletier C, Peltonen M, Pereira AC, Pérez RM, Peters A, Petkeviciene J, Pham ST, Pigeot I, Pikhart H, Pilav A, Pilotto L, Pitakaka F, Plans-Rubió P, Polakowska M, Polašek O, Porta M, Portegies MLP, Pourshams A, Pradeepa R, Prashant M, Price JF, Puiu M, Punab M, Qasrawi RF, Qorbani M, Radic I, Radisauskas R, Rahman M, Raitakari O, Raj M, Rao SR, Ramachandran A, Ramos E, Rampal S, Reina DAR, Rasmussen F, Redon J, Reganit PFM, Ribeiro R, Riboli E, Rigo F, de Wit TFR, Ritti-Dias RM, Robinson SM, Robitaille C, Rodríguez-Artalejo F, Rodriguez-Perez del Cristo M, Rodríguez-Villamizar LA, Rojas-Martinez R, Rosengren A, Rubinstein A, Rui O, Ruiz-Betancourt BS, Horimoto ARVR, Rutkowski M, Sabanayagam C, Sachdev HS, Saidi O, Sakarya S, Salanave B, Salazar Martinez E, Salmerón D, Salomaa V, Salonen JT, Salvetti M, Sánchez-Abanto J, Sans S, Santos D, Santos IS, dos Santos RN, Santos R, Saramies JL, Sardinha LB, Margolis GS, Sarrafzadegan N, Saum K-U, Savva SC, Scazufca M, Schargrodsky H, Schneider IJ, Schultsz C, Schutte AE, Sen A, Senbanjo IO, Sepanlou SG, Sharma SK, Shaw JE, Shibuya K, Shin DW, Shin Y, Siantar R, Sibai AM, Silva DAS, Simon M, Simons J, Simons LA, Sjöström M, Skovbjerg S, Slowikowska-Hilczer J, Slusarczyk P, Smeeth L, Smith MC, Snijder MB, So H-K, Sobngwi E, Söderberg S, Solfrizzi V, Sonestedt E, Song Y, Sørensen TIA, Jérome CS, Soumare A, Staessen JA, Starc G, Stathopoulou MG, Stavreski B, Steene-Johannessen J, Stehle P, Stein AD, Stergiou GS, Stessman J, Stieber J, Stöckl D, Stocks T, Stokwiszewski J, Stronks K, Strufaldi MW, Sun C-A, Sundström J, Sung Y-T, Suriyawongpaisal P, Sy RG, Tai ES, Tammesoo M-L, Tamosiunas A, Tang L, Tang X, Tanser F, Tao Y, Tarawneh MR, Tarqui-Mamani CB, Taylor A, Theobald H, Thijs L, Thuesen BH, Tjonneland A, Tolonen HK, Tolstrup JS, Topbas M, Topór-Madry R,

Tormo MJ, Torrent M, Traissac P, Trichopoulos D, Trichopoulou A, Trinh OTH, Trivedi A, Tshepo L, Tulloch-Reid MK, Tuomainen T-P, Tuomilehto J, Turley ML, Tynelius P, Tzourio C, Ueda P, Ugel E, Ulmer H, Uusitalo HMT, Valdivia G, Valvi D, van der Schouw YT, Van Herck K, van Rossem L, van Valkengoed IGM, Vanderschueren D, Vanuzzo D, Vatten L, Vega T, Velasquez-Melendez G, Veronesi G, Verschuren WMM, Verstraeten R, Victora CG, Viet L, Viikari-Juntura E, Vineis P, Vioque J, Virtanen JK, Visvikis-Siest S, Viswanathan B, Vollenweider P, Voutilainen S, Vrdoljak A, Vrijheid M, Wade AN, Wagner A, Walton J, Mohamud WNW, Wang M-D, Wang Q, Wang YX, Wannamethee SG, Wareham N, Wederkopp N, Weerasekera D, Whincup PH, Widhalm K, Widyahening IS, Wiecek A, Wijga AH, Wilks RJ, Willeit J, Willeit P, Williams EA, Wilsgaard T, Wojtyniak B, Wong TY, Wong-McClure RA, Woo J, Woodward M, Wu AG, Wu FC, Wu SL, Xu H, Yan W, Yang X, Ye X, Yiallouros PK, Yoshihara A, Younger-Coleman NO, Yusoff AF, Yusoff MFM, Zambon S, Zdrojewski T, Zeng Y, Zhao D, Zhao W, Zheng Y, Zhu D, Zimmermann E, Zuñiga Cisneros J. Worldwide trends in blood pressure from 1975 to 2015: a pooled analysis of 1479 population-based measurement studies with 19·1 million participants. Lancet. 2017;389:37–55.

Zink LM, Hake S. Histone variants: nuclear function and disease. Curr Opin Genet Dev. 2016;37:82–9.

Chapter 8
Progress of Genomics in Atherosclerosis-Coronary Heart Disease and Myocardial Infarction

Jinqing Yuan and Yue Liu

Abstract Coronary artery disease (CAD) has long been the leading cause of death in the world. It produces great burden on public health and individual life quality. Albeit a complete picture of its biological nature hasn't been clarified yet, it is acknowledged that CAD, as one of common complex disorders, results from a combination of heritable and environmental factors. In the past decades, the advent and rapid growth of genome-wide association studies (GWASs) for the analysis between phenotypes and genotypes shed light on the heritability underlying CAD. Dozens of risk variants associated with CAD have been identified, and a minority of them present connection with known biological pathways. Nevertheless, it is kept in dispute with respect to whether these substantial and costly projects truly contribute to basic research and clinical utility. Hereinafter, we summarized major findings in genetic architecture of CAD, challenges, and prospects of GWASs.

Keywords Atherosclerosis · Coronary heart disease · Myocardial infarction · GWAS

Abbreviations

CAD	Coronary artery disease
CARDIoGRAM	Coronary artery disease genome-wide replication and meta-analysis
CHD	Coronary heart disease
CNV	Copy-number variations
CVD	Cardiovascular disease

J. Yuan (✉) · Y. Liu
State Key Laboratory of Cardiovascular Disease, Fuwai Hospital, National Center for Cardiovascular Diseases, Chinese Academy of Medical Sciences and Peking Union Medical College, Beijing, People's Republic of China

H. Jiang, M. Liu (eds.), *Heart Genomics*, Translational Bioinformatics 16,
https://doi.org/10.1007/978-981-13-1429-2_8

EWAS	Epigenome-wide association study
FDR	False discovery rate
FH	Familial hypercholesterolemia
FRS	Framingham risk score
GRS	Genetic risk score
GWAS	Genome-wide association study
IHD	Ischemic heart disease
LD	Linkage disequilibrium
LDL	Low-density lipoprotein
MAF	Minor allele frequency
MI	Myocardial infarction
MR	Mendelian randomization
NGS	Next-generation sequencing
PCSK9	Proprotein convertase subtilisin/kexin type 9
PRS	Polygenic risk score
RRS	Reynolds risk score
SNP	Single-nucleotide polymorphism
WES	Whole exome sequencing
WGS	Whole-genome sequencing
WTCCC	Wellcome Trust Case Control Consortium

Coronary artery disease (CAD) also called coronary heart disease (CHD), ischemic heart disease (IHD), is a common term for a group of diseases including angina pectoris, myocardial infarction, and sudden cardiac death. Albeit various types of classifications, all feature atherosclerotic plaque in coronary artery that leads to short supply of blood flow.

1 Overview of Coronary Artery Disease: Epidemiology and Etiology

CAD has long been the number one killer in the world. It produces great burden on public health and individual life quality. The Global Burden of Disease 2016 Study (GBD 2016) announced over 153 million people with prevalent CAD and 19% increase in total deaths from CAD, rising from 7.96 million in 2006 to 9.48 million in 2016 (GBD, C.O.D.C. 2017; GBD, D.A.I.I. 2017). According to American Heart Association, 80% mortality from cardiovascular disease (CVD) occurred in low- and middle-income countries, and 16.5 million Americans suffered from coronary heart disease (myocardial infarction and angina pectoris) (Benjamin et al. 2017). Furthermore, epidemiological evidence also shows increased prevalence of CAD globally, which is caused by mixed factors such as aggravating trend of aging population, adoption of a sedentary lifestyle and Western diets, and so on. To be note, substantial decrease in incidence is found in the developed world (Assimes and Roberts 2016). This again underscores that strategies of prevention and management matter a lot.

Current scientific knowledge has not hitherto clarified the underlying etiology of CAD completely. It is generally accepted that CAD is characterized by lifelong exposure to an atherogenic environment in the context of heritable susceptibility. Traditional risk factors for CAD include dyslipidemia, hypertension, tobacco use, diabetes, etc. Meanwhile, a family history of CAD particularly premature CAD also acts as an independent and strong risk factor (Chow et al. 2011; Sivapalaratnam et al. 2010). It is said that genetic variance accounts for as much as 40–60% of the variation in risk of CAD incidence (Björkegren et al. 2015).

2 Genetic Variation

Human genome consists of about 22,500 genes and 3.2 billion base pairs of DNA (Pertea and Salzberg 2010). No two humans share exactly identical genetic information for nearly 0.5% of bases (1.5 million) are completely different. The term "genetic variation" is used to describe all DNA sequence variants that occur between subjects, including single-nucleotide polymorphism (SNP) and structural variation (Johansen et al. 2010). It can disrupt the integrity of genes and thereafter alters their expressions or functions sometimes. SNP denotes a substitution of one base for another, and it occurs frequently, whereas the latter one refers to a change of DNA generally ranging from 1 to 5000 kb in size. Structural variations can be either unbalanced (e.g., copy-number variations) or balanced (e.g., genomic inversions, reciprocal translocations). It is now believed that copy-number variations (CNVs) present more prevalent than SNPs. Deletion, duplications, insertion, triplication, and translocation can all result in CNVs. CNVs can further affect the proper dosage of genes, as they can contain entire genes or regulatory elements (Johansen et al. 2010; Stankiewicz and Lupski 2010). Nevertheless, SNPs analysis has been the predominant approach to unveil the association between genetics/genomes with complex diseases so far. Hybrid genotyping platforms have recently been developed to genotype SNPs and CNVs simultaneously (McCarroll et al. 2008).

3 Genome-Wide Association Study (GWAS): Rationale and Development

Scientists believe that mutations that contribute to susceptibility for common diseases shall themselves be common ["common disease, common variant" (CDCV) hypothesis] (Altshuler et al. 2008). To be specific, many genetic variants predispose individuals to common complex disorders additively, though each with a relatively weak effect. Based on this rationale, GWASs were launched in the last decades to elucidate associations between genetic variants and traits in a large scale of population. GWAS is a high-throughput genotyping way that is capable of testing hundreds to thousands of SNPs. Its aim is to elicit disease-related loci by comparing SNPs frequencies among affected and unaffected individuals on the whole-genome level, so there is no need to assume "candidate genes" in advance (Altshuler et al. 2008). Two main advances in genetics and technology expedite the feasibility of conducting GWAS. First, the principle of linkage disequilibrium (LD) at the population level, which depicts a nonrandom correlation among alleles probably inherited altogether. The International HapMap project, choosing 269 DNA samples from four populations, identified almost ten million common SNPs (typically above 5% frequency) within haplotype blocks (LD structure), facilitating a reduced number of SNPs for covering complete whole genome (International HapMap Consortium 2005). And the second one is due to the development of DNA microarrays in parallel with reduced costs, allowing eligibility of larger-scale genotyping.

Since the first GWASs' results were published in 2005–2007, thousands of loci have been identified to be linked with a large range of traits, including common diseases such as type 2 diabetes and schizophrenia, metabolic processes known as risk biomarkers like blood lipids, and individual traits such as height (Visscher et al. 2012). The Wellcome Trust Case Control Consortium (WTCCC), one of the first GWASs of CAD including 1926 case subjects, discovered the first loci at chromosome 9 (9p21.3) that was strongly associated with CAD (WTCCC 2007). Later, to better detect true associations for CAD which requires larger samples, the Coronary Artery Disease Genome-wide Replication and Meta-Analysis (CARDIoGRAM) consortium was set up (Preuss et al. 2010). This project with collaboration of 14 GWAS involved constitutes 86,995 white individuals (22,233 cases and 64,762 controls) in total. Former GWASs data were limited in their ability to shed light on the genetic architecture of CAD in non-European populations. Nowadays, several GWASs concerning both Eastern Asian and African subjects have been carried out (Lee et al. 2013; Nikpay et al. 2015). These data further identified novel loci in non-European decedents to help provide a complete picture of CAD etiology.

To date, GWASs have yielded a valuable repository of over 200 genome-wide significant or suggestive risk loci for CAD, the majority of them having small effect sizes with odd ratio <1.25 (Nikpay et al. 2015). Most of them are in noncoding regions and may involve in disease process via regulation of gene expression (Brænne et al. 2015). Meta-analyses only interpreted causality of a few loci related to traditional risk factors (lipid, blood pressure, inflammation, blood glucose, ABO system, etc.), of which an association with lipid metabolism constitutes the majority (Deloukas et al. 2013). There still hangs a long list of loci that failed to show any relevance with known biomarkers for CAD. Therefore, it is logic to infer the possibility of other mechanisms underlying the pathogenesis of CAD that is worthy of exploration.

4 Genetic Variants Associated with Coronary Artery Disease

4.1 9p21 (CDKN2B-CDKN2A-ANRIL)

The locus on chromosome 9 region 21, a 53 kb LD block with multiple highly correlated SNPs, was the first common variant locus of CAD identified and shows the robust and most consistent relation with CAD in replicable studies (Samani et al. 2007). This allele has a highly prevalent minor allele frequency (MAF) of 50% in European decedents. About 25% of Europeans were homozygous for the risk allele and had a 40% increased risk of CAD (McPherson et al. 2007). Comparable results were later confirmed in other ethnic groups like East-Asian, Hispanic, Han-Chinese, except Africans (Zhou et al. 2008; Teo et al. 2010; Assimes et al. 2008). Interestingly, its effect for CAD via mechanisms independent of known risk factors still remains to be discussed.

Several genes at this locus are investigated such as CDKN2B, CDKN2A, and ANRIL (CDKN2B-AS1). CDKN2B and CDKN2A are genes encoding protein as cyclin-dependent kinase inhibitors. These proteins are involved in the regulation of cell cycle and apoptosis, stem cell renewal, and maintenance of cellular homeostasis (Assimes and Roberts 2016). The expressions of CDKN2A and CDKN2B have been found in human vasculature and atherosclerotic plaques. Deletion of 9p21 in mice confirmed decreased CDKN2A and CDKN2B, causing excessively proliferative capacity in aortic smooth muscle cells (Visel et al. 2010). ANRIL (CDKN2B-AS1) encodes long noncoding RNA (lnc-RNA) that can alter gene expression of neighboring genes by RNA interference or gene silencing, which may participate in the regulation of CDKN2A and CDKN2B expression (Broadbent et al. 2008). Its expression in vascular endothelium, monocyte-derived macrophages and vascular smooth muscle cells implicates its role in atherosclerosis (Hannou et al. 2015).

4.2 1p13 (CELSR2-PSRC1-SORT1)

The strongest risk loci for CAD second only to 9p21 is the CELSR2-PSRC1-SORT1 gene cluster (Samani et al. 2007). These three genes are located in chromosome 1 region 13. 1p13.3 is found to be highly related with both coronary artery disease and low-density lipoprotein-cholesterol (LDL-C) concentrations. Musunuru et al. demonstrated that rs12740374 at intergenic noncoding region functions as a causative SNP that creates a C/EBP transcription factor binding site and consequently decreases hepatic gene expression SORT1 (Musunuru et al. 2010). Complex effects of SORT1 on lipid metabolism were found under different conditions. Sortilin, production of SORT1, is known as a neurotensin receptor. Kjolby et al. noticed a reduction of cholesterol levels in SORT1-deficient mice (Kjolby et al. 2010). Reasonable explanation is that sortilin can bind apoB100 in the Golgi and facilitate its secretion from the liver. Another experiment supported that sortilin increases LDL concentrations by promotion of PCSK9 secretion and downregulation of LDL receptors (Gustafsen et al. 2014). Moreover, another experiment noticed the direct pro-atherosclerotic effect of sortilin in immune cells (Patel et al. 2011). Above all, it implies that sortilin is engaged in multiple pathways of atherosclerosis.

4.3 19p13 (LDLR) and 1p32 (PCSK9)

Both LDLR (LDL receptor) and PCSK9 (proprotein convertase subtilisin/kexin type 9) are also LDL-C-associated loci that involve in the occurrence of CAD and MI (Kathiresan et al. 2009). The risk alleles are, respectively, located in the 19p13 and 1p32. Both of them are confirmed as independent risk alleles across multiple populations and function in the metabolism of LDL particles (Johansen et al. 2010). Rare variants in LDLR and PCSK9 cause a monogenic disease familial hypercholesterolemia (FH) characterized by lipid accumulation in tendons, the cornea, and arterial wall (Hegele 2009; Yamamoto et al. 1989). In untreated heterozygous FH patients, about 50% of males and 30% of females will develop CVD before the age of 60 years (Stone et al. 1974).

4.4 6q25 (LPA)

In the GWA studies, another highly significant signal at 6q25 at the lipoprotein(a) (LPA) was explored. Encoded by LPA, apolipoprotein little a [Apo(a)], together with ApoB in covalent link forms lipoprotein(a) [Lp(a)] (Schmidt et al. 2016). Prior study has pointed out that genetic variation across the entire LPA locus cumulatively explains approximately 91% of the variation in plasma Lp(a) concentrations (Boerwinkle et al. 1992).

Common genetic variation at the LPA includes various promoter polymorphisms, nonsynonymous variants, and the structure Kringle IV type 2 (KIV-2) which leads to the multiallelic (1 to >40 copies) intragenic copy-number variation (CNV) known as the KIV-2 CNV (Schmidt et al. 2016).

It is well acknowledged that there is an inverse correlation of CNV length with Lp(a) levels. Short KIV-2 CNV alleles, as a consequence, are risk factors for CAD at elevated Lp(a) levels in large multiethnic sample (Lanktree et al. 2010). Lp(a) has become a heated nontraditional biomarker of CAD lately and manifests predictive value in large prospective researches and meta-analyses, though the physiological function of Lp(a) is still ambiguous (Erqou et al. 2009; Kamstrup et al. 2008). Several mechanisms have been proposed in vitro experiments. It suggested that Lp(a) may serve as a link between the cholesterol transport and the fibrinolytic system and modulates blood clotting and fibrinolytic processes (Schmidt et al. 2016). Additive hypotheses include involvement in wound healing and tissue repair and oxidized phospholipids (OxPLs) transport (Bergmark et al. 2008), etc.

4.5 10q11 (CXCL12)

The novel risk alleles of interest, rs1746048 and rs501120, reside in 10q11 region at downstream of CXCL12 (chemokine [C-X-C-motif] ligand 12) gene. Two SNPs were both associated with increased plasma CXCL12 levels, and rs1746048 appeared to have higher transcript levels of CXCL12 in human natural killer cells and the liver (Mehta et al. 2011). Controversial results in vivo and vitro experiments might baffle the understanding of exact action of CXCL12. Cumulative findings indicated that CXCL12 is a cardioprotective molecule under pre- and postischemic conditions, through involvement in several biofunctions as activating growth and survival signaling pathways in cardiomyocytes or stimulating cardiac regeneration by recruiting hematopoietic stem and progenitor cells and so on (Veldkamp et al. 2009). While other experiments found it central to both the progress of arteriosclerosis and the rupture of artery atherosclerotic plaque by platelet activation and leukocyte migration (Zernecke and Weber 2010). Epidemiological trials have reported that higher plasma CXCL12 in patients with chronic kidney disease was associated with CVD risk factors and increased incidence of CVD, MI, and death (Mehta et al. 2014). Meta-analysis revealed that rs1746048 increase a 12% incidence rate of coronary heart disease (95% CI 1.09–1.15, $P < 0.0001$) across various populations (Huang et al. 2013).

4.6 Risk Variants Related to Myocardial Infarction

Myocardial infarction, as a serious sequelae of CAD, has been previously treated as the same phenotype as CAD to understand genetic work on coronary atherosclerosis. Apart from the common pathophysiology of endothelial dysfunction, monocyte adhesion, lipid deposition, and inflammatory response, plaque vulnerability and thrombosis are distinct pathways involved in the incidence of MI. A complex interaction of factors influences whether or not an advanced plaque ruptures, including the extent and degree of lipid-core necrosis sustained by proliferating macrophages within the plaque, de novo monocyte migration/emigration, the degree of luminal stenosis, plaque burden, vessel remodeling, and the thickness of the fibrous cap (Bentzon et al. 2014). Contents within the plaque are free into blood flow and then spur thrombus formation to occlude vessels. Yet whether a susceptibility locus contributes to atherosclerosis or to plaque rupture, thrombosis causing myocardial infarction is hard to tell from previous findings.

One way to handle it to a certain extent is to compare subjects with prior MIs to those without prior MIs among subjects with existing CAD (Assimes and Roberts 2016). The ABO locus has been identified in this way as the top MI-predisposing locus in the presence of coronary atherosclerosis (Reilly et al. 2011). The ABO gene encodes proteins transferase A and transferase B in slight difference that transfer a carbohydrate onto von Willebrand Factor (vWF). The blood group O gene also codes a protein with no enzymatic activity in that a deletion of guanine. Plasma vWF increase 25% in people with non-O blood groups compared with blood group O subjects (Jenkins and O'Donnell 2006) and in turn increase the risk of MI (Reilly et al. 2011; Dentali et al. 2014). Other GWASs also found ABO as a locus for LDL-C metabolism, type 2 diabetes, and biomarkers like ICAM-1 and E-selectin (Qi et al. 2010; Chasman et al. 2009; Barbalic et al. 2010), suggesting multiplex mechanisms of ABO leading to atherosclerosis and thrombosis. Analysis from a GWAS added evidence of linking ABO locus with MI after adjustments for SNPs related to those traits above. But it still needs further investigation.

5 Lessons and Challenges

Certainly, the sheer number of uncovered loci delivers new clues to the landscape of pathophysiology for this common, complex disease. Yet achievements have been attaining paralleled to controversies.

5.1 *Missing Heritability*

In the vast majority of cases, the estimated effect sizes are small to modest, mostly increases in risk by a factor of 1.1–1.5 per associated allele (Nikpay et al. 2015). Needless to say, despite the inclusion of remarkably large numbers of subjects, the bulk of the heritability of CAD cannot be ascribed to loci that have emerged from GWASs, as these loci collectively explaining less than 20% of genetic variance in CAD (Nikpay et al. 2015).

Several explanations may shed light on its intricate nature. The prevailing view pointed to those additional to-be-explored variants and enduring debate has focused on whether these additional variants are common alleles (frequency ≥1%) with moderate-to-small effects or rare alleles (frequency <1%) with large effects. Moreover, genetic interactions, gene-environment interactions, as well as transgenerational epigenetic changes might also contribute to this enigma (Prins et al. 2012).

5.2 *Evidence of Rare Variants' Influence on Risk of CAD*

While GWASs have unmasked growing number of common variants, its coverage dramatically constrained for low-frequency and rare alleles. Such lower-frequency alleles may be somewhat of great importance underlying complex disorders for natural selection acts in order to suppress its damaging influence widespread (Altshuler et al. 2008). At present our knowledge of the impact of rare variants on CAD is still limited. So numerous and individually rare are they, it is not possible to create a complete catalog in the general population. Instead, they must be identified by sequencing among cases and controls in each study. Moreover, because each variant is too rare to prove statistical evidence of association, the mutations must be aggregated as a class (called burden test) to compare the overall frequency of cases versus controls. Recent advances in high-throughput sequencing technology pave the way toward the discovery of rare risk alleles. Rare mutation probably plays an essential role in common diseases as well.

The example of rare mutations in lipoprotein lipase (LPL) gene illustrates the impact of rare variants underpinning CAD heritability. LPL is a rate-determining enzyme in the clearance of circulating triglyceride-rich lipoproteins (Eckel 1989). A recent study performed whole-exome sequencing in over 22,000 participants and found 123 loss-of-function or missense variants in LPL with minor allele frequency less than 1%. It showed that heterozygous carriers of an LPL damaging SNP had higher plasma triglyceride levels and higher risk of CAD (odds ratio 1.84; 95% CI, 1.35–2.51; $P < 0.001$) (Khera et al. 2017).

Though effect sizes of rare variants on intermediate traits and risk of CAD are larger compared to those of common variants, it also became apparent that low-frequency polymorphisms collectively still only explain a very small proportion of the heritability for CAD (Benson et al. 2018), where a significant fraction of heritability of CAD lie seems far beyond explanation.

5.3 SNPs in Noncoding Regions

It is noteworthy that GWAS tags SNPs not particular gene and associations detected by GWASs provide little causality evidence unless the loci are translated into genes and pathways. Until now, only a handful of SNPs found from GWASs are located in exons and cause non-synonymous missense mutations of certain genes, directly changing the amino acid sequence. Most of the genetic risk variants for CAD were found in either introns or intergenic regions, which may affect the expression of neighboring (or remote) genes through intergenic transcription factor binding sites, enhancer regions, or noncoding RNAs (Barth and Tomaselli 2016). What biological function those associated SNPs perform warrants thorough consideration via integrative tools (Fig. 8.1).

5.4 Inconformity Among Races and Repeatability

To date, most ongoing analyses are based on the datasets established in the population of European ancestry. CAD-associated regions were generally found to be concordant between subjects of European or Asian ancestry. Whereas most loci failed to replicate or showed considerably reduced effect sizes in African ancestry populations (Teo et al. 2010), one example is the difference in representative risk loci 9p21. It is a pressing need for additional GWASs at full scale.

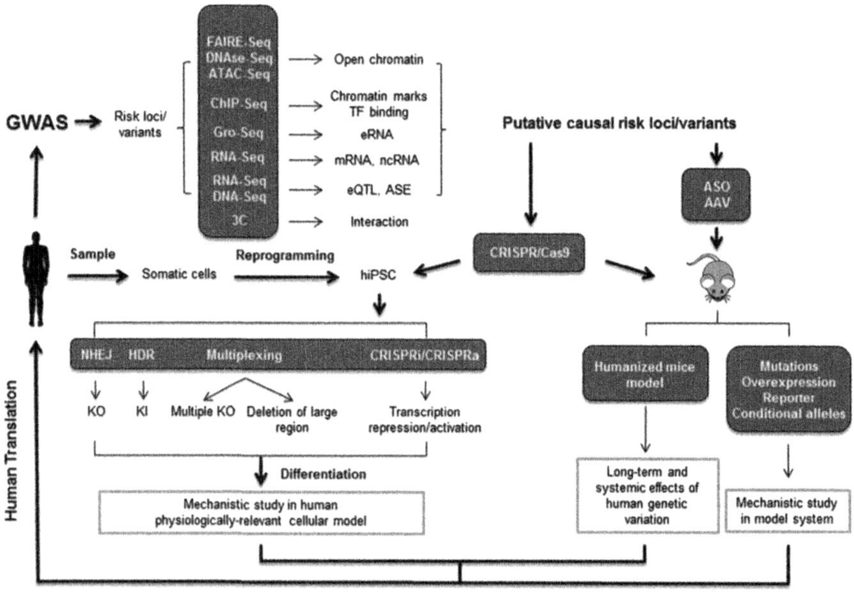

Fig. 8.1 Experimental tools for genome-wide association studies (GWAS) functional follow-up studies. GWAS findings can be functionally annotated using genome-wide methods, which help prioritize loci with likely biological function. These putative risk loci can be further interrogated by genome editing using the clustered regularly interspaced short palindromic repeats/clustered regularly interspaced short palindromic repeat-associated 9 (CRISPR/Cas9) system in vitro and in vivo. Additionally, adeno-associated virus (AAV) and antisense oligos (ASO) can be used to study candidate gene knockdown and overexpression in the mouse model. *ASE* indicates allele-specific expression, *ATAC-Seq* assay for transposase-accessible chromatin using sequencing, *ChIP-Seq* chromatin immunoprecipitation with massively parallel sequencing, *eRNA* enhancer RNA, *eQTL* expression quantitative trait loci, *FAIRE-Seq* formaldehyde-assisted isolation of regulatory elements, *HDR* homology-directed repair, *iPSC* induced pluripotent stem cells, *KO* knockout, *KI* knockin, *ncRNA* noncoding RNA, *NHEJ* nonhomologous end joining, *TF* transcription factor

6 Post-GWAS Era: New Strategies and Prospects

6.1 Replication of Larger-Scale, Multi-ethnical GWA Studies

Growing number of genetic variants has been reported gradually in parallel to expansion of sample size and improved analytic strategies. Yet a finite amount of genetic information shed light on CAD etiology as noted above. On the one hand, previous GWASs had low statistical power and as a result left many loci with common variants of similar and smaller effects undetected. Needless to say, lower-frequency common variants (0.5–1%) and CNVs were also missed. On the other hand, in the majority of studies only enrolling European subjects, their results would likely be not matched in non-European populations, especially Africans. Take all into consideration, GWASs of larger sample size and multiple ethnicities in more cohorts are demanded to evaluate the identified loci as well as newly associated ones. It is estimated that a GWAS incorporating one million subjects may necessitate it to yield a clinically useful CAD genetic predictor (Chatterjee et al. 2013). For now, the UK Biobank excels in its unprecedented scale with 1.5 million sample enrolled, which seems to be a leaping advance toward the needed sample sizes (Chasman 2018).

6.2 Next-Generation Sequencing

As mentioned above, taking root in "common disease, common variance," GWASs in CAD have explored dozens of common variants in human genome and established many associations with modest to low effect size on its risk. But a disappointing fact comes out that there is no major breakthrough on the horizon as primarily envisioned. All discovered risk loci collectively illuminate only a small fraction of inherited risk factors of CAD. It is hypothesized that there may exist low-frequency and rare variants in complex diseases and effect sizes of which are generally greater than more common variants (Altshuler et al. 2008). Thus it calls for replaceable methods to launch more profound discovery.

Next-generation sequencing (NGS), including whole-exome and whole-genome sequencing, emerges as a methodological leap forward in genotyping and genetic mapping. High cost-effectiveness and sensitivity distinguished it as an invaluable technique, for both increase in throughput by several orders of magnitude and simultaneous sequencing of tremendous number of samples (Huang et al. 2009). NGS promises to advance our full understanding of CAD heritability via identification of additional common, rare, and structural variants at newly identified loci. Meanwhile, analyzing specific functional variants or variant accumulation in specific genes with CAD may help distinguish causative genes in a gene-rich locus (Johansen et al. 2010).

Exome is composed of an entire set of exons which can be transcribed to messenger RNA (mRNA) without being removed by RNA splicing. Though accounting for about 1% of the human genome, mutations in the exome are thought to harbor 85% of mutations that have a large effect on disease (Choi et al. 2009). Whole-exome sequencing is a new tool combining exon trapping and high-throughput sequencing. With only exome targeted, WES bridges the gap where GWASs genotype SNPs generally in intergenic region and intron. It promotes analytic sensitivity of low-frequency and rare variants associated with disease as well as common ones (McPherson and Tybjaerg-Hansen 2016).

Here we exhibit some examples to illustrate how it works in the field of genetic architecture of CAD. Khera et al. performed WES and reported that the presence of rare damaging mutations in lipoprotein lipase (LPL) was significantly associated with higher triglyceride levels and susceptibility of CAD (Khera et al. 2017) (see Sect. 5.2 above).

Lu et al. carried out a meta-analysis of exome-wide association studies in East-Asian population. They identified 12 novel loci and 31 protein-altering variants pointing to 25 functional lipid genes (Lu et al. 2017). Further investigation revealed that most coding variants on modulating serum lipid levels of low-density lipoprotein, triglyceride, and total cholesterol except high-density lipoprotein were correlated with CAD. Moreover, all discovered genetic variants, low-frequency or rare (MAF, 4.21–0.03%) in this sample, appear to be population specific, which restated the necessity of enrollments from multiethnical backgrounds.

Clonal hematopoiesis of indeterminate potential (CHIP) is a common phenomenon among the elderly, defined as an expanded somatic blood-cell clone in individuals without other hematologic abnormalities. Previous studies reported its impact on the risk of hematologic cancer. Incorporating 4726 participants with coronary heart disease and 3529 controls, Jaiswal et al. used WES to detect related genetic mutations and further assessed the function of a risk allele in vitro (Jaiswal et al. 2017). It was found that (1) CHIP carriers had 1.9 times higher risk of coronary heart disease than noncarriers, and the incidence of early-onset myocardial infarction quadrupled as compared to noncarriers. (2) Mutations in DNMT3A, TET2, ASXL1, and JAK2 genes were identified. (3) Histocytological experiments observed accelerated atherosclerosis in mice with Tet2 deficiency and increase in expression of inflammatory chemokines in macrophages. In conclusion, this research supplied evidence that CHIP may contribute causally to CAD and its medical importance as a modified risk factor remains to be discussed.

NGS seeks to further identify all low-frequency and rare variants either amid exome or across the entire genome; thus it holds the promise to acquire comprehensive catalogs of genomic variants and in turn functional characterization of those genes. With regard to clinical application, it will increase the power of risk prediction algorithms and facilitate the realization of precision medicine in terms of prediction, diagnosis and risk stratification, and individualized treatment by genotyping of each patient. What is likely to change in the near future is that GWASs by SNP arrays will be gradually replaced by GWASs by WGS, particularly for quantitative traits and very common diseases.

However, myriads of additional data produced by NGS technology are challenging current computational capacity. Strategies must be figured out to improve storage and software in order to effectively handle data.

6.3 Mendelian Randomization

Observational study and experimental trial are two major medical methods to verify whether exposures, also named risk factors or intermediate phenotypes, account for a certain consequence as incidence and outcome of diseases and effect of treatments. Randomized controlled trial (RCT) is the gold standard to assess causality between a risk factor and an outcome. It refers to a precise intervention to ensure that all variables maintain comparable except the factor to be investigated among study groups. Then the differences in the outcome are directly on account of the effect of the exposure. In some instances, RCTs are not feasible, owing to their limitations of excessive cost of time and expense, ethical confliction, and ineffaceable confounders (Lawlor et al. 2008).

Mendelian randomization (MR) is an emerging methodology based on the Mendelian genetic law, which is called "Parental alleles that randomly assigned to the offspring." If a genetic variant proved to be associated with both a phenotype and a clinical endpoint, then the genotype can be used as instrumental variable (IV) to infer the causal relationship between the latter ones (Thomas and Conti 2004). Three key prerequisites for successful MR studies follows: (1) genetic variant is strongly associated with the exposure of interest; (2) genetic variant isn't associated with confounders; and (3) genetic variant must lead to the outcome through effects on the risk factor rather than any other way (Sekula et al. 2016). With its strength in avoidance of confounding, reverse causation and other forms of bias, as well as rapid development of various biomedical techniques and availability of large GWASs statistics, MR studies are widely used in recent years.

MR studies have provided some intriguing insights in the causality of a variety of cardiometabolic risk factors. Some traditional risk factors have been revalidated. A statement from the European Atherosclerosis Society Consensus Panel supported that variants in over 50 genes in relation with lower LDL-C levels could reduce risk of CAD under MR trials. And a time-cumulative and log-linear association is set between LDL-C change and the risk of cardiovascular disease. Furthermore, meta-analyses of MR studies involving more than 300,000 subjects and 80,000 CAD ones further provide robust proof that LDL-C plays a causal role in atherosclerotic cardiovascular disease (ASCVD) (Ference et al. 2017). However, some risk markers like high-density lipoprotein cholesterol and C-reactive protein that were found relevant with CAD before turning out have no causality with CAD or MI in MR studies (Elliott et al. 2009; Voight et al. 2012). Therefore, increasing HDL cholesterol concentrations or lowering CRP concentrations are unlikely to represent effective therapeutic approaches to reduce CVD risk. A latest Mendelian randomization has supported previous view that elevated serum calcium levels may increase the risk of CAD and MI. Six SNPs derived from two GWASs related to serum calcium levels and without pleiotropic associations with potential confounders were analyzed. With per 0.5 mg/dL increase in serum calcium levels predicted by six SNPs altogether, the odds ratios of CAD and MI were 1.25 (95% CI 1.08-1.45, $p = 0.003$) and 1.24 (95% CI 1.05–1.46, $p = 0.009$), respectively (Larsson et al. 2017).

6.4 Epigenetics and Epigenomics

The study of a complete set of epigenetic modifications in an organism, namely, epigenomics, comes out as an intriguing field for further dissection of pathophysiology. Epigenetics is defined as heritable changes in gene function that do not involve changes in the DNA sequence (Dupont et al. 2009). It consists of DNA methylation, histone modification, microRNA (miRNA) long noncoding RNA (lncRNA), and so on.

DNA methylation refers to the addition of methyl groups onto the cytosine of 5′-CpG-3′ dinucleotide, which is catalyzed by DNA methyltransferase (DNMT). Upregulation of methylation causes the DNA structure to linked tightly so that the gene cannot be transcribed. Thus, average gene methylation level tested was inversely associated with the expression of the gene. While methylated CpGs are mainly located at nonregulatory areas in the genome, demethylated CpGs aggregating at certain segments, namely, CpG islands, always appear at the regions of promoters and act as essential regulatory regions of DNA transcription (Smith et al. 2012).

Two methods for analyzing methylation in vivo follows: (1) compare genome-wide methylation between CAD patients and control group by human methylation array or sequencing. (2) Detect methylation of specific critical genes by pyrosequencing or methylation-specific PCR (MSP). Publications of epigenome-wide association studies (EWASs) have soared in recent years, helping revealing novel molecular mechanisms of disease. From epigenome-wide association studies, it is consistent that men with CAD or a history of MI exhibit less variable DNA methylation levels. Besides a growing number of critical genes and sequences, such as ATP-binding cassette A1 (ABCA1), long interspersed element 1 (LINE-1 and Alu have been suggested to show their linkage of methylation levels on CAD (Duan et al. 2018). Take ABCA1 as an example; its product catalyzes cholesterol transfer from peripheral cells to nascent HDL particles. In patients with familial hypercholesterolemia, ABCA1 DNA methylation levels were negatively associated with serum HDL-C levels and higher prevalence of CAD (Guay et al. 2012).

Histone modification consists of methylation, phosphorylation, acetylation, ubiquitylation, and sumoylation of histone proteins (Strahl and Allis 2000). It alters chromatin tightness and in turn allows or inhibits binding of transcription factors onto DNA. Several researches focused on the activities of histone acetyltransferases (HATs) and histone deacetylases (HDACs) and have showed their effect on signaling modification and transcriptional regulation, which are essential for CAD (Duan et al. 2017). In another experiment, utilizing a model of double knockout (DKO) mice with both HDAC9 and LDLr deficiency showed that HDAC9 deficiency resulted in macrophage polarization for promoting inflammation resolution and reverse cholesterol transport, which could put the brakes on progression and promote lesion regression (Cao et al. 2014). In addition, the work of HDAC9 inhibitors implicates a potential drug target of interest.

Both miRNAs and lncRNAs enjoy overwhelming high profiles today. They belong to a group of noncoding RNA molecules with distinct lengths that work in broad gene regulation. Whereas miRNAs contain about 22 nucleotides and the latter ones greater than 200 nucleotides. Not only are they of note in the elucidation of physiopathological pathways of CAD progression, they also show strengths in clinical use as biomarkers and potential treatment targets (Duan et al. 2017).

These patterns regulate gene expression dynamically, ranging from short to long term. Furthermore, they turn out to be interfered by environmental variables like lifestyle and perhaps therapeutic approaches. It is fair to say that their intricate characteristics present a formidable challenge to unravel the etiology of certain disease. From another perspective, epigenetic modifications are supposed to be promising markers in diagnosis and treatment evaluation, worthy of a huge amount of work. However, there are some pending challenges in the interpretation of tentative associations about EWAS findings, partly due to cellular heterogeneity, reverse causality, and replication of identified DNA methylation variable positions (MVPs) (Paul and Beck 2014).

6.5 Systems Genetics

With tons of data in various omics available to the public, their true significance is far beyond what we thought. Target tissues and organs, biological processes, and molecular pathways through which risk loci affect CAD etiology have been largely unsolved. To perform laboratory experiments for so many loci will be certainly time-consuming and expensive. Additively, the vast majority of identified loci show small or modest effects, strangling implements of molecular biology experiments.

System genetics, derived from the idea of systems biology, aims to elucidate what genetic variants do at different molecular levels. It pools resources ranging from genetic variance, intermediate phenotypes such as transcript levels, protein levels and metabolite levels, to gene-gene and gene-environment interactions in all relevant tissues (Nadeau and Dudley 2011). With profound progress in multiple omics techniques, it is now eligible and promising to integrate multidimensional datasets and survey thousands of biological molecules quantitatively. Eventually, it is supposed to pave the way for unraveling fully spectrum of molecular processes driving complex diseases.

Utilizing systems genetics approach, Lempiäinen et al. investigated 68 prioritized CAD genes for their regulatory roles in over 200 metabolic and vascular tissue gene-protein sub-networks named "modules" (Lempiäinen et al. 2018). The modules and genes within were scored for CAD druggability and animal experiment. Twenty-five top gene/protein modules were identified by calculating a CAD-feasibility score. Biological processed where they engaged included extracellular matrix organization, coagulation, platelet activation, innate immune response, complement activation, leukocyte migration, and multiplex cellular signaling mechanisms. Though many genes/proteins were found druggable, they didn't coincide with assumed target tissues and pathways mostly, inferring unknown pleiotropic effects of cardiometabolic drugs. Moreover, beyond consistency in established and revealed druggable targets, target genes/proteins for non-cardiometabolic drugs (antineoplastic and immune-modulating agents) were also noticed. This research presents a paradigm of network analysis to provide a more integrative insight on the effect of genetic variance on process and treatment of CAD.

Another research that integrated genomics and proteomics profiling data also suggested the future direction. After identifying 153 proteins strongly associated with the net Framingham risk score (FRS) proteomic platform, they performed GWASs and exome array analyses and successfully uncovered dozens of novel genetic determinants of these FRS-related proteins (Benson et al. 2018). This work again underlined the significance of integrative strategy to freshen insight into pathophysiology.

7 Genetic/Polygenic Risk Score (GFS/PFS): Clinical Value in Primary and Secondary Prevention

As mentioned above, GWASs have yielded an increasing list of multiple risk variants (mostly common variants) regarding the heritability of CAD. It is suggested that the majority of these variants associated with CAD outside conventional risk factors. Therefore, there is a pressing need to explore potential novel mechanisms underpinning CAD etiology on the one hand. On the other hand, these variants may be applied to facilitate prediction and prognosis of CAD in the real world.

According to the "common disease, common variant" hypothesis, an individual predisposes to CAD progress through an aggregate effect of risk variants carried in total, not to mention that environmental factors and gene-gene and gene-environment interactions all contribute to CAD prevalence within populations. The Framingham risk score (FRS) and Reynolds risk score (RRS) are major acknowledged risk prediction algorithms to estimate 10-year likelihood of developing CHD in an individual: while the latter added high sensitivity C-reactive protein and a family history of early-onset heart attack, two scoring models share variables as age, sex, blood pressure, total cholesterol, high-sensitivity lipoprotein, and smoking (Wilson et al. 1998; Ridker et al. 2007, 2008). Though these well-established algorithms prove to be of great value in primary prevention, ongoing efforts are taken for refinement with the advent of increasing novel biomarkers. It is reasonable to embrace a more patient-centered approach to risk assessment with more comprehensive and accurate risk factor evaluation.

By contrast to conventional risk factors ranging from signs, biochemical factors, to imaging information, genotype may enjoy a more promising option. Less susceptible to biological interference or estimate error as it is, it remains consistent over a lifetime compared to quantitative traits such as lipoproteins (Johansen et al. 2010). Therefore, genotype data can be obtained at any point in a patient's life and seem unbiased to predict susceptibility to CAD even for lifelong period.

The prevailing view is to integrate several risk variants as a single variable, called genetic risk score (GRS) or polygenic risk score (PRS), with established score systems like FRS (Goldstein et al. 2014). A pertinent research selected 152 SNPs (including 49 genome-wide significant SNPs) related with CAD from the CARDIoGRAMplusC4D consortium to form GRSs (de Vries et al. 2015). Here came a conclusion that GRSs including either 49 SNPs or total SNPs were significantly associated with incidence of CAD after adjustment for traditional risk factors and family history, though they didn't improve 10-year risk prediction of CHD probably owing to incidence-prevalence bias. Besides, their work also lighted a hint that adding more suggestively associated SNPs (with a false discovery rate (FDR) <10%) would offer incremental association with incidence. Another similar study with over 4000 people in CoLaus study supported the significant link of GRS to incident CAD. And both discrimination [C-index improvement, 0.016; $p = 0.048$] and reclassification [net reclassification improvement, 8.6%; $p = 0.027$] were modestly improved (Antiochos et al. 2016). Likewise, in this research a weighted, polygenic GRS including SNPs at MAF <5% outperformed GRSs including merely SNPs that achieved statistical significance.

Previously, great attention was paid on pinpointing susceptible alleles that in turn can be used for discerning people at high risk of CAD. After numerous risk variants being gauged, some focus on whether these affecting risk of CAD influence clinical outcomes as well. Mega et al. conducted a research including several trials (cohort and randomized control trials) to evaluate the usefulness of GRS in both incident and recurrent CHD (Mega et al. 2015). Apart from revalidation of its work in primary prevention, they concluded that genetic risk score, based on 27 validated CHD-risk SNPs, is associated with recurrent CHD independent of established clinical biomarkers. Interestingly, their study further revealed that those in higher genetic risk categories enjoyed greater clinical benefit as risk reduction in events from statin therapy, which implicates its potential utility of therapeutic management.

It is hypothesized that susceptibility loci for CAD predispose an individual to clinical complications of CAD throughout one's life course. Given the nature of genetic variants, some views that using PRS as a lifelong risk prediction system may sound reasonable. Salfati et al. integrated high-risk alleles into a GRS and scrutinized the degree of subclinical coronary atherosclerosis by staining and direct visual inspection of the right coronary artery in the youngest group and scanning for coronary artery calcification in others (Salfati et al. 2015). It did show a significant positive connection between GRS and risk of subclinical coronary atherosclerosis. At present, whether GRS can be used in lifetime risk prediction and guide early intervention still lacks of elaborate evidence.

Genotyping is only the first step toward etiological understanding, to bring genomics findings to the real world for promotion of clinical management calls for more interest. Although a GRS aggregating the effects of all loci was statistically significant, it doesn't show much improvement of any established risk score. GRS to some extent can help, quote, "explain a sufficient proportion of variation (between 1% and 15% at present for highly polygenic traits without a major gene) to separate groups, for example, samples with the highest and lowest risk" (Visscher et al. 2017). In the future, replicated studies will help elaborate the real potential of GRSs as markers or treatment indices.

8 Conclusion (Fig. 8.2)

This chapter provides an overview of major GWASs focusing on CAD and several typical validated CAD-risk variants within. In parallel to fruitful discoveries, challenges as well as new biology and emerging translational outcomes for CAD have also been described.

The past decade has witnessed so many genetic discoveries that were generated via the experimental design of GWASs in human populations. Discovery of identified CAD-risk common variants, either validated ($p < 5 * 10^{-8}$) or only suggestive, led to soaring amount of information on risk genes, signaling pathways, and intermediate causal factors, providing a wealth of novel insights on genetic architecture and biological nature of CAD. To date, with ever-increasing sample size, there seems no trend reaching a plateau of the number of risk loci associated with CAD.

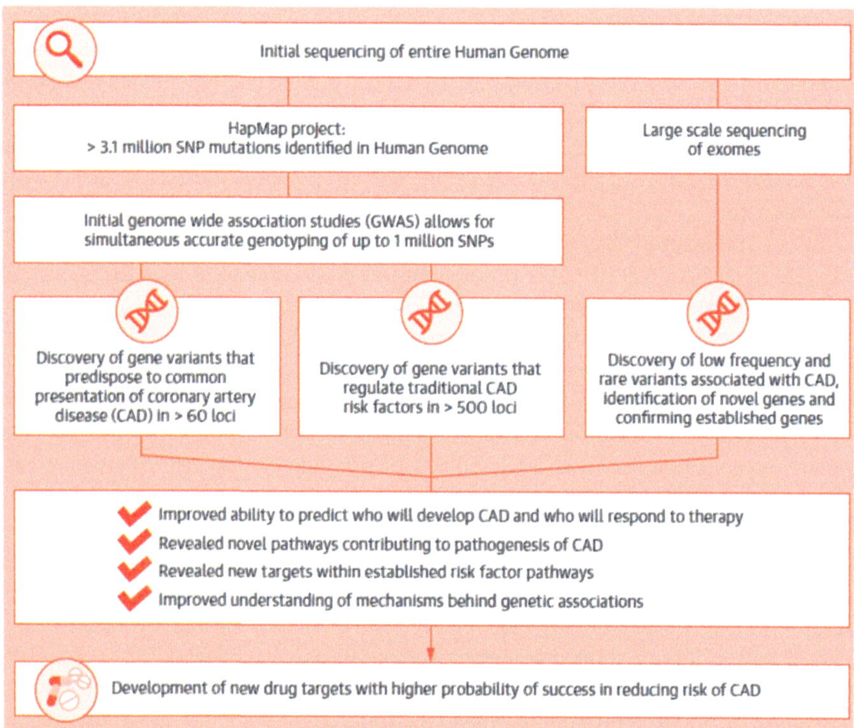

Fig. 8.2 Progress and future directions of GWASs in CAD (Assimes, T.L. et al. Genetics: Implications for Prevention and Management of Coronary Artery Disease. doi: https://doi.org/10.1016/j.jacc.2016.10.039). *CAD* coronary artery disease, *GWAS* genome-wide association studies, *PCSK9* proprotein convertase subtilisin/kexin type 9, *SNP* single-nucleotide polymorphism

GWASs based on SNP arrays are designed to tag common variants in the genome, which limits its power to cover all genetic variants in the population. Cumulating analyses have emphasized the significance underlying the identification of lower-frequency alleles with large effect size as well as CNVs. Upgrade of alternative mechanical and computational tools and expansion of sample size is proposed as eligible strategies. As an illustration, next-generation sequencing enables the intensive sequencing of entire genomes or exomes beyond common variants with the strengths of time- and cost-effectiveness; thus it may be eligible as a replacement for SNP microarray. It is reasonable to infer that future technological advances will surely revolutionize the current knowledge of CAD physiopathology.

The emphasis in research will need to convert from gene discovery to interpretation into biological understanding and clinical practicality. Given the fact that the vast majority of genetic variants' influences remain in the fog, what pathways lie in initiation, progression of disease and drug metabolism behind these variants require enduring studies. A sea of variants in GWASs datasets coupled with multiplex-omics data will be granted as a preferable treasury for mining clinical utility of prevention or early intervention of disease, development and repurposing of candidate therapeutics, etc.

References

Altshuler D, Daly MJ, Lander ES. Genetic mapping in human disease. Science. 2008;322(5903):881–8.

Antiochos P, et al. Association between parental history and genetic risk scores for coronary heart disease prediction: the population-based CoLaus study. Atherosclerosis. 2016;244:59–65.

Assimes TL, Roberts R. Genetics: implications for prevention and management of coronary artery disease. J Am Coll Cardiol. 2016;68(25):2797–818.

Assimes TL, et al. Susceptibility locus for clinical and subclinical coronary artery disease at chromosome 9p21 in the multi-ethnic ADVANCE study. Hum Mol Genet. 2008;17(15):2320–8.

Barbalic M, et al. Large-scale genomic studies reveal central role of ABO in sP-selectin and sICAM-1 levels. Hum Mol Genet. 2010;19(9):1863–72.

Barth AS, Tomaselli GF. Gene scanning and heart attack risk. Trends Cardiovasc Med. 2016;26(3):260–5.

Benjamin EJ, et al. Heart disease and stroke statistics-2017 update: a report from the American Heart Association. Circulation. 2017;135(10):e146–603.

Benson MD, et al. Genetic architecture of the cardiovascular risk proteome. Circulation. 2018;137(11):1158–72.

Bentzon JF, et al. Mechanisms of plaque formation and rupture. Circ Res. 2014;114(12):1852–66.

Bergmark C, et al. A novel function of lipoprotein [a] as a preferential carrier of oxidized phospholipids in human plasma. J Lipid Res. 2008;49(10):2230–9.

Björkegren J, et al. Genome-wide significant loci: how important are they? Systems genetics to understand heritability of coronary artery disease and other common complex disorders. J Am Coll Cardiol. 2015;65(8):830–45.

Boerwinkle E, et al. Apolipoprotein(a) gene accounts for greater than 90% of the variation in plasma lipoprotein(a) concentrations. J Clin Invest. 1992;90(1):52–60.

Brænne I, et al. Prediction of causal candidate genes in coronary artery disease loci. Arterioscler Thromb Vasc Biol. 2015;35(10):2207–17.

Broadbent HM, et al. Susceptibility to coronary artery disease and diabetes is encoded by distinct, tightly linked SNPs in the ANRIL locus on chromosome 9p. Hum Mol Genet. 2008;17(6):806–14.
Cao Q, et al. HDAC9 represses cholesterol efflux and generation of alternatively activated macrophages in atherosclerosis development. Arterioscler Thromb Vasc Biol. 2014;34(9):1871–9.
Chasman DI. Peering into the future of CAD genomics. Circ Res. 2018;122(3):391–3.
Chasman DI, et al. Forty-three loci associated with plasma lipoprotein size, concentration, and cholesterol content in genome-wide analysis. PLoS Genet. 2009;5(11):e1000730.
Chatterjee N, et al. Projecting the performance of risk prediction based on polygenic analyses of genome-wide association studies. Nat Genet. 2013;45(4):400–5. 405e1–3
Choi M, et al. Genetic diagnosis by whole exome capture and massively parallel DNA sequencing. Proc Natl Acad Sci U S A. 2009;106(45):19096–101.
Chow CK, et al. Parental history and myocardial infarction risk across the world: the INTERHEART study. J Am Coll Cardiol. 2011;57(5):619–27.
de Vries PS, et al. Incremental predictive value of 152 single nucleotide polymorphisms in the 10-year risk prediction of incident coronary heart disease: the Rotterdam study. Int J Epidemiol. 2015;44(2):682–8.
Deloukas P, et al. Large-scale association analysis identifies new risk loci for coronary artery disease. Nat Genet. 2013;45(1):25–33.
Dentali F, et al. ABO blood group and vascular disease: an update. Semin Thromb Hemost. 2014;40(1):49–59.
Duan L, et al. Epigenetic mechanisms in coronary artery disease: the current state and prospects. Trends Cardiovasc Med. 2017. https://doi.org/10.1016/j.tcm.2017.12.012.
Duan L, et al. The role of DNA methylation in coronary artery disease. Gene. 2018;646:91–7.
Dupont C, Armant DR, Brenner CA. Epigenetics: definition, mechanisms and clinical perspective. Semin Reprod Med. 2009;27(5):351–7.
Eckel RH. Lipoprotein lipase. A multifunctional enzyme relevant to common metabolic diseases. N Engl J Med. 1989;320(16):1060–8.
Elliott P, et al. Genetic loci associated with C-reactive protein levels and risk of coronary heart disease. JAMA. 2009;302(1):37–48.
Erqou S, et al. Lipoprotein(a) concentration and the risk of coronary heart disease, stroke, and nonvascular mortality. JAMA. 2009;302(4):412–23.
Ference BA, et al. Low-density lipoproteins cause atherosclerotic cardiovascular disease. 1. Evidence from genetic, epidemiologic, and clinical studies. A consensus statement from the European Atherosclerosis Society Consensus Panel. Eur Heart J. 2017;38(32):2459–72.
GBD, C.O.D.C. Global, regional, and national age-sex specific mortality for 264 causes of death, 1980–2016: a systematic analysis for the Global Burden of Disease Study 2016. Lancet (London, England). 2017;390(10100):1151–210.
GBD, D.A.I.I. Global, regional, and national incidence, prevalence, and years lived with disability for 328 diseases and injuries for 195 countries, 1990–2016: a systematic analysis for the Global Burden of Disease Study 2016. Lancet. 2017;390(10100):1211–59.
Goldstein BA, et al. Simple, standardized incorporation of genetic risk into non-genetic risk prediction tools for complex traits: coronary heart disease as an example. Front Genet. 2014;5:254.
Guay SP, et al. ABCA1 gene promoter DNA methylation is associated with HDL particle profile and coronary artery disease in familial hypercholesterolemia. Epigenetics. 2012;7(5):464–72.
Gustafsen C, et al. The hypercholesterolemia-risk gene SORT1 facilitates PCSK9 secretion. Cell Metab. 2014;19(2):310–8.
Hannou SA, et al. Functional genomics of the CDKN2A/B locus in cardiovascular and metabolic disease: what have we learned from GWASs? Trends Endocrinol Metab. 2015;26(4):176–84.
Hegele RA. Plasma lipoproteins: genetic influences and clinical implications. Nat Rev Genet. 2009;10(2):109–21.
Huang X, et al. High-throughput genotyping by whole-genome resequencing. Genome Res. 2009;19(6):1068–76.
Huang Y, et al. Relationship between chemokine (C-X-C motif) ligand 12 gene variant (rs1746048) and coronary heart disease: case-control study and meta-analysis. Gene. 2013;521(1):38–44.

International HapMap Consortium. A haplotype map of the human genome. Nature. 2005;437(7063):1299–320.

Jaiswal S, et al. Clonal hematopoiesis and risk of atherosclerotic cardiovascular disease. N Engl J Med. 2017;377(2):111–21.

Jenkins PV, O'Donnell JS. ABO blood group determines plasma von Willebrand factor levels: a biologic function after all? Transfusion. 2006;46(10):1836–44.

Johansen CT, Lanktree MB, Hegele RA. Translating genomic analyses into improved management of coronary artery disease. Futur Cardiol. 2010;6(4):507–21.

Kamstrup PR, et al. Extreme lipoprotein(a) levels and risk of myocardial infarction in the general population: the Copenhagen City Heart Study. Circulation. 2008;117(2):176–84.

Kathiresan S, et al. Genome-wide association of early-onset myocardial infarction with single nucleotide polymorphisms and copy number variants. Nat Genet. 2009;41(3):334–41.

Khera AV, et al. Association of rare and common variation in the lipoprotein lipase gene with coronary artery disease. JAMA. 2017;317(9):937–46.

Kjolby M, et al. Sort1, encoded by the cardiovascular risk locus 1p13.3, is a regulator of hepatic lipoprotein export. Cell Metab. 2010;12(3):213–23.

Lanktree MB, et al. Comprehensive analysis of genomic variation in the LPA locus and its relationship to plasma lipoprotein(a) in South Asians, Chinese, and European Caucasians. Circ Cardiovasc Genet. 2010;3(1):39–46.

Larsson SC, Burgess S, Michaëlsson K. Association of genetic variants related to serum calcium levels with coronary artery disease and myocardial infarction. JAMA. 2017;318(4):371–80.

Lawlor DA, et al. Mendelian randomization: using genes as instruments for making causal inferences in epidemiology. Stat Med. 2008;27(8):1133–63.

Lee JY, et al. A genome-wide association study of a coronary artery disease risk variant. J Hum Genet. 2013;58(3):120–6.

Lempiäinen H, et al. Network analysis of coronary artery disease risk genes elucidates disease mechanisms and druggable targets. Sci Rep. 2018;8(1):3434.

Lu X, et al. Exome chip meta-analysis identifies novel loci and East Asian-specific coding variants that contribute to lipid levels and coronary artery disease. Nat Genet. 2017;49(12):1722–30.

McCarroll SA, et al. Integrated detection and population-genetic analysis of SNPs and copy number variation. Nat Genet. 2008;40(10):1166–74.

McPherson R, Tybjaerg-Hansen A. Genetics of coronary artery disease. Circ Res. 2016;118(4):564–78.

McPherson R, et al. A common allele on chromosome 9 associated with coronary heart disease. Science. 2007;316(5830):1488–91.

Mega JL, et al. Genetic risk, coronary heart disease events, and the clinical benefit of statin therapy. Lancet (London, England). 2015;385(9984):2264–71.

Mehta NN, et al. The novel atherosclerosis locus at 10q11 regulates plasma CXCL12 levels. Eur Heart J. 2011;32(8):963–71.

Mehta NN, et al. Higher plasma CXCL12 levels predict incident myocardial infarction and death in chronic kidney disease: findings from the Chronic Renal Insufficiency Cohort study. Eur Heart J. 2014;35(31):2115–22.

Musunuru K, et al. From noncoding variant to phenotype via SORT1 at the 1p13 cholesterol locus. Nature. 2010;466(7307):714–9.

Nadeau JH, Dudley AM. Genetics. Systems genetics. Science. 2011;331(6020):1015–6.

Nikpay M, et al. A comprehensive 1,000 genomes-based genome-wide association meta-analysis of coronary artery disease. Nat Genet. 2015;47(10):1121–30.

Patel KM, et al. Macrophage sortilin promotes LDL uptake, foam cell formation, and atherosclerosis. Circ Res. 2011;116(5):789.

Paul DS, Beck S. Advances in epigenome-wide association studies for common diseases. Trends Mol Med. 2014;20(10):541–3.

Pertea M, Salzberg SL. Between a chicken and a grape: estimating the number of human genes. Genome Biol. 2010;11(5):206.

Preuss M, et al. Design of the Coronary ARtery DIsease Genome-Wide Replication And Meta-Analysis (CARDIoGRAM) study: a genome-wide association meta-analysis involving more than 22 000 cases and 60 000 controls. Circ Cardiovasc Genet. 2010;3(5):475–83.
Prins BP, et al. Genetics of coronary artery disease: genome-wide association studies and beyond. Atherosclerosis. 2012;225(1):1–10.
Qi L, et al. Genetic variants in ABO blood group region, plasma soluble E-selectin levels and risk of type 2 diabetes. Hum Mol Genet. 2010;19(9):1856–62.
Reilly MP, et al. Identification of ADAMTS7 as a novel locus for coronary atherosclerosis and association of ABO with myocardial infarction in the presence of coronary atherosclerosis: two genome-wide association studies. Lancet. 2011;377(9763):383–92.
Ridker PM, et al. Development and validation of improved algorithms for the assessment of global cardiovascular risk in women: the Reynolds Risk Score. JAMA. 2007;297(6):611–9.
Ridker PM, et al. C-reactive protein and parental history improve global cardiovascular risk prediction: the Reynolds Risk Score for men. Circulation. 2008;118(22):2243–51. 4p following 2251
Salfati E, et al. Susceptibility loci for clinical coronary artery disease and subclinical coronary atherosclerosis throughout the life-course. Circ Cardiovasc Genet. 2015;8(6):803–11.
Samani NJ, et al. Genome wide association analysis of coronary artery disease. N Engl J Med. 2007;357(5):443–53.
Schmidt K, et al. Structure, function, and genetics of lipoprotein (a). J Lipid Res. 2016;57(8):1339–59.
Sekula P, et al. Mendelian randomization as an approach to assess causality using observational data. J Am Soc Nephrol. 2016;27(11):3253–65.
Sivapalaratnam S, et al. Family history of premature coronary heart disease and risk prediction in the EPIC-Norfolk prospective population study. Heart. 2010;96(24):1985–9.
Smith ZD, et al. A unique regulatory phase of DNA methylation in the early mammalian embryo. Nature. 2012;484(7394):339–44.
Stankiewicz P, Lupski JR. Structural variation in the human genome and its role in disease. Annu Rev Med. 2010;61:437–55.
Stone NJ, et al. Coronary artery disease in 116 kindred with familial type II hyperlipoproteinemia. Circulation. 1974;49(3):476–88.
Strahl BD, Allis CD. The language of covalent histone modifications. Nature. 2000;403(6765):41–5.
Teo YY, Small KS, Kwiatkowski DP. Methodological challenges of genome-wide association analysis in Africa. Nat Rev Genet. 2010;11(2):149–60.
Thomas DC, Conti DV. Commentary: the concept of 'Mendelian Randomization'. Int J Epidemiol. 2004;33(1):21–5.
Veldkamp CT, et al. Monomeric structure of the cardioprotective chemokine SDF-1/CXCL12. Protein Sci. 2009;18(7):1359–69.
Visel A, et al. Targeted deletion of the 9p21 non-coding coronary artery disease risk interval in mice. Nature. 2010;464(7287):409–12.
Visscher PM, et al. Five years of GWAS discovery. Am J Hum Genet. 2012;90(1):7–24.
Visscher PM, et al. 10 years of GWAS discovery: biology, function, and translation. Am J Hum Genet. 2017;101(1):5–22.
Voight BF, et al. Plasma HDL cholesterol and risk of myocardial infarction: a mendelian randomisation study. Lancet. 2012;380(9841):572–80.
Wilson PW, et al. Prediction of coronary heart disease using risk factor categories. Circulation. 1998;97(18):1837–47.
WTCCC. Genome-wide association study of 14,000 cases of seven common diseases and 3,000 shared controls. Nature. 2007;447(7145):661–78.
Yamamoto A, et al. Clinical features of familial hypercholesterolemia. Arteriosclerosis. 1989;9(1 Suppl):I66.
Zernecke A, Weber C. Chemokines in the vascular inflammatory response of atherosclerosis. Cardiovasc Res. 2010;86(2):192–201.
Zhou L, et al. Associations between single nucleotide polymorphisms on chromosome 9p21 and risk of coronary heart disease in Chinese Han population. Arterioscler Thromb Vasc Biol. 2008;28(11):2085–9.

Chapter 9
Progress of Genomics in Cardiac Conduction and Rhythm Disorders

Zhenyan Xu, Qinmei Xiong, Yang Shen, Jinzhu Hu, and Kui Hong

Abstract Genes and genetic loci linked thus far to cardiac electrical function and arrhythmia have been primarily uncovered through numerous studies. The contribution of epigenetics to heart rhythm disorders has also recently been highlighted in common pathologies. This chapter provides an overview to recent progress of genomics in cardiac conduction and rhythm disorders, including long QT syndrome (LQTS), Brugada syndrome (BrS), short QT syndrome (SQT), catecholaminergic polymorphic ventricular tachycardias (CPVT), early repolarization syndrome (ERS), atrial fibrillation (AF), idiopathic ventricular fibrillation (IVF), progressive cardiac conduction defect (PCCD), and sick sinus syndrome (SSS). After a brief introduction of each disorder, the disease-associated implication of genetic variations will be illustrated comprehensively. It can be found that there is marked genotype heterogeneity and cardiac phenotype heterogeneity in this field. Finally, the impact of genetics on clinical management of cardiac conduction and rhythm

Z. Xu
Department of Cardiovascular Medicine, The Second Affiliated Hospital of Nanchang University, Nanchang of Jiangxi, China

Department of Genetic Medicine, The Second Affiliated Hospital of Nanchang University, Nanchang of Jiangxi, China

Q. Xiong · J. Hu
Department of Cardiovascular Medicine, The Second Affiliated Hospital of Nanchang University, Nanchang of Jiangxi, China

Y. Shen
Department of Genetic Medicine, The Second Affiliated Hospital of Nanchang University, Nanchang of Jiangxi, China

Jiangxi Key Laboratory of Molecular Medicine, Nanchang of Jiangxi, China

K. Hong (✉)
Department of Cardiovascular Medicine, The Second Affiliated Hospital of Nanchang University, Nanchang of Jiangxi, China

Department of Genetic Medicine, The Second Affiliated Hospital of Nanchang University, Nanchang of Jiangxi, China

Jiangxi Key Laboratory of Molecular Medicine, Nanchang of Jiangxi, China

H. Jiang, M. Liu (eds.), *Heart Genomics*, Translational Bioinformatics 16,
https://doi.org/10.1007/978-981-13-1429-2_9

disorders will be summarized, including elucidation of an individual's genetic predisposition and personalized approaches to treatment.

Keywords Long QT syndrome · Brugada syndrome · Short QT syndrome · Catecholaminergic polymorphic ventricular tachycardias · Early repolarization syndrome · Atrial fibrillation · Idiopathic ventricular fibrillation · Progressive cardiac conduction defect · Sick sinus syndrome · Genetic variation · Genomics

1 Long QT Syndrome (LQTS)

1.1 Introduction

Long QT syndrome (LQTS) is an inheritable cardiac electric disease characterized by a prolonged QT interval on 12-lead electrogram (ECG) and susceptibility to develop polymorphic ventricular tachycardia known as torsads de pointes with an estimated incidence of at least 1 in 2500 people (Schwartz et al. 2009). It is considered as a common inherited channelopathy responsible for syncope or sudden cardiac death in individuals with structurally normal hearts (Jons et al. 2010).

The discovery of LQTS disease genes is a milestone research in the field of cardiovascular disease genetic research, representing a crucial first step in developing an understanding of the molecular basis for abnormal cardiac repolarization. These findings could contribute to identify new therapies in LQTS and help to reveal important insights into the mechanisms of arrhythmias and provide a theoretical basis for the most effective treatments and preventive measures (Nakano and Shimizu 2016).

A personal history of syncope, aborted cardiac arrest, or sudden death in the child or young adult period, and family history, may lead to suspicion of LQTS. The syncope is typically precipitous and without warning, thus differing from the common vasovagal and orthostatic forms of syncope in which presyncope and other warning symptoms occur. The absence of aura, incontinence, and postictal findings helps differentiate LQTS-associated syncope from seizures.

1.2 Genotype

Over the past 2 dozens of years, we have witnessed remarkable advances in our understanding of LQTS molecular genetic mechanisms. Since the first discovery of LQTS-causative gene, a total of 15 genes have been reported in autosomal dominant forms of Romano-Ward-type congenital LQTS and 2 genes for autosomal recessive forms of the Jervell and Lange-Nielsen syndrome (Table 9.1). Among of them, three LQTS-causative, KCNH2-encoded Kv11.1 (LQT2) genes including KCNQ1-encoded Kv7.1 channel and SCN5A-encoded Nav1.5 (LQT3) result in LQTS-associated mutations in 75% of LQTS.

Table 9.1 Congenital LQTS genes and affected ion currents

Locus name	Gene	Chromosome locus	Protein	Current
LQT1	KCNQ1	11p15.5-p15.4	Potassium voltage- gated channel subfamily KQT member 1	K^+ (I_{Ks})
LQT2	KCNH2	7q36.1	Potassium voltage-gated channel subfamily H member 2	K^+ (I_{Kr})
LQT3	SCN5A	3p22.2	Sodium channel protein type 5 subunit alpha	Na^+ (I_{Na})
LQT4	ANK2	4q25-q26	Ankyrin-2	Na^+ (I_{Na})
LQT5	KCNE1	21q22.12	Potassium voltage-gated channel subfamily E member 1	K^+ (I_{Ks})
LQT6	KCNE2	21q22.11	Potassium voltage-gated channel subfamily E member 2	K^+ (I_{Kr})
LQT7	KCNJ2	17q24.3	Inward rectifier potassium channel 2	K^+ ($K_{ir2.1}$)
LQT8	CACNA1C	12p13.33	Voltage-dependent L-type calcium channel subunit alpha-1C	Ca^{2+}($I_{Ca\text{-}L}$)
LQT9	CAV3	3p25.3	Caveolin-3	Na^+ (I_{Na})
LQT10	SCN4B	11q23.3	Sodium channel subunit beta-4	Na^+ (I_{Na})
LQT11	AKAP9	7q21.2	A-kinase anchor protein 9	K^+ (I_{Ks})
LQT12	SNTA1	20q11.21	Alpha-1-syntrophin	Na^+ (I_{Na})
LQT13	KCNJ5	11q24.3	G-protein-activated inward rectifier potassium channel 4	K^+ (K_{ir})
LQT14	CALM1	14q32.11	Calmodulin	Dysfunctional Ca^{2+} signaling
LQT15	CALM2	2p21	Calmodulin	Dysfunctional Ca^{2+} signaling

1.3 Mutations in Potassium Channel Genes

LQTS subtypes associated with slow activation of delayed rectifier potassium current (IKs) dysfunction include LQT1(KCNQ1), LQT5(KCNE1), LQT11(AKP-9) (Chen et al. 2007), JLN1(KCNQ1), and JLN2(KCNE2). Mutations in KCNE1 are associated with LQT5 which cause defective trafficking of IKs channel and reduce amplitude of IKs current (Harmer et al. 2010). A mutation in AKAP-9 gene is associated with LQT11 which has been shown to reduce the interaction between Kv7.1 and A-kinase anchor protein 9 and reduce the cyclic AMP-induced phosphorylation of the channel resulting in the prolongation of action potential in ventricular cardiomyocyte (Chen et al. 2007).

LQTS subtype associated with rapid activation of delayed rectifier potassium current (IKr) dysfunction includes LQT2 (KCNH2) and LQT6 (KCNE2). Most of mutations in KCNH2 gene disrupt the maturation and trafficking process before reducing the number of function in ion channels. Mutations in KCNE2 (β-subunit of Kv11.1) have been reported to modulate Kv11.1 channel gate and currents.

1.4 Mutation in Sodium Channel Genes

LQTS subtypes with dysfunction of late-activating sodium channels (I_{Na}) contain LQT3 (SCN5A), LQT9 (CAV3) (Vatta et al. 2006), LQT10 (SCN4b), and LQT12 (SNTA1). Phenotypic overlap of LQT3 with Brugada syndrome is also shown in SCN5A mutations. Mutation in SCN4B which encodes subunit of sodium channel has been shown to increase the persistence of I_{Na} with a positive shift toward inactivation (Medeiros-Domingo et al. 2007). Mutation in CAV3 which acts in conjunction to increase peak I_{Na} through a cyclic AMP-independent pathway results in QT prolongation (Vatta et al. 2006). ANTA1 encoding a-1 syntrophin is associated with Nav1.5 channel. Mutation in SNTA1 is associated with an increase in peak of I_{Na}.

1.5 Mutations in Other Genes

Another subtype of LQTS includes LQT7 (KCNJ2), LQT8 (CACNA1C), LQT14 (CALM1), and LQT15 (CALM2). LQT7, also known as Andersen-Tawil syndrome, is characterized by ventricular arrhythmias and skeletal development abnormalities. KCNJ2 mutation resulting in a reduction in Kir2.1 prolongs the terminal phase of the cardiac action potential. KCNJ2 mutations also were identified in patients with CPVT phenotype, so differential diagnosis between LQT7 and CPVT through genotyping becomes more challenging. Timothy syndrome (LQT8) is the most severe phenotypic form of LQTS, which is associated with mutations in CACNA1C. These gene mutations slowed an inactivation of CaV1.2 that increased the influx of calcium, prolonged the cardiac action potential, and resulted in the lethal arrhythmias. De novo mutations in CALM1 and CALM2 encoding calmodulin, which disrupted calcium-ion binding to the protein, are responsible for LQT14 and LAT15 (Crotti et al. 2013; Makita et al. 2014), respectively.

1.6 Genotype-Phenotype Correlations

Understanding the pathogenetic link between genotype and clinical phenotype is vital to the diagnosis and treatment of inherited arrhythmias. For each phenotype, environmental factors (e.g., autonomic nerve activity, ion concentrations) and other genetic factors might affect the phenotypes. Importantly, insights into these relationships might point the way to improve treatment of disease. Analysis of genotype and phenotype has shown that mutation location and type influence the variability of symptoms. LQTS associated with biallelic pathogenic variants or heterozygosity for pathogenic variants in two different genes is generally associated with a more severe phenotype with a longer QTc interval and a higher incidence of cardiac events. The percentage of genetically affected individuals with a normal QTc was higher in the LQTS type 1 group (36%) than in the LQTS type 2 group (19%) or the type 3 group (10%) (Goldenberg et al. 2011). At least 37% of individuals with the LQTS type 1 phenotype, 54% with the type 2 phenotype, and 82% with the type 3 phenotype remain asymptomatic.

Patients with mutations in the transmembrane region of KCNQ1 (IKs) have a greater probability of presenting arrhythmic events than those with mutations in the C-terminal region (Shimizu 2005). Missense cytoplasmic-loop mutations have been reported to cause a longer QT interval at enrolment and an increased risk of ACA/SCD (Barsheshet et al. 2012). The same is true for patients with mutations in the pore region of KCNH2 or HERG (Crozier et al. 1989), as compared to those with mutations in the N- or C-terminal regions (Moss et al. 2002). In LQT2 patients, pore-loop mutations, especially missense mutations, show a higher incidence of ACA/SCD compared with non-pore-loop mutations (Shimizu et al. 2009).

With LQT1, and subjects with a history of exercise-induced syncope, swimming and diving are generally contraindicated. Exercise precautions are more imperative with LQT1, or in those who have already experienced events during exercise, than in LQT2 and LQT3. With LQT2, or those with a history of auditory-evoked events, removing loud alarm clocks and turning down the volume on the phone at night are advisable.

1.7 Gene Screening

Genetic testing is a valuable tool for managing inherited arrhythmias. The identification of LQTS disease genes represents a crucial first step in developing an understanding of the molecular basis, which could contribute to gene diagnosis and help to identify new potential therapies. Different LQTS genotypes with different genetic backgrounds have great differences in the clinical penetrance and expression. The goal of identifying the genetic basis of the disease is to individualize and optimize treatment strategies. The ultimate goal is to prevent future adverse clinical outcomes and SCD events in surviving relatives (Angrist et al. 2010; Zumhagen et al. 2012).

LQTS genetic testing contributes to not only diagnosis but also gene-specific and mutation-specific risk stratification and clinical management. The genetic screening from consensus report recommended that comprehensive or specific genetic testing for KCNQ1, KCNH2, and SCN5A should be performed for the patients who fulfill the criteria. Genetic information will help to identify the risk stratification of LQTS. Genotype-phenotype data demonstrated that those with mutations in the transmembrane region of Kv7.1 had greater risk of arrhythmic events than those with other mutations (Moss et al. 2007) and had a greater sensitivity to sympathetic stimulation than those with C-terminal mutations (Shimizu et al. 2004). Subsequently, another study reported that patients with C-loop missense mutations in the KCNQ1 gene exhibited a higher risk for life-threatening events (Shimizu et al. 2009).

2 Brugada Syndrome (BrS)

2.1 Introduction

Brugada syndrome (BrS) is an inherited cardiac arrhythmia characterized by distinct coved-type ST segment elevation in the right precordial leads on ECG and high risk for sudden cardiac death in young adults (Antzelevitch et al. 2005). ECG prevalence of BrS in the general population varies significantly among continents, countries, and ethnic groups. Southeast Asia has the highest prevalence of BrS, estimated to be greater than 5 per 10,000 inhabitants (Brugada et al. 2014). The true prevalence of BrS in the general population is difficult to estimate because the ECG pattern is often concealed.

BrS is a channelopathy with an autosomal dominant mode of transmission. To date, more than 450 variants have been identified in 24 genes encoding sodium, potassium, and calcium channels or associated proteins (ABCC9, CACNA1C, CACNA2D1, CACNB2, FGF12, GPD1L, HCN4, HEY2, KCND2, KCND3, KCNE3, KCNE5, KCNH2, KCNJ8, PKP2, RANGRF, SCN10A, SCN1B, SCN2B, SCN3B, SCN5A, SEMA3A, SLMAP, and TRPM4) as associated with the Brugada phenotype (Table 9.2). These mutations cause either a decrease in inward sodium or calcium current or an increase in outward potassium currents, resulting in an outward shift in the balance of current active during the early phases of the action potential.

Table 9.2 Genes associated with Brugada syndrome (BrS)

Sodium channel ($I_{Na\downarrow}$)	Calcium channel ($I_{Ca\downarrow}$)	Potassium channel ($I_{K\uparrow}$)	Proteins that interact with ion channel
SCN5A	CACNA1C	KCND2	RANGRF ($I_{Na\downarrow}$)
SCN1B	CACNB2	KCND3	GPD1L ($I_{Na\downarrow}$)
SCN2B	CACNA2D1	KCNE3	SLMAP ($I_{Na\downarrow}$)
SCN3B		KCNE5	PKP2 ($I_{Na\downarrow}$)
SCN10A		KCNJ8	TRPM4 ($I_{Na\downarrow}$)
		ABCC9	FGF12 ($I_{Na\downarrow}$)
		KCNH2	HEY2 (I_{Na}?)
			HCN4 (?)
			SEMA3A ($I_{to\uparrow}$)

2.2 Sodium Channel Mutations

The first gene to be linked to BrS is SCN5A, encoding the α-subunit of the voltage-gated cardiac sodium channel, Nav1.5 (Chen et al. 1998). To date, more than 300 BrS-related mutations in SCN5A have been reported. Approximately 20–25% of BrS patients are genetically diagnosed with pathogenic variations in SCN5A. Several of these mutations have been functionally expressed and shown to cause loss of function of sodium channel current.

Besides SCN5A, pathogenic variations in SCN1B (Watanabe et al. 2008), SCN2B (Riuro et al. 2013), and SCN3B (Hu et al. 2009), which encode the β1, β2, and β3 subunits of the Nav1.5 sodium channel, respectively, and SCN10A (Hu et al. 2014), encoding the neuronal sodium channel Nav1.8, have been discovered to modify the sodium channel's function. Watanabe et al. (2008) identified that mutant SCN1B caused loss of function of sodium channel peak current. Co-expression of mutant SCN1B with WT-SCN5A and WT-KCND3 (separately) induced 55.6% decrease of sodium channel peak current and 70.6% gain of function in Ito, which elucidate the pathogenesis of BrS. The mutant SCN2B causes a significant reduction in sodium current density due to a decreased Nav1.5 cell surface expression (Riuro et al. 2013). The mutant SCN3B leads to a decrease in peak sodium current density, accelerated inactivation, and slowed reactivation, due to an impaired intracellular transport and cell surface expression of the cardiac sodium channel (Hu et al. 2009; Ishikawa et al. 2013). Co-expression of SCN5A-WT with SCN10A-mutant leads to a major loss of function of I_{Na}, contributing to the manifestation of BrS (Hu et al. 2014). GWAS studies have identified SCN10A variants that lead to the altered cardiac function and arrhythmogenesis, indicating that Nav1.8 plays an important role in the electrical function of the heart in both health and disease (121–129). While mutations in SCN1B, SCN2B, and SCN3B are rare, several studies have reported that SCN10A mutations account for 2.5–16% of BrS patients (Hu et al. 2014; Behr et al. 2015; Le Scouarnec et al. 2015; Fukuyama et al. 2016).

2.3 Calcium Channel Mutations

Calcium channels (CACNA1C, CANB2b, and CACNA2D1) have also been reported as BrS-susceptibility genes. Pathogenic variants in CACNA1C and CACNB2, which encode the α1c and β-2b subunits of the L-type cardiac Ca2+ channel, respectively, cause a decrease in I_{Ca} channels and make up nearly 11.5% of BrS cases in which patients present with a syndrome overlapping the typical BrS ECG pattern with a short QT interval (Antzelevitch et al. 2007; Cordeiro et al. 2009). CACNA2D1, which encodes the α-2/δ subunit of the L-type cardiac Ca^{2+} channel, regulates the current density and activation/inactivation kinetics of the Ca^{2+} channel and is a gene associated with BrS but in a low frequency (Burashnikov et al. 2010).

2.4 Potassium Channel Mutations

Putative gain-of-function mutations in genes encoding channels that conduct outward potassium currents have also been reported in a few BrS cases. Gain-of-function mutations in KCND3 encoding Kv4.3, the α-subunit of the Ito channel (Giudicessi et al. 2011), have been implicated in BrS (Giudicessi et al. 2011) with an enhanced Ito current gradient within the right ventricle where KCND3 expression is the highest. Gain-of-function mutations in KCND2 encoding KV4.2, the potassium voltage-gated channel subfamily D member 2, have been associated with BrS (Perrin et al. 2014). Gain-of-function mutations in KCNE3, encoding MiRP2, have been linked to BrS. The functional role of MiRP2 is modulation of several cardiac potassium currents, including Ito and IKs. Co-expression of KCNE3 mutations with WT-KCND3 leads to a gain-of-function and accelerated kinetics of Ito (Delpon et al. 2008). KCNE5 is located on the X chromosome and encodes a regulatory β-subunit of Ito and IKs channels. Mutations in KCNE5 cause modifications to potassium channels that lead to BrS by increasing Ito (Ohno et al. 2011). Mutations in KCNJ8, encoding Kir6.1, the pore-forming subunit of adenosine triphosphate (ATP)-sensitive potassium channel (KATP), have been reported to result in gain of function of IKATP. KATP channel is closed under normoxic conditions. Gain of function of IKATP leads to accentuation of the action potential notch as well as depression of the plateau, leading to BrS. Mutations in ABCC9, encoding SUR2A, the sulfonylurea receptors of KATP channels, have been recently reported to cause a severe arrhythmic phenotype typical of BrS (Barajas-Martinez et al. 2012). Mutations in ABCC9 can reduce the ATP sensitivity of the receptor complex, resulting in gain of function of KATP channel, due to reduced tonic inhibitory effect of ATP. Mutations in KCNH2, encoding the α-subunit of the rapid delayed rectifier potassium channel (hERG), result in increased I_{Kr}, leading to short QT interval and Brugada ECG (Kawamura et al. 2009; Wang et al. 2014). Evidence is currently insufficient to demonstrate causality.

2.5 Ion Channel-Related Mutated Proteins

Mutations in GPD1L, encoding the glycerol-3-phosphat dehydrogenase 1-like protein, which has been found to be in close structural and functional association with Nav1.5, may affect trafficking of the Nav1.5 to the cell surface and decrease cardiac sodium current (Shy et al. 2013) but has only been implied in a small number of BrS cases (London et al. 2007). Missense mutations in RANGRF, encoding MOG1, a protein known to modulate Nav1.5, can impair the trafficking of Nav1.5 to the membrane, leading to I_{Na} reduction and clinical manifestation of BrS (Kattygnarath et al. 2011). Mutations in SLMAP, encoding the sarcolemmal membrane-associated protein, a component of T tubules and sarcoplasmic reticulum, are known to cause BrS by impairing the intracellular trafficking of the Nav1.5 and causing loss of function in I_{Na} (Ishikawa et al. 2012). Mutations in PKP2, encoding the desmosomal protein plakophillin-2, the primary gene responsible for arrhythmogenic right ventricular cardiomyopathy, have also been recently reported to cause loss of function of sodium channel function and lead to BrS (Cerrone et al. 2014; Cerrone and Delmar 2014). Pathogenic mutations in TRPM4, encoding a calcium-activated nonselective cation channel, alter the resting potential of the membrane, and changes in the function of this channel may reduce the availability of sodium channels, ultimately leading to BrS (Liu et al. 2013). FGF12 encodes for a fibroblast growth factor homologous factor (FHF-1), which expressed in the human ventricle. FHFs exert modulatory effects on cardiac sodium and calcium channels (Wang et al. 2011; Hennessey et al. 2013a, b). Missense mutation in FGF12 has been shown to reduce I_{Na} and is associated with BrS (Hennessey et al. 2013a, b). HEY2, encoding the transcriptional repressor hairy/enhancer of split related with YRPW motif protein 2, has been found to play a role in the regulation of SCN5A expression and conduction velocity in the heart, indicating that BrS may originate from altered transcriptional programming during cardiac development (Bezzina et al. 2013). Loss-of-function mutations in HCN4 encoding the hyperpolarization-activated cyclic nucleotide-gated potassium channel 4, a voltage-gated ion channel mediating the pacemaker current in the heart, have been reported in a few patients with BrS (Ueda et al. 2009). The causative role remains unclear. Mutation in SEMA3A gene encoding semaphorin-3A protein is associated with BrS through a KV4.3 gain-of-function mechanism (Boczek et al. 2014). WT semaphorin-3A protein can inhibit the function of KV4.3.

It is notable that except for SCN5A and GPD1L, most of the genes associated with BrS susceptibility have been identified only in single patients, in a few unrelated patients, or in small families through candidate gene analysis. Therefore, further investigation is needed before genes are implicated in the pathogenesis of BrS in order to avoid false-positive reports of causality in the genetic counseling (Wilde et al. 2010). Current clinical guidelines only recommend genetic analysis of the SCN5A gene (Priori et al. 2016). However, known BrS-susceptibility genes can only explain 30–35% of clinically diagnosed patients, indicating that 65–70% of BrS patients remain genetically unsolved (Brugada et al. 2014).

3 Short QT Syndrome (SQTS)

3.1 Introduction

Short QT syndrome (SQTS) is an inherited arrhythmia syndrome characterized by abbreviated short QT interval as measured on ECG (<350 ms), which can cause malignant ventricular arrhythmias and sudden cardiac death. To date, seven genes have been associated with SQTS (SQTS1-7, Table 9.3). Genetic studies have identified mutations in genes encoding in potassium channels that are also causative

Table 9.3 Genetic defects associated with SQTS

Type	Locus	Gene/protein	Function
SQT1	7q35-36	*KCNH2*, HERG	$I_{Kr\uparrow}$
SQT 2	11p15.5	*KCNQ1*, KVLQT1	$I_{ks\uparrow}$
SQT 3	17q23.1-24.2	*KCNJ2*, Kir2.1	$I_{K1\uparrow}$
SQT 4	12p13.3	*CACNA1C, Cav1.2*	$I_{Ca\downarrow}$
SQT 5	10p12	*CACNB2b, Cavβ2b*	$I_{Ca\downarrow}$
SQT 6	7q21-22	*CACNA2D1*, $Ca_v\alpha_2\delta$-1	$I_{Ca\downarrow}$
SQT 7	2q35	SLC4A3,anion exchanger (AE3)	Altered anion transport

genes for LQTS (KCNH2, KCNJ2, and KCNQ1). Mutations in genes encoding calcium channels (CACNA1C and CACNB2) cause either SQTS or an overlapping phenotype that combines an abbreviated QT interval and a Brugada ECG phenotype. Other disease-causing genes include CACNA2D1 (L-type calcium channel subunit) and SLC4A3 gene encoding anion exchanger (AE3). Gain-of-function mutations in the KCNH2 (I_{Kr}), KCNQ1 (I_{Ks}), and KCNJ2 (I_{K1}) genes account for SQT1, SQT2, and SQT3, respectively. Loss-of-function mutations in the CACNA1C, CACNB2, and CACNA2D1 genes encoding for the α1, β, and a2δ-1 subunits of the cardiac L-type calcium channel account for SQT4, SQT5, and SQT6, respectively (Khera and Jacobson 2016; Thorsen et al. 2017). SLC4A3 is a new pathogenic gene account for SQT7.

3.2 *KCNH2, KCNQ1, and KCNJ2 Genes*

SQTS-causing genes in potassium channels exert a gain-of-function effect on the affected channel and lead to a shortened cardiac repolarization. KCNH2 (human ether-a-go-go related gene [HERG]) locates on chromosome 7q35 and encodes the pore-forming (α) subunit of the IKr (the rapidly activating delayed rectifier potassium channels). In 2004, Brugada et al. (2004) firstly reported the KCNH2 gene linked to SQTS and ventricular arrhythmias. The article also reported two missense mutations (C.1764C>A,C.1764C>G) which both lead to an identical amino acid substitution (N588 K) in the KCNH2 gene of two different families and demonstrated the link of SQTS to gain of function of IKr for the first time. Subsequently, Hong et al. (2005a) identified the same locus mutation in the KCNH2 gene in a family presenting with SQTS and atrial fibrillation. Hereafter, additional gain-of-function mutations, including the S631A, T618I, and R1135H mutations in KCNH2, were gradually identified (McPate et al. 2008; Itoh et al. 2009; Sun et al. 2011). These gain-of-function mutations lead to increased current flow through IKr and shortened action potential duration and refractoriness. Thus the mutations result in the shortening of the QT interval to form the subtype named SQT1.

KCNQ1 (KVLQT1) channel subunits is assembled with the accessory β-subunit KCNE1 to mediate the slowly activating delayed rectifier potassium channels (I_{Ks}).

In 2004, Bellocq et al. (2004) firstly reported the mutation (V307L) in KCNQ1, in a 70-year-old man with idiopathic ventricular fibrillation and markedly shorten QT interval. The mutation V307L leads to the substitution of the valine at position 307 by leucine (V307L). The activation of V307L was shifted toward more negative potentials, which suggested that the kinetics of activation in the mutant was faster than that of the WT. All of these data indicate a gain-of-function mechanism. The V141M mutations were identified to be associated with SQT2 (Hong et al. 2005b). The gain of function of V141M mutation caused IKs to activate faster and at more negative potentials, which make channels constitutively open during the cardiac action potential. In 2015, Wu et al. (2015) reported the R259H mutation in Chinese people. Compared with the former V307L and V141M mutations, the R259H mutation displayed an increased current density and markedly slower deactivation, which means no change in steady-state activation (SSA) and steady-state inactivation (SSI).The mutations in the KCNQ1 gene are linked to a gain of function and responsible for SQT2.

KCNJ2 is implicated in the Kir2.1 channels that underlie the inward rectifier potassium currents (IK1), contributing to a stable resting membrane potential and phase 3 repolarization of the cardiac action potential. In 2005, Priori et al. (2005) reported the mutation D172N in the KCNJ2 gene in a patient with SQTS. Changes in IK1 due to the Kir2.1 D172N mutation demonstrated a larger outward IK1 and rightward voltage shifts of peak-repolarizing current, which results in abbreviating the ventricular action potential duration (APD) and effective refractory period (ERP). The gain of function in the KCNJ2 mutation is named as SQT3. In 2012, Hattori et al. (2012) published a case of a novel M301K mutation in the KCNJ2 gene and described possible mechanisms for the associated markedly shortened QT interval. Furthermore, the mutation E299V in KCNJ2 gene was detected in an 11-year-old boy with a remarkably short QT interval and atrial fibrillation (AF) in 2013 (Deo et al. 2013). The next year, Ambrosini et al. observed a lysine-to-threonine substitution at position 346 (K346T) of the Kir2.1 on monozygotic twins displaying a short QT interval on electrocardiogram recordings and autism-epilepsy phenotype. Both gain-of-function mutations, E299V and K346T, caused an increased outward current to induce type III short QT syndrome (SQT3) (Whittaker et al. 2017).

Genetic and heterologous expression studies revealed gain-of-function missense mutations in KCNH2, KCNJ2, and KCNQ1, which, encoding potassium channels, were associated with SQTS.

3.3 CACNAlC, CACNB2b, and CACNA2D1 Genes

Mutations affecting the L-type calcium channel (LTCC) subunits linked to SQTS are expected to cause a loss of channel function and abbreviation of AP. The LTCC is composed of four subunits: the main pore-forming α1 (Cav1.2) subunit, encoded by CACNAlC, which determines the main biophysical and pharmacologic properties of the channel, and three auxiliary subunits, including a cytoplasmic β-subunit, encoded by CACNB; α2δ, encoded by CACNA2D; and a γ-subunit, which is not expressed in the heart. CACNAlC, CACNB2b, and CACNA2D1 genes locate on chromosomes 12p13.3, 10p12.33, and 7q21-q22, respectively (Pereira et al. 2017). Functional analysis showed that loss-of-function mutations in these genes, especially, lead to shorter QTc intervals by decreasing the amplitude of inward calcium current. In 2007, research group from Antzelevitch (Antzelevitch et al. 2007) firstly reported that loss-of-function missense mutations of the CACNA1C (A39V and G490R) and CACNB2 (S481L) genes are associated with a familial sudden cardiac death syndrome. The findings revealed that the loss-of-function mutation leads to a significantly reduced inward calcium flow during phase 2 of action potential (AP) which induces AP repolarization time and ERP shortening. A39V mutation in CACNAIC causes a defect in trafficking of mature Cav1.2 channels, whereas G490R Cav1.2 or S481L Cavβ2b subunits transport normally. Subsequently, study from the same group[17] reported that mutations in CACNA1C and CACNB2b cause a major loss of function in calcium channel activity and lead to a decrease in inwardly directed calcium currents, displaying an abbreviated QTc. Recently, Chen et al. (2017) reported the novel mutations (c.700G>A/p.E234K in DES, c.2966G>A/p.R989H in MYPN, and c.5918G>C/p.R1973P in CACNA1C) detected in a Chinese family of obstructive HCM with early repolarization (ER) and SQTS. The characteristics of short QT interval might be attributed to the mutation p.R1973P in CACNA1C. Meanwhile, accelerated repolarization caused by the mutation could lead to a subsequent short QT interval on the surface ECG. A new variant of SQTS caused by a mutation in the gene encoding the Cava2d-1 subunit of the L-type calcium channel (CACNA2D1) was reported by Templin et al. (2011). This study identifies that a pathogenic variant in the CACNA2D1 gene leads to SQTS, defining SQTS6, during the loss of function of LTCC.

3.4 SCN5A and Other Genes

The role of SCN5A has been firmly established over the past decade. In 2012, Hong group (2012) first described a novel mutation of SCN5A in a Chinese man with the overlapping phenotypes of short QT interval and Brugada-like ECG. The mutation decreased I_{Na}, which might result in the reduction of cardiac excitability and electrical conduction velocity. This study revealed that loss-of-function mutation in the SCN5A gene (R689H) caused short QT interval. It is worthy to know that the particular expressed phenotype often depends on the genetic background, including modulating rare and common variants, as well as epigenetic and environmental factors. The tools for identifying novel mutations and genetic modifiers that change phenotypic expression and disease outcome are extremely urgent. Recently, Thorsen et al. (2017) identify a missense mutation in the anion exchanger (AE3)-encoding SLC4A3 gene in two unrelated families with SQTS. The mutation causes a reduced surface expression of AE3 and membrane bicarbonate transport. Slc4a3 knockdown in zebrafish causes an increased cardiac pHi, short QTc, and reduced systolic duration, which is rescued by wild-type but not mutated SLC4A3. Mechanistic analyses suggest that an increase in pHi and a decrease in [Cl-] shortened the AP duration. However, other mechanisms may also play a role. Altered anion transport producing development of arrhythmia might provide new therapeutic possibilities.

3.5 Genetic Screening

A recent study conducted in 45 probands with SQTS identified mutations in only 14% of cases, despite familial disease being present in almost half (Priori et al. 2013). Data from the same study on 73 patients suggested that SQTS is highly lethal with SCD often being the first manifestation of the disease.

4 Catecholaminergic Polymorphic Ventricular Tachycardias (CPVT)

4.1 Introduction

Catecholaminergic polymorphic ventricular tachycardia (CPVT) is an inherited cardiac arrhythmia disorder that is characterized by episodic syncope occurring during exercise or acute emotion in individuals without structural cardiac abnormalities (Lieve et al. 2016). CPVT is a rare disease with an estimated prevalence of 1:10,000 but with a life-threatening potential because CPVT plays an important role in syncope and sudden cardiac death (SCD) in the young. The majority of patients present around the age of 10 years with exercise-induced syncopal episodes (Roston et al. 2015).

Patients with CPVT have a normal 12-lead ECG at rest; the occurrence of a bidirectional ventricular tachycardia (VT) is highly suggestive of the diagnosis CPVT but is infrequently observed (Priori et al. 2013). Exercise testing is the most helpful clinical tool to diagnose CPVT. In asymptomatic relatives of CPVT patients, the exercise test has a specificity of 97% and a sensitivity of 50% for predicting the presence of the familial CPVT-associated mutation (Hayashi et al. 2012).

4.2 Genetic Basis of CPVT

CPVT has two inheritance model, autosomal dominant inheritance and autosomal recessive inheritance. The familial occurrence of CPVT was already recognized in 1960, which described three sisters suffering from syncopal episodes during exercise or emotional stress (Berg 1960). Genetic linkage studies in two unrelated

Table 9.4 Genes associated with CPVT

Gene	Full name	Gene ID	Location	Inheritance pattern
RYR2	Ryanodine receptor 2	6262	1q43	AD
CASQ2	Calsequestrin 2	845	1p13.1	AR
TRDN	Triadin	10345	6q22.31	AR
KCNJ2	Potassium voltage-gated channel subfamily J member 2	3759	17q24.3	AD
ANK2	Ankyrin 2	287	4q25-q26	AD
CALM1	Calmodulin 1	801	14q32.11	AD

Finnish families revealed a disease-causing locus with an autosomal dominant inheritance pattern on chromosome 1q42–q43 (Swan et al. 1999). The gene encoding the cardiac ryanodine receptor (RYR2) was identified as the first disease-causing gene residing on this locus (Priori et al. 2001), and five more genes were found to be associated with CPVT later (Table 9.4).

Mutations in RyR2 account for approximately 65% of the CPVT cases (Ackerman et al. 2011). RYR2 mainly expresses in the sarcoplasmic reticulum of cardiomyocytes; it is involved in intracellular calcium hemostasis and plays a role in the excitation-contraction coupling of the heart. Mutations in RyR2 cause a diastolic calcium leakage, with a subsequent increase in the cytosolic calcium concentration (Jiang et al. 2004). This increase in cytosolic calcium concentration ultimately activates the electrogenic sodium-calcium exchanger, leading to a transient inward current, resulting in delayed after depolarizations that in turn can lead to triggered arrhythmias. The diastolic calcium leakage becomes more pronounced in the setting of high β-adrenergic tone. More than 200 gain-of-function RyR2 mutations have been described in patients with CPVT (Roston et al. 2018). However, loss-of-function mutations in RyR2, such as A4860G, also exist (Jiang et al. 2007). A deletion of exon 3 of RyR2 can cause a distinct subtype of CPVT characterized by sinoatrial node and atrioventricular node dysfunction, supraventricular arrhythmias, and dilated cardiomyopathy (Bhuiyan et al. 2007).

Mutations in CASQ2 account for approximately 2–5% of the CPVT cases (Ackerman et al. 2011), which are autosomal recessive inherited. CASQ2 encodes cardiac calsequestrin, which is a calcium-buffering protein within the sarcoplasmic reticulum with an inhibitory effect on RyR2. Mutations in CASQ2 gene disrupt the handling of Ca2+ within myocytes. During exercise or emotional stress, impaired Ca2+ regulation in the heart can lead to VT in people with CPVT. Genetic information proved to be important in the prediction of risk on lethal ventricular arrhythmias of affected individuals (Wilde and Amin 2017).

Mutations in the gene encoding triadin (TRDN) were identified in the probands of two families of CPVT of an autosomal recessive inheritance pattern (Roux-Buisson et al. 2012). TRDN mutations might cause CPVT by an impaired interaction between FKBP12.6 and RyR2 or by a reduction of CASQ2. CALM1 encodes calmodulin; its mutation was revealed in CPVT cases. Calmodulin is a calcium-binding protein that stabilizes RyR2 and reduces the probability of its opening during diastole (Nyegaard et al. 2012). Finally, mutations in the genes ANK2, encoding ankyrin-B, and KCNJ2, encoding the potassium inwardly rectifying channel Kir2.1, are generally associated with the congenital long QT syndrome types 4 and 7, respectively, but may phenocopy CPVT (Pérez-Riera et al. 2018).

5 Early Repolarization Syndrome (ERS)

5.1 Introduction

Early repolarization (ER) is characterized by an overt J-wave or J-point elevation, notch or slur of the terminal part of the QRS, and ST elevation in at least two contiguous inferior and/or lateral leads of a standard 12-lead electrocardiogram (ECG) (Priori et al. 2013). For decades, the ER pattern was historically thought to be a benign ECG variant in the general population, with a prevalence ranging from 1% to 13% depending on the population studied (Mizusawa and Bezzina 2014). However, accumulating evidence from a variety of case-control and population-based studies has shown that ER is linked with an increased risk of arrhythmic death including ventricular tachycardia (VT), ventricular fibrillation (VF), and sudden cardiac death (SCD). The reported association between ER pattern and VT/VF in these studies has resulted in the recognition of a clinical syndrome, referring as the early repolarization syndrome (ERS). Despite the underlying arrhythmic substrates of ERS are currently not well known, there is an increasing evidence that the gene mutations in the cardiac ion channels play a key pathogenic role in the development of ERS. It is widely believed that ERS is a type of inherited arrhythmogenic disease. Thus, a better understanding of the genetic basis linking ERS and ventricular arrhythmogenesis would have great significance for early diagnosis, risk stratification, and possible treatment. To this end, this chapter reviews current progress in the mechanistic and genetic research in ERS.

5.2 Potassium Channel Genes

Information relative to the genetic basis for ERS is relatively limited, however, growing evidence indicating that ERS is related to mutations in a number of cardiac ion channel genes (Table 9.5). Of these, KCNJ8, encoding the Kir6.1 subunit of cardiac adenosine triphosphate (ATP)-sensitive potassium channel (KATP), is the most frequently reported ERS-related gene. It has been demonstrated that gain-of-function mutations in KCNJ8 in patients with idiopathic VT/VF are associated with ERS (Delpon et al. 2008; Haissaguerre et al. 2009; Medeiros-Domingo et al. 2010; Barajas-Martinez et al. 2012; Hu et al. 2014). Haissaguerre et al. firstly reported

Table 9.5 Genetic defects associated with ERS

Type	Locus	Gene/protein	Ion channel	% of probands
ERS1	12p11.23	*KCNJ8, Kir6.1*	$I_{K\text{-}ATP}\uparrow$	
ERS2	12p13.3	*CACNA1C, Ca$_v$1.2*	$I_{Ca}\downarrow$	4.1%
ERS3	10p12.33	*CACNB2b, Ca$_v\beta$2b*	$I_{Ca}\downarrow$	8.3%
ERS4	7q21.11	*CACNA2D1*, $Ca_v\alpha_2\delta$-1	$I_{Ca}\downarrow$	4.1%
ERS5	12p12.1	*ABCC9, SUR2A*	$I_{K\text{-}ATP}\uparrow$	Rare
ERS6	3p21	*SCN5A, Na$_v$1.5*	$I_{Na}\downarrow$	Rare
ERS7	3p22.2	*SCN10A, Na$_v$1.8*	$I_{Na}\downarrow$	

KCNJ8 variant in ERS in the case of a young women with electrical storm (Haissaguerre et al. 2009). However, functional expression data for the KCNJ8 were unavailable to verify the link of such mutation with the phenotype. Subsequently, Medeiros-Domingo et al. reported an association between the S422L missense mutation in KCNJ8 and ERS (Medeiros-Domingo et al. 2010). They genetically screened a S422L-KCNJ8 mutation in 1 out of 14 probands with ERS. Functional characterization of the mutation revealed a gain of function in the KATP. The identical KCNJ8 S422L mutation was also reported by Barajas-Martínez et al. They found the mutation in 4 out of 204 J-wave syndrome prohands and showed a two-fold gain of function of KATP current when KCNJ8-S422L was co-expressed with SUR2A wild type (Barajas-Martinez et al. 2012). More recently, mutations in ABCC9, which is responsible for the ATP-binding cassette transporter of KATP, have been identified in ERS patients. Hu et al. screened a cohort of 150 patients with ERS and identified four ERS patients with ABCC9 mutation (Hu et al. 2014). Functional analysis revealed that the mutation indirectly enhances the KATP current (Hu et al. 2014).

The Ito channel is of great importance for the early phase of repolarization and reported to be the predominant contributor to phase I of the cardiac action potential (Delpon et al. 2008). Genetic variants in genes encoding Ito have been implicated as an important contributor to ERS. Perrin et al. performed genetic screening in a cohort of 31 ERS patients for mutations in genes encoding Ito channel subunits (KCNA4, KCND2, and KCND3). A single mutation in KCND2 – D612N – was identified in a patient with an unusual J-wave syndrome with evidence of QRS notching in the anterior leads. Functional characterization revealed a gain-of-function effect with an increased Ito current density (Perrin et al. 2014). Sinner MF et al. performed a meta-analysis of genome-wide association studies (GWAS) in almost 15,000 individuals with ER from community-based cohorts of European ancestry and reported that genetic variants in KCND3 possibly lead to altered channel properties that could underlie an increased arrhythmogenic potential (Sinner et al. 2012).

5.3 Calcium Channel Genes

Mutations in ion channels that encode inward currents have also been involved in the development of ERS. Several L-type calcium channel mutations have been associated with ERS. Loss-of-function mutations in the subunits of the cardiac L-type calcium channel genes (CACNA2D1, CACNA1C, and CACNB2) have been identified in ERS patients (Burashnikov et al. 2010; Napolitano and Antzelevitch 2011; Chen et al. 2017; Liu et al. 2017). Liu X et al. reported an association of ERS with a CACNA1C-Q1916R mutation in a large Chinese family with nocturnal sudden cardiac death. Functional studies using the exogenous expression approach revealed that the mutation decreased L-type calcium current and the protein expression, underlying ERS-related sudden cardiac death (Liu et al. 2017).

5.4 Sodium Channel Genes

In addition, the mutations in sodium channel have also been associated with ERS. Watanabe et al. identified mutations in the SCN5A gene in three ER patients after screening a cohort of 26 patients. Functional studies demonstrated a loss-of-function effect, with all three mutations failing to generate an INa current (Watanabe et al. 2011). Guo qi et al. characterized a novel SCN5A mutation that was associated with ERS. They reported a novel heterozygous missense mutation of A1055G in the SCN5A gene that led to loss of function of sodium channels accounting for the arrhythmogenic characteristics of ERS (Guo et al. 2016). Interestingly, a recent Brugada syndrome study has reported mutations in SCN10A were associated with ERS suggesting that SCN10A is a potential susceptibility gene for ERS (Hu et al. 2014).

The ankyrin-2 gene (ANK2) encodes ankyrin-B, a molecule critical for the membrane targeting of key cardiac ion channels, transporters, and signaling proteins. Loss-of-function variants in the ANK2 gene can cause a variety of cardiac rhythm abnormalities including sinus node dysfunction, atrial fibrillation, and ventricular arrhythmias. Recently, Broendberg AK et al., for the first time, reported a potential recognized association between ANK2 variants and ERS in two patients with idiopathic VF storm (Krogh et al. 2015). Although conclusion should be made with caution, it was suggested that these ANK2 variants may be capable of modulating the substrate for the development of ERS.

6 Atrial Fibrillation (AF)

6.1 *Introduction*

Atrial fibrillation (AF) is known as the most common cardiac supraventricular arrhythmia. The incidence of the familial form of AF remains unknown; however, recent studies suggest that up to 30% of AF patients without an identified cause have a family history of this condition. Over the past decade, the genetic substance of AF susceptibility has been increasingly uncovered through studies. Numerous genetic rare variations have been identified in AF families and sporadic cases, and common risk variants linked to AF susceptibility have been mapped in general population. These findings have established an important role of genetic variations in generating a substrate for AF. To date, AF-associated genes have predominantly encoded proteins including cardiac ion channels, myocardial structural components, and transcription factors closely related to the process of cardiac development.

6.2 *Cardiac Ion Channels Related Genes*

In 1997, the chromosomal locus 10q22-q24 was firstly found in three families with autosomal dominant AF using linkage analysis (Brugada et al. 1997). Afterward, the first mutation gene KCNQ1 that encodes a voltage-gated potassium (K+) channel was identified to be responsible for AF in 2003 (Chen et al. 2003). These findings have launched a cascade of candidate gene screening studies on other cardiac ion channels. As exemplified by the findings of Hong et al., a de novo missense p.V141M mutation in KCNQ1 was identified in a baby girl who was diagnosed with AF with slow ventricular response and a short QT interval (Hong et al. 2005a, b). Sequential researches on genetic variations responsible for AF have shown that almost all of the potassium channel and sodium channel genes are associated with the development of AF. To date, all disease genes encoding cardiac ion channels linked to AF have been listed in Table 9.6. Further functional analyses of the ion channel gene mutation have discovered that gain-of-function effects on potassium current and loss-of-function effects on sodium current can generally be responsible for the pathogenesis of AF, with some exceptional cases.

The SCN5A mutation p.M1875T associated with familial AF displayed a gain-of-function-type modulation of cardiac Na+ channels, which is a novel mechanism predisposing to increased atrial excitability and familial AF (Makiyama et al. 2008). Mutations in KCNA5, encoding the ultrarapid delayed rectifier potassium current (IKur), can result in a loss-of-function effect on IKur current and eventually lead to AF (Olson et al. 2006; Yang et al. 2009, 2010).

Meanwhile, mutations in the ryanodine receptor 2 (RYR2) gene, contributing to abnormal Ca2+ handling in cardiomyocytes, have been detected to be associated

Table 9.6 Cardiac ion channel-related genes responsible for AF

Gene	Protein
ABCC9	ATP-binding cassette transporter; I_{KATP}
HCN4	Pacemaker current; I_F
JPH2	Intracellular Ca^{2+} signaling
KCNA5	Voltage-gated K^+ channel, Kv1.5; I_{Kur}
KCND3	Voltage-gated K^+ channel, Kv4.3; I_{to}
KCNE1	β-subunit; I_{Ks}, I_{Kr}
KCNE2	β-subunit; I_{Ks}, I_{Kr}
KCNE3	β-subunit; I_{Ks}
KCNE4	β-subunit; I_{Ks}
KCNE5	β-subunit; I_{Ks}
KCNH2	Voltage-gated K^+ channel, Kv11.1; I_{Kr}
KCNJ2	Inwardly rectifying K^+ channel, Kir2.1; I_{K1}
KCNJ5	Inwardly rectifying K^+ channel, Kir3.4; I_{KAch}
KCNJ8	Inwardly rectifying K^+ channel, Kir6.1; I_{KATP}
KCNK3	Two-pore domain K^+ channel, TASK1; I_{leak}
KCNN2	Small conductance, Ca^{2+}-activated K^+ channel, KCa2.2; I_{KCa}
KCNN3	Small conductance, Ca^{2+}-activated K^+ channel, KCa2.2; I_{KCa}
KCNQ1	Voltage-gated K^+ channel,Kv7.1; I_{Ks}
RYR2	Ryanodine receptor 2, Ca^{2+} release
SCN1B	β-subunit; I_{Na}
SCN2B	β-subunit; I_{Na}
SCN3B	β-subunit; I_{Na}
SCN4B	β-subunit; I_{Na}
SCN5A	Voltage-gated Na^+ channel, Nav1.5; I_{Na}
SCN10A	Voltage-gated Na^+ channel, Nav1.8; $I_{Na\text{-}L}$

with clinical phenotype characterized by AF (Kazemian et al. 2011; Zhabyeyev et al. 2013). Junctophilin-2 (JPH2) is believed to have an important role in sarcoplasmic reticulum (SR) Ca2+ handling and modulation of RYR2. Beavers et al. screened 203 unrelated hypertrophic cardiomyopathy patients and uncovered a novel JPH2 missense mutation E169K in two patients with juvenile-onset paroxysmal AF (Beavers et al. 2013). Further analysis suggested that JPH2-mediated destabilization of RyR2 due to loss-of-function mutation can promote SR Ca2+ leak and lead to CPVT and AF; these data may underscore the importance of Ca2+ dysregulation as a fundamental mechanism for both atrial and ventricular tachyarrhythmias, representing a potential novel therapeutic target for AF.

6.3 Non-ion Channel Genes

Recently, genetic variants in a broad spectrum of non-ion channel genes responsible for AF have been found through studies, which predominantly encode cardiac transcription factors, myocardial structural components, and signaling proteins (Table 9.7).

The Pitx2 (paired-like homeodomain transcription factor 2) deficiency has been established as a plausible mechanism for AF in a series of studies on Pitx2 knockout

Table 9.7 Non-cardiac ion channel-related genes responsible for AF

Gene	Protein
Transcription factors	
CUX2	Homeobox protein Cux-2
GATA4	GATA-binding protein 4
GATA5	GATA-binding protein 5
GATA6	GATA-binding protein 6
HAND2	Heart- and neural crest derivative-expressed protein 2
NKX2-5	Homeobox protein NKX2-5
NKX2-6	Homeobox protein NKX2-6
PITX2	Paired-like homeodomain protein 2
PRRX1	Paired related homeobox protein 1
SHOX2	Short stature homeobox protein 2
SOX5	SRY-box 5
TBX5	T-box protein 5
ZFHX3	Zinc finger homeobox protein 3
Myocardial structural components	
CAV1	Caveolae protein, caveolin 1
GJA1	Gap junction protein, connexin 43
GJA5	Gap junction protein, connexin 40
LMNA	Nuclear envelope protein, lamin A/C
MYH6	Sarcomere protein, myosin heavy chain 6
MYL4	Sarcomere protein, myosin light chain 4
NEBL	Nebulette
SYNE2	Nuclear envelope protein, spectrin repeat containing nuclear envelope protein 2
SYNPO2L	Z-disc protein, synaptopodin 2-like
Signaling, protein turnover, other	
CAND2	TATA-binding protein, Cullin associated and neddylation disassociated 2
CEP68	Centrosomal protein 68
GREM2	BMP antagonist, gremlin 2
NEURL	Neutralized E3 ubiquitin protein ligase 1
NPPA	Natriuretic peptide precursor A
PPF1A4	Liprin-alpha 4
SH3PXD2A	Tyrosine kinase substrate, SH3 and PX domains 2

mice (Chinchilla et al. 2011). Then the role for transcription factors in the pathogenesis for AF was brought to attention. Firstly, T-box transcription factor 5 (TBX5) is expressed in the embryonic heart and regulates transcription of downstream genes such as the atrial natriuretic factor (NPPA) and fibroblast growth factor 10 (FGF10) by the binding to T-box-binding elements, often in combination with the NKX2-5 transcription factor. A gain-of-function TBX5 gene mutation G125R has been found in a large atypical Holt-Oram syndrome (HOS) family with mild skeletal deformations and paroxysmal AF (Postma et al. 2008). The potential linkage of AF and some cardiogenesis genes including GATA4, GATA5, and GATA6 has been investigated by Yang et al. (2011, 2012a, b). Mutations in these genes have been causally implicated in AF and congenital heart diseases, which may suggest a novel insight into the underlying mechanism in the pathogenesis of AF.

Connexins are key members of gap junctional proteins, including connexin43 (Cx43) and connexin40 (Cx40), highly expressed in the atrial tissues. Genetic studies have suggested that mutations in GJA1 and GJA5, the genes encoding Cx43 and Cx40, respectively, also are involved in the pathogenesis of AF. A recent study have shown that a novel nonsense mutation (c.G1494A, p.Trp498Ter) in the LMNA gene can be identified as the candidate mutation in a Chinese AF family using exome sequencing and Sanger sequencing (Zhao et al. 2016).

6.4 *AF-Associated GWAS Loci*

Genome-wide association studies (GWAS) have yielded variants at more than 30 loci that associate with atrial fibrillation (AF). The first major GWAS in AF was reported in 2007, with a significant locus identified in an intergenic noncoding region on chromosome 4q25 (Gudbjartsson et al. 2007). Olesen et al. replicated the GWAS associations of SNPs in three loci on chromosomes 4q25, 7p31, and 12p12 in a population of patients with early-onset lone AF (Olesen et al. 2012). Very recently, a large-scale, trans-ancestry meta-analyses of common and rare variant association studies including over 240,000 individuals with AF identified 12 new genetic loci, involved in cardiac electrical and structural remodeling (Christophersen et al. 2017). Meanwhile, Low and colleagues from Japan described a GWAS in over 8000 AF cases and identified six loci, many of which were specific to individuals of Asian ancestry (Low et al. 2017). These GWAS findings provide further insights into the molecular basis of atrial fibrillation and may facilitate the identification of new potential targets for drug discovery.

Table 9.8 IVF-related genes

Gene	Description	Location
SCN5A	Sodium voltage-gated channel alpha subunit 5	Chromosome 3
SCN3B	Sodium voltage-gated channel beta subunit 3	Chromosome 11
RYR2	Ryanodine receptor 2	Chromosome 1
CALM1	Calmodulin 1	Chromosome 14
KCNE5	Potassium voltage-gated channel subfamily E regulatory subunit 5	Chromosome X
DPP6	Dipeptidyl peptidase like 6	Chromosome 7

7 Idiopathic Ventricular Fibrillation (IVF)

7.1 Introduction

Ventricular fibrillation is considered as a leading cause of sudden cardiac death worldwide. However, 5–12% of these cases, with no demonstrable cardiac or non-cardiac disorders to account for the episode, have been classified as idiopathic ventricular fibrillation (IVF). The etiology of IVF remains incompletely illustrated since its first description in 1927. To date, previous genetic studies have shown that causative gene mutations can be detected among IVF patients, indicating IVF as a rare primary cardiac arrhythmia syndrome with inherited characteristics.

Although the etiology of IVF remains incompletely understood yet, several lines of evidence have indicated that this disorder is a channelopathy involving hereditary malfunction of cardiac ion-channel genes, including sodium channels, potassium channels, or calcium channels. The IVF-related genes have been shown in Table 9.8. Mutations in these genes could contribute to the risk of developing IVF.

7.2 Sodium Channel-Associated Genes

SCN5A has been identified as one of the responsible genes for IVF patients for over a decade. In 2000, Akai J et al. found a novel SCN5A missense mutation S1710L in a symptomatic IVF patient that did not exhibit the typical Brugada ECG by performing genetic screenings (Akai et al. 2000). Heterologously expressed S1710L channels showed a marked acceleration in the current decay together with a large hyperpolarizing shift of steady-state inactivation and depolarizing shift of activation. These findings have suggested that SCN5A is one of the responsible genes for IVF patients in the absence of Brugada syndrome phenotype. In addition, a novel mutation in the SCN3B-encoded sodium channel beta subunit Navbeta3 that causes a loss of function of Nav1.5 channels has been found in a case with IVF (Valdivia et al. 2010).

7.3 Calcium Channel-Associated Genes

Marsman et al. carried out exome sequencing on a family presenting IVF and sudden death and identified a mutation in CALM1 gene encoding calmodulin underlying IVF manifesting in childhood and adolescence (Marsman et al. 2014). Recently, a novel CALM gene mutation F90L was identified in a family with IVF and early-onset sudden cardiac death (Nomikos et al. 2014). It is known that calmodulin associated with the cardiac muscle ryanodine receptor (RyR2) regulates excitation-contraction coupling. This F90L mutation can lead to dysregulation of RyR2-mediated Ca(2+) release, indicating a potential mechanism that underlies familial IVF. Additionally, two novel, functional heterozygous mutations, c.6224T>C and c.13781A>G, in the RyR2 gene have been found in a family with IVF (Paech et al. 2014). These findings suggest that mutations in CALM and RYR2 genes are potentially causative factors responsible for IVF.

7.4 Potassium Channel-Associated Genes

KCNE5 (KCNE1L) is located in the X chromosome and encodes an auxiliary β-subunit for K channels. It can modify the transient outward current (Ito), which plays a key role in determining the repolarization process in the myocardium. When KCNE5 variants screening was conducted among 205 Japanese patients with BrS or IVF with negative SCN5A mutation, two novel KCNE5 variants were identified to cause IVF, especially BrS, through gain-of-function effects on Ito (Ohno et al. 2011).

7.5 Other Genes

By performing a genome-wide haplotype-sharing analysis, Alders et al. identified a haplotype, on chromosome 7q36 (involving the arrhythmia gene DPP6), in IVF patients. Hence, DPP6 was proposed as a gene for IVF, and increased DPP6 expression was considered as the pathogenetic mechanism (Alders et al. 2009). A recent study expanded our knowledge of IVF linked to the DPP6 locus by analyzing the detailed clinical characterization of 286 DPP6 risk-haplotype-positive (haplotype-positive) and 315 DPP6 risk-haplotype-negative (haplotype-negative) individuals (Ten et al. 2016). The results have shown that median survival was 70 vs. 93 years for haplotype positives vs. haplotype negatives, with a worse phenotype in males.

7.6 Genetic Testing

Currently, knowledge concerning the functional meaning of genetic testing for IVF is growing. It is known that except for arrhythmia, IVF does not manifest with identifiable clinical abnormalities. Therefore, genetic testing plays an important role as it enables the identification of individuals at risk of IVF. Additionally, genetic testing and a mechanistic understanding of IVF may lead to rational therapies for IVF.

8 Progressive Cardiac Conduction Defect (PCCD)

8.1 Introduction

Progressive cardiac conduction disease (PCCD) is an inherited cardiac channelopathy which is clinically characterized by a progressive slow conduction through the His-Purkinje system, with right and/or left bundle branch block and widening of the QRS complex, leading to complete atrioventricular node block. The electrocardiogram (ECG) was mainly manifested in abnormal PR, atrioventricular block, left or right bundle, atrial static, sinus node dysfunction, and sinus atrioventricular block. So it is one of the major causes of permanent pacemaker implantation in developing countries. When the patient progresses to intermittent or chronic high- or third-degree atrioventricular block, it can cause syncope and sudden cardiac death (SCD).

PCCD often involving left and right bundle branch fibrosis, also can be associated with calcification, atrophy, and diffuse involvement of the conduction system which is gradually replaced by fibrous tissue, but the adjacent tissue was not affected. PCCD may be associated with structural heart disease or may present as a primary electrical disease or channelopathy.

8.2 Genetic Basis of PCCD

PCCD is mostly autosomal dominant inherited, with some of which are recessive inherited or sporadic cases. In 1901, Morquio first reported a familial aggregation of cardiac conduction disturbances (Baruteau et al. 2015). With the development of molecular genetic technology, 24 related genes have been reported, 4 of which are identified as disease-causing genes (Table 9.9).

SCN5A encodes the α-subunit of cardiac sodium ion channel Nav1.5 protein, forms the ion-conducting pore, and participates in the formation of myocardial action potential (Liu et al. 2014). SCN5A is highly expressed in human heart and is known as the first disease-causing gene of PCCD. Schott first found that SCN5A loss-of-function mutation G514C can lead to PCCD in a French family (Schott et al. 1999). The myocardial sodium current Nav1.5 density did not change, but it can increase the sodium channel activation voltage, make the sodium channel inactivation

Table 9.9 Genes associated with PCCD

Gene	Full name	Gene ID	Location	Inheritance pattern
SCN5A	Sodium voltage-gated channel alpha subunit 5	6331	3p22.2	AD
TRPM4	Transient receptor potential cation channel subfamily M member 4	54795	19q13.33	AD
LMNA	Lamin A/C	4000	1q22	AD
NKX2.5	NK2 homeobox 5	1482	5q35.1	AD
SCN1B	Sodium voltage-gated channel beta subunit 1	6324	19q13.11	AD
PRKAG2	Protein kinase AMP-activated non-catalytic subunit gamma 2	51422	7q36.1	AD
HCN4	Hyperpolarization-activated cyclic nucleotide-gated potassium channel 4	10021	15q24.1	AD
SCN10A	Sodium voltage-gated channel alpha subunit 10	6336	3p22.2	AD
DES	Desmin	1674	2q35	AR, AD
DSP	Desmoplakin	1832	6p24.3	AR
CACNB2	Calcium voltage-gated channel auxiliary subunit beta 2	783	10p12.33-p12.31	
TBX5	T-box 5	6910	12q24.21	AD
TBX3	T-box 3	6926	12q24.21	AD
GATA4	GATA-binding protein 4	2626	8p23.1	AD
CAP2	Cyclase-associated actin cytoskeleton regulatory protein 2	10486	6p22.3	
LDB3	LIM domain binding 3	11155	10q23.2	
ABCC9	ATP-binding cassette subfamily C member 9	10060	12p12.1	
GJC1	Gap junction protein gamma 1	10052	17q21.31	
CLCA2	Chloride channel accessory 2	9635	1p22.3	
GJA5	Gap junction protein alpha 5	2702	1q21.2	
PDYN	Prodynorphin	5173	20p13	
KCNQ1	Potassium voltage-gated channel subfamily Q member 1	3784	11p15.5-p15.4	
KCNH2	Potassium voltage-gated channel subfamily H member 2	3757	7q36.1	
KCNK17	Potassium two-pore domain channel subfamily K member 17	89822	6p21.2	

faster, and reduce the number of activated sodium channels. This mutation affects the rapid influx of Na+, thus changing the 0 phase of action potential, affecting the rapid response of cardiac myocyte electrical activity conduction. In addition, recent studies have also found that mutated sodium channel proteins in patients with PCCD can't be transported to the plasma membrane after being modified (Yu et al. 2017). In heterozygous SCN5A knockout mice model, age-related progressive development of the conduction defect was recorded, including progressive impairment with aging of atrial and ventricular conduction associated with myocardial rearrangements and fibrosis (Royer et al. 2005). Mutations in the SCN5A gene lead to a broad spectrum of hereditary arrhythmias with variable phenotypic expression, so the clinical phenotypes of PCCD caused by SCN5A mutation are often overlapped with Brugada syndrome, long QT syndrome type 3, and sick sinus syndrome (Abriel 2010).

The SCN1B gene encodes the β-subunit of the cardiac sodium channel Nav1,5, which is involved in the regulation of sodium channel expression and function. Watanabe et al. found a causative mutation in the SCN1B gene in a pedigree with PCCD; co-expression of mutant SCN1B with Nav1.5 reduced the number of sodium channels in the cell membrane (Watanabe et al. 2008). Genome-wide association study found that SCN10A gene, which encodes the cardiac sodium channel Nav1.8, its enhancer interacts with the promoter of SCN5A, thus explaining the reason that SCN10A gene mutation may affect the cardiac conduction system (Pfeufer et al. 2010). The expression of Nav1.8 in the heart is still under debate, so the mechanism that causes PCCD is not yet clear.

Alterations in the expression of cardiac connexins may cause abnormal activation to spread through the myocardium and lead to conduction disorders. Cx45 mutant R75H is responsible for a novel disease entity of PCCD associated with craniofacial and dentodigital malformation (Seki et al. 2017). Heterologous expression of Cx40-Q58L in connexin-deficient cells resulted in a marked reduction in junctional conductance and the diffuse localization of Cx40 proteins in the vicinity of the plasma membrane, without the formation of gap junctions (Makita et al. 2012). It emphasizes the importance of Cx40 in the normal propagation of the electrical impulse in the specialized cardiac conduction system.

The transient receptor potential melastatin 4 (TRPM4) cation channel is abundantly expressed in the heart and mediates a Ca2+-activated nonselective cationic current (Kruse and Pongs 2014). In pedigree carrying the TRPM4-I376T mutation, the functional analysis showed that the mutation increased the expression of TRPM4 in the cell membrane and increased the current density (Daumy et al. 2016). E7K mutation of TRPM4 gene was found to be pathogenic in a PCCD family and was found that it affects the endocytosis and stability of the membrane channel proteins and produces greater currents (Kruse et al. 2009). Functional experiments expressing three variants of TRPM4 suggested a similar gain-of-function phenomenon related to altered deSUMOylation (Liu et al. 2010). The above study shows that TRPM4 plays a key role in the genetic pathogenesis of PCCD. It may be that gain-of-function mutations lead to depolarization of the cell membrane of the conduction system, thereby reducing the number of available sodium channels.

PCCD may be associated with structural heart disease, including dilated cardiomyopathy and sporadic congenital heart diseases. It has been reported that several genes are associated with this kind of PCCD, such as LMNA, Nkx2.5, and TBX5. In an autosomal dominant PCCD family, clinical phenotypes include atrioventricular block, sinus bradycardia, and sudden death; a complex rearrangement of the LMNA gene which encodes the nuclear membrane protein A/C was detected (Marsman et al. 2011). Patients with mutations in the LMNA often have a high risk of sudden death, and guidelines recommend that the patients be considered for ICD implantation (Priori et al. 2015). The correlation between NKX2.5 and PCCD was first found in four families with atrial septal defect (Schott et al. 1998). In addition, the mutation of NKX2.5 gene can also result in the insufficiency of noncompaction of ventricular myocardium and sudden death. Cardiac transcription factor TBX5 gene mutation can lead to Holt-Oram syndrome, consisting of bilateral symmetric limb malformations, congenital cardiac defects, and rapidly progressive cardiac conduction disease (Bogarapu et al. 2014).

9 Sick Sinus Syndrome (SSS)

9.1 Introduction

Sick sinus syndrome (SSS), also known as sinus node dysfunction (SND), is a disorder of the sinoatrial (SA) node caused by impaired pacemaker function and impulse transmission producing a constellation of abnormal rhythms. It may result in palpitations and tissue under-perfusion leading to fatigue, lightheadedness, pre-syncope, and syncope (Monfredi and Boyett 2015). In 1906, the phenomenon of atrial wave shedding on the electrocardiogram was first discovered by Karel Frederik Wenckebach and described as sinoatrial block (Schamroth and Dove 1966). The prevalence of sick sinus syndrome in patients with heart disease over the age of 65 is approximately 1/600 (Rodriguez and Schocken 1990), and the incidence is approximately 0.8/1000 per year in the normal population (Jensen et al. 2014). Inherited SSS, also known as congenital SSS or familial SSS, refers to the genetic cause of cardiac sinoatrial node dysfunction and leads to a variety of arrhythmias and clinical symptoms of the syndrome. Hereditary SSS can be found in fetuses, infants, or children with or without cardiac structural abnormalities and the incidence of obvious familial tendency but with the lack of specific epidemiological data.

9.2 Genetic Basis of SSS

SSS is inherited in autosomal dominant and autosomal recessive patterns; to date, ten pathogenic genes have been reported to be associated with inherited SSS (Table 9.10).

SCN5A, which encodes cardiac sodium channel gene, is the first pathogenic gene of hereditary SSS. There are about 24 pathogenic mutations of SCN5A gene, which can cause SSS in autosomal dominant or autosomal recessive pattern. The molecular mechanism of dominant familial inheritance due to heterozygous SCN5A mutations is the abnormal conduction of the sinus node resulting from the loss of

Table 9.10 Genes associated with SSS

Gene	Full name	Gene ID	Location	Inheritance pattern
SCN5A	Sodium voltage-gated channel alpha subunit 5	6331	3p22.2	AD, AR
HCN4	Hyperpolarization-activated cyclic nucleotide-gated potassium channel 4	10021	15q24.1	AD
MYH6	Myosin heavy chain 6	4624	14q11.2	AD
ANK2	Ankyrin 2	287	4q25-q26	AD
GNB2	G-protein subunit beta 2	2783	7q22.1	
KCNQ1	Potassium voltage-gated channel subfamily Q member 1	3784	11p15.5-p15.4	
CACNA1D	Calcium voltage-gated channel subunit alpha1 D	776	3p21.1	
LMNA	Lamin A/C	4000	1q22	
CAV3	Caveolin-3	859	3p25.3	
PRKAG2	Protein kinase AMP-activated non-catalytic subunit gamma 2	51422	7q36.1	

function of SCN5A (Milanesi et al. 2015). It is noteworthy that since the SCN5A-encoded Na+ channel is an important protein involved in the action potential of cardiomyocytes, some patients with inherited SSS caused by SCN5A mutation also have other arrhythmia phenotypes, including Brugada syndrome, long QT syndrome, and progressive cardiac conduction disease (Smits et al. 2005).

HCN4 encodes hyperpolarization-activated cyclic nucleotide-gated potassium channel, which is key ion channel proteins that generate electrical impulses in the sinus node tissue. The loss-of-function mutations can cause autosomal dominant inherited SSS. So far, about ten HCN4 mutations have been reported to be associated with SSS (Milanesi et al. 2015). The main pathogenesis is that HCN4 gene mutations lead to sinus node cell action potential spontaneous disorder.

Ankyrin 2 is a structural connexin that is encoded by ANK2 gene and is mainly expressed in myocardium. Its function is to help immobilize and connect ion channel proteins with large structural proteins and maintain the normal distribution of ion channels in cells. Hereditary SSS caused by ANK2 follows an autosomal dominant inheritance pattern. There are two known pathogenic mutations associated with hereditary SSS, which are reported from two independent families (Le Scouarnec et al. 2008; Robaei et al. 2015).

MYH6, which encodes the human α-myosin heavy chain family, is the only non-ion channel gene that causes hereditary SSS and is autosomal dominant inherited. To date, only there MYH6 mutations have been reported to be associated with hereditary SSS. The first SSS-associated MYH6 gene variant was derived from a study of 38,384 people in genome-wide association studies (GWAS) (Holm et al. 2011). Subsequently, a proband screening study and one pedigree study found one pathogenic mutation, further confirming the correlation between the MYH6 gene and hereditary SSS (Ishikawa et al. 2015; Lam et al. 2015).

In addition to the four pathogenic genes mentioned above, the other six genes associated with hereditary SSS are reported from rare families or sporadic cases. So more genetic studies are still needed to confirm its pathogenicity in the future.

References

Abriel H. Cardiac sodium channel Na(v)1.5 and interacting proteins: physiology and pathophysiology. J Mol Cell Cardiol. 2010;48(1):2–11.

Ackerman MJ, Priori SG, Willems S, Berul C, Brugada R, Calkins H, Camm AJ, Ellinor PT, Gollob M, Hamilton R, Hershberger RE, Judge DP, Le Marec H, McKenna WJ, Schulze-Bahr E, Semsarian C, Towbin JA, Watkins H, Wilde A, Wolpert C, Zipes DP. HRS/EHRA expert consensus statement on the state of genetic testing for the channelopathies and cardiomyopathies this document was developed as a partnership between the Heart Rhythm Society (HRS) and the European Heart Rhythm Association (EHRA). Heart Rhythm. 2011;8(8):1308–39.

Akai J, Makita N, Sakurada H, Shirai N, Ueda K, Kitabatake A, Nakazawa K, Kimura A, Hiraoka M. A novel SCN5A mutation associated with idiopathic ventricular fibrillation without typical ECG findings of Brugada syndrome. FEBS Lett. 2000;479(1–2):29–34.

Alders M, Koopmann TT, Christiaans I, Postema PG, Beekman L, Tanck MW, Zeppenfeld K, Loh P, Koch KT, Demolombe S, Mannens MM, Bezzina CR, Wilde AA. Haplotype-sharing analysis implicates chromosome 7q36 harboring DPP6 in familial idiopathic ventricular fibrillation. Am J Hum Genet. 2009;84(4):468–76.

Angrist M, Chandrasekharan S, Heaney C, Cook-Deegan R. Impact of gene patents and licensing practices on access to genetic testing for long QT syndrome. Genet Med. 2010;12(4):S111–54.

Antzelevitch C, Brugada P, Borggrefe M, Brugada J, Brugada R, Corrado D, Gussak I, LeMarec H, Nademanee K, Perez RA, Shimizu W, Schulze-Bahr E, Tan H, Wilde A. Brugada syndrome: report of the second consensus conference. Heart Rhythm. 2005;2(4):429–40.

Antzelevitch C, Pollevick GD, Cordeiro JM, Casis O, Sanguinetti MC, Aizawa Y, Guerchicoff A, Pfeiffer R, Oliva A, Wollnik B, Gelber P, Bonaros EJ, Burashnikov E, Wu Y, Sargent JD, Schickel S, Oberheiden R, Bhatia A, Hsu LF, Haissaguerre M, Schimpf R, Borggrefe M, Wolpert C. Loss-of-function mutations in the cardiac calcium channel underlie a new clinical entity characterized by ST-segment elevation, short QT intervals, and sudden cardiac death. Circulation. 2007;115(4):442–9.

Barajas-Martinez H, Hu D, Ferrer T, Onetti CG, Wu Y, Burashnikov E, Boyle M, Surman T, Urrutia J, Veltmann C, Schimpf R, Borggrefe M, Wolpert C, Ibrahim BB, Sanchez-Chapula JA, Winters S, Haissaguerre M, Antzelevitch C. Molecular genetic and functional association of Brugada and early repolarization syndromes with S422L missense mutation in KCNJ8. Heart Rhythm. 2012;9(4):548–55.

Barsheshet A, Goldenberg I, O-Uchi J, Moss AJ, Jons C, Shimizu W, Wilde AA, McNitt S, Peterson DR, Zareba W, Robinson JL, Ackerman MJ, Cypress M, Gray DA, Hofman N, Kanters JK, Kaufman ES, Platonov PG, Qi M, Towbin JA, Vincent GM, Lopes CM. Mutations in cytoplasmic loops of the KCNQ1 channel and the risk of life-threatening events: implications for mutation-specific response to beta-blocker therapy in type 1 long-QT syndrome. Circulation. 2012;125(16):1988–96.

Baruteau AE, Probst V, Abriel H. Inherited progressive cardiac conduction disorders. Curr Opin Cardiol. 2015;30(1):33–9.

Beavers DL, Wang W, Ather S, Voigt N, Garbino A, Dixit SS, Landstrom AP, Li N, Wang Q, Olivotto I, Dobrev D, Ackerman MJ, Wehrens X. Mutation E169K in junctophilin-2 causes atrial fibrillation due to impaired RyR2 stabilization. J Am Coll Cardiol. 2013;62(21):2010–9.

Behr ER, Savio-Galimberti E, Barc J, Holst AG, Petropoulou E, Prins BP, Jabbari J, Torchio M, Berthet M, Mizusawa Y, Yang T, Nannenberg EA, Dagradi F, Weeke P, Bastiaenan R, Ackerman

MJ, Haunso S, Leenhardt A, Kaab S, Probst V, Redon R, Sharma S, Wilde A, Tfelt-Hansen J, Schwartz P, Roden DM, Bezzina CR, Olesen M, Darbar D, Guicheney P, Crotti L, Jamshidi Y. Role of common and rare variants in SCN10A: results from the Brugada syndrome QRS locus gene discovery collaborative study. Cardiovasc Res. 2015;106(3):520–9.

Bellocq C, van Ginneken AC, Bezzina CR, Alders M, Escande D, Mannens MM, Baro I, Wilde AA. Mutation in the KCNQ1 gene leading to the short QT-interval syndrome. Circulation. 2004;109(20):2394–7.

Berg KJ. Multifocal ventricular extrasystoles with Adams-Stokes syndrome in siblings. Am Heart J. 1960;60(6):965–70.

Bezzina CR, Barc J, Mizusawa Y, Remme CA, Gourraud JB, Simonet F, Verkerk AO, Schwartz PJ, Crotti L, Dagradi F, Guicheney P, Fressart V, Leenhardt A, Antzelevitch C, Bartkowiak S, Borggrefe M, Schimpf R, Schulze-Bahr E, Zumhagen S, Behr ER, Bastiaenen R, Tfelt-Hansen J, Olesen MS, Kaab S, Beckmann BM, Weeke P, Watanabe H, Endo N, Minamino T, Horie M, Ohno S, Hasegawa K, Makita N, Nogami A, Shimizu W, Aiba T, Froguel P, Balkau B, Lantieri O, Torchio M, Wiese C, Weber D, Wolswinkel R, Coronel R, Boukens BJ, Bezieau S, Charpentier E, Chatel S, Despres A, Gros F, Kyndt F, Lecointe S, Lindenbaum P, Portero V, Violleau J, Gessler M, Tan HL, Roden DM, Christoffels VM, Le Marec H, Wilde AA, Probst V, Schott JJ, Dina C, Redon R. Common variants at SCN5A-SCN10A and HEY2 are associated with Brugada syndrome, a rare disease with high risk of sudden cardiac death. Nat Genet. 2013;45(9):1044–9.

Bhuiyan ZA, van den Berg MP, van Tintelen JP, Bink-Boelkens MT, Wiesfeld AC, Alders M, Postma AV, van Langen I, Mannens MM, Wilde AA. Expanding spectrum of human RYR2-related disease: new electrocardiographic, structural, and genetic features. Circulation. 2007;116(14):1569–76.

Boczek NJ, Ye D, Johnson EK, Wang W, Crotti L, Tester DJ, Dagradi F, Mizusawa Y, Torchio M, Alders M, Giudicessi JR, Wilde AA, Schwartz PJ, Nerbonne JM, Ackerman MJ. Characterization of SEMA3A-encoded semaphorin as a naturally occurring Kv4.3 protein inhibitor and its contribution to Brugada syndrome. Circ Res. 2014;115(4):460–9.

Bogarapu S, Bleyl SB, Calhoun A, Viskochil D, Saarel EV, Everitt MD, Frank DU. Phenotype of a patient with contiguous deletion of TBX5 and TBX3: expanding the disease spectrum. Am J Med Genet A. 2014;164A(5):1304–9.

Brugada R, Tapscott T, Czernuszewicz GZ, Marian AJ, Iglesias A, Mont L, Brugada J, Girona J, Domingo A, Bachinski LL, Roberts R. Identification of a genetic locus for familial atrial fibrillation. N Engl J Med. 1997;336(13):905–11.

Brugada R, Hong K, Dumaine R, Cordeiro J, Gaita F, Borggrefe M, Menendez TM, Brugada J, Pollevick GD, Wolpert C, Burashnikov E, Matsuo K, Wu YS, Guerchicoff A, Bianchi F, Giustetto C, Schimpf R, Brugada P, Antzelevitch C. Sudden death associated with short-QT syndrome linked to mutations in HERG. Circulation. 2004;109(1):30–5.

Brugada R, Campuzano O, Sarquella-Brugada G, Brugada J, Brugada P. Brugada syndrome. Methodist Debakey Cardiovasc J. 2014;10(1):25–8.

Burashnikov E, Pfeiffer R, Barajas-Martinez H, Delpon E, Hu D, Desai M, Borggrefe M, Haissaguerre M, Kanter R, Pollevick GD, Guerchicoff A, Laino R, Marieb M, Nademanee K, Nam GB, Robles R, Schimpf R, Stapleton DD, Viskin S, Winters S, Wolpert C, Zimmern S, Veltmann C, Antzelevitch C. Mutations in the cardiac L-type calcium channel associated with inherited J-wave syndromes and sudden cardiac death. Heart Rhythm. 2010;7(12):1872–82.

Cerrone M, Delmar M. Desmosomes and the sodium channel complex: implications for arrhythmogenic cardiomyopathy and Brugada syndrome. Trends Cardiovasc Med. 2014;24(5):184–90.

Cerrone M, Lin X, Zhang M, Agullo-Pascual E, Pfenniger A, Chkourko GH, Novelli V, Kim C, Tirasawadichai T, Judge DP, Rothenberg E, Chen HS, Napolitano C, Priori SG, Delmar M. Missense mutations in plakophilin-2 cause sodium current deficit and associate with a Brugada syndrome phenotype. Circulation. 2014;129(10):1092–103.

Chen Q, Kirsch GE, Zhang D, Brugada R, Brugada J, Brugada P, Potenza D, Moya A, Borggrefe M, Breithardt G, Ortiz-Lopez R, Wang Z, Antzelevitch C, O'Brien RE, Schulze-Bahr E, Keating

MT, Towbin JA, Wang Q. Genetic basis and molecular mechanism for idiopathic ventricular fibrillation. Nature. 1998;392(6673):293–6.

Chen YH, Xu SJ, Bendahhou S, Wang XL, Wang Y, Xu WY, Jin HW, Sun H, Su XY, Zhuang QN, Yang YQ, Li YB, Liu Y, Xu HJ, Li XF, Ma N, Mou CP, Chen Z, Barhanin J, Huang W. KCNQ1 gain-of-function mutation in familial atrial fibrillation. Science. 2003;299(5604):251–4.

Chen L, Marquardt ML, Tester DJ, Sampson KJ, Ackerman MJ, Kass RS. Mutation of an A-kinase-anchoring protein causes long-QT syndrome. Proc Natl Acad Sci U S A. 2007;104(52):20990–5.

Chen Y, Barajas-Martinez H, Zhu D, Wang X, Chen C, Zhuang R, Shi J, Wu X, Tao Y, Jin W, Wang X, Hu D. Novel trigenic CACNA1C/DES/MYPN mutations in a family of hypertrophic cardiomyopathy with early repolarization and short QT syndrome. J Transl Med. 2017;15(1):78.

Chinchilla A, Daimi H, Lozano-Velasco E, Dominguez JN, Caballero R, Delpon E, Tamargo J, Cinca J, Hove-Madsen L, Aranega AE, Franco D. PITX2 insufficiency leads to atrial electrical and structural remodeling linked to arrhythmogenesis. Circ Cardiovasc Genet. 2011;4(3):269–79.

Christophersen IE, Rienstra M, Roselli C, Yin X, Geelhoed B, Barnard J, Lin H, Arking DE, Smith AV, Albert CM, Chaffin M, Tucker NR, Li M, Klarin D, Bihlmeyer NA, Low SK, Weeke PE, Muller-Nurasyid M, Smith JG, Brody JA, Niemeijer MN, Dorr M, Trompet S, Huffman J, Gustafsson S, Schurmann C, Kleber ME, Lyytikainen LP, Seppala I, Malik R, Horimoto A, Perez M, Sinisalo J, Aeschbacher S, Theriault S, Yao J, Radmanesh F, Weiss S, Teumer A, Choi SH, Weng LC, Clauss S, Deo R, Rader DJ, Shah SH, Sun A, Hopewell JC, Debette S, Chauhan G, Yang Q, Worrall BB, Pare G, Kamatani Y, Hagemeijer YP, Verweij N, Siland JE, Kubo M, Smith JD, Van Wagoner DR, Bis JC, Perz S, Psaty BM, Ridker PM, Magnani JW, Harris TB, Launer LJ, Shoemaker MB, Padmanabhan S, Haessler J, Bartz TM, Waldenberger M, Lichtner P, Arendt M, Krieger JE, Kahonen M, Risch L, Mansur AJ, Peters A, Smith BH, Lind L, Scott SA, Lu Y, Bottinger EB, Hernesniemi J, Lindgren CM, Wong JA, Huang J, Eskola M, Morris AP, Ford I, Reiner AP, Delgado G, Chen LY, Chen YI, Sandhu RK, Li M, Boerwinkle E, Eisele L, Lannfelt L, Rost N, Anderson CD, Taylor KD, Campbell A, Magnusson PK, Porteous D, Hocking LJ, Vlachopoulou E, Pedersen NL, Nikus K, Orho-Melander M, Hamsten A, Heeringa J, Denny JC, Kriebel J, Darbar D, Newton-Cheh C, Shaffer C, Macfarlane PW, Heilmann-Heimbach S, Almgren P, Huang PL, Sotoodehnia N, Soliman EZ, Uitterlinden AG, Hofman A, Franco OH, Volker U, Jockel KH, Sinner MF, Lin HJ, Guo X, Dichgans M, Ingelsson E, Kooperberg C, Melander O, Loos R, Laurikka J, Conen D, Rosand J, van der Harst P, Lokki ML, Kathiresan S, Pereira A, Jukema JW, Hayward C, Rotter JI, Marz W, Lehtimaki T, Stricker BH, Chung MK, Felix SB, Gudnason V, Alonso A, Roden DM, Kaab S, Chasman DI, Heckbert SR, Benjamin EJ, Tanaka T, Lunetta KL, Lubitz SA, Ellinor PT. Large-scale analyses of common and rare variants identify 12 new loci associated with atrial fibrillation. Nat Genet. 2017;49(6):946–52.

Cordeiro JM, Marieb M, Pfeiffer R, Calloe K, Burashnikov E, Antzelevitch C. Accelerated inactivation of the L-type calcium current due to a mutation in CACNB2b underlies Brugada syndrome. J Mol Cell Cardiol. 2009;46(5):695–703.

Crotti L, Johnson CN, Graf E, De Ferrari GM, Cuneo BF, Ovadia M, Papagiannis J, Feldkamp MD, Rathi SG, Kunic JD, Pedrazzini M, Wieland T, Lichtner P, Beckmann BM, Clark T, Shaffer C, Benson DW, Kaab S, Meitinger T, Strom TM, Chazin WJ, Schwartz PJ, George AJ. Calmodulin mutations associated with recurrent cardiac arrest in infants. Circulation. 2013;127(9):1009–17.

Crozier IG, Loughnan A, Dow LJ, Low CJ, Ikram H. Congenital long QT syndrome in adults. N Z Med J. 1989;102(871):340–1.

Daumy X, Amarouch MY, Lindenbaum P, Bonnaud S, Charpentier E, Bianchi B, Nafzger S, Baron E, Fouchard S, Thollet A, Kyndt F, Barc J, Le Scouarnec S, Makita N, Le Marec H, Dina C, Gourraud JB, Probst V, Abriel H, Redon R, Schott JJ. Targeted resequencing identifies TRPM4 as a major gene predisposing to progressive familial heart block type I. Int J Cardiol. 2016;207:349–58.

Delpon E, Cordeiro JM, Nunez L, Thomsen PE, Guerchicoff A, Pollevick GD, Wu Y, Kanters JK, Larsen CT, Hofman-Bang J, Burashnikov E, Christiansen M, Antzelevitch C. Functional effects of KCNE3 mutation and its role in the development of Brugada syndrome. Circ Arrhythm Electrophysiol. 2008;1(3):209–18.

Deo M, Ruan Y, Pandit SV, Shah K, Berenfeld O, Blaufox A, Cerrone M, Noujaim SF, Denegri M, Jalife J, Priori SG. KCNJ2 mutation in short QT syndrome 3 results in atrial fibrillation and ventricular proarrhythmia. Proc Natl Acad Sci U S A. 2013;110(11):4291–6.

Fukuyama M, Ohno S, Makiyama T, Horie M. Novel SCN10A variants associated with Brugada syndrome. Europace. 2016;18(6):905–11.

Giudicessi JR, Ye D, Tester DJ, Crotti L, Mugione A, Nesterenko VV, Albertson RM, Antzelevitch C, Schwartz PJ, Ackerman MJ. Transient outward current (I(to)) gain-of-function mutations in the KCND3-encoded Kv4.3 potassium channel and Brugada syndrome. Heart Rhythm. 2011;8(7):1024–32.

Goldenberg I, Horr S, Moss AJ, Lopes CM, Barsheshet A, McNitt S, Zareba W, Andrews ML, Robinson JL, Locati EH, Ackerman MJ, Benhorin J, Kaufman ES, Napolitano C, Platonov PG, Priori SG, Qi M, Schwartz PJ, Shimizu W, Towbin JA, Vincent GM, Wilde AA, Zhang L. Risk for life-threatening cardiac events in patients with genotype-confirmed long-QT syndrome and normal-range corrected QT intervals. J Am Coll Cardiol. 2011;57(1):51–9.

Gudbjartsson DF, Arnar DO, Helgadottir A, Gretarsdottir S, Holm H, Sigurdsson A, Jonasdottir A, Baker A, Thorleifsson G, Kristjansson K, Palsson A, Blondal T, Sulem P, Backman VM, Hardarson GA, Palsdottir E, Helgason A, Sigurjonsdottir R, Sverrisson JT, Kostulas K, Ng MC, Baum L, So WY, Wong KS, Chan JC, Furie KL, Greenberg SM, Sale M, Kelly P, MacRae CA, Smith EE, Rosand J, Hillert J, Ma RC, Ellinor PT, Thorgeirsson G, Gulcher JR, Kong A, Thorsteinsdottir U, Stefansson K. Variants conferring risk of atrial fibrillation on chromosome 4q25. Nature. 2007;448(7151):353–7.

Guo Q, Ren L, Chen X, Hou C, Chu J, Pu J, Zhang S. A novel mutation in the SCN5A gene contributes to arrhythmogenic characteristics of early repolarization syndrome. Int J Mol Med. 2016;37(3):727–33.

Haissaguerre M, Chatel S, Sacher F, Weerasooriya R, Probst V, Loussouarn G, Horlitz M, Liersch R, Schulze-Bahr E, Wilde A, Kaab S, Koster J, Rudy Y, Le Marec H, Schott JJ. Ventricular fibrillation with prominent early repolarization associated with a rare variant of KCNJ8/KATP channel. J Cardiovasc Electrophysiol. 2009;20(1):93–8.

Harmer SC, Wilson AJ, Aldridge R, Tinker A. Mechanisms of disease pathogenesis in long QT syndrome type 5. Am J Phys Cell Phys. 2010;298(2):C263–73.

Hattori T, Makiyama T, Akao M, Ehara E, Ohno S, Iguchi M, Nishio Y, Sasaki K, Itoh H, Yokode M, Kita T, Horie M, Kimura T. A novel gain-of-function KCNJ2 mutation associated with short-QT syndrome impairs inward rectification of Kir2.1 currents. Cardiovasc Res. 2012;93(4):666–73.

Hayashi M, Denjoy I, Hayashi M, Extramiana F, Maltret A, Roux-Buisson N, Lupoglazoff JM, Klug D, Maury P, Messali A, Guicheney P, Leenhardt A. The role of stress test for predicting genetic mutations and future cardiac events in asymptomatic relatives of catecholaminergic polymorphic ventricular tachycardia probands. Europace. 2012;14(9):1344–51.

Hennessey JA, Marcou CA, Wang C, Wei EQ, Wang C, Tester DJ, Torchio M, Dagradi F, Crotti L, Schwartz PJ, Ackerman MJ, Pitt GS. FGF12 is a candidate Brugada syndrome locus. Heart Rhythm. 2013a;10(12):1886–94.

Hennessey JA, Wei EQ, Pitt GS. Fibroblast growth factor homologous factors modulate cardiac calcium channels. Circ Res. 2013b;113(4):381–8.

Holm H, Gudbjartsson DF, Sulem P, Masson G, Helgadottir HT, Zanon C, Magnusson OT, Helgason A, Saemundsdottir J, Gylfason A, Stefansdottir H, Gretarsdottir S, Matthiasson SE, Thorgeirsson GM, Jonasdottir A, Sigurdsson A, Stefansson H, Werge T, Rafnar T, Kiemeney LA, Parvez B, Muhammad R, Roden DM, Darbar D, Thorleifsson G, Walters GB, Kong A, Thorsteinsdottir U, Arnar DO, Stefansson K. A rare variant in MYH6 is associated with high risk of sick sinus syndrome. Nat Genet. 2011;43(4):316–20.

Hong K, Bjerregaard P, Gussak I, Brugada R. Short QT syndrome and atrial fibrillation caused by mutation in KCNH2. J Cardiovasc Electrophysiol. 2005a;16(4):394–6.

Hong K, Piper DR, Diaz-Valdecantos A, Brugada J, Oliva A, Burashnikov E, Santos-de-Soto J, Grueso-Montero J, Diaz-Enfante E, Brugada P, Sachse F, Sanguinetti MC, Brugada R. De novo KCNQ1 mutation responsible for atrial fibrillation and short QT syndrome in utero. Cardiovasc Res. 2005b;68(3):433–40.

Hong K, Hu J, Yu J, Brugada R. Concomitant Brugada-like and short QT electrocardiogram linked to SCN5A mutation. Eur J Hum Genet. 2012;20(11):1189–92.

Hu D, Barajas-Martinez H, Burashnikov E, Springer M, Wu Y, Varro A, Pfeiffer R, Koopmann TT, Cordeiro JM, Guerchicoff A, Pollevick GD, Antzelevitch C. A mutation in the beta 3 subunit of the cardiac sodium channel associated with Brugada ECG phenotype. Circ Cardiovasc Genet. 2009;2(3):270–8.

Hu D, Barajas-Martinez H, Pfeiffer R, Dezi F, Pfeiffer J, Buch T, Betzenhauser MJ, Belardinelli L, Kahlig KM, Rajamani S, DeAntonio HJ, Myerburg RJ, Ito H, Deshmukh P, Marieb M, Nam GB, Bhatia A, Hasdemir C, Haissaguerre M, Veltmann C, Schimpf R, Borggrefe M, Viskin S, Antzelevitch C. Mutations in SCN10A are responsible for a large fraction of cases of Brugada syndrome. J Am Coll Cardiol. 2014;64(1):66–79.

Ishikawa T, Sato A, Marcou CA, Tester DJ, Ackerman MJ, Crotti L, Schwartz PJ, On YK, Park JE, Nakamura K, Hiraoka M, Nakazawa K, Sakurada H, Arimura T, Makita N, Kimura A. A novel disease gene for Brugada syndrome: sarcolemmal membrane-associated protein gene mutations impair intracellular trafficking of hNav1.5. Circ Arrhythm Electrophysiol. 2012;5(6):1098–107.

Ishikawa T, Takahashi N, Ohno S, Sakurada H, Nakamura K, On YK, Park JE, Makiyama T, Horie M, Arimura T, Makita N, Kimura A. Novel SCN3B mutation associated with Brugada syndrome affects intracellular trafficking and function of Nav1.5. Circ J. 2013;77(4):959–67.

Ishikawa T, Jou CJ, Nogami A, Kowase S, Arrington CB, Barnett SM, Harrell DT, Arimura T, Tsuji Y, Kimura A, Makita N. Novel mutation in the alpha-myosin heavy chain gene is associated with sick sinus syndrome. Circ Arrhythm Electrophysiol. 2015;8(2):400–8.

Itoh H, Sakaguchi T, Ashihara T, Ding WG, Nagaoka I, Oka Y, Nakazawa Y, Yao T, Jo H, Ito M, Nakamura K, Ohe T, Matsuura H, Horie M. A novel KCNH2 mutation as a modifier for short QT interval. Int J Cardiol. 2009;137(1):83–5.

Jensen PN, Gronroos NN, Chen LY, Folsom AR, DeFilippi C, Heckbert SR, Alonso A. Incidence of and risk factors for sick sinus syndrome in the general population. J Am Coll Cardiol. 2014;64(6):531–8.

Jiang D, Xiao B, Yang D, Wang R, Choi P, Zhang L, Cheng H, Chen SR. RyR2 mutations linked to ventricular tachycardia and sudden death reduce the threshold for store-overload-induced Ca2+ release (SOICR). Proc Natl Acad Sci U S A. 2004;101(35):13062–7.

Jiang D, Chen W, Wang R, Zhang L, Chen SR. Loss of luminal Ca2+ activation in the cardiac ryanodine receptor is associated with ventricular fibrillation and sudden death. Proc Natl Acad Sci U S A. 2007;104(46):18309–14.

Jons C, Moss AJ, Goldenberg I, Liu J, McNitt S, Zareba W, Qi M, Robinson JL. Risk of fatal arrhythmic events in long QT syndrome patients after syncope. J Am Coll Cardiol. 2010;55(8):783–8.

Kattygnarath D, Maugenre S, Neyroud N, Balse E, Ichai C, Denjoy I, Dilanian G, Martins RP, Fressart V, Berthet M, Schott JJ, Leenhardt A, Probst V, Le Marec H, Hainque B, Coulombe A, Hatem SN, Guicheney P. MOG1: a new susceptibility gene for Brugada syndrome. Circ Cardiovasc Genet. 2011;4(3):261–8.

Kawamura M, Ozawa T, Yao T, Ashihara T, Sugimoto Y, Yagi T, Itoh H, Ito M, Makiyama T, Horie M. Dynamic change in ST-segment and spontaneous occurrence of ventricular fibrillation in Brugada syndrome with a novel nonsense mutation in the SCN5A gene during long-term follow-up. Circ J. 2009;73(3):584–8.

Kazemian P, Gollob MH, Pantano A, Oudit GY. A novel mutation in the RYR2 gene leading to catecholaminergic polymorphic ventricular tachycardia and paroxysmal atrial fibrillation: dose-dependent arrhythmia-event suppression by beta-blocker therapy. Can J Cardiol. 2011;27(6):870–7.

Khera S, Jacobson JT. Short QT syndrome in current clinical practice. Cardiol Rev. 2016;24(4):190–3.

Krogh BA, Pedersen LN, Nielsen JC, Jensen HK. Ankyrin-2 variants associated with idiopathic ventricular fibrillation storm in patients with intermittent early repolarization pattern. Heart Rhythm Case Rep. 2015;1(5):337–41.

Kruse M, Pongs O. TRPM4 channels in the cardiovascular system. Curr Opin Pharmacol. 2014;15:68–73.

Kruse M, Schulze-Bahr E, Corfield V, Beckmann A, Stallmeyer B, Kurtbay G, Ohmert I, Schulze-Bahr E, Brink P, Pongs O. Impaired endocytosis of the ion channel TRPM4 is associated with human progressive familial heart block type I. J Clin Invest. 2009;119(9):2737–44.

Lam L, Ingles J, Turner C, Kilborn M, Bagnall RD, Semsarian C. Exome sequencing identifies a novel mutation in the MYH6 gene in a family with early-onset sinus node dysfunction, ventricular arrhythmias, and cardiac arrest. Heart Rhythm Case Rep. 2015;1(3):141–5.

Le Scouarnec S, Bhasin N, Vieyres C, Hund TJ, Cunha SR, Koval O, Marionneau C, Chen B, Wu Y, Demolombe S, Song LS, Le Marec H, Probst V, Schott JJ, Anderson ME, Mohler PJ. Dysfunction in ankyrin-B-dependent ion channel and transporter targeting causes human sinus node disease. Proc Natl Acad Sci U S A. 2008;105(40):15617–22.

Le Scouarnec S, Karakachoff M, Gourraud JB, Lindenbaum P, Bonnaud S, Portero V, Duboscq-Bidot L, Daumy X, Simonet F, Teusan R, Baron E, Violleau J, Persyn E, Bellanger L, Barc J, Chatel S, Martins R, Mabo P, Sacher F, Haissaguerre M, Kyndt F, Schmitt S, Bezieau S, Le Marec H, Dina C, Schott JJ, Probst V, Redon R. Testing the burden of rare variation in arrhythmia-susceptibility genes provides new insights into molecular diagnosis for Brugada syndrome. Hum Mol Genet. 2015;24(10):2757–63.

Lieve KV, van der Werf C, Wilde AA. Catecholaminergic polymorphic ventricular tachycardia. Circ J. 2016;80(6):1285–91.

Liu H, El ZL, Kruse M, Guinamard R, Beckmann A, Bozio A, Kurtbay G, Megarbane A, Ohmert I, Blaysat G, Villain E, Pongs O, Bouvagnet P. Gain-of-function mutations in TRPM4 cause autosomal dominant isolated cardiac conduction disease. Circ Cardiovasc Genet. 2010;3(4):374–85.

Liu H, Chatel S, Simard C, Syam N, Salle L, Probst V, Morel J, Millat G, Lopez M, Abriel H, Schott JJ, Guinamard R, Bouvagnet P. Molecular genetics and functional anomalies in a series of 248 Brugada cases with 11 mutations in the TRPM4 channel. PLoS One. 2013;8(1):e54131.

Liu M, Yang KC, Dudley SJ. Cardiac sodium channel mutations: why so many phenotypes? Nat Rev Cardiol. 2014;11(10):607–15.

Liu X, Shen Y, Xie J, Bao H, Cao Q, Wan R, Xu X, Zhou H, Huang L, Xu Z, Zhu W, Hu J, Cheng X, Hong K. A mutation in the CACNA1C gene leads to early repolarization syndrome with incomplete penetrance: a Chinese family study. PLoS One. 2017;12(5):e177532.

London B, Michalec M, Mehdi H, Zhu X, Kerchner L, Sanyal S, Viswanathan PC, Pfahnl AE, Shang LL, Madhusudanan M, Baty CJ, Lagana S, Aleong R, Gutmann R, Ackerman MJ, McNamara DM, Weiss R, Dudley SJ. Mutation in glycerol-3-phosphate dehydrogenase 1 like gene (GPD1-L) decreases cardiac Na+ current and causes inherited arrhythmias. Circulation. 2007;116(20):2260–8.

Low SK, Takahashi A, Ebana Y, Ozaki K, Christophersen IE, Ellinor PT, Ogishima S, Yamamoto M, Satoh M, Sasaki M, Yamaji T, Iwasaki M, Tsugane S, Tanaka K, Naito M, Wakai K, Tanaka H, Furukawa T, Kubo M, Ito K, Kamatani Y, Tanaka T. Identification of six new genetic loci associated with atrial fibrillation in the Japanese population. Nat Genet. 2017;49(6):953–8.

Makita N, Seki A, Sumitomo N, Chkourko H, Fukuhara S, Watanabe H, Shimizu W, Bezzina CR, Hasdemir C, Mugishima H, Makiyama T, Baruteau A, Baron E, Horie M, Hagiwara N, Wilde AA, Probst V, Le Marec H, Roden DM, Mochizuki N, Schott JJ, Delmar M. A connexin40 mutation associated with a malignant variant of progressive familial heart block type I. Circ Arrhythm Electrophysiol. 2012;5(1):163–72.

Makita N, Yagihara N, Crotti L, Johnson CN, Beckmann BM, Roh MS, Shigemizu D, Lichtner P, Ishikawa T, Aiba T, Homfray T, Behr ER, Klug D, Denjoy I, Mastantuono E, Theisen D, Tsunoda T, Satake W, Toda T, Nakagawa H, Tsuji Y, Tsuchiya T, Yamamoto H, Miyamoto

Y, Endo N, Kimura A, Ozaki K, Motomura H, Suda K, Tanaka T, Schwartz PJ, Meitinger T, Kaab S, Guicheney P, Shimizu W, Bhuiyan ZA, Watanabe H, Chazin WJ, George AJ. Novel calmodulin mutations associated with congenital arrhythmia susceptibility. Circ Cardiovasc Genet. 2014;7(4):466–74.

Makiyama T, Akao M, Shizuta S, Doi T, Nishiyama K, Oka Y, Ohno S, Nishio Y, Tsuji K, Itoh H, Kimura T, Kita T, Horie M. A novel SCN5A gain-of-function mutation M1875T associated with familial atrial fibrillation. J Am Coll Cardiol. 2008;52(16):1326–34.

Marsman RF, Bardai A, Postma AV, Res JC, Koopmann TT, Beekman L, van der Wal AC, Pinto YM, Lekanne DR, Wilde AA, Jordaens LJ, Bezzina CR. A complex double deletion in LMNA underlies progressive cardiac conduction disease, atrial arrhythmias, and sudden death. Circ Cardiovasc Genet. 2011;4(3):280–7.

Marsman RF, Barc J, Beekman L, Alders M, Dooijes D, van den Wijngaard A, Ratbi I, Sefiani A, Bhuiyan ZA, Wilde AA, Bezzina CR. A mutation in CALM1 encoding calmodulin in familial idiopathic ventricular fibrillation in childhood and adolescence. J Am Coll Cardiol. 2014;63(3):259–66.

McPate MJ, Duncan RS, Hancox JC, Witchel HJ. Pharmacology of the short QT syndrome N588K-hERG K+ channel mutation: differential impact on selected class I and class III antiarrhythmic drugs. Br J Pharmacol. 2008;155(6):957–66.

Medeiros-Domingo A, Kaku T, Tester DJ, Iturralde-Torres P, Itty A, Ye B, Valdivia C, Ueda K, Canizales-Quinteros S, Tusie-Luna MT, Makielski JC, Ackerman MJ. SCN4B-encoded sodium channel beta4 subunit in congenital long-QT syndrome. Circulation. 2007;116(2):134–42.

Medeiros-Domingo A, Tan BH, Crotti L, Tester DJ, Eckhardt L, Cuoretti A, Kroboth SL, Song C, Zhou Q, Kopp D, Schwartz PJ, Makielski JC, Ackerman MJ. Gain-of-function mutation S422L in the KCNJ8-encoded cardiac K(ATP) channel Kir6.1 as a pathogenic substrate for J-wave syndromes. Heart Rhythm. 2010;7(10):1466–71.

Milanesi R, Bucchi A, Baruscotti M. The genetic basis for inherited forms of sinoatrial dysfunction and atrioventricular node dysfunction. J Interv Card Electrophysiol. 2015;43(2):121–34.

Mizusawa Y, Bezzina CR. Early repolarization pattern: its ECG characteristics, arrhythmogenicity and heritability. J Interv Card Electrophysiol. 2014;39(3):185–92.

Monfredi O, Boyett MR. Sick sinus syndrome and atrial fibrillation in older persons – a view from the sinoatrial nodal myocyte. J Mol Cell Cardiol. 2015;83:88–100.

Moss AJ, Zareba W, Kaufman ES, Gartman E, Peterson DR, Benhorin J, Towbin JA, Keating MT, Priori SG, Schwartz PJ, Vincent GM, Robinson JL, Andrews ML, Feng C, Hall WJ, Medina A, Zhang L, Wang Z. Increased risk of arrhythmic events in long-QT syndrome with mutations in the pore region of the human ether-a-go-go-related gene potassium channel. Circulation. 2002;105(7):794–9.

Moss AJ, Shimizu W, Wilde AA, Towbin JA, Zareba W, Robinson JL, Qi M, Vincent GM, Ackerman MJ, Kaufman ES, Hofman N, Seth R, Kamakura S, Miyamoto Y, Goldenberg I, Andrews ML, McNitt S. Clinical aspects of type-1 long-QT syndrome by location, coding type, and biophysical function of mutations involving the KCNQ1 gene. Circulation. 2007;115(19):2481–9.

Nakano Y, Shimizu W. Genetics of long-QT syndrome. J Hum Genet. 2016;61(1):51–5.

Napolitano C, Antzelevitch C. Phenotypical manifestations of mutations in the genes encoding subunits of the cardiac voltage-dependent L-type calcium channel. Circ Res. 2011;108(5):607–18.

Nomikos M, Thanassoulas A, Beck K, Vassilakopoulou V, Hu H, Calver BL, Theodoridou M, Kashir J, Blayney L, Livaniou E, Rizkallah P, Nounesis G, Lai FA. Altered RyR2 regulation by the calmodulin F90L mutation associated with idiopathic ventricular fibrillation and early sudden cardiac death. FEBS Lett. 2014;588(17):2898–902.

Nyegaard M, Overgaard MT, Sondergaard MT, Vranas M, Behr ER, Hildebrandt LL, Lund J, Hedley PL, Camm AJ, Wettrell G, Fosdal I, Christiansen M, Borglum AD. Mutations in calmodulin cause ventricular tachycardia and sudden cardiac death. Am J Hum Genet. 2012;91(4):703–12.

Ohno S, Zankov DP, Ding WG, Itoh H, Makiyama T, Doi T, Shizuta S, Hattori T, Miyamoto A, Naiki N, Hancox JC, Matsuura H, Horie M. KCNE5 (KCNE1L) variants are novel modulators of Brugada syndrome and idiopathic ventricular fibrillation. Circ Arrhythm Electrophysiol. 2011;4(3):352–61.

Olesen MS, Holst AG, Jabbari J, Nielsen JB, Christophersen IE, Sajadieh A, Haunso S, Svendsen JH. Genetic loci on chromosomes 4q25, 7p31, and 12p12 are associated with onset of lone atrial fibrillation before the age of 40 years. Can J Cardiol. 2012;28(2):191–5.

Olson TM, Alekseev AE, Liu XK, Park S, Zingman LV, Bienengraeber M, Sattiraju S, Ballew JD, Jahangir A, Terzic A. Kv1.5 channelopathy due to KCNA5 loss-of-function mutation causes human atrial fibrillation. Hum Mol Genet. 2006;15(14):2185–91.

Paech C, Gebauer RA, Karstedt J, Marschall C, Bollmann A, Husser D. Ryanodine receptor mutations presenting as idiopathic ventricular fibrillation: a report on two novel familial compound mutations, c.6224T>C and c.13781A>G, with the clinical presentation of idiopathic ventricular fibrillation. Pediatr Cardiol. 2014;35(8):1437–41.

Pereira R, Campuzano O, Sarquella-Brugada G, Cesar S, Iglesias A, Brugada J, Cruz FF, Brugada R. Short QT syndrome in pediatrics. Clin Res Cardiol. 2017;106(6):393–400.

Pérez-Riera AR, Barbosa-Barros R, de Rezende Barbosa MPC, Daminello-Raimundo R, de Lucca AA Jr, de Abreu LC. Catecholaminergic polymorphic ventricular tachycardia, an update. Ann Noninvasive Electrocardiol. 2018 Jul;23(4):e12512.

Perrin MJ, Adler A, Green S, Al-Zoughool F, Doroshenko P, Orr N, Uppal S, Healey JS, Birnie D, Sanatani S, Gardner M, Champagne J, Simpson C, Ahmad K, van den Berg MP, Chauhan V, Backx PH, van Tintelen JP, Krahn AD, Gollob MH. Evaluation of genes encoding for the transient outward current (Ito) identifies the KCND2 gene as a cause of J-wave syndrome associated with sudden cardiac death. Circ Cardiovasc Genet. 2014;7(6):782–9.

Pfeufer A, van Noord C, Marciante KD, Arking DE, Larson MG, Smith AV, Tarasov KV, Muller M, Sotoodehnia N, Sinner MF, Verwoert GC, Li M, Kao WH, Kottgen A, Coresh J, Bis JC, Psaty BM, Rice K, Rotter JI, Rivadeneira F, Hofman A, Kors JA, Stricker BH, Uitterlinden AG, van Duijn CM, Beckmann BM, Sauter W, Gieger C, Lubitz SA, Newton-Cheh C, Wang TJ, Magnani JW, Schnabel RB, Chung MK, Barnard J, Smith JD, Van Wagoner DR, Vasan RS, Aspelund T, Eiriksdottir G, Harris TB, Launer LJ, Najjar SS, Lakatta E, Schlessinger D, Uda M, Abecasis GR, Muller-Myhsok B, Ehret GB, Boerwinkle E, Chakravarti A, Soliman EZ, Lunetta KL, Perz S, Wichmann HE, Meitinger T, Levy D, Gudnason V, Ellinor PT, Sanna S, Kaab S, Witteman JC, Alonso A, Benjamin EJ, Heckbert SR. Genome-wide association study of PR interval. Nat Genet. 2010;42(2):153–9.

Postma AV, van de Meerakker JB, Mathijssen IB, Barnett P, Christoffels VM, Ilgun A, Lam J, Wilde AA, Lekanne DR, Moorman AF. A gain-of-function TBX5 mutation is associated with atypical Holt-Oram syndrome and paroxysmal atrial fibrillation. Circ Res. 2008;102(11):1433–42.

Priori SG, Napolitano C, Tiso N, Memmi M, Vignati G, Bloise R, Sorrentino V, Danieli GA. Mutations in the cardiac ryanodine receptor gene (hRyR2) underlie catecholaminergic polymorphic ventricular tachycardia. Circulation. 2001;103(2):196–200.

Priori SG, Pandit SV, Rivolta I, Berenfeld O, Ronchetti E, Dhamoon A, Napolitano C, Anumonwo J, di Barletta MR, Gudapakkam S, Bosi G, Stramba-Badiale M, Jalife J. A novel form of short QT syndrome (SQT3) is caused by a mutation in the KCNJ2 gene. Circ Res. 2005;96(7):800–7.

Priori SG, Wilde AA, Horie M, Cho Y, Behr ER, Berul C, Blom N, Brugada J, Chiang CE, Huikuri H, Kannankeril P, Krahn A, Leenhardt A, Moss A, Schwartz PJ, Shimizu W, Tomaselli G, Tracy C. HRS/EHRA/APHRS expert consensus statement on the diagnosis and management of patients with inherited primary arrhythmia syndromes: document endorsed by HRS, EHRA, and APHRS in May 2013 and by ACCF, AHA, PACES, and AEPC in June 2013. Heart Rhythm. 2013;10(12):1932–63.

Priori SG, Blomstrom-Lundqvist C, Mazzanti A, Blom N, Borggrefe M, Camm J, Elliott PM, Fitzsimons D, Hatala R, Hindricks G, Kirchhof P, Kjeldsen K, Kuck KH, Hernandez-Madrid A, Nikolaou N, Norekval TM, Spaulding C, Van Veldhuisen DJ. ESC guidelines for the management of patients with ventricular arrhythmias and the prevention of sudden cardiac death: the task force for the management of patients with ventricular arrhythmias and the prevention of sudden cardiac death of the European Society of Cardiology (ESC). Endorsed by: Association for European Paediatric and Congenital Cardiology (AEPC). Eur Heart J. 2015;36(41):2793–867.

Priori SG, Blomstrom-Lundqvist C, Mazzanti A, Blom N, Borggrefe M, Camm J, Elliott PM, Fitzsimons D, Hatala R, Hindricks G, Kirchhof P, Kjeldsen K, Kuck KH, Hernandez-Madrid A, Nikolaou N, Norekval TM, Spaulding C, Van Veldhuisen DJ. ESC guidelines for the management of patients with ventricular arrhythmias and the prevention of sudden cardiac death. Rev Esp Cardiol (Engl Ed). 2016;69(2):176.

Riuro H, Beltran-Alvarez P, Tarradas A, Selga E, Campuzano O, Verges M, Pagans S, Iglesias A, Brugada J, Brugada P, Vazquez FM, Perez GJ, Scornik FS, Brugada R. A missense mutation in the sodium channel beta2 subunit reveals SCN2B as a new candidate gene for Brugada syndrome. Hum Mutat. 2013;34(7):961–6.

Robaei D, Ford T, Ooi SY. Ankyrin-B syndrome: a case of sinus node dysfunction, atrial fibrillation and prolonged QT in a young adult. Heart Lung Circ. 2015;24(2):e31–4.

Rodriguez RD, Schocken DD. Update on sick sinus syndrome, a cardiac disorder of aging. Geriatrics. 1990;45(1):26–30. 33–6

Roston TM, Vinocur JM, Maginot KR, Mohammed S, Salerno JC, Etheridge SP, Cohen M, Hamilton RM, Pflaumer A, Kanter RJ, Potts JE, LaPage MJ, Collins KK, Gebauer RA, Temple JD, Batra AS, Erickson C, Miszczak-Knecht M, Kubus P, Bar-Cohen Y, Kantoch M, Thomas VC, Hessling G, Anderson C, Young ML, Cabrera OM, Lau YR, Johnsrude CL, Fournier A, Kannankeril PJ, Sanatani S. Catecholaminergic polymorphic ventricular tachycardia in children: analysis of therapeutic strategies and outcomes from an international multicenter registry. Circ Arrhythm Electrophysiol. 2015;8(3):633–42.

Roston TM, Yuchi Z, Kannankeril PJ, Hathaway J, Vinocur JM, Etheridge SP, Potts JE, Maginot KR, Salerno JC, Cohen MI, Hamilton RM, Pflaumer A, Mohammed S, Kimlicka L, Kanter RJ, LaPage MJ, Collins KK, Gebauer RA, Temple JD, Batra AS, Erickson C, Miszczak-Knecht M, Kubus P, Bar-Cohen Y, Kantoch M, Thomas VC, Hessling G, Anderson C, Young ML, Choi S, Cabrera OM, Lau YR, Johnsrude CL, Fournier A, Van Petegem F, Sanatani S. The clinical and genetic spectrum of catecholaminergic polymorphic ventricular tachycardia: findings from an international multicentre registry. Europace. 2018;20(3):541–7.

Roux-Buisson N, Cacheux M, Fourest-Lieuvin A, Fauconnier J, Brocard J, Denjoy I, Durand P, Guicheney P, Kyndt F, Leenhardt A, Le Marec H, Lucet V, Mabo P, Probst V, Monnier N, Ray PF, Santoni E, Tremeaux P, Lacampagne A, Faure J, Lunardi J, Marty I. Absence of triadin, a protein of the calcium release complex, is responsible for cardiac arrhythmia with sudden death in human. Hum Mol Genet. 2012;21(12):2759–67.

Royer A, van Veen TA, Le Bouter S, Marionneau C, Griol-Charhbili V, Leoni AL, Steenman M, van Rijen HV, Demolombe S, Goddard CA, Richer C, Escoubet B, Jarry-Guichard T, Colledge WH, Gros D, de Bakker JM, Grace AA, Escande D, Charpentier F. Mouse model of SCN5A-linked hereditary Lenegre's disease: age-related conduction slowing and myocardial fibrosis. Circulation. 2005;111(14):1738–46.

Schamroth L, Dove E. The Wenckebach phenomenon in sino-atrial block. Br Heart J. 1966;28(3):350–8.

Schott JJ, Benson DW, Basson CT, Pease W, Silberbach GM, Moak JP, Maron BJ, Seidman CE, Seidman JG. Congenital heart disease caused by mutations in the transcription factor NKX2-5. Science. 1998;281(5373):108–11.

Schott JJ, Alshinawi C, Kyndt F, Probst V, Hoorntje TM, Hulsbeek M, Wilde AA, Escande D, Mannens MM, Le Marec H. Cardiac conduction defects associate with mutations in SCN5A. Nat Genet. 1999;23(1):20–1.

Schwartz PJ, Stramba-Badiale M, Crotti L, Pedrazzini M, Besana A, Bosi G, Gabbarini F, Goulene K, Insolia R, Mannarino S, Mosca F, Nespoli L, Rimini A, Rosati E, Salice P, Spazzolini C. Prevalence of the congenital long-QT syndrome. Circulation. 2009;120(18):1761–7.

Seki A, Ishikawa T, Daumy X, Mishima H, Barc J, Sasaki R, Nishii K, Saito K, Urano M, Ohno S, Otsuki S, Kimoto H, Baruteau AE, Thollet A, Fouchard S, Bonnaud S, Parent P, Shibata Y, Perrin JP, Le Marec H, Hagiwara N, Mercier S, Horie M, Probst V, Yoshiura KI, Redon R,

Schott JJ, Makita N. Progressive atrial conduction defects associated with bone malformation caused by a Connexin-45 mutation. J Am Coll Cardiol. 2017;70(3):358–70.

Shimizu W. The long QT syndrome: therapeutic implications of a genetic diagnosis. Cardiovasc Res. 2005;67(3):347–56.

Shimizu W, Horie M, Ohno S, Takenaka K, Yamaguchi M, Shimizu M, Washizuka T, Aizawa Y, Nakamura K, Ohe T, Aiba T, Miyamoto Y, Yoshimasa Y, Towbin JA, Priori SG, Kamakura S. Mutation site-specific differences in arrhythmic risk and sensitivity to sympathetic stimulation in the LQT1 form of congenital long QT syndrome: multicenter study in Japan. J Am Coll Cardiol. 2004;44(1):117–25.

Shimizu W, Moss AJ, Wilde AA, Towbin JA, Ackerman MJ, January CT, Tester DJ, Zareba W, Robinson JL, Qi M, Vincent GM, Kaufman ES, Hofman N, Noda T, Kamakura S, Miyamoto Y, Shah S, Amin V, Goldenberg I, Andrews ML, McNitt S. Genotype-phenotype aspects of type 2 long QT syndrome. J Am Coll Cardiol. 2009;54(22):2052–62.

Shy D, Gillet L, Abriel H. Cardiac sodium channel NaV1.5 distribution in myocytes via interacting proteins: the multiple pool model. Biochim Biophys Acta. 2013;1833(4):886–94.

Sinner MF, Porthan K, Noseworthy PA, Havulinna AS, Tikkanen JT, Muller-Nurasyid M, Peloso G, Ulivi S, Beckmann BM, Brockhaus AC, Cooper RR, Gasparini P, Hengstenberg C, Hwang SJ, Iorio A, Junttila MJ, Klopp N, Kahonen M, Laaksonen MA, Lehtimaki T, Lichtner P, Lyytikainen LP, Martens E, Meisinger C, Meitinger T, Merchant FM, Nieminen MS, Peters A, Pietila A, Perz S, Oikarinen L, Raitakari O, Reinhard W, Silander K, Thorand B, Wichmann HE, Sinagra G, Viikari J, O'Donnell CJ, Ellinor PT, Huikuri HV, Kaab S, Newton-Cheh C, Salomaa V. A meta-analysis of genome-wide association studies of the electrocardiographic early repolarization pattern. Heart Rhythm. 2012;9(10):1627–34.

Smits JP, Koopmann TT, Wilders R, Veldkamp MW, Opthof T, Bhuiyan ZA, Mannens MM, Balser JR, Tan HL, Bezzina CR, Wilde AA. A mutation in the human cardiac sodium channel (E161K) contributes to sick sinus syndrome, conduction disease and Brugada syndrome in two families. J Mol Cell Cardiol. 2005;38(6):969–81.

Sun Y, Quan XQ, Fromme S, Cox RH, Zhang P, Zhang L, Guo D, Guo J, Patel C, Kowey PR, Yan GX. A novel mutation in the KCNH2 gene associated with short QT syndrome. J Mol Cell Cardiol. 2011;50(3):433–41.

Swan H, Piippo K, Viitasalo M, Heikkila P, Paavonen T, Kainulainen K, Kere J, Keto P, Kontula K, Toivonen L. Arrhythmic disorder mapped to chromosome 1q42-q43 causes malignant polymorphic ventricular tachycardia in structurally normal hearts. J Am Coll Cardiol. 1999;34(7):2035–42.

Templin C, Ghadri JR, Rougier JS, Baumer A, Kaplan V, Albesa M, Sticht H, Rauch A, Puleo C, Hu D, Barajas-Martinez H, Antzelevitch C, Luscher TF, Abriel H, Duru F. Identification of a novel loss-of-function calcium channel gene mutation in short QT syndrome (SQTS6). Eur Heart J. 2011;32(9):1077–88.

Ten SJ, Postema PG, Boekholdt SM, Tan HL, van der Heijden JF, de Groot NM, Volders PG, Zeppenfeld K, Boersma LV, Nannenberg EA, Christiaans I, Wilde AA. Detailed characterization of familial idiopathic ventricular fibrillation linked to the DPP6 locus. Heart Rhythm. 2016;13(4):905–12.

Thorsen K, Dam VS, Kjaer-Sorensen K, Pedersen LN, Skeberdis VA, Jurevicius J, Treinys R, Petersen I, Nielsen MS, Oxvig C, Morth JP, Matchkov VV, Aalkjaer C, Bundgaard H, Jensen HK. Loss-of-activity-mutation in the cardiac chloride-bicarbonate exchanger AE3 causes short QT syndrome. Nat Commun. 2017;8(1):1696.

Ueda K, Hirano Y, Higashiuesato Y, Aizawa Y, Hayashi T, Inagaki N, Tana T, Ohya Y, Takishita S, Muratani H, Hiraoka M, Kimura A. Role of HCN4 channel in preventing ventricular arrhythmia. J Hum Genet. 2009;54(2):115–21.

Valdivia CR, Medeiros-Domingo A, Ye B, Shen WK, Algiers TJ, Ackerman MJ, Makielski JC. Loss-of-function mutation of the SCN3B-encoded sodium channel {beta}3 subunit associated with a case of idiopathic ventricular fibrillation. Cardiovasc Res. 2010;86(3):392–400.

Vatta M, Ackerman MJ, Ye B, Makielski JC, Ughanze EE, Taylor EW, Tester DJ, Balijepalli RC, Foell JD, Li Z, Kamp TJ, Towbin JA. Mutant caveolin-3 induces persistent late sodium current and is associated with long-QT syndrome. Circulation. 2006;114(20):2104–12.

Wang C, Hennessey JA, Kirkton RD, Wang C, Graham V, Puranam RS, Rosenberg PB, Bursac N, Pitt GS. Fibroblast growth factor homologous factor 13 regulates Na+ channels and conduction velocity in murine hearts. Circ Res. 2011;109(7):775–82.

Wang QI, Ohno S, Ding WG, Fukuyama M, Miyamoto A, Itoh H, Makiyama T, Wu J, Bai J, Hasegawa K, Shinohara T, Takahashi N, Shimizu A, Matsuura H, Horie M. Gain-of-function KCNH2 mutations in patients with Brugada syndrome. J Cardiovasc Electrophysiol. 2014;25(5):522–30.

Watanabe H, Koopmann TT, Le Scouarnec S, Yang T, Ingram CR, Schott JJ, Demolombe S, Probst V, Anselme F, Escande D, Wiesfeld AC, Pfeufer A, Kaab S, Wichmann HE, Hasdemir C, Aizawa Y, Wilde AA, Roden DM, Bezzina CR. Sodium channel beta1 subunit mutations associated with Brugada syndrome and cardiac conduction disease in humans. J Clin Invest. 2008;118(6):2260–8.

Watanabe H, Nogami A, Ohkubo K, Kawata H, Hayashi Y, Ishikawa T, Makiyama T, Nagao S, Yagihara N, Takehara N, Kawamura Y, Sato A, Okamura K, Hosaka Y, Sato M, Fukae S, Chinushi M, Oda H, Okabe M, Kimura A, Maemura K, Watanabe I, Kamakura S, Horie M, Aizawa Y, Shimizu W, Makita N. Electrocardiographic characteristics and SCN5A mutations in idiopathic ventricular fibrillation associated with early repolarization. Circ Arrhythm Electrophysiol. 2011;4(6):874–81.

Whittaker DG, Ni H, El HA, Hancox JC, Zhang H. Atrial arrhythmogenicity of KCNJ2 mutations in short QT syndrome: insights from virtual human atria. PLoS Comput Biol. 2017;13(6):e1005593.

Wilde A, Amin A. Channelopathies, genetic testing and risk stratification. Int J Cardiol. 2017;237:53–5.

Wilde AA, Postema PG, Di Diego JM, Viskin S, Morita H, Fish JM, Antzelevitch C. The pathophysiological mechanism underlying Brugada syndrome: depolarization versus repolarization. J Mol Cell Cardiol. 2010;49(4):543–53.

Wu ZJ, Huang Y, Fu YC, Zhao XJ, Zhu C, Zhang Y, Xu B, Zhu QL, Li Y. Characterization of a Chinese KCNQ1 mutation (R259H) that shortens repolarization and causes short QT syndrome 2. J Geriatr Cardiol. 2015;12(4):394–401.

Yang Y, Li J, Lin X, Yang Y, Hong K, Wang L, Liu J, Li L, Yan D, Liang D, Xiao J, Jin H, Wu J, Zhang Y, Chen YH. Novel KCNA5 loss-of-function mutations responsible for atrial fibrillation. J Hum Genet. 2009;54(5):277–83.

Yang T, Yang P, Roden DM, Darbar D. Novel KCNA5 mutation implicates tyrosine kinase signaling in human atrial fibrillation. Heart Rhythm. 2010;7(9):1246–52.

Yang YQ, Wang MY, Zhang XL, Tan HW, Shi HF, Jiang WF, Wang XH, Fang WY, Liu X. GATA4 loss-of-function mutations in familial atrial fibrillation. Clin Chim Acta. 2011;412(19–20):1825–30.

Yang YQ, Li L, Wang J, Zhang XL, Li RG, Xu YJ, Tan HW, Wang XH, Jiang JQ, Fang WY, Liu X. GATA6 loss-of-function mutation in atrial fibrillation. Eur J Med Genet. 2012a;55(10):520–6.

Yang YQ, Wang J, Wang XH, Wang Q, Tan HW, Zhang M, Shen FF, Jiang JQ, Fang WY, Liu X. Mutational spectrum of the GATA5 gene associated with familial atrial fibrillation. Int J Cardiol. 2012b;157(2):305–7.

Yu R, Fan XF, Chen C, Liu ZH. Whole exome sequencing identifies a novel mutation (R367G) in SCN5A to be associated with familial cardiac conduction disease. Mol Med Rep. 2017;16(1):410–4.

Zhabyeyev P, Hiess F, Wang R, Liu Y, Wayne CS, Oudit GY. S4153R is a gain-of-function mutation in the cardiac Ca(2+) release channel ryanodine receptor associated with catecholaminergic polymorphic ventricular tachycardia and paroxysmal atrial fibrillation. Can J Cardiol. 2013;29(8):993–6.

Zhao J, Yao H, Li Z, Wang L, Liu G, Wang DW, Wang DW, Liang Z. A novel nonsense mutation in LMNA gene identified by exome sequencing in an atrial fibrillation family. Eur J Med Genet. 2016;59(8):396–400.

Zumhagen S, Stallmeyer B, Friedrich C, Eckardt L, Seebohm G, Schulze-Bahr E. Inherited long QT syndrome. Herzschrittmacherther Elektrophysiol. 2012;23(3):211–9.

Chapter 10
Progress of Epigenetic Changes in Heart Failure

Peng Yu, Ming Liu, and Hong Jiang

Abstract Cardiac function requires sophisticated regulatory networks to orchestrate environmental adaptation. These regulatory networks in heart remodeling to maintain its function are composed of transcriptional control and signaling systems. Epigenetic processes play vital roles in the process. These include DNA and histone modification and classes of regulatory RNAs, such as microRNAs and long noncoding RNAs. Here, we review this emerging landscape to summarize their functions and roles in cardiac remodeling and to assist in exploring this previously hidden layer of gene regulation that may underpin heart adaptation in remodeling.

Keywords Heart failure · Epigenetics · DNA modification · Histone modification · Noncoding RNAs

Heart failure (HF), a leading cause of morbidity and mortality, is a growing public health concern, affecting at least 20% individuals at some point in their lifetime and contributing to 11% of deaths, with an incidence that is expected to rise by 46% from 2012 to 2030 (Benjamin et al. 2017; Rau and Vondriska 2017). As a terminal stage of a set of cardiovascular diseases, heart failure can be contributed by coronary artery disease, hypertension, arrhythmia, valvular disease, etc. which are defined as stage A heart failure. The stage B is pre-heart failure, meaning the heart was under remodeling, in which cardiomyocytes and fibroblasts reprogrammed the transcriptome and introduced new phenotype. If the remodeling was not treated appropriately, then the patients could develop into stage C heart failure who have (currently) or had (previously) signs and symptoms of the condition. The process mentioned above implies that comprehension of the cardiac remodeling could lead

P. Yu · H. Jiang (✉)
Shanghai Institute of Cardiovascular Diseases, Clinical Science Institute, Zhongshan Hospital, Shanghai Medical College of Fudan University, Shanghai, China
e-mail: jiang.hong@zs-hospital.sh.cn

M. Liu
Department of General Practice, Zhongshan Hospital, Shanghai Medical College of Fudan University, Shanghai, China

H. Jiang, M. Liu (eds.), *Heart Genomics*, Translational Bioinformatics 16,
https://doi.org/10.1007/978-981-13-1429-2_10

to new insight and treatment of heart failure. Cardiac genomic reprogramming, consisting of upregulation and downregulation in particular sets of genes, is critical in modulating heart function and has a fundamental role in pathology of cardiac hypertrophy and heart failure (Greco and Condorelli 2015).

In 1977 Fred Sanger and Alan R. Coulson introduced the rapid determination of DNA sequence, which lead to transform biology as a whole by providing a tool for deciphering complete genes and later entire genomes (Sanger et al. 1977; Schuster 2008). Next-generation sequencing has demonstrated its huge potential for anyone who is devoted in the life sciences, at a time when many believed the arrival of post-genomic age. The accessibility of next-generation sequencing facilitates comprehensive investigation of disease process for its high-throughput property (Schuster 2008).

Epigenetic mechanisms, emerged as a key regulatory element in cardiac physiological or pathological state, included four main categories on the basis of their mechanism of action: DNA modification, histone modifications, ATP-dependent chromatin remodeling, and regulation by noncoding RNAs, including microRNAs (miRNAs) and long noncoding RNAs (LncRNA) (Papait et al. 2013). Some of the latest advances in the study of gene expression regulation have been in the fields of epigenomic modification and noncoding (nc) RNAs. The former comprises secondary chemical alterations of DNA (methylation of cytosines) and proteins (posttranslational modification of histones), inducing changes to the state of chromatin. Transcription is modulated via transformation of the genetic regulatory elements more or less permissive to the activity of transcription factors. Conversely, ncRNAs, diverse groups of RNA, do not encode proteins, which are now recognized to be a fundamental part of gene expression regulatory mechanisms (Greco and Condorelli 2015). In this chapter, we would outline the analysis of genomic changes presented in remodeling hearts.

1 DNA Modification

Gene expression levels of failing cardiac myocyte genes mainly regulate sarcomere function, excitation-contraction coupling, the cytoskeleton, and cardiac energetics (Mann and Bristow 2005). Some prior studies have identified the important role for activation of specific DNA-binding transcription factors leading to the dysregulated gene expression profile in heart failure; more recent studies have suggested that histone modifications and changes in chromatin structure may play equally vital roles in regulating the transcriptional dynamics of the failing heart (Anand et al. 2013; Haldar and McKinsey 2014; Greco et al. 2016).

The modification of DNA is an important epigenetic mechanism that affects the spatial and temporal regulation of gene expression (Pai et al. 2011). Modifications to the genome that are persistent and maintained through cellular process but do not involve sequence changes, notably DNA methylation, add another layer of complexity in the form of epigenetic variability (Rau and Vondriska 2017).

Methylation patterns have been found to be involved in transposon silencing and genomic imprinting (Smith and Meissner 2013). Although stable, DNA methylation is not immutable. Environmental factors can change methylation patterns, making it correlated with phenotype, a moving target. Meanwhile, DNA methylation is an appealing molecular integrator of genetics and exposure (Rau and Vondriska 2017). Regarding the heart, whole-genome bisulfite sequencing (WGBS), which allows genome-wide DNA methylation profiling, revealed the methylation of DNA in cardiac diseases (Smith and Meissner 2013; Greco and Condorelli 2015; Meder et al. 2017).

Medar et al. (2017) carried out transcriptome-, methylome-, and epigenome-wide association studies in the blood from heart failure patients caused by dilated cardiomyopathy (DCM) and control patients. Furthermore, the studies were repeated in multiple replication cohorts, increasing the likelihood of relevance in larger populations and diminishing concerns of bias resulting from batch effects, sample collection, and overfitting of models in data analysis. The whole-genome sequencing was used to rule out sequence variation as the cause of epigenetic changes, an important step toward definitive demonstration of links between DNA methylation and phenotypes (可能有重复) (Meder et al. 2017). They examined the methylomes of peripheral blood cells in addition to cardiac cells and identified 217 methylation sites where altered methylation between control patients and patients with DCM is conserved between these two tissues, including sites located near the disease-associated genes *NPPA* and *NPPB*. Strikingly, these methylation marks hold across blood and heart cells from patients undergoing transplantation, although these cells have distinct genomes and the heart had a different epigenome when transplanted.

However, 5-hydroxymethylcytosine (5-hmC)—an oxidation product of methylation of DNA catalyzed by the ten-eleven translocation (TET) family proteins—were shown to act as a relatively stable epigenetic mark, regulating gene transcription dependent on its genomic location, rather than an intermediate product of an active demethylation process (Ficz et al. 2011; Pastor et al. 2011). Greco and colleagues (2016) revealed DNA hydroxymethylation is critical for the gene expression program of cardiac cells. DNA hydroxymethylation at the gene body and enhancer and and regulation of gene expression was in different profile in fetal, adult, and remodeling cardiac cells. The findings give new insights into mechanisms governing cardiac gene expression (Greco et al. 2016).

2 Acetylation and Methylation of Histones

The nucleosome is the fundamental unit of chromatin and is made up of a core of eight histones around which a double strand of DNA is wrapped. The amino-terminal tails of all eight core histones protrude through DNA and subject to a variety of posttranslational covalent modifications on specific residues, including acetylation, methylation, ubiquitylation, sumoylation, and phosphorylation (Papait et al. 2013).

In the last few years, it has been demonstrated that acetylation and methylation are important epigenetic mechanisms involved in regulating key cellular processes, such as gene transcription (Bernstein et al. 2007).

Histone acetylation occurs on Lys residues, mainly presented in the tails of histone proteins H2B, H3, and H4. It signals for transcription activation, on the condition that hypoacetylated histone proteins are found in transcriptionally inactive regions. Acetylation is a dynamic process, under the control of two families of enzymes: histone acetyltransferases (HATs) and histone deacetylases (HDACs), both of which include multiple enzyme classes whose expression and activity are subtly regulated (Wang and Hayes 2008; Papait et al. 2013).

Indeed, many proteins are acetylated and deacetylated by HATs and HDACs in the cell. Similar to phosphorylation, acetylation of a protein, especially the histones, can have a wide range of effects including altering stability of the protein or facilitating new protein–protein interactions. A number of transcription factors are acetylated, as well as nuclear import factors, cell cycle regulators, and structural proteins. The attenuation of hypertrophy by HDACs inhibitors could be due to increased acetylation of one or a combination of these protein substrates. However, some controversial evidence exists. For example, high-mobility group box 1 (HMGB1) working as architectural protein that can bend DNA in the nucleus could be hyperacetylated by the inhibition of HDAC4 during hepatic ischemia and reperfusion injury (Evankovich et al. 2010). When HMGB1 is hyperacetylated, it redirects from nucleus into secretion (Bonaldi et al. 2003). We have previously demonstrated that secreted extracellular HMGB1 could exacerbate cardiac remodeling to heart failure (Zhang et al. 2016).

HDACs are the main or well-studied enzymes that control histone acetylation as stress-responsive regulators of gene expression in the heart (McKinsey and Olson 2005). In in vitro study, pharmacological inhibition of HDACs blocks hypertrophic growth of cardiac myocytes (Antos et al. 2003), and some evidence suggests inhibition of HDACs block hypertrophy in vivo (Kook et al. 2003). Kong et al. (2006) demonstrated the compounds inhibiting HDACs impacted hypertrophic growth of the heart in a physiologically relevant model of pressure overload in vivo. The study elucidated that two broad-spectrum HDAC inhibitors are capable of blunting hypertrophic growth in a model of cardiac hypertrophy, and they could preserve systolic function, blunt stress-induced fibrogenesis, inhibit collagen biosynthesis in cardiac fibroblasts, and blunt pathological switching of MHC isoforms, which is marked as cardiac remodeling and heart failure (Kong et al. 2006). The role of acetylation of histones in heart failure is supported by studies demonstrating that spectrum inhibitors of HDACs—such as trichostatin A (TSA) and sodium butyrate (NaB)—are able to prevent cardiac hypertrophy in vitro and in vivo (Antos et al. 2003; Kee et al. 2006; Kong et al. 2006; Gallo et al. 2008).

Unlike acetylation, histone methylation can modulate either activation or repression of transcription, depending on the particular methylated Lys or Arg position and the degree of methylation (mono-, di-, and tri-methylation) (Kouzarides 2002). Methylated modification of histones is not immutable, which is regulated by two classes of enzymes: histone methyltransferases (HMTs) and histone demethylases

(HDMTs) (Cloos et al. 2008). Combinations of DNA methylation and histone modifications create an "epigenetic code" that is an important aspect of epigenetic regulation, which results in different functional genomic regions: chromatin structure that is poor in methylated DNA and rich in H3 histone acetylated and methylated on Lys 4 and Lys 36 is transcriptionally active genomic regions; in contrast, regions characterized by a high level of methylated DNA and H3 histone that is deacetylated and methylated on Lys 9 and Lys 27 are transcriptionally inactive (Turner 2007).

Recent studies suggest that histone methylation is involved in the transcriptional regulation of cardiac hypertrophy. Keneda et al. investigated the genome-wide distribution of the two types of histone methylation—tri-methylated histone H3H4 and tri-methylated histone H3K9—in rat and human heart. They found these epigenetic marks are altered in heart failure (Kaneda et al. 2009). A more recent report showed that histone demethylase is involved in cardiac hypertrophy. Studies in mice have shown that a reduction of expression of histone demethylase in cardiomyocytes manifests resistance to pressure overload induced by transverse aortic constriction (TAC), whereas the transgenic mice display cardiac hypertrophy increased (Zhang et al. 2011; Papait et al. 2013).

These studies support a possible role for changes of histone methylation in gene transcriptional regulation underlying heart failure. However, whether the epigenetic marks are involved in regulating gene expression in cardiac hypertrophy or at which stage of the pathology its involvement is needs further study to elucidate (Papait et al. 2013).

3 The Chromatin Remodeling

That chromatin remodeling regulates gene expression changes underlying heart failure is supported by recent studies on Brg1, an ATPase subunit of BAF (Hang et al. 2010; Han et al. 2014; He et al. 2014). Mice lacking Brg1 are more resistant to a pro-hypertrophy stimulus, such as TAC, in comparison with wild-type mice. Moreover, Brg1 was found upregulated in the heart of certain hypertrophic cardiomyopathy patients, whose expression level is correlated with disease severity. Brg1 plays a key role in the switch from Myh6 to Myh7 isoform during cardiac hypertrophy. Brg1 is expressed in embryonic heart rather than in adult cardiomyocytes. When challenged with pro-hypertrophy stimuli, Brg1 re-expressed in the heart and then interacted with its embryonic partners HDAC and PARP, forming molecular complexes leading to the pathological shift from adult to fetal MHC isoforms (from Myh6 to Myh7) (Papait et al. 2013; Han et al. 2014).

Rosa-Garrido and colleagues (2017) examined changes in myocardial chromatin configuration isolated from mouse hearts subjected to tamoxifen-inducible cardiac-specific CCCTC-binding factor (CTCF) knockout, which is a zinc finger protein and ubiquitous chromatin structural protein. Genome-wide chromatin conformation capture and DNA sequencing were performed, and deep RNA sequencing, as well, was used as a functional readout for relevant epigenetic changes. Cardiac-specific deletion of CTCF was sufficient to induce a dilated cardiac phenotype, showing LV ejection fraction reduced and myocyte hypertrophy and myocardial fibrosis increased. Meanwhile, expression of CTCF increased in human failing hearts when supported with a LV assist device. High-resolution chromatin mapping showed the cardiac-specific CTCF deletion mice had changes in the chromatin architecture. In cardiac-specific CTCF deletion mouse model, looping structure of chromatin was altered, which were enriched with cardiac functional genes. Meanwhile, enrichment of aberrant genes coincided with the loss of chromatin loops. The study also showed that expression levels of several target genes associated with the development of heart failure (e.g., Nppa [atrial natriuretic factor]) changed, accompanied by changes in the local chromatin microenvironment (Mann 2017; Rosa-Garrido et al. 2017).

Recent study uses a proteomic approach to further identify the range of chromatin remodelers playing a role in cardiac remodeling. In the study, chromatin isolated from control mice, and mice in the hypertrophy or failure phase in response to pressure overload, manifested many different groups of proteins that showed differential responses to cardiac pressure overload (Franklin et al. 2012). One of the proteins further studied included the high-mobility group protein box 2 (HMGB2), which is associated with chromatin remodeling and described as a protein that can bend DNA and could facilitate various complexes (promoter and enhancer regions) assembled with their respective interacting partners to regulate gene transcription. Mehta and colleagues (2015) have shown Mitf works in concert with Brg1 in cardiac transcription factor Gata4 activation. In vivo study, the combined recruitment of Mitf and Brg1 appeared to be crucial for a normal hypertrophy response to neurohormonal or mechanical cardiac pressure overload.

Overall, these studies point to an important role for chromatin remodeling in accessibility of transcription factors to activate the cardiac hypertrophy gene program (Franklin et al. 2012; Anand et al. 2013; Mehta et al. 2015; van Berlo 2015). To what extent cardiomyocyte chromatin displays plasticity and the gene-specific regulations of chromatin allow activation of the cardiac hypertrophy program is not fully understood. Future studies focus on the chromatin changes that permit cardiac hypertrophy to improve our understanding of the underlying mechanism (van Berlo 2015).

4 miRNAs Are Key Regulators of Heart Failure

Abundant studies have suggested the important role for a family of tiny regulatory RNAs, known as microRNAs (miRNAs or miRs), in the control of diverse aspects of cardiac function. With respect to modification of DNA and histones, the role of miRNAs in heart failure has been more extensively investigated. A handful of reports have demonstrated an important role of miRNAs in heart failure (Divakaran and

Mann 2008). Many miRNAs has been found with altered expression in animal models of heart failure and in human cardiac patients (Latronico and Condorelli 2009). Interestingly, the miRNA signature of failing heart shows a similarity to fetal cardiac tissue, which is in accordance with the notion that the fetal-gene expression program is reactivated during cardiac remodeling (Papait et al. 2013). miRNA signature may be specific for different pathologies. A genome-wide study on miRNA expression demonstrated that even in heart failure, for different pathologies—aortic stenosis, dilated cardiomyopathy, and ischemic cardiomyopathy—43 out of 87 tested miRNAs expressed differentially in more than one of these disease groups (Ikeda et al. 2007; Papait et al. 2013).

Pathological stress can activate signaling pathways that trigger cardiomyocyte hypertrophy. In vitro and in vivo studies have demonstrated the key role of miRNAs in driving gene expression change during heart failure. One of the first studies mechanistically investigating the importance of a microRNA in cardiac hypertrophy demonstrated miR-208 crucially involved in hypertrophic signaling (van Rooij et al. 2007). Another founder study demonstrated that miR-133 levels were downregulated in the heart from pressure-overloaded mice; simultaneously, inhibition of miR-133 with an anti-miR-133 oligonucleotide resulted in cardiac hypertrophy in vivo (Care et al. 2007). miR-1 could attenuate cardiomyocyte hypertrophy in cultured cardiomyocytes and in the intact adult heart by regulation of myocardium growth signaling pathway via modulation of calcium signaling components such as calmodulin (Ikeda et al. 2009). Overexpression of miR-23a, miR-23b, miR-24, miR-195, or miR-214, whose profile showed it upregulated with heart failure, could individually induce cardiac hypertrophy in neonatal cardiomyocytes (Papait et al. 2013).

Cardiac fibrosis is caused by exaggerated activation of mostly resident cardiac fibroblasts, which secreted excessive extracellular matrix proteins, growth factors, cytokines, and genetic material (Thum and Condorelli 2015). Enrichment of several microRNAs in cardiac fibroblasts has been shown to modulate various cellular processes powerfully, including proliferation and growth factor secretion. For instance, miR-133 and miR-30, which control expression of connective tissue growth factor, are usually downregulated under cardiac stress (Duisters et al. 2009). miR-21 was shown to be involved in fibrosis development via silencing the ERK–MAP kinase inhibitor sprouty-1 (Thum et al. 2008). Paradoxically, miR-21 knockout mice developed cardiac fibrosis under cardiac stress, whereas pharmacological inhibition of miR-21 ameliorated fibrosis in the heart (Thum et al. 2008; Patrick et al. 2010).

Interestingly, some reports have shown miRNAs could act as promising biomarkers. For example, circulating levels of miR-29 were shown to be directly correlated with hypertrophy and fibrosis in patients with hypertrophic cardiomyopathy, (Roncarati et al. 2014), whereas other circulating miRNA patterns may discriminate between patients with heart failure (Watson et al. 2015). Altogether, these studies suggest that miRNAs represent a good therapeutic target and/or have a diagnostic significance for heart failure.

5 lncRNAs in Heart Failure

Advanced high-throughput sequencing techniques unveiled the ubiquitous expression of lncRNAs in various cells (Ounzain et al. 2015). Most of the lncRNAs proceed specific expression changes in pathological heart conditions and experimentally induced cardiac stress in humans and rodent models, respectively. Consequently, lncRNA may work as key regulators of homeostasis and disease processes. It has been reported lncRNAs were involved in cardiac development and disorders (Batista and Chang 2013; Klattenhoff et al. 2013; Han et al. 2014; Ounzain et al. 2015). Therefore, lncRNAs may act as therapeutic targets for the treatment of cardiac disease in the future (Bär et al. 2016). Although research of lncRNAs is still in its infancy, pharmacological manipulation of several lncRNAs in the pathologic mouse heart has demonstrated promising results (Bär et al. 2016; Wang et al. 2016).

In a pioneer study, Han et al. identified a cardioprotective antisense lncRNAs expressed in murine heart specifically in the myosin heavy chain 7 (Myh7) locus, which was termed myosin heavy chain-associated RNA transcripts or Mhrt (Han et al. 2014). It is downregulated when the heart is challenged with pressure overload, and transgenic overexpression of Mhrt protects mice from pathological hypertrophy following heart failure. The nuclear-localized Mhrt binds the chromatin-remodeling factor BRG1 under homeostatic conditions to deter its binding with chromatin, acting as decoy nucleotide target to prevent BRG1 binding and epigenetic repression of its targeted chromatin, whose role in cardiac remodeling elucidated above.

In contrast with the cardioprotective role of Mhrt, Viereck et al. recently identified a pro-hypertrophic lncRNA named cardiac hypertrophy-associated transcript, Chast (Viereck et al. 2016). It is modulated by the NFAT and is upregulated in response to hypertrophic stimuli specifically in myocardium. Overexpression of Chast results in myocardium hypertrophy in vitro and in vivo. Meanwhile, using pharmacological inhibition ameliorates cardiac remodeling and hypertrophy. This study provides the first proof of principle that lncRNA-targeting therapy is used in the treatment of cardiac hypertrophy. Chast has a homologue in human heart, the expression of which is elevated in human ESC-derived myocardium on exposure to hypertrophic stimuli. And its elevation is validated in hearts of patients with aortic stenosis compared with healthy hearts (Bär et al. 2016; Viereck et al. 2016). Similarly, Wang et al. discovered lncRNA Chaer, the abbreviation of cardiac-hypertrophy-associated epigenetic regulator, another lncRNA required for the development of cardiac hypertrophy, which directly interacts with PRC2 and consequently inhibits the formation of transcriptionally silent chromatin at pro-hypertrophic loci (Bär et al. 2016; Wang et al. 2016).

In mice, Wang et al. discovered lncRNA Chrf (cardiac-hypertrophy-related factor) which acts as a competing endogenous RNA sequestering miR-489. miR-489 targets the mRNA of myeloid differentiation primary response gene 88 (Myd88), whose upregulation is known to result in hypertrophy (Wang et al. 2014; Thum and Condorelli 2015).

In summary, several human and murine lncRNAs associated with cardiac development and remodeling have been identified and partially characterized. Given the rapidly evolving research field, the list of lncRNAs implicated in the processes of cardiac hypertrophy and heart failure is expected to grow exponentially in the future. As a result, improvement of knowledge in lncRNAs will potentially contribute to facilitating cardiovascular prevention and intervention (Bär et al. 2016).

6 Perspective

The dynamic aspects of epigenetics not only provide more accurate evidence to the role of changing environmental factors in the process of heart failure, associating the environment with the genome, but also offer novel ways to reactivate or silence the pathologic activation of genes. While pharmacogenetics varies individually, epigenomics offers more information that may help develop more precise therapy. In the field of oncology, epigenetic drugs have already been translated into clinical (Delmore et al. 2011; Duygu et al. 2013; Coppede 2014).

It is clear that epigenetic modifications including DNA modifications, histone modifications, and noncoding RNA-based mechanisms are the molecular targets for environmental stimuli and may lead to the onset and aggravation of heart failure. To elucidate the role of epigenetic mechanisms in cardiac hypertrophy and heart failure could provide a basis for the development of new tools against these pathologies involved in heart failure. Additional research is obviously warranted to further clarify the impaction of epigenetic mechanisms on the onset and development of heart disease and to eventually identify new pharmacologic targets in heart failure (Papait and Condorelli 2010; Duygu et al. 2013).

References

Anand P, Brown JD, Lin CY, Qi J, Zhang R, Artero PC, Alaiti MA, Bullard J, Alazem K, Margulies KB, Cappola TP, Lemieux M, Plutzky J, Bradner JE, Haldar SM. BET bromodomains mediate transcriptional pause release in heart failure. Cell. 2013;154:569–82.

Antos CL, McKinsey TA, Dreitz M, Hollingsworth LM, Zhang CL, Schreiber K, Rindt H, Gorczynski RJ, Olson EN. Dose-dependent blockade to cardiomyocyte hypertrophy by histone deacetylase inhibitors. J Biol Chem. 2003;278:28930–7.

Bär C, Chatterjee S, Thum T. Long noncoding RNAs in cardiovascular pathology, diagnosis, and therapy. Circulation. 2016;134:1484–99.

Batista PJ, Chang HY. Long noncoding RNAs: cellular address codes in development and disease. Cell. 2013;152:1298–307.

Benjamin EJ, Blaha MJ, Chiuve SE, Cushman M, Das SR, Deo R, de Ferranti SD, Floyd J, Fornage M, Gillespie C, Isasi CR, Jiménez MC, Jordan LC, Judd SE, Lackland D, Lichtman JH, Lisabeth L, Liu S, Longenecker CT, Mackey RH, Matsushita K, Mozaffarian D, Mussolino ME, Nasir K, Neumar RW, Palaniappan L, Pandey DK, Thiagarajan RR, Reeves MJ, Ritchey M, Rodriguez CJ, Roth GA, Rosamond WD, Sasson C, Towfighi A, Tsao CW, Turner MB,

Virani SS, Voeks JH, Willey JZ, Wilkins JT, Wu JH, Alger HM, Wong SS, Muntner P. Heart disease and stroke statistics—2017 update: a report from the American Heart Association. Circulation. 2017;135:e146–603.
Bernstein BE, Meissner A, Lander ES. The mammalian epigenome. Cell. 2007;128:669–81.
Bonaldi T, Talamo F, Scaffidi P, Ferrera D, Porto A, Bachi A, Rubartelli A, Agresti A, Bianchi ME. Monocytic cells hyperacetylate chromatin protein HMGB1 to redirect it towards secretion. EMBO J. 2003;22:5551–60.
Care A, Catalucci D, Felicetti F, Bonci D, Addario A, Gallo P, Bang ML, Segnalini P, Gu Y, Dalton ND, Elia L, Latronico MV, Hoydal M, Autore C, Russo MA, Dorn GN, Ellingsen O, Ruiz-Lozano P, Peterson KL, Croce CM, Peschle C, Condorelli G. MicroRNA-133 controls cardiac hypertrophy. Nat Med. 2007;13:613–8.
Cloos PA, Christensen J, Agger K, Helin K. Erasing the methyl mark: histone demethylases at the center of cellular differentiation and disease. Genes Dev. 2008;22:1115–40.
Coppede F. Epigenetic biomarkers of colorectal cancer: focus on DNA methylation. Cancer Lett. 2014;342:238–47.
Delmore JE, Issa GC, Lemieux ME, Rahl PB, Shi J, Jacobs HM, Kastritis E, Gilpatrick T, Paranal RM, Qi J, Chesi M, Schinzel AC, McKeown MR, Heffernan TP, Vakoc CR, Bergsagel PL, Ghobrial IM, Richardson PG, Young RA, Hahn WC, Anderson KC, Kung AL, Bradner JE. Mitsiades CS. BET bromodomain inhibition as a therapeutic strategy to target c-Myc. Cell. 2011;146:904–17.
Divakaran V, Mann DL. The emerging role of MicroRNAs in cardiac remodeling and heart failure. Circ Res. 2008;103:1072–83.
Duisters RF, Tijsen AJ, Schroen B, Leenders JJ, Lentink V, van der Made I, Herias V, van Leeuwen RE, Schellings MW, Barenbrug P, Maessen JG, Heymans S, Pinto YM, Creemers EE. miR-133 and miR-30 regulate connective tissue growth factor: implications for a role of microRNAs in myocardial matrix remodeling. Circ Res. 2009;104:170–8. 6p–178p
Duygu B, Poels EM, Da CMP. Genetics and epigenetics of arrhythmia and heart failure. Front Genet. 2013;4:219.
Evankovich J, Cho SW, Zhang R, Cardinal J, Dhupar R, Zhang L, Klune JR, Zlotnicki J, Billiar T, Tsung A. High mobility group box 1 release from hepatocytes during ischemia and reperfusion injury is mediated by decreased histone deacetylase activity. J Biol Chem. 2010;285:39888–97.
Ficz G, Branco MR, Seisenberger S, Santos F, Krueger F, Hore TA, Marques CJ, Andrews S, Reik W. Dynamic regulation of 5-hydroxymethylcytosine in mouse ES cells and during differentiation. Nature. 2011;473:398–402.
Franklin S, Chen H, Mitchell-Jordan S, Ren S, Wang Y, Vondriska TM. Quantitative analysis of the chromatin proteome in disease reveals remodeling principles and identifies high mobility group protein B2 as a regulator of hypertrophic growth. Mol Cell Proteomics. 2012;11:M111–14258.
Gallo P, Latronico MV, Gallo P, Grimaldi S, Borgia F, Todaro M, Jones P, Gallinari P, De Francesco R, Ciliberto G, Steinkuhler C, Esposito G, Condorelli G. Inhibition of class I histone deacetylase with an apicidin derivative prevents cardiac hypertrophy and failure. Cardiovasc Res. 2008;80:416–24.
Greco CM, Condorelli G. Epigenetic modifications and noncoding RNAs in cardiac hypertrophy and failure. Nat Rev Cardiol. 2015;12:488–97.
Greco CM, Kunderfranco P, Rubino M, Larcher V, Carullo P, Anselmo A, Kurz K, Carell T, Angius A, Latronico MV, Papait R, Condorelli G. DNA hydroxymethylation controls cardiomyocyte gene expression in development and hypertrophy. Nat Commun. 2016;7:12418.
Haldar SM, McKinsey TA. BET-ting on chromatin-based therapeutics for heart failure. J Mol Cell Cardiol. 2014;74:98–102.
Han P, Li W, Lin C, Yang J, Shang C, Nurnberg ST, Jin KK, Xu W, Lin C, Lin C, Xiong Y, Chien H, Zhou B, Ashley E, Bernstein D, Chen P, Chen HV, Quertermous T, Chang C. A long noncoding RNA protects the heart from pathological hypertrophy. Nature. 2014;514:102–6.
Hang CT, Yang J, Han P, Cheng H, Shang C, Ashley E, Zhou B, Chang C. Chromatin regulation by Brg1 underlies heart muscle development and disease. Nature. 2010;466:62–7.

He L, Tian X, Zhang H, Hu T, Huang X, Zhang L, Wang Z, Zhou B. BAF200 is required for heart morphogenesis and coronary artery development. PLoS One. 2014;9:e109493.

Ikeda S, Kong SW, Lu J, Bisping E, Zhang H, Allen PD, Golub TR, Pieske B, Pu WT. Altered microRNA expression in human heart disease. Physiol Genomics. 2007;31:367–73.

Ikeda S, He A, Kong SW, Lu J, Bejar R, Bodyak N, Lee KH, Ma Q, Kang PM, Golub TR, Pu WT. MicroRNA-1 negatively regulates expression of the hypertrophy-associated calmodulin and Mef2a genes. Mol Cell Biol. 2009;29:2193–204.

Kaneda R, Takada S, Yamashita Y, Choi YL, Nonaka-Sarukawa M, Soda M, Misawa Y, Isomura T, Shimada K, Mano H. Genome-wide histone methylation profile for heart failure. Genes Cells. 2009;14:69–77.

Kee HJ, Sohn IS, Nam KI, Park JE, Qian YR, Yin Z, Ahn Y, Jeong MH, Bang YJ, Kim N, Kim JK, Kim KK, Epstein JA, Kook H. Inhibition of histone deacetylation blocks cardiac hypertrophy induced by angiotensin II infusion and aortic banding. Circulation. 2006;113:51–9.

Klattenhoff CA, Scheuermann JC, Surface LE, Bradley RK, Fields PA, Steinhauser ML, Ding H, Butty VL, Torrey L, Haas S, Abo R, Tabebordbar M, Lee RT, Burge CB, Boyer LA. Braveheart, a long noncoding RNA required for cardiovascular lineage commitment. Cell. 2013;152:570–83.

Kong Y, Tannous P, Lu G, Berenji K, Rothermel BA, Olson EN, Hill JA. Suppression of class I and II histone deacetylases blunts pressure-overload cardiac hypertrophy. Circulation. 2006;113:2579–88.

Kook H, Lepore JJ, Gitler AD, Lu MM, Wing-Man YW, Mackay J, Zhou R, Ferrari V, Gruber P, Epstein JA. Cardiac hypertrophy and histone deacetylase-dependent transcriptional repression mediated by the atypical homeodomain protein Hop. J Clin Invest. 2003;112:863–71.

Kouzarides T. Histone methylation in transcriptional control. Curr Opin Genet Dev. 2002;12:198–209.

Latronico MV, Condorelli G. MicroRNAs and cardiac pathology. Nat Rev Cardiol. 2009;6:419–29.

Mann DL. High-resolution chromatin mapping in heart failure. Circulation. 2017;136:1626–8.

Mann DL, Bristow MR. Mechanisms and models in heart failure: the biomechanical model and beyond. Circulation. 2005;111:2837–49.

McKinsey TA, Olson EN. Toward transcriptional therapies for the failing heart: chemical screens to modulate genes. J Clin Invest. 2005;115:538–46.

Meder B, Haas J, Sedaghat-Hamedani F, Kayvanpour E, Frese K, Lai A, Nietsch R, Scheiner C, Mester S, Bordalo DM, Amr A, Dietrich C, Pils D, Siede D, Hund H, Bauer A, Holzer DB, Ruhparwar A, Mueller-Hennessen M, Weichenhan D, Plass C, Weis T, Backs J, Wuerstle M, Keller A, Katus HA, Posch AE. Epigenome-wide association study identifies cardiac gene patterning and a novel class of biomarkers for heart failure. Circulation. 2017;136:1528–44.

Mehta G, Kumarasamy S, Wu J, Walsh A, Liu L, Williams K, Joe B, de la Serna IL. MITF interacts with the SWI/SNF subunit, BRG1, to promote GATA4 expression in cardiac hypertrophy. J Mol Cell Cardiol. 2015;88:101–10.

Ounzain S, Micheletti R, Beckmann T, Schroen B, Alexanian M, Pezzuto I, Crippa S, Nemir M, Sarre A, Johnson R, Dauvillier J, Burdet F, Ibberson M, Guigo R, Xenarios I, Heymans S, Pedrazzini T. Genome-wide profiling of the cardiac transcriptome after myocardial infarction identifies novel heart- specific long non-coding RNAs. Eur Heart J. 2015;36:353–68.

Pai AA, Bell JT, Marioni JC, Pritchard JK, Gilad Y. A Genome-wide study of DNA methylation patterns and gene expression levels in multiple human and chimpanzee tissues. PLoS Genet. 2011;7:e1001316.

Papait R, Condorelli G. Epigenetics in heart failure. Ann N Y Acad Sci. 2010;1188:159–64.

Papait R, Greco C, Kunderfranco P, Latronico MVG, Condorelli G. Epigenetics: a new mechanism of regulation of heart failure? Basic Res Cardiol. 2013;108:361.

Pastor WA, Pape UJ, Huang Y, Henderson HR, Lister R, Ko M, McLoughlin EM, Brudno Y, Mahapatra S, Kapranov P, Tahiliani M, Daley GQ, Liu XS, Ecker JR, Milos PM, Agarwal S, Rao A. Genome- wide mapping of 5-hydroxymethylcytosine in embryonic stem cells. Nature. 2011;473:394–7.

Patrick DM, Montgomery RL, Qi X, Obad S, Kauppinen S, Hill JA, van Rooij E, Olson EN. Stress-dependent cardiac remodeling occurs in the absence of microRNA-21 in mice. J Clin Invest. 2010;120:3912–6.

Rau CD, Vondriska TM. DNA methylation and human heart failure. Circulation. 2017;136:1545–7.

Roncarati R, Viviani AC, Losi MA, Papa L, Cavarretta E, Da CMP, Contaldi C, Saccani JG, Franzone A, Galastri L, Latronico MV, Imbriaco M, Esposito G, De Windt L, Betocchi S, Condorelli G. Circulating miR-29a, among other up-regulated microRNAs, is the only biomarker for both hypertrophy and fibrosis in patients with hypertrophic cardiomyopathy. J Am Coll Cardiol. 2014;63:920–7.

Rosa-Garrido M, Chapski DJ, Schmitt AD, Kimball TH, Karbassi E, Monte E, Balderas E, Pellegrini M, Shih T, Soehalim E, Liem D, Ping P, Galjart NJ, Ren S, Wang Y, Ren B, Vondriska TM. High- resolution mapping of chromatin conformation in cardiac myocytes reveals structural remodeling of the epigenome in heart failure. Circulation. 2017;136:1613–25.

Sanger F, Air GM, Barrell BG, Brown NL, Coulson AR, Fiddes CA, Hutchison CA, Slocombe PM, Smith M. Nucleotide sequence of bacteriophage phi X174 DNA. Nature. 1977;265:687–95.

Schuster SC. Next-generation sequencing transforms today's biology. Nat Methods. 2008;5:16–8.

Smith ZD, Meissner A. DNA methylation: roles in mammalian development. Nat Rev Genet. 2013;14:204–20.

Thum T, Condorelli G. Long noncoding RNAs and microRNAs in cardiovascular pathophysiology. Circ Res. 2015;116:751–62.

Thum T, Gross C, Fiedler J, Fischer T, Kissler S, Bussen M, Galuppo P, Just S, Rottbauer W, Frantz S, Castoldi M, Soutschek J, Koteliansky V, Rosenwald A, Basson MA, Licht JD, Pena JT, Rouhanifard SH, Muckenthaler MU, Tuschl T, Martin GR, Bauersachs J, Engelhardt S. MicroRNA-21 contributes to myocardial disease by stimulating MAP kinase signalling in fibroblasts. Nature. 2008;456:980–4.

Turner BM. Defining an epigenetic code. Nat Cell Biol. 2007;9:2–6.

van Berlo JH. Chromatin remodeling permits cardiac hypertrophy to develop. J Mol Cell Cardiol. 2015;89:119–21.

van Rooij E, Sutherland LB, Qi X, Richardson JA, Hill J, Olson EN. Control of stress-dependent cardiac growth and gene expression by a microRNA. Science. 2007;316:575–9.

Viereck J, Kumarswamy R, Foinquinos A, Xiao K, Avramopoulos P, Kunz M, Dittrich M, Maetzig T, Zimmer K, Remke J, Just A, Fendrich J, Scherf K, Bolesani E, Schambach A, Weidemann F, Zweigerdt R, de Windt LJ, Engelhardt S, Dandekar T, Batkai S, Thum T. Long noncoding RNA Chast promotes cardiac remodeling. Sci Transl Med. 2016;8:322r–6r.

Wang X, Hayes JJ. Acetylation mimics within individual core histone tail domains indicate distinct roles in regulating the stability of higher-order chromatin structure. Mol Cell Biol. 2008;28:227–36.

Wang K, Liu F, Zhou LY, Long B, Yuan SM, Wang Y, Liu CY, Sun T, Zhang XJ, Li PF. The long noncoding RNA CHRF regulates cardiac hypertrophy by targeting miR-489. Circ Res. 2014;114:1377–88.

Wang Z, Zhang XJ, Ji YX, Zhang P, Deng KQ, Gong J, Ren S, Wang X, Chen I, Wang H, Gao C, Yokota T, Ang YS, Li S, Cass A, Vondriska TM, Li G, Deb A, Srivastava D, Yang HT, Xiao X, Li H, Wang Y. The long noncoding RNA Chaer defines an epigenetic checkpoint in cardiac hypertrophy. Nat Med. 2016;22:1131–9.

Watson CJ, Gupta SK, O'Connell E, Thum S, Glezeva N, Fendrich J, Gallagher J, Ledwidge M, Grote- Levi L, McDonald K, Thum T. MicroRNA signatures differentiate preserved from reduced ejection fraction heart failure. Eur J Heart Fail. 2015;17:405–15.

Zhang QJ, Chen HZ, Wang L, Liu DP, Hill JA, Liu ZP. The histone trimethyllysine demethylase JMJD2A promotes cardiac hypertrophy in response to hypertrophic stimuli in mice. J Clin Invest. 2011;121:2447–56.

Zhang L, Liu M, Jiang H, Yu Y, Yu P, Tong R, Wu J, Zhang S, Yao K, Zou Y, Ge J. Extracellular high- mobility group box 1 mediates pressure overload-induced cardiac hypertrophy and heart failure. J Cell Mol Med. 2016;20:459–70.

Chapter 11
Progress of Genetics in Inherited Cardiomyopathies-Induced Heart Failure

Baoli Zhang, Xue Yang, Ning Feng, and Hong Jiang

Abstract Heart failure (HF) has been an increasingly growing public health problem and social burden. One prevalent cause of HF is familial cardiomyopathy, such as the dilated, hypertrophic, restrictive, or arrhythmogenic cardiomyopathy and left ventricular non-compaction. With the help of the genome-wide association study and progress in genetic screening techniques such as next-generation sequencing, virtually 100 disease-associated genes have been continuously identified for cardiomyopathies. Recent genetic researches of single gene variation and genotype-phenotype relationships have provided profound insights into the complicated mechanism of cardiomyopathies. These findings greatly promoted the understanding of cardiomyopathies with different etiologies and facilitated the genetic counselling, screening inchoate diagnosis, and better management and prognostics of inherited cardiomyopathies and HF. In this chapter, the major genetic causes of inherited cardiomyopathies causing HF and the cellular functions of the genetic factors are mainly discussed.

Keywords Heart failure · Inherited cardiomyopathies · Genetics · DCM · HCM · RCM · ARVC · LVNC

1 Introduction

Heart failure (HF) is a predominant cause of morbidity and mortality in the world, and in recent decades, its prevalence has been rising in industrialized countries with the population aging, as well as growing prevalence of hypertension, diabetes

B. Zhang · X. Yang · H. Jiang (✉)
Shanghai Institute of Cardiovascular Diseases, Clinical Science Institute, Zhongshan Hospital, Shanghai Medical College of Fudan University, Shanghai, China
e-mail: jiang.hong@zs-hospital.sh.cn

N. Feng
Division of Cardiology, Heart and Vascular Institute, University of Pittsburgh School of Medicine, Pittsburgh, PA, USA

H. Jiang, M. Liu (eds.), *Heart Genomics*, Translational Bioinformatics 16,
https://doi.org/10.1007/978-981-13-1429-2_11

mellitus, and chronic kidney diseases. Although advancements in science, technology, and public awareness improve the survival time of cardiovascular diseases, diabetes, and chronic kidney diseases, HF is still a big challenge and results in extremely high health-care-related costs, heavily burdening either on patients or on society. Studies indicate an overall HF prevalence of 11.8% (range 4.7–13.3%) in the general population (van Riet et al. 2016) and about 50% survival average 5 years after diagnosis (Bhatt and Butler 2018), which shows a poor prognosis for HF.

In HF, the symptoms are dyspnea, fatigue, and fluid retention with signs of cardiac structural anomalies, while the stroke volume is decompensated by neurohumoral regulation, incapable of fulfilling the need of the organism. In spite of the majority of cases of HF due to coronary heart disease, hypertension, diabetes, etc. (Dickstein et al. 2008), epidemiology studies reveal that the risk of HF is involved in genetic predisposition, even in these acquired conditions. It is recognized that one of the most prevalent causes of HF are cardiomyopathies, like genetic cardiomyopathies, and concretely, HF could be triggered by Mendelian genetic disorders in a crucial proportion of cases. In accordance with structural and functional abnormalities in heart muscle, the various clinical phenotypes are classified into dilated cardiomyopathy (DCM), hypertrophic cardiomyopathy (HCM), arrhythmogenic right ventricular cardiomyopathy (ARVC), restrictive cardiomyopathy (RCM), and unclassified cardiomyopathies (Elliott et al. 2008). Studying single gene variants that lead to cardiomyopathy offers an opportunity for elucidating critical molecules implicated in the process of remodeling and for perception of the responses caused by gene mutations liable to and also relevant to common etiologies of HF. The genotype-phenotype relationships of the cardiomyopathies primarily include gene mutations involved in force generation and transmission, energy production and regulation, calcium cycling, ion channels, and transcriptional regulators, which have been discussed previously (Morita et al. 2005; Lopes and Elliott 2013; Maron and Maron 2013; Perez-Serra et al. 2016; Weintraub et al. 2017).

2 Features and Epidemiology

The inherited cardiomyopathies leading to HF are composed of five main types: DCM, HCM, RCM, ARVC, and left ventricular non-compaction (LVNC). Foremost, DCM, characterized by dilatation of chamber size, ventricular wall thinning, and systolic deterioration of left or both ventricles, is one of the prevalent causes of HF and responsible for around 55% of cardiomyopathies (Maron et al. 2006), and the annual incidence is 2–8 cases per 100,000 with an approximated prevalence of 1/2500 population (Maron et al. 2006; Towbin et al. 2006; Elliott et al. 2014).More importantly, 30–50% of cases link to genetic etiologies (Dec and Fuster 1994; Fatkin 2011; Givertz and Mann 2013). Of note, HCM is the most common inherited cardiovascular disorders (Sabater-Molina et al. 2018) and defined phenotypically by asymmetric or concentric wall thickening with or without outflow tract obstruction, in the absence of other cardiac conditions or potential systemic diseases

(Marian and Braunwald 2017), with a prevalence of 1/500 in adults (Maron et al. 2006; Jacoby and McKenna 2012; Elliott et al. 2014). Furthermore, restrictive cardiomyopathy (RCM) is a sparse form of the myocardial disorders (Jacoby and McKenna 2012) and described by normal or reduced sized ventricles with biatrial enlargement, worsened ventricular stiffness with normal, or increased ventricular wall thickness and systolic function (Maron et al. 2006; Muchtar et al. 2017). Also, ARVC is mostly first presented with arrhythmia, characterized by regional or global right ventricular dysfunction; histologically it is the myocardium replaced by fibro-fatty and may also affect the left ventricle (Meyer et al. 2014; El-Battrawy et al. 2018). The prevalence of ARVC is approximately 1:2000–1:5000 and more prevalent in males (2–3:1) (Elliott et al. 2014; Meyer et al. 2014; Camargo-Ariza et al. 2018; El-Battrawy et al. 2018; Shah and Bhimji 2018). Ultimately, LVNC is considered an unclassified cardiomyopathy (Elliott et al. 2008) and characterized by prominent trabeculations and recesses in the left ventricular walls (Meyer et al. 2014), with the approximated prevalence ranging from 4.5 to 26 per 10,000 adult patients (Elliott et al. 2008).

3 Inheritance Patterns

The inheritance patterns of the inherited cardiomyopathies consist of autosomal dominant, recessive, X-linked, or mitochondrial-transmitted inheritance. Particularly, autosomal dominant inheritance is the major pattern of transmission in inherited, familial DCM (FDCM), with 30–50% of cases of DCM in autosomal dominant inheritance, while the cases of autosomal recessive, X-linked, and mitochondrial inheritance account for a small part (Dec and Fuster 1994; Fatkin 2011; Meyer et al. 2014). Generally, HCM has an autosomal dominant pattern of inheritance and about 60% of cases with it show an obviously identifiable familial disease, though autosomal recessive and X-linked modes of inheritance have been depicted but are exiguous (Branzi et al. 1985; Hartmannova et al. 2013; Marian and Braunwald 2017). Then, RCM is much less common and usually related to systemic disorders (Brieler et al. 2017), and it is inherited as an autosomal dominant trait, whereas autosomal recessive, X-linked, and mitochondrial-transmitted disease also exists. Moreover, ARVC has usually an autosomal dominant trait with incomplete penetrance (Hamid et al. 2002; Maron et al. 2006; Ellinor et al. 2010; Corrado et al. 2017), and analyses of first- and second-degree relatives of patients with ARVC indicate that equivalent to 50% of ARVC cases are familial (Hamid et al. 2002; Maron et al. 2006; Ellinor et al. 2010) and although two autosomal recessive forms have been described (Protonotarios and Tsatsopoulou 2004). Eventually, LVNC frequently has autosomal dominant inheritance or X-linked recessive, with reports showing 70% having autosomal dominant and 30% X-linked inheritance (Ichida et al. 1999; Towbin 2010), while infrequent cases of autosomal recessive and mitochondrial inheritance also emerge (Towbin 2010).

4 Molecular Genetics and Genome-Wide Association Study(GWAS)

Molecular genetics provide significant insights into the pathogenesis of the inherited cardiomyopathies, which gives novel outlooks for the diagnosis and management of affected patients. And GWAS offer an unbiased strategy to research the correlation of clinical genotype and phenotype, with identifying single nucleotide polymorphism (SNP) and other mutations in DNA correlated with a disease. For instance, GWAS identified new risk loci for HF induced by DCM (Villard et al. 2011; Meder et al. 2014). Additionally, for the acquired HF, the GWAS in advanced all-cause HF found regions containing HSPB7 and FRMD4B as new susceptibility loci (Cappola et al. 2010), rs28714259 as a validated SNP related to anthracycline-induced HF (Wells et al. 2017), and locus on chromosome 3p22 in an intron of CKLF-like MARVEL transmembrane domain-containing 7 (CMTM7) as an all-cause mortality predictor in HF (Morrison et al. 2010).

5 Next-Generation Sequencing (NGS)

Fundamentally, employing of next-generation sequencing (NGS) platforms promotes the development of GWAS and has multiplied the number of genes analyzed related to the inherited cardiomyopathies such as DCM, HCM, ARVC, etc. The technique has revolutionized the clinical screening in genetics arising from making concrete analyses of substantial genes in a short time and in a cost-effective way (Sikkema-Raddatz et al. 2013; Jamuar and Tan 2015). The sum of data and the identification of genes using NGS, however, has caused a genetic overlap between DCM and other cardiomyopathies such as HCM and ARVC (Biswas et al. 2014). In addition, this high-throughput analysis frequently identifies novel DNA sequence mutations for which pathogenicity or significance is not unraveled (Maron et al. 2012) and these ambiguous mutations have practically discombobulated family screening and boost puzzles in explanation of genetic testing results, and therefore, this is a big challenge for geneticist and clinicians. Nevertheless, the pathogenicity of such ambiguous variants can be tested in computational biology tools and be predicted in silico.

6 Genetics in Inherited Cardiomyopathies-Induced HF

Collectively, the known mutated genes of different types of inherited cardiomyopathies inducing HF have been categorized in terms of the likely or related functional processes or outcomes of variants or the components of the cardiomyocytes or extracellular matrix such as force generation and transmission, energy production and modulation or metabolism, calcium cycling and ion channels, transcriptional regulation, and other cytological processes.

6.1 *Mutations of Genes in Force Generation and Transmission*

The essential functions of cardiomyocytes are generation of contractile force by the sarcomere as well as its propagation to the extracellular matrix. However, inappropriate execution of each component furthers the remodeling in the heart, hypertrophy or dilatation, and causes HF. There are a multitude of components in the process of force production and transmission such as sarcomere, Z-disk, cytoskeleton, nuclear envelope and desmosome, etc. The sarcomere is a vital part for the contraction in heart and skeleton muscle limited by Z-bands, and muscle contraction is triggered by interplay between thick filaments and thin filaments coupling with titin and other proteins. Hence, pathogenic mutations in sarcomere encoding genes are related to downward contractile function (Kamisago et al. 2000). Notably, as a symbol of the region between neighboring sarcomeres and attaching to actin filaments and titin, the Z-disk is a dark thin protein band in striated muscle, and quite a few regulatory proteins propagating the generated force of muscle contraction to neighboring sarcomeres are involved in the Z-disk (Perez-Serra et al. 2016). Also, cytoskeleton extending from the nucleus to the cell membrane in the muscle cell and composed of intermediate filaments, etc. is a sophisticated network which connects filaments and tubules that spread around the cytoplasm, and its primary function would help the true function of muscle contraction. Moreover, intermediate filaments as cytoskeletal proteins connect the Z-disk with the sarcolemma. In addition, the nuclear envelope with many small holes consists of the inner nuclear membrane and the outer nuclear membrane and exists as an interplay between a couple of proteins regulating the nucleon-cytoskeleton interaction. Finally, desmosome is a component necessary for mechanical intercellular junctions and is capable of sustaining the mechanical and electrical integrity of the myocytes. Consequently, multiple gene variants are implicated in the process of force generation and transmission rather than other processes or classes.

6.1.1 Mutations in Genes Associated with DCM

A large number of variants in the genes of the components in the process of force generation and transmission are involved in the pathogenesis of inherited DCM, like mutations in the gene ACTC1 (Olson et al. 1998; Lakdawala et al. 2013; Chow et al. 2014; Pugh et al. 2014), ACTN2 (Andreasen et al. 2013; Ng et al. 2013; Pugh et al. 2014), ANKRD1 (Duboscq-Bidot et al. 2009; Moulik et al. 2009; Ng et al. 2013), BAG3 (Norton et al. 2011; Feldman et al. 2014; Franaszczyk et al. 2014), CRYAB (Inagaki et al. 2006; Pilotto et al. 2006), CSRP3 (Hershberger et al. 2008), CTF1 (Erdmann et al. 2000), DES (Andreasen et al. 2013; Tse et al. 2013; Zhao et al. 2015; Brodehl et al. 2016), DMD (Milasin et al. 1996; Ferlini et al. 1998; Kimura et al. 2007; Diegoli et al. 2011), DSC2 (Elliott et al. 2010; Garcia-Pavia et al. 2011; Andreasen et al. 2013), DSG2 (Elliott et al. 2010; Garcia-Pavia et al. 2011; Ng et al. 2013), DSP (Garcia-Pavia et al. 2011), EMD (Bione et al. 1995; Zhao et al. 2015), FHL2 (Arimura et al. 2007), FKRP (Muller et al. 2005), FKTN

(Murakami et al. 2006; Hobbiebrunken et al. 2016), LAMA4 (Knoll et al. 2007; Andreasen et al. 2013), LDB3 (Vatta et al. 2003; Arimura et al. 2009a, b; Andreasen et al. 2013), LMNA (Sebillon et al. 2003; Parks et al. 2008; Perez-Serra et al. 2015), MURC (Rodriguez et al. 2011; Andreasen et al. 2013), MYBPC3 (Ehlermann et al. 2008), MYH6 (Carniel et al. 2005; Merlo et al. 2013; Zhao et al. 2015), MYH7 (Hershberger et al. 2008; Lakdawala et al. 2012; Lamont et al. 2014), MYPN (Duboscq-Bidot et al. 2008), NEBL (Andreasen et al. 2013; Maiellaro-Rafferty et al. 2013), NEXN (Hassel et al. 2009), OBSCN (Marston et al. 2015), PDLIM3 (Arola et al. 2007), PKP2 (Elliott et al. 2010), PLEC (Bolling et al. 2010), RBM20 (Brauch et al. 2009; Guo et al. 2013; Khan et al. 2016), SGCD (Tsubata et al. 2000; Karkkainen et al. 2003), SYNE1 (Puckelwartz et al. 2010; Zhou et al. 2017), SYNE2 (Duong et al. 2014), TCAP (Hershberger et al. 2008; Hirtle-Lewis et al. 2013), TMPO (Taylor et al. 2005), TNNC1 (Mogensen et al. 2004; Hershberger et al. 2010; Zhao et al. 2015), TNNI3 (Carballo et al. 2009; Murakami et al. 2010; Memo et al. 2013), TNNT2 (Hershberger et al. 2008; Zimmerman et al. 2010; Pugh et al. 2014), TPM1 (Zimmerman et al. 2010; Pugh et al. 2014), TTN (Gerull et al. 2002; Yoskovitz et al. 2012; Norton et al. 2013), VCL (Pugh et al. 2014), etc. (Table 11.1). The majority of the genes are inherited as an autosomal dominant trait, whereas only a few follow the autosomal recessive or X-linked inheritance. The major genes related to a predominant cardiac phenotype include TTN and LMNA, which are responsible for up to 25% and 5% of all cases of autosomal dominant DCM, respectively (Fatkin et al. 1999; Gerull et al. 2002; Parks et al. 2008). DCM mutations in other genes are far less common.

In DCM, genetic heterogeneity presents in autosomal dominant DCM with not less than 40 genes recognized for DCM (Ackerman et al. 2011; Cahill et al. 2013; Mestroni and Taylor 2013; Morales and Hershberger 2013; Teekakirikul et al. 2013; Weintraub et al. 2017), with substantial genes encoding sarcomere, Z-disk, as well as cytoskeletal proteins, while mutations in a smaller number of genes which encoded nuclear envelop, desmosome and basal lamina proteins, etc. have also been identified (Table 11.1). These genes mainly encode sarcomere or related sarcomere organization proteins (actin, alpha cardiac muscle 1 (ACTC1), caveolae-associated protein 4 (MURC), myosin-binding protein C (MYBPC3), myosin-6 (MYH6), myosin-6 (MYH7), RNA-binding protein 20 (RBM20), troponin C (TNNC1), troponin I (TNNI3), troponin T (TNNT2), tropomyosin alpha-1 chain (TPM1), titin (TTN)), Z-disk (alpha-actinin-2 (ACTN2), ankyrin repeat domain-containing protein 1 or CARD (ANKRD1), Bcl-2-associated athanogene 3 (BAG3), cysteine- and glycine-rich protein 3 or cardiac LIM protein (CSRP3), LIM domain-binding protein 3 (LDB3), myopalladin (MYPN), nebulette (NEBL), nexilin or F-actin-binding protein (NEXN), PDZ and LIM domain protein 3 (PDLIM3), telethonin or titin-cap protein (TCAP)), and cytoskeleton proteins (desmin (DES), fukutin-related protein (FKRP), fukutin (FKTN), plectin (PLEC), delta-sarcoglycan (SGCD), vinculin (VCL)), whereas genes coding nuclear membrane proteins (prelamin-A/C or lamin A/C (LMNA), nesprin-1 (SYNE1), nesprin-2(SYNE2), lamina-associated polypeptide 2, isoform alpha (TMPO)), desmosomal proteins (desmoplakin (DSP), desmocollin-2 (DSC2), desmoglein-2 (DSG2), plakophilin-2(PKP2)), basal lamina protein

Table 11.1 Genetics in force generation and transmission

Gene[a]	Protein[a]	OMIM[a]	Inheritance[a]
Dilated cardiomyopathy			
ACTC1	Actin, alpha cardiac muscle 1	102540	AD[b]
ACTN2	Alpha-actinin-2	102573	AD
ANKRD1	Ankyrin repeat domain-containing protein 1 (CARD)	609599	
BAG3	Bcl-2-associated athanogene 3	603883	AD
CRYAB	Alpha-crystallin B chain	123590	AD
CSRP3	Cysteine- and glycine-rich protein 3(cardiac LIM protein)	600824	
CTF1	Cardiotrophin-1	600435	
CFL2	Cofilin-2	601443	
DES	Desmin	125660	
DMD	Dystrophin	300377	XL[b]
DSC2	Desmocollin-2	125645	
DSG2	Desmoglein-2	125671	
DSP	Desmoplakin	125647	AD, AR[b]
EMD	Emerin	300384	XLR[b]
FHL2	Four and a half LIM domains protein 2	602633	
FKRP	Fukutin-related protein	606596	
FKTN	Fukutin	607440	AR
ILK	Integrin-linked protein kinase	602366	
LAMA4	Laminin subunit alpha-4	600133	AD
LDB3	LIM domain-binding protein 3	605906	AD
LMNA	Prelamin-A/C (lamin A/C)	150330	AD
MYBPC3	Myosin-binding protein C, cardiac-type	600958	AD
MYH6	Myosin-6 (Alpha-myosin heavy chain 6)	160710	
MYH7	Myosin-7 (Beta-myosin heavy chain 7)	160760	AD
MURC	Caveolae-associated protein 4 (muscle-related coiled-coil protein)	617714	
MYPN	Myopalladin	608517	AD
NEBL	Nebulette	605491	
NEXN	Nexilin (F-actin-binding protein)	613121	AD
PDLIM3	PDZ and LIM domain protein 3	605889	
OBSCN	Obscurin	608616	
PKP2	Plakophilin-2	602861	
PLEC	Plectin	601282	
RBM20	RNA-binding protein 20	613171	AD
SGCD	Delta-sarcoglycan	601411	
SYNE1	Nesprin-1	608441	
SYNE2	Nesprin-2	608442	
TCAP	Telethonin (titin-cap protein)	604488	
TMPO	Lamina-associated polypeptide 2, isoform alpha	188380	
TNNC1	Troponin C, slow skeletal and cardiac muscles	191040	

(continued)

Table 11.1 (continued)

Gene[a]	Protein[a]	OMIM[a]	Inheritance[a]
TNNI3	Troponin I, cardiac muscle	613932	AD
		191044	AR
TNNT2	Troponin T, cardiac muscle	191045	AD
TPM1	Tropomyosin alpha-1 chain	191010	AD
TTN	Titin	188840	
VCL	Vinculin	193065	
Hypertrophic cardiomyopathy			
ACTC1	Actin, alpha cardiac muscle 1	102540	AD
ACTN2	Alpha-actinin-2	102573	AD
ANKRD1	Ankyrin repeat domain-containing protein 1 (CARD)	609599	
CAV3	Caveolin-3	601253	AD
CRYAB	Alpha-crystallin B chain	123590	
CSRP3	Cysteine and glycine-rich protein 3	600824	AD
DES	Desmin	125660	
DSC2	Desmocollin-2	125645	
FHL1	Four and a half LIM domains protein 1	300163	
FLNC	Filamin-C	102565	
LDB3	LIM domain-binding protein 3	605906	AD
MYBPC3	Myosin-binding protein C, cardiac-type	600958	AD
MYH6	Myosin-6	160710	AD
MYH7	Myosin-7	160760	AD
MYL2	Myosin regulatory light chain 2, ventricular/cardiac muscle isoform	160781	AD
MYL3	Myosin light chain 3	160790	AD
MYLK2	Myosin light chain kinase 2, skeletal/cardiac muscle	606566	AD
MYO6	Unconventional myosin-VI	600970	AD
MYOZ2	Myozenin-2	605602	AD
MYPN	Myopalladin	608517	AD
NEXN	Nexilin	613121	AD
TCAP	Telethonin	604488	AD
TNNC1	Troponin C, slow skeletal and cardiac muscles	191040	AD
TNNI3	Troponin I, cardiac muscle	191044	AD
TNNT2	Troponin T, cardiac muscle	191045	AD
TPM1	Tropomyosin alpha-1 chain	191010	AD
TTN	Titin	188840	AD
VCL	Vinculin	193065	AD
Restrictive cardiomyopathy			
ACTC1	Actin, alpha cardiac muscle 1	102540	
BAG3	Bcl-2-associated athanogene 3	603883	
DES	Desmin	125660	
FLNC	Filamin-C	102565	AD
MYH7	Myosin-7	160760	

(continued)

Table 11.1 (continued)

Gene[a]	Protein[a]	OMIM[a]	Inheritance[a]
MYL2	Myosin regulatory light chain 2, ventricular/cardiac muscle isoform	160781	
MYL3	Myosin light chain 3	160790	
TNNCI	Troponin C, slow skeletal and cardiac muscles	191040	
MYPN	Myopalladin	608517	AD
TNNI3	Troponin I, cardiac muscle	191044	AD
TNNT2	Troponin T, cardiac muscle	191045	AD
TPM1	Tropomyosin alpha-1 chain	191010	
TTN	Titin	188840	
Arrhythmogenic right ventricular cardiomyopathy			
CTNNA3	Catenin alpha-3	607667	AD
DES	Desmin	125660	
DSC2	Desmocollin-2	125645	AD, AR
DSG2	Desmoglein-2	125671	AD
DSP	Desmoplakin	125647	AD
JUP	Junction plakoglobin	173325	AD
LDB3	LIM domain-binding protein 3	605906	
LMNA	Prelamin-A/C	150330	
PKP2	Plakophilin-2	602861	AD
PKP4	Plakophilin-4	604276	
STRN	Striatin	614765	
TMEM43	Transmembrane protein 43	612048	AD
TTN	Titin	188840	
Left ventricular non-compaction			
ACTC1	Actin, alpha cardiac muscle 1	102540	AD
ACTN2	Alpha-actinin-2	102573	
DMD	Dystrophin	300377	
DTNA	Dystrobrevin alpha	601239	AD
LDB3	LIM domain-binding protein 3	605906	AD
LMNA	Prelamin-A/C	150330	
MYBPC3	Myosin-binding protein C, cardiac-type	600958	AD
MYH7	Myosin-7	160760	AD
TNNI3	Troponin I, cardiac muscle	191044	
TNNT2	Troponin T, cardiac muscle	191045	AD
TPM1	Tropomyosin alpha-1 chain	191010	AD
TTN	Titin	188840	

[a]Data are edited from the following standard references: genes from HGNC, proteins from UniProt, and OMIM numbers and inheritance from OMIM

[b]*AD* autosomal dominant, *AR* autosomal recessive, *XL* X-linked, *XLR* X-linked recessive

(laminin subunit alpha-4 (LAMA4)), and other proteins (alpha-crystallin B chain (CRYAB), four and a half LIM domains protein 2 (FHL2)) have also been indicated to lead to DCM phenotypes (Table 11.1). Mechanistically, defects of power production have been concluded to be correlated with sarcomere protein-induced DCM, whereas cytoskeletal proteins are speculated to cause deficiencies of power propagation conducing to the DCM phenotype (Ehler and Perriard 2000; Debold et al. 2007). Besides, the abnormal desmosomal proteins seem to undermine the relationship between sarcomere, the intercalated disk, and Z-disk (Basso et al. 2006; Yang et al. 2006).

It is substantiated that anomalies of sarcomeres in force generation and transmission are considered to conduce to the development of DCM (Ehler and Perriard 2000; Fatkin and Graham 2002; Debold et al. 2007; McNally et al. 2013). Defects in the TTN gene are the major reason of DCM, and it encodes a pivotal and giant sarcomeric cytoskeletal protein, titin. Titin interacts with thin and thick filaments and plays an essential role in sarcomere assembly or in force generation and transmission, tendering a passive power and flexibility to maintain systolic and diastolic function. Pathogenic mutations in the TTN gene are primarily truncation variants (Herman et al. 2012), while the accurate pathogenic role of missense variants in TTN is not completely elucidated. Next, actin, alpha cardiac muscle 1 encoded by ACTC1, is a thin filament protein in sarcomere. It interplays with the troponin complex and tropomyosin and relates the sarcomere to the sarcolemma by combining to dystrophin. DCM-causing actin variations result in force transmission deficiencies via perturbing dystrophin-actin binding (Kamisago et al. 2000). Furthermore, actin interacts with myosin-7 and troponin T where variants lead to DCM via the abnormalities of force generation because of perturbing the actin-myosin interplay or changing cross-bridge activity during contraction and TNNT2 variants perturb calcium-sensitive troponin C binding (Olson et al. 1998; Kamisago et al. 2000; McNally et al. 2013). Likewise, pathogenic variants in MYH6 are correlated with DCM (Carniel et al. 2005; Merlo et al. 2013; Zhao et al. 2015). Also, tropomyosin alpha-1 chain variants destabilize actin interplay and damage force transmission to adjacent sarcomeres (Olson et al. 2001). Additionally, MYBPC3, associated with modulation of cardiac contraction, is expressed exclusively in myocardium and the mutation renders DCM (Ehlermann et al. 2008; Pugh et al. 2014; Wang et al. 2014). Prominently, troponin I encoded by TNNI3 is the inhibitory subunit in troponin complex and prohibit actin-myosin interplay, modulating the relaxation of myocardium, and it causes DCM when mutated (Carballo et al. 2009; Murakami et al. 2010; Memo et al. 2013). However, troponin C bound to calcium suppresses the inhibitory action of troponin I, therefore permitting the interplay of actin with myosin, the hydrolysis of ATP, and the production of tension. Similarly, mutations in TNNC1 contribute to the development of DCM (Mogensen et al. 2004; Dweck et al. 2008; Hershberger et al. 2010; Zhao et al. 2015). Additionally, RNA-binding protein 20 is crucial for normal splicing of many cardiac genes, playing an important role in regulating circular RNA production from the titin gene (Khan et al. 2016), and abnormalities in it lead to DCM (Brauch et al. 2009). Universally, obscurin encoded by OBSCN is showed to participate in the connection between the M-line of the sarcomere, the sarcoplasmic reticulum and an element of the sarcomere. It has been suggested to interact with titin, small ankyrin1, and myomesin (Fukuzawa et al. 2008; Kontrogianni-Konstantopoulos et al. 2009; Gautel 2011). And Steven Marston et al. have reported that OBSCN mutations are related to DCM due to haploinsufficiency (Marston et al. 2015). In addition, the MURC gene encodes caveolae-associated protein 4, promoting myofibrillar organization, and pathogenic mutations regarding it have been linked to DCM (Rodriguez et al. 2011; Andreasen et al. 2013). Ultimately, cofilin-2 encoded by CFL2 modulates actin-filament dynamics. In 2015, Subramanian et al. indicated that abnormal cofilin-2 is related to DCM (Subramanian et al. 2015).

Z-disk, which is important, is another molecular level target for DCM (Knoll et al. 2002) and symbolizes the boundaries of the sarcomeres, where actin filaments, nebulette filaments, and titin are anchored. Accordingly, one terminal of every actin thin filament can be fixed, and the sarcomeres are eligible for the contraction generation. In principle, variants in the protein Bcl-2-associated athanogene 3 encoded by BAG3 are linked to DCM (Norton et al. 2011; Villard et al. 2011; Feldman et al. 2014; Franaszczyk et al. 2014), and Fang et al. have identified that the E455K mutation in BAG3 perturb the interplay between Bcl-2-associated athanogene 3 and heat shock protein 70(HSP70), suggesting that the interaction between BAG3 and HSP70 is critical for BAG3 to stabilize HSPs and sustain the homeostasis of myocyte proteins (Fang et al. 2017). Further, the CARP protein encoded by ANKRD1 interacts with titin or myopalladin, and it is implicated in the myofibrillar stretch-sensor system. Also, it is a potential marker for disease development in idiopathic DCM or cardiac remodeling (Bogomolovas et al. 2015) and pathogenic variants in ANKRD1 are related to DCM (Duboscq-Bidot et al. 2009; Moulik et al. 2009; Ng et al. 2013). Again, LIM domain-binding protein 3 encoded by LDB3 plays an essential role in maintaining the structural integrity of Z-disk in heart muscle (Lin et al. 2014), and pathogenic variants in it are associated with DCM (Vatta et al. 2003; Andreasen et al. 2013). Intriguingly, variants in cysteine- and glycine-rich protein 3 encoded by CSRP3 perturb its interaction with telethonin and its stretch-sensor activity (Knoll et al. 2002). Nonetheless, mutations in CSRP3 result in aberrance in Z-disk architecture and T-tubule system (Mohapatra et al. 2003). Further, alpha-actinin-2 encoded by ACTN2 is the unique muscle isoform of alpha-actinin exclusively expressed in striated muscle. Variants in the protein undermine CSPR3 binding or stretch sensing and are involved in DCM (Andreasen et al. 2013; Ng et al. 2013; Pugh et al. 2014). In addition, nexilin encoded by NEXN is a filamentous actin-binding protein and is probably implicated in cell adhesion and migration and PDZ and LIM domain protein 3 encoded by the PDLIM3 gene functions in cytoskeletal assembly. Remarkably, pathogenic variants in these genes result in DCM (Arola et al. 2007; Hassel et al. 2009; Andreasen et al. 2013; Pugh et al. 2014). Next, nebulette encoded by NEBL is a Z-disk protein substantially expressed in myocardia and plays a critical role in mechanic sensing or power production through its interplay with tropomyosin-troponin complex and actin. And myopalladin encoded by MYPN interplays with nebulette as well as α-actinin and has a regulatory or structural function. Ultimately, telethonin encoded by TCAP is expressed in cardiac and skeletal muscle, and it is thought to be crucial to sarcomere assembly. In summary, pathogenic mutations in these genes are associated with DCM (Hirtle-Lewis et al. 2013; Maiellaro-Rafferty et al. 2013).

It is surmised that cytoskeletal proteins cause deficiencies of power propagation inducing the DCM phenotype. Among them, intermediate filaments play an important role in cytoskeleton connecting the Z-disk with the sarcolemma and dystrophin-glycoprotein complex (DGC) function in the connection between the F-actin cytoskeleton and the extracellular matrix. What's more, defects of proteins connecting the sarcomere with the extracellular matrix are susceptible to impair force transmission. In detail, desmin encoded by DES is a muscle-specific class III intermediate filament protein, and pathogenic variants in it have been shown to be related to DCM (Andreasen et al. 2013; Zhao et al. 2015; Brodehl et al. 2016). Then, delta-sarcoglycan encoded by SGCD is expressed considerably in striated muscle and is one of the components of the sarcoglycan complex, a subcomplex of DGC. And pathogenic mutations in delta-sarcoglycan have an important relationship with the onset of DCM (Tsubata et al. 2000; Karkkainen et al. 2003; Pugh et al. 2014). Also, vinculin (expressed ubiquitously) and its splice variant, metavinculin (primarily expressed in smooth and cardiac muscle), both of them encoded by VCL, are cytoskeletal proteins linked to cell-cell and cell-matrix junctions and are pivotal constituents of multiple protein assemblies connecting the extracellular matrix to the actin cytoskeleton. They offer attachment of thin filaments to the cell membrane and critically, function in propagating productive force. Therefore, the aberrance in VCL gene has also been associated with DCM (Olson et al. 2002; Pugh et al. 2014). Further, alpha-crystallin B chain encoded by the CRYAB gene participates in the intracellular architecture and pathogenic mutations in the gene have been related to DCM (Inagaki et al. 2006; Pilotto et al. 2006; Andreasen et al. 2013). Still, the LAMA4 gene encodes laminin subunit alpha-4, one of the members in laminins family, being recognized to be extracellular matrix glycoproteins and the main non-collagenous component of basement membranes, and the pathogenic defects in the gene have been indicated to render the development of DCM (Knoll et al. 2007; Andreasen et al. 2013). At last, four and a half LIM domains protein 2 encoded by FHL2 is considered to have an influence on the assembly of extracellular membranes and malignant mutations in this gene could be a cause of DCM through the impairment of recruiting metabolic enzymes to the sarcomere (Arimura et al. 2007).

In the nuclear envelope, mutations of some genes encoding nuclear membrane proteins have been associated with the pathogenesis of DCM. The LMNA gene encodes prelamin-A/C, one of the components of the inner nuclear membrane protein complex. The interactions between the lamin and integral proteins of the inner membrane maintain the integrity or stability of nuclear envelop. Markedly, LMNA is the second gene accounting for DCM (Hershberger and Siegfried 2011), and it is inherited primarily in an autosomal dominant inheritance pattern. In addition, the extraordinary importance of the encoded protein in electrophysiologic cells is recognized by the prominent electrophysiologic anomalies (progressive atrioventricular block and atrial arrhythmias) detected in variants of prelamin-A/C. Then, lamina-associated polypeptide 2, isoform alpha encoded by TMPO, plays a crucial role in preserving the structural organization of the nuclear membrane. Finally, nesprin-1 encoded by SYNE1 and nesprin-2 encoded by SYNE2 concentrate on the nuclear membrane. They are constituents of the linker of nucleoskeleton and cytoskeleton (LINC) complexes, constituting a physical connection between the nucleus and the cytoskeleton. As a general rule, the pathogenic mutations of the proteins in the nuclear envelope related to DCM is infrequent (Taylor et al. 2005; Puckelwartz et al. 2010).

Several pathogenic variants of DSC2, DSG2, DSP, PKP2, etc. (Table 11.1) in desmosomes, assisting in withstanding shearing forces as cell-cell junctions and observed in convergence in cells susceptible to mechanical stress, lead to familial DCM (Elliott et al. 2010; Garcia-Pavia et al. 2011; Andreasen et al. 2013; Ng et al. 2013). Concretely, desmocollin-2 (encoded by DSC2) and desmoglein-2 (encoded by DSG2) are predominant constituents of the desmosome. In common, desmoplakin is encoded by DSP, and the C-terminal region of it connects with intermediate filaments. Conversely, the N-terminus of it interplays with the N-terminus of plakoglobin or plakophilin 1 and is necessary for localizing to the desmosome. Further, the desmoplakin-intermediate filament junction regulates cell mechanics (Broussard et al. 2017). Of note, several mutations in DSP have been correlated with DCM (Garcia-Pavia et al. 2011). Eventually, plakophilin-2 (encoded by PKP2) lying in desmosomes and nuclei of cells makes a prominent effect on connecting cadherins to intermediate filaments in the cytoskeleton.

Curiously, the inheritance patterns of a small part of DCM cases are elucidated as an autosomal recessive pattern, like variants in gene TNNI3, DES, DSP, FKRP, FKTN, PLEC, etc. In advance, TNNI3 and DES (mentioned before) have both autosomal dominant and autosomal recessive pattern (Murphy et al. 2004; McLaughlin et al. 2013) in DCM. Next, desmoplakin encoded by DSP is one of the main elements of desmosomes and recessive mutations have an impact on DCM (Norgett et al. 2000; Ramoglu et al. 2017). Then, the FKRP gene encodes fukutin-related protein, essential for posttranslational modification of dystroglycan. Pathogenic mutations in the gene, as a general rule, have been correlated with limb-girdle muscular dystrophy. Nonetheless, a family suffering from DCM and without muscular abnormality has also been reported (Muller et al. 2005). Further, mutations in fukutin (encoded by FKTN) have also been associated with DCM (Murakami et al. 2006). Lastly, plectin encoded by PLEC functions to preserve cell or tissue integrity as well as coordinating dynamic alterations in cytoarchitecture or cell shape and can connect different elements of the cytoskeleton. Interestingly, variants in PLEC is also related to DCM with epidermolysis bullosa simplex and muscular dystrophy (Bolling et al. 2010).

Few DCM cases following an X-linked inheritance pattern have been published, like mutations in gene DMD and EMD. Principally, dystrophin encoded by DMD connects the sarcomere to the sarcolemma and extracellular matrix (ECM), and when mutated, it leads to DCM (Milasin et al. 1996; Rimessi et al. 2005; Kimura et al. 2007; Diegoli et al. 2011). Factually, this gene was the first delineated linked to DCM (Muntoni et al. 1993). Besides, emerin encoded by EMD is one of the members of the inner nuclear membrane protein complex and plays an essential part in membrane anchoring to the cytoskeleton. Deficiencies in emerin result in X-linked Emery-Dreifuss muscular dystrophy and DCM (Mook et al. 2013; Zhao et al. 2015), etc.

6.1.2 Variations in Genes Leading to HCM

Genetic causes for HCM have thrown light on more than 20 genes disclosed in force generation and transmission, mostly affecting the sarcomere proteins (actin, alpha cardiac muscle 1 (ACTC1), filamin-C (FLNC), myosin-binding protein C (MYBPC3), myosin-6 (MYH6), myosin-7 (MYH7), myosin regulatory light chain 2 (MYL2), myosin light chain 3 (MYL3), unconventional myosin-6 (MYO6), myopalladin (MYPN), troponin C (TNNC1), troponin I (TNNI3), troponin T (TNNT2), tropomyosin alpha-1 chain (TPM1), titin (TTN) and Z-disk proteins (alpha-actinin-2 (ACTN2), ankyrin repeat domain-containing protein 1 (ANKRD1), cysteine- and glycine-rich protein 3 (CSRP3), LIM domain-binding protein 3 (LDB3), myozenin-2 (MYOZ2), nexilin (NEXN), and telethonin (TCAP)); nevertheless, defects in genes encoding proteins of the cytoskeleton (caveolin-3 (CAV3), vinculin (VCL)), or desmosome (desmocollin-2, DSC2) have also been recognized (Table 11.1). HCM is often depicted as a disease of the sarcomere and frequently inherited as an autosomal dominant trait.

More than ten causative genes encode sarcomere proteins, with pathogenic mutations observed in nearly all sarcomere proteins required for generating the molecular force of contraction in cardiomyocytes, predominantly the ones encoded by MYBPC3 and MYH7 genes (each up to 40% in HCM frequency) (Sabater-Molina et al. 2018). However, the pathogenic variants in other genes containing MYL2, MYL3, ACTC1, TNNI3, TNNT2, TPM1, etc. are responsible for a greatly smaller proportion of cases causing HCM (Sabater-Molina et al. 2018). The introductions of proteins encoded by ACTC1, MYBPC3, MYH6, MYH7, TNNI3, TTN, etc. have been mentioned before. Initially, filamins encoded by FLNC are actin-binding proteins participating in reshaping of the cytoskeleton. Then, myosin regulatory light chain 2 encoded by MYL2 plays a critical part in stabilizing the long alpha helical neck of the myosin head and myosin light chain 3 encoded by MYL3 is a structural component of the myosins. Commonly, unconventional myosin-VI encoded by MYO6 is an actin-based molecular motor or unconventional myosins implicated in organelle transport and intracellular vesicle. Additionally, myopalladin encoded by MYPN is an element of the sarcomere that tethers nebulin in muscle. Obviously, pathogenic mutations in sarcomere encoding genes (Table 11.1) are associated with the pathogenesis of HCM (Valdes-Mas et al. 2014; Mogensen et al. 2015; Towe et al. 2015; Ntusi et al. 2016; Chen et al. 2017; Marian and Braunwald 2017; Mendes et al. 2017; Walsh et al. 2017; Frustaci et al. 2018; Sabater-Molina et al. 2018).

In addition, variants in Z-disk and other non-sarcomere encoding genes have been identified in HCM cases, such as in ACTN2, ANKRD1, CAV3, CRYAB, CSRP3, DES, DSC2, FHL1, LDB3, MYOZ2, NEXN, TCAP, VCL, etc. (Table 11.1). Definitely, caveolin-3 encoded by CAV3 is converged in subsarcolemmal regions essential for the electrical transmission of the contractile impulse and belongs to the dystrophin-glycoprotein complex, conferring stability to the muscle cell membrane. Again, myozenin-2 encoded by MYOZ2 interacts and colocalize with alpha-actinin and thus attach calcineurin to the sarcomere in striated muscle. Dramatically, pathogenic mutations in these genes (Table 11.1), likewise, are related to HCM (Bos et al. 2006; Osio et al. 2007; Arimura et al. 2009a, b; Fratev et al. 2014; Girolami et al. 2014; Haywood et al. 2016; Rubattu et al. 2016; San et al. 2016; Marian and Braunwald 2017; Mendes et al. 2017; Walsh et al. 2017; Sabater-Molina et al. 2018).

6.1.3 Defects in Genes Related to RCM

Ordinarily, RCM can be an inherited disease or an acquired disorder and frequently a combination with a systemic condition (Muchtar et al. 2017). In inherited RCM, pathogenic mutations are described in the sarcomere protein genes ACTC1 (Kaski et al. 2008), FLNC (Tucker et al. 2017), MYH7, TNNI3 (Rai et al. 2009), MYL2, MYL3, TPM1 (Caleshu et al. 2011), MYPN, TNNT2, and TTN (Peled et al. 2014) or non-sarcomere protein genes BAG3 (Schanzer et al. 2018), DES, etc. (Table 11.1). Notably, RCM can manifest as desmin-related cardiomyopathy attributable to variants in the gene DES. RCM often has autosomal dominant inheritance; however, autosomal recessive pattern occurs (Ploski et al. 2016).

6.1.4 Deficiencies in Genes Causing ARVC

ARVC, one of the hereditary cardiomyopathies, is frequently inherited as an autosomal dominant pattern, whereas recessive dominant pattern of the disorder equally happens (McKoy et al. 2000). The pathogenesis of ARVC is mainly attributed to mutations in the genes encoding desmosomal proteins (like DSP, DSG2, DSC2, PKP2, PKP4) (Castelletti et al. 2017; Liu et al. 2017; Tsatsopoulou 2017) and other cell-to-cell junctional apparatus proteins (like CTNNA3, JUP, STRN); however, sparse pathogenic defects in genes unrelated to cell-to-cell junctional apparatus such as DES, LDB3, TTN, LMNA, and TMEM43 (Table 11.1) also contribute to the development of ARVC (Lazzarini et al. 2015; Lopez-Ayala et al. 2015; AbdelWahab et al. 2018). The majority of the genes have been introduced before. Particularly, plakophilin-4 encoded by PKP4 is one of the members in plakophilin subfamily. Then, catenin alpha-3 encoded by CTNNA3 is a constituent of a unique hybrid adhering junction that involves proteins related to either adherent junctions or desmosomes and an adhesion molecule in intercalated disks of the heart (Li et al. 2012). Fundamentally, junction plakoglobin encoded by JUP is an essential component of the desmosomes and the intermediate junctions. Mutations in the gene were the initial disease-causing variants to be observed in patients with Naxos disease. And the characteristics of ARVC as a cell adhesion condition were initially proposed by the patients with Naxos disease, an autosomal recessive cardiocutaneous disease (McKoy et al. 2000). And yet, mutations in the JUP gene can be inherited as an autosomal dominant pattern in ARVC. Moreover, striatin encoded by STRN is a component of cell junction and transmembrane protein 43 encoded by TMEM43 is a greatly conserved nuclear membrane protein. In general, pathogenic variants in the foregoing genes are correlated with ARVC.

6.1.5 Variations in Genes Associated with LVNC

Rudimentarily, LVNC can be a distinct condition as well as be in association with other cardiomyopathies or many congenital disorders (Arbustini et al. 2016; Weir-McCall et al. 2016). The disease-causing genes of LVNC basically include sarcomere encoding genes (e.g., ACTC1, MYBPC3, MYH7, TNNI3, TNNT2, TPM1, TTN) (Finsterer and Stollberger 2015; Yang et al. 2015; Yoshida et al. 2016; Miszalski-Jamka et al. 2017; van Waning et al. 2018), and non-sarcomeric genes also play a part in LVNC (e.g., ACTN2, LDB3, DMD, DTNA, LMNA, EMD) (Bagnall et al. 2014; Parent et al. 2015; Liu et al. 2016; Shan et al. 2017) (Table 11.1). Of note, mutations in MYH7, MYBPC3, and TTN are the most prevalent mutations (71%) observed in genetic LVNC, and variants are more common in children (44%) than in adults (30%) (van Waning et al. 2018). Concretely, dystrobrevin alpha encoded by DTNA is homologous to some extent to dystrophin, and defects in it have been associated with LVNC (Cao et al. 2017). And evidently, LVNC is prevalent in patients with Duchenne or Becker muscular dystrophy because of mutations in DMD gene (Kimura et al. 2013).

6.2 Defects of Genes in Energy Production and Modulation or Metabolism

6.2.1 Mutations in Genes Related to DCM

In energy production and modulation or metabolism, mutations in several genes encoding proteins like DNAJC19, DNM1L (Ashrafian et al. 2010), LAMP2, SDHA, TAZ, TTR (Parry et al. 2015), and TXNRD2 (Table 11.2) are linked to inherited DCM. Originally, mutations in the DNAJC19 gene promote the development of DCM associated with ataxia syndrome (Al et al. 2016; Ucar et al. 2017). Defects in DNAJC19 encoding protein impair the energy production and mitochondrial survival as result of changing the protein transportation in mitochondria. Also, dynamin-1-like protein encoded by DNM1L is one of the proteins in the dynamin superfamily of GTPases and plays a crucial role in modulating mitochondrial and peroxisomal fission. Next, defects in thioredoxin reductase 2 encoded by TXNRD2 result in disturbed mitochondrial redox homeostasis contributing to DCM (Horstkotte et al. 2011; Sibbing et al. 2011). Additionally, succinate dehydrogenase [ubiquinone] flavoprotein subunit, mitochondrial encoded by SDHA is localized in the inner membrane of the mitochondria and is one of the components of

Table 11.2 Genetics in energy production and modulation or metabolism

Gene[a]	Protein[a]	OMIM[a]	Inheritance[a]
Dilated cardiomyopathy			
DNAJC19	DnaJ homolog subfamily C member 19	608977	
DNM1L	Dynamin-1-like protein	603850	
LAMP2	Lysosome-associated membrane glycoprotein 2	309060	
SDHA	Succinate dehydrogenase [ubiquinone] flavoprotein subunit, mitochondrial	600857	
TAZ	Tafazzin	300394	XLR[b]
TTR	Transthyretin	176300	
TXNRD2	Thioredoxin reductase 2	606448	
Hypertrophic cardiomyopathy			
COX15	Cytochrome c oxidase assembly protein COX15 homolog	603646	
GLA	Alpha-galactosidase A	300644	XL[b]
LAMP2	Lysosome-associated membrane glycoprotein 2	309060	
PRKAG2	5′-AMP-activated protein kinase subunit gamma-2	602743	AD[b]
TAZ	Tafazzin	300394	XLR
TTR	Transthyretin	176300	
Restrictive cardiomyopathy			
GLA	Alpha-galactosidase A	300644	XL
TTR	Transthyretin	176300	
Left ventricular non-compaction			
NNT	Nicotinamide nucleotide transhydrogenase	607878	
NUBPL	Nucleotide-binding protein-like protein	613621	
TAZ	Tafazzin	300394	XLR

[a]Data are edited from the following standard references: genes from HGNC, proteins from UniProt, and OMIM numbers and inheritance from OMIM

[b]*AD* autosomal dominant, *XL* X-linked, *XLR* X-linked recessive

succinate-ubiquinone oxidoreductase. Deficiencies in SDHA have been related to familial neonatal DCM (Levitas et al. 2010). Then, lysosome-associated membrane glycoprotein 2 encoded by LAMP2 is one of the members in the lysosomal membrane glycoproteins and makes a critical impact on conservation, defense, and adhesion of the lysosome. Defects in this gene may cause Danon disease, an X-linked dominant disorder, manifested as failure of heart and skeletal muscles. And DCM is usually a phenotype in Danon disease (Taylor et al. 2007), and variants in LAMP2 have also been found in DCM patients (Mook et al. 2013) Ultimately, tafazzin encoded by TAZ can maintain structure and play an important role in energy metabolism in cell via catalyzing a phospholipid synthesis in the inner mitochondrial membrane (Houtkooper et al. 2009). Pathogenic mutations in this gene alter the metabolism of phospholipid, thereby changing structure of the mitochondrial membrane and functions, and have been published correlated with DCM (Man et al. 2013), as the gene is generally related to Barth syndrome which can present as DCM, LVNC, or, less common, HCM (Baksiene et al. 2016).

6.2.2 Variations in Genes Associated with HCM

It is corroborated that a couple of gene mutations in metabolism are etiologies of hypertrophic cardiac remodeling and HF in HCM, such as COX15, GLA, LAMP2 (Fu et al. 2016), PRKAG2, TAZ (Zhao et al. 2017), etc. (Table 11.2). Basically, 5′-AMP-activated protein kinase subunit gamma-2 encoded by PRKAG2 is one of the component of AMP-activated protein kinase (AMPK), activated by different cellular stresses. When activated, AMPK turns on catabolic pathways and turns off a number of ATP-consuming processes. Hence, AMPK is recognized as a metabolic stress sensor and regulates a variety of proteins regarding cell energy and metabolism. Mutations in PRKAG2 are associated with HCM (Calore 2017). In addition, alpha-galactosidase A encoded by GLA is a lysosomal hydrolase, and the majority of GLA variants may lead to Fabry disease, a multisystemic disorder which can have a phenotype of cardiac hypertrophy (Adalsteinsdottir et al. 2017), but some deficiencies conduce to HCM (Marian and Braunwald 2017; Sabater-Molina et al. 2018). Eventually, cytochrome c oxidase assembly protein COX15 homolog encoded by COX15, one of the COX assembly factors found in yeast, functions in the mitochondrial heme biosynthetic pathway. One study has established COX15 is an additional cause of fatal infantile HCM, linked to isolated COX deficiency (Antonicka et al. 2003).

6.2.3 Defects in Genes Leading to RCM

Some gene mutations in energy production and regulation or metabolism are correlated with RCM, like GLA (Muchtar et al. 2017) and TTR (Table 11.2). Specifically, transthyretin is encoded by TTR, and mutant TTR gene brings about misfolded proteins (amyloid deposition) leading to extracellular deposits impairing organ function. Functionally, infiltrative amyloid deposition induces RCM (diastolic dysfunction), and subsequently systolic HF and TTR are linked to amyloid-related RCM (Mueller et al. 2010; Sen-Chowdhry et al. 2010; Muchtar et al. 2017).

6.2.4 Deficiencies in Genes Linked to LVNC

Nonetheless, variants in several genes in energy production and modulation or metabolism have a relationship with LVNC, such as NNT (Bainbridge et al. 2015), NUBPL, and TAZ (Miszalski-Jamka et al. 2017) (Table 11.2). On the whole, NNT encodes nicotinamide nucleotide transhydrogenase, localized in inner mitochondrial membrane, and these proteins make an influence on catalyzing the transfer of a hydride ion in the energy-transfer system of the respiratory chain. Curiously, nucleotide-binding protein-like protein encoded by NUBPL plays a crucial part in the assembly of respiratory complex I, a large mitochondrial inner membrane enzyme, and the mutations in NUBPL may contribute to the development of LVNC (Zhou et al. 2018).

6.3 Variations of Genes in Calcium Cycling and Ion Channels

It is illuminated that anomalies in calcium homeostasis and defects in cardiac ion channel proteins are related to the etiologies of inherited cardiomyopathies, such as DCM, HCM, ARVC, and LVNC. What's more, quite a lot studies suggest that dysfunction of cardiac calcium cycling and ion channel proteins to be a common and fundamental mechanism for HF and several mutations in the genes encoding cardiac calcium cycling and ion channel proteins have been found associated with inherited cardiomyopathies. Commonly, via the voltage-gated calcium channels in the sarcolemma, calcium ion gets into the cardiomyocytes, and sarcomere contraction is induced by this influx of calcium proceeding with the opening of the ryanodine receptor (RyR) calcium channel in the sarcoplasmic reticulum (SR). On the contrary, the relaxation of the myocardia is initiated, when the released cytoplasmic calcium ion is reabsorbed into SR through the SR calcium-ATPase pump (SERCA2a). The SR or endoplasmic reticulum (ER), a network of tubules and sacs in striated muscle, functions in muscular contraction and relaxation in virtue of releasing and storing calcium. Besides, ion channels, furthering the diffusion of ions across the membranes of the organelles and cells (e.g., myocardial cells), are protein complexes embedded in the membranes. Significantly, pathogenic defects in calcium cycling and ion channel proteins in cardiomyocytes participate in the disturbance of muscular contraction.

6.3.1 Defects in Genes Linked to DCM

A couple of variants in the genes regarding calcium cycling and ion channels are related to DCM, like ABCC9, CLIC2, KCNQ1 (Xiong et al. 2015), PLN, PSEN1, PSEN2, and SCN5A (Table 11.3).Firstly, ABCC9 encodes ATP-binding cassette subfamily C member 9, which is one of the members of ATP-binding cassette (ABC) transport proteins, transporting different molecules between intra- and extracellular membranes, and pathogenic mutations in the gene are correlated with DCM (Bienengraeber et al. 2004; Pugh et al. 2014). Secondarily, potassium voltage-gated channel subfamily KQT member 1 encoded by KCNQ1 is essential for the repolarization in the heart. Next, cardiac phospholamban encoded by PLN is localized in SR and a key regulator of cardiac diastolic function. Definitely, cardiac relaxation occurs with the decline of intracellular concentrations of calcium because reuptake into the SR occurs through SERCA2a. Then phospholamban whose activity is prohibited by protein kinase A (PKA)-mediated phosphorylation regulates calcium recycling by suppressing SERCA2a. As a rule, the pathogenic mutations in PLN are related to DCM (Schmitt et al. 2003; Ng et al. 2013; Pugh et al. 2014). Again, PSEN1 encodes presenilin-1 and PSEN2 encodes presenilin-2. The presenilins are expressed in the heart and play an essential part in cardiac development. Mechanistically, PSEN1 coimmunoprecipitates with SERCA2a, and the PSEN1 oligomer interplays straight with the calcium channel (Gianni et al. 2010). And

Table 11.3 Genetics in calcium cycling and ion channels

Gene[a]	Protein[a]	OMIM[a]	Inheritance[a]
Dilated cardiomyopathy			
ABCC9	ATP-binding cassette subfamily C member 9	601439	
CLIC2	Chloride intracellular channel protein 2	300138	
KCNQ1	Potassium voltage-gated channel subfamily KQT member 1	607542	
PLN	Cardiac phospholamban	172405	
PSEN1	Presenilin-1	104311	AD[b]
PSEN2	Presenilin-2	600759	AD
SCN5A	Sodium channel protein type 5 subunit alpha	600163	AD
Hypertrophic cardiomyopathy			
JPH2	Junctophilin-2	605267	AD
PLN	Cardiac phospholamban	172405	AD
Arrhythmogenic right ventricular cardiomyopathy			
KCNQ1	Potassium voltage-gated channel subfamily KQT member 1	607542	
RYR2	Ryanodine receptor 2	180902	AD
Left ventricular non-compaction			
ABCC9	ATP-binding cassette subfamily C member 9	601439	
CASQ2	Calsequestrin-2	114251	
HCN4	Potassium/sodium hyperpolarization-activated cyclic nucleotide-gated channel 4	605206	
KCNH2	Potassium voltage-gated channel subfamily H member 2	152427	
RYR2	Ryanodine receptor 2	180902	
SCN5A	Sodium channel protein type 5 subunit alpha	600163	

[a]Data are edited from the following standard references: genes from HGNC, proteins from UniProt, and OMIM numbers and inheritance from OMIM
[b]*AD* autosomal dominant

mutations in these genes contribute to the pathogenesis of DCM and HF (Li et al. 2006; Song et al. 2018). In addition, SCN5A encodes sodium channel protein type 5 subunit alpha, and sodium channels play a pivotal role in the fast depolarization of the heart muscles and in the conservation of impulse conduction. Pathogenic mutations in this gene contribute to ventricular remodeling and DCM (Nguyen et al. 2008; Weintraub et al. 2017) by inducing an anomaly in the currents of the various ions implicated in the contraction of the myocardial cells. Finally, chloride intracellular channel protein 2 encoded by CLIC2 participates in the modulation of calcium homeostasis in myocardial cells by interplaying with the cardiac ryanodine receptor 2 (RYR2) (Molina-Navarro et al. 2013).

6.3.2 Mutations in Genes Correlated with HCM

Involved in myocyte calcium homeostasis and ion channels, defects in some genes are linked to HCM, such as JPH2 (Landstrom et al. 2007; Marian and Braunwald 2017) and PLN (Landstrom et al. 2011; Marian and Braunwald 2017) (Table 11.3). Specifically, JPH2 encodes junctophilin-2, one of the subtypes in junctophilins which are constituents of the junctional complexes. Junctional complexes between the cell membrane and SR/ER are a characteristic of all excitable cells, and it is physically close to the plasmalemmal L-type calcium channel or SR.

6.3.3 Deficiencies in Genes Causing ARVC

Certain variants in the genes RYR2 result in ARVC (Roux-Buisson et al. 2014), and KCNQ1 may be a possible causal gene for ARVC (Xiong et al. 2015). Particularly, RYR2 encodes ryanodine receptor 2, the main source of calcium essential for excitation-contraction coupling in the myocardia. And the phosphorylation of RYR2 separates FKBP12.6 and modulates the channel open probability.

6.3.4 Variations in Genes Leading to LVNC

It is deemed that defects in some genes (e.g., ABCC9 (Waldmuller et al. 2015), CASQ2 (Egan et al. 2013), HCN4, KCNH2, RYR2 (Ohno et al. 2014), SCN5A (Shan et al. 2008) (Table 11.3)) are related to LVNC. Explicitly, CASQ2 encodes calsequestrin-2, an isoform of calsequestrin in SR of the cardiac muscles and a component of a protein complex that involves the ryanodine receptor. Calsequestrin 2 functions as the main Ca^{2+} reservoir in SR of the myocardial cells. Then, HCN4 encodes potassium/sodium hyperpolarization-activated cyclic nucleotide-gated channel 4, expressed in the sinoatrial node and producing the cardiac pacemaker current or regenerating cardiac action potentials, and some of the HCN4 mutations may play a part on LVNC (Milano et al. 2014; Schweizer et al. 2014), albeit a research indicated that mutations HCN4 (G811E) may not be a monogenic factor to lead to the cardiac conditions (Yokoyama et al. 2018). Lastly, KCNH2 encodes potassium voltage-gated channel subfamily H member 2, having a marked impact on the ultimate repolarization of the ventricular action potential, and the mutations in KCNH2 may cause LVNC regardless of the mechanism remaining unclear (Miszalski-Jamka et al. 2017).

6.4 Defects in Genes in Transcriptional Regulation and Other Cytological Processes

In a general way, several genes encoding cardiac transcription factors (TFs) play an essential role in cardiogenesis and structural remodeling. Frequently, defects in these genes cause cardiac malformations, and the aberrant heart remodeling contributes to the development of inherited cardiomyopathies and HF. For the most part, pathogenic mutations in the genes of transcriptional regulation and other cytological processes (e.g., cellular differentiation) make a difference in some types of inherited cardiomyopathies, such as DCM, HCM, ARVC, and LVNC.

6.4.1 Variations in Genes Leading to DCM

Deficiencies in EYA4, FOXD4, GATA4, GATA5, GATA6, GATAD1, NKX2-5, RAF1, RXRA (RXRα), TBX5, and TBX20 are correlated with familial or sporadic DCM (Table 11.4). Initially, the EYA4 gene encodes eyes absent homolog 4, one of the members in eyes absent (EYA) family. Crucially, this protein functions as a transcriptional activator via its protein phosphatase activity, and it is essential for the development of the eyes or organ of Corti. From OMIM, the pattern of the gene is autosomal dominant inheritance in DCM. Defects in the EYA4 gene lead to DCM and sensorineural hearing loss (Schonberger et al. 2005), and it also actuates hypertrophy through modulation of p27kip1 in DCM (Williams et al. 2015). Next, forkhead box protein D4 encoded by the FOXD4 gene, one of the members in winged-helix transcription factors family of TFs, regulating embryonal development and tissue differentiation, and a pathogenic variant, is related to inherited DCM with obsessive-compulsive disorder and suicidality (Minoretti et al. 2007). Furthermore, transcription factor GATA-4 encoded by GATA4, transcription factor GATA-5 encoded by GATA5, and transcription factor GATA-6 encoded by GATA6 involve zinc finger motifs and modulate the expression of structural and regulatory genes in the heart.TF GATA-4 controls gene expression crucial for myocardial differentiation and function, containing sarcomere proteins and brain-type natriuretic factor. Noticeably, several variants in GATA4, GATA5, and GATA6 make a critical

Table 11.4 Genetics in transcriptional regulation and other cytological processes

Gene[a]	Protein[a]	OMIM[a]	Inheritance[a]
Dilated cardiomyopathy			
EYA4	Eyes absent homolog 4	603550	AD[b]
FOXD4	Forkhead box protein D4	601092	
GATA4	Transcription factor GATA-4	600576	
GATA5	Transcription factor GATA-5	611496	
GATA6	Transcription factor GATA-6	601656	
GATAD1	GATA zinc finger domain-containing protein 1	614518	AR[b]
NKX2-5	Homeobox protein Nkx-2.5	600584	
RAF1	RAF proto-oncogene serine/threonine-protein kinase	164760	AD
RXRA	Retinoic acid receptor RXR -alpha	180245	
TBX5	T-box transcription factor TBX5-A	601620	
TBX20	T-box transcription factor TBX20	606061	
Hypertrophic cardiomyopathy			
RAF1	RAF proto-oncogene serine/threonine-protein kinase	164760	
TRIM63	E3 ubiquitin-protein ligase TRIM63	606131	
Arrhythmogenic right ventricular cardiomyopathy			
TGFB3	Transforming growth factor beta-3	190230	AD
Left ventricular non-compaction			
MIB1	E3 ubiquitin-protein ligase MIB1	608677	AD

[a]Data are edited from the following standard references: genes from HGNC, proteins from UniProt, and OMIM numbers and inheritance from OMIM

[b]*AD* autosomal dominant, *AR* autosomal recessive

influence on the pathogenesis of DCM (Li et al. 2013, 2014; Xu et al. 2014; Zhao et al. 2014; Zhang et al. 2015a, b). Again, RAF proto-oncogene serine/threonine-protein kinase is encoded by RAF1, one of the genes in rat sarcoma-mitogen-activated protein kinase (RAS-MAPK) pathway, playing a significant role in myocardial biology, cell division, and differentiation (Sala et al. 2012), and RAF1 mutations cause childhood-onset DCM (Dhandapany et al. 2014). What's more, NKX2-5, TBX5, and TBX20 genes encode TFs which induce the transcription of a series of genes making a key influence on cardiac development and structural remodeling, and equally, a couple of pathogenic variants contribute to the development of DCM (Yuan et al. 2015; Zhang et al. 2015a, b; Zhou et al. 2015, 2016; Hanley et al. 2016; Zhao et al. 2016; Xu et al. 2017). Additionally, RXRA encodes retinoic acid receptor RXR alpha, a member of the retinoid X receptors, retinoid-dependent transcriptional regulators, playing a critical role in embryonic and fetal development. Predominantly, RXRα overexpression in myocardial cells leads to DCM (Subbarayan et al. 2000). Ultimately, GATA zinc finger domain-containing protein 1 encoded by GATAD1 modulates gene expression via combining with a histone modification site. A pathogenic variant in GATAD1 is correlated with familial DCM in an autosomal recessive inheritance pattern (Theis et al. 2011).

6.4.2 Defects in Genes Linked to HCM

Variations in the gene RAF1 (Wang et al. 2016; Thompson et al. 2017) and TRIM63 conduce to HCM. Usually, TRIM63 encodes E3 ubiquitin-protein ligase TRIM63, namely, muscle-specific ring finger protein 1(MURF1), expressed selectively in striated muscles such as myocardia. The RING finger motif, inducing protein-DNA and protein-protein interplay, participates in a multitude of functions, like transcriptional repression, ubiquitination, etc. TRIM63 is able to mediate polyubiquitination and degradation of some thick filament proteins (Kedar et al. 2004; Clarke et al. 2007), and the mutation in TRIM63 causes human HCM (Chen et al. 2012).

6.4.3 Deficiencies in Genes Correlated with AC

TGFB3 encodes transforming growth factor beta-3, a cytokine member that participates in multiple physiological processes, such as cardiac development, proliferation, differentiation, etc., and it regulates cardiac fibrosis and remodeling. What's more, variants in TGFB3 play an important role in the development of AC (Tamargo 2012).

6.4.4 Mutations in Genes Related to LVNC

E3 ubiquitin-protein ligase MIB1 encoded by MIB1 is an E3 ubiquitin ligase that makes an essential effect on modulating the endocytosis of Notch ligands, necessary for the activation of Notch signaling, controlling cell fate specification, differentiation, and patterning. Arrestingly, expression of MIB2 mutations influenced Notch signaling in primary rat cardiomyocytes (Piccolo et al. 2017), and defects in MIB1 are responsible for LVNC in an autosomal dominant pattern (Luxan et al. 2013).

7 Conclusion

HF is a life-threatening disorder and a main cause of death worldwide. Despite acquired conditions, HF could be induced by inherited cardiomyopathies, such as DCM, HCM, RCM, ARVC, and LVNC. Profound insight into genetic background and genotype-phenotype relationships of the inherited cardiomyopathies may shed further light on the pathogenic pathways or pathophysiology of HF and contribute to the innovation of therapeutics. Furthermore, complete profiling of the fundamental prevalent and sparse genetic variation of inherited forms in HF will improve diagnosis, therapies, prognosis, and personalized counselling. Dramatically, GWAS has facilitated the development of molecular genetics in a variety of diseases, and employment of NGS has identified multiple novel genes, beneficial to elucidate the pathophysiological pathways. In the future, a multitude of further exploration of the molecular pathways and the complicated mechanisms of gene expression modulation are essential for the diagnosis and management of HF.

References

AbdelWahab A, Gardner M, Parkash R, Gray C, Sapp J. Ventricular tachycardia ablation in arrhythmogenic right ventricular cardiomyopathy patients with TMEM43 gene mutations. J Cardiovasc Electrophysiol. 2018;29(1):90–7.

Ackerman MJ, Priori SG, Willems S, Berul C, Brugada R, Calkins H, Camm AJ, Ellinor PT, Gollob M, Hamilton R, Hershberger RE, Judge DP, Le Marec H, McKenna WJ, Schulze-Bahr E, Semsarian C, Towbin JA, Watkins H, Wilde A, Wolpert C, Zipes DP. HRS/EHRA expert consensus statement on the state of genetic testing for the channelopathies and cardiomyopathies this document was developed as a partnership between the Heart Rhythm Society (HRS) and the European Heart Rhythm Association (EHRA). Heart Rhythm. 2011;8(8):1308–39.

Adalsteinsdottir B, Palsson R, Desnick RJ, Gardarsdottir M, Teekakirikul P, Maron M, Appelbaum E, Neisius U, Maron BJ, Burke MA, Chen B, Pagant S, Madsen CV, Danielsen R, Arngrimsson R, Feldt-Rasmussen U, Seidman JG, Seidman CE, Gunnarsson GT. Fabry disease in families with hypertrophic cardiomyopathy: clinical manifestations in the classic and later-onset phenotypes. Circ Cardiovasc Genet. 2017;10(4):e001639.

Al TA, Siriwardena K, George K, Mital S, Mercimek-Mahmutoglu S. Progressive cerebellar atrophy and a novel homozygous pathogenic DNAJC19 variant as a cause of dilated cardiomyopathy ataxia syndrome. Pediatr Neurol. 2016;62:58–61.

Andreasen C, Nielsen JB, Refsgaard L, Holst AG, Christensen AH, Andreasen L, Sajadieh A, Haunso S, Svendsen JH, Olesen MS. New population-based exome data are questioning the pathogenicity of previously cardiomyopathy-associated genetic variants. Eur J Hum Genet. 2013;21(9):918–28.

Antonicka H, Mattman A, Carlson CG, Glerum DM, Hoffbuhr KC, Leary SC, Kennaway NG, Shoubridge EA. Mutations in COX15 produce a defect in the mitochondrial heme biosynthetic pathway, causing early-onset fatal hypertrophic cardiomyopathy. Am J Hum Genet. 2003;72(1):101–14.

Arbustini E, Favalli V, Narula N, Serio A, Grasso M. Left ventricular noncompaction: a distinct genetic cardiomyopathy? J Am Coll Cardiol. 2016;68(9):949–66.

Arimura T, Hayashi T, Matsumoto Y, Shibata H, Hiroi S, Nakamura T, Inagaki N, Hinohara K, Takahashi M, Manatsu SI, Sasaoka T, Izumi T, Bonne G, Schwartz K, Kimura A. Structural analysis of four and half LIM protein-2 in dilated cardiomyopathy. Biochem Biophys Res Commun. 2007;357(1):162–7.

Arimura T, Bos JM, Sato A, Kubo T, Okamoto H, Nishi H, Harada H, Koga Y, Moulik M, Doi YL, Towbin JA, Ackerman MJ, Kimura A. Cardiac ankyrin repeat protein gene (ANKRD1) mutations in hypertrophic cardiomyopathy. J Am Coll Cardiol. 2009a;54(4):334–42.

Arimura T, Inagaki N, Hayashi T, Shichi D, Sato A, Hinohara K, Vatta M, Towbin JA, Chikamori T, Yamashina A, Kimura A. Impaired binding of ZASP/cypher with phosphoglucomutase 1 is associated with dilated cardiomyopathy. Cardiovasc Res. 2009b;83(1):80–8.

Arola AM, Sanchez X, Murphy RT, Hasle E, Li H, Elliott PM, McKenna WJ, Towbin JA, Bowles NE. Mutations in PDLIM3 and MYOZ1 encoding myocyte Z line proteins are infrequently found in idiopathic dilated cardiomyopathy. Mol Genet Metab. 2007;90(4):435–40.

Ashrafian H, Docherty L, Leo V, Towlson C, Neilan M, Steeples V, Lygate CA, Hough T, Townsend S, Williams D, Wells S, Norris D, Glyn-Jones S, Land J, Barbaric I, Lalanne Z, Denny P, Szumska D, Bhattacharya S, Griffin JL, Hargreaves I, Fernandez-Fuentes N, Cheeseman M, Watkins H, Dear TN. A mutation in the mitochondrial fission gene Dnm1l leads to cardiomyopathy. PLoS Genet. 2010;6(6):e1001000.

Bagnall RD, Molloy LK, Kalman JM, Semsarian C. Exome sequencing identifies a mutation in the ACTN2 gene in a family with idiopathic ventricular fibrillation, left ventricular noncompaction, and sudden death. BMC Med Genet. 2014;15:99.

Bainbridge MN, Davis EE, Choi WY, Dickson A, Martinez HR, Wang M, Dinh H, Muzny DM, Pignatelli R, Katsanis N, Boerwinkle E, Gibbs RA, Jefferies JL. Loss of function mutations in NNT are associated with left ventricular noncompaction. Circ Cardiovasc Genet. 2015;8(4):544–52.

Baksiene M, Benusiene E, Morkuniene A, Ambrozaityte L, Utkus A, Kucinskas V. A novel intronic splice site tafazzin gene mutation detected prenatally in a family with Barth syndrome. Balkan J Med Genet. 2016;19(2):95–100.

Basso C, Czarnowska E, Della BM, Bauce B, Beffagna G, Wlodarska EK, Pilichou K, Ramondo A, Lorenzon A, Wozniek O, Corrado D, Daliento L, Danieli GA, Valente M, Nava A, Thiene G, Rampazzo A. Ultrastructural evidence of intercalated disc remodelling in arrhythmogenic right ventricular cardiomyopathy: an electron microscopy investigation on endomyocardial biopsies. Eur Heart J. 2006;27(15):1847–54.

Bhatt KN, Butler J. Myocardial energetics and heart failure: a review of recent therapeutic trials. Curr Heart Fail Rep. 2018;15:191.

Bienengraeber M, Olson TM, Selivanov VA, Kathmann EC, O'Cochlain F, Gao F, Karger AB, Ballew JD, Hodgson DM, Zingman LV, Pang YP, Alekseev AE, Terzic A. ABCC9 mutations identified in human dilated cardiomyopathy disrupt catalytic KATP channel gating. Nat Genet. 2004;36(4):382–7.

Bione S, Small K, Aksmanovic VM, D'Urso M, Ciccodicola A, Merlini L, Morandi L, Kress W, Yates JR, Warren ST, et al. Identification of new mutations in the Emery-Dreifuss muscular dystrophy gene and evidence for genetic heterogeneity of the disease. Hum Mol Genet. 1995;4(10):1859–63.

Biswas A, Rao VR, Seth S, Maulik SK. Next generation sequencing in cardiomyopathy: towards personalized genomics and medicine. Mol Biol Rep. 2014;41(8):4881–8.

Bogomolovas J, Brohm K, Celutkiene J, Balciunaite G, Bironaite D, Bukelskiene V, Daunoravicus D, Witt CC, Fielitz J, Grabauskiene V, Labeit S. Induction of Ankrd1 in dilated cardiomyopathy correlates with the heart failure progression. Biomed Res Int. 2015;2015:273936.

Bolling MC, Pas HH, de Visser M, Aronica E, Pfendner EG, van den Berg MP, Diercks GF, Suurmeijer AJ, Jonkman MF. PLEC1 mutations underlie adult-onset dilated cardiomyopathy in epidermolysis bullosa simplex with muscular dystrophy. J Invest Dermatol. 2010;130(4):1178–81.

Bos JM, Poley RN, Ny M, Tester DJ, Xu X, Vatta M, Towbin JA, Gersh BJ, Ommen SR, Ackerman MJ. Genotype-phenotype relationships involving hypertrophic cardiomyopathy-associated mutations in titin, muscle LIM protein, and telethonin. Mol Genet Metab. 2006;88(1):78–85.

Branzi A, Romeo G, Specchia S, Lolli C, Binetti G, Devoto M, Bacchi M, Magnani B. Genetic heterogeneity of hypertrophic cardiomyopathy. Int J Cardiol. 1985;7(2):129–38.

Brauch KM, Karst ML, Herron KJ, de Andrade M, Pellikka PA, Rodeheffer RJ, Michels VV, Olson TM. Mutations in ribonucleic acid binding protein gene cause familial dilated cardiomyopathy. J Am Coll Cardiol. 2009;54(10):930–41.

Brieler J, Breeden MA, Tucker J. Cardiomyopathy: an overview. Am Fam Physician. 2017;96(10):640–6.

Brodehl A, Dieding M, Biere N, Unger A, Klauke B, Walhorn V, Gummert J, Schulz U, Linke WA, Gerull B, Vorgert M, Anselmetti D, Milting H. Functional characterization of the novel DES mutation p.L136P associated with dilated cardiomyopathy reveals a dominant filament assembly defect. J Mol Cell Cardiol. 2016;91:207–14.

Broussard JA, Yang R, Huang C, Nathamgari S, Beese AM, Godsel LM, Hegazy MH, Lee S, Zhou F, Sniadecki NJ, Green KJ, Espinosa HD. The desmoplakin-intermediate filament linkage regulates cell mechanics. Mol Biol Cell. 2017;28(23):3156–64.

Cahill TJ, Ashrafian H, Watkins H. Genetic cardiomyopathies causing heart failure. Circ Res. 2013;113(6):660–75.

Caleshu C, Sakhuja R, Nussbaum RL, Schiller NB, Ursell PC, Eng C, De Marco T, McGlothlin D, Burchard EG, Rame JE. Furthering the link between the sarcomere and primary cardiomyopathies: restrictive cardiomyopathy associated with multiple mutations in genes previously associated with hypertrophic or dilated cardiomyopathy. Am J Med Genet A. 2011;155A(9):2229–35.

Calore M. The PRKAG2 gene and hypertrophic cardiomyopathy: an energetically imbalanced relationship. Am J Physiol Heart Circ Physiol. 2017;313(2):H248–50.

Camargo-Ariza WA, Galvis-Blanco SJ, Camacho-Enciso T, Quiroz-Romero CA, Bermudez-Echeverry JJ. Arrhythmogenic right ventricular cardiomyopathy/dysplasia. Literature review and case report. Arch Cardiol Mex. 2018;88(1):51–9.

Cao Q, Shen Y, Liu X, Yu X, Yuan P, Wan R, Liu X, Peng X, He W, Pu J, Hong K. Phenotype and functional analyses in a transgenic mouse model of left ventricular noncompaction caused by a DTNA mutation. Int Heart J. 2017;58(6):939–47.

Cappola TP, Li M, He J, Ky B, Gilmore J, Qu L, Keating B, Reilly M, Kim CE, Glessner J, Frackelton E, Hakonarson H, Syed F, Hindes A, Matkovich SJ, Cresci S, Dorn GN. Common variants in HSPB7 and FRMD4B associated with advanced heart failure. Circ Cardiovasc Genet. 2010;3(2):147–54.

Carballo S, Robinson P, Otway R, Fatkin D, Jongbloed JD, de Jonge N, Blair E, van Tintelen JP, Redwood C, Watkins H. Identification and functional characterization of cardiac troponin I as a novel disease gene in autosomal dominant dilated cardiomyopathy. Circ Res. 2009;105(4):375–82.

Carniel E, Taylor MR, Sinagra G, Di Lenarda A, Ku L, Fain PR, Boucek MM, Cavanaugh J, Miocic S, Slavov D, Graw SL, Feiger J, Zhu XZ, Dao D, Ferguson DA, Bristow MR, Mestroni L. Alpha-myosin heavy chain: a sarcomeric gene associated with dilated and hypertrophic phenotypes of cardiomyopathy. Circulation. 2005;112(1):54–9.

Castelletti S, Vischer AS, Syrris P, Crotti L, Spazzolini C, Ghidoni A, Parati G, Jenkins S, Kotta MC, McKenna WJ, Schwartz PJ, Pantazis A. Desmoplakin missense and non-missense mutations in arrhythmogenic right ventricular cardiomyopathy: genotype-phenotype correlation. Int J Cardiol. 2017;249:268–73.

Chen SN, Czernuszewicz G, Tan Y, Lombardi R, Jin J, Willerson JT, Marian AJ. Human molecular genetic and functional studies identify TRIM63, encoding Muscle RING Finger Protein 1, as a novel gene for human hypertrophic cardiomyopathy. Circ Res. 2012;111(7):907–19.

Chen Y, Barajas-Martinez H, Zhu D, Wang X, Chen C, Zhuang R, Shi J, Wu X, Tao Y, Jin W, Wang X, Hu D. Novel trigenic CACNA1C/DES/MYPN mutations in a family of hypertrophic cardiomyopathy with early repolarization and short QT syndrome. J Transl Med. 2017;15(1):78.

Chow ML, Shaffer JF, Harris SP, Dawson JF. Altered interactions between cardiac myosin binding protein-C and alpha-cardiac actin variants associated with cardiomyopathies. Arch Biochem Biophys. 2014;550–551:28–32.

Clarke BA, Drujan D, Willis MS, Murphy LO, Corpina RA, Burova E, Rakhilin SV, Stitt TN, Patterson C, Latres E, Glass DJ. The E3 ligase MuRF1 degrades myosin heavy chain protein in dexamethasone-treated skeletal muscle. Cell Metab. 2007;6(5):376–85.

Corrado D, Link MS, Calkins H. Arrhythmogenic right ventricular cardiomyopathy. N Engl J Med. 2017;376(1):61–72.

Debold EP, Schmitt JP, Patlak JB, Beck SE, Moore JR, Seidman JG, Seidman C, Warshaw DM. Hypertrophic and dilated cardiomyopathy mutations differentially affect the molecular force generation of mouse alpha-cardiac myosin in the laser trap assay. Am J Physiol Heart Circ Physiol. 2007;293(1):H284–91.

Dec GW, Fuster V. Idiopathic dilated cardiomyopathy. N Engl J Med. 1994;331(23):1564–75.

Dhandapany PS, Razzaque MA, Muthusami U, Kunnoth S, Edwards JJ, Mulero-Navarro S, Riess I, Pardo S, Sheng J, Rani DS, Rani B, Govindaraj P, Flex E, Yokota T, Furutani M, Nishizawa T, Nakanishi T, Robbins J, Limongelli G, Hajjar RJ, Lebeche D, Bahl A, Khullar M, Rathinavel A, Sadler KC, Tartaglia M, Matsuoka R, Thangaraj K, Gelb BD. RAF1 mutations in childhood-onset dilated cardiomyopathy. Nat Genet. 2014;46(6):635–9.

Dickstein K, Cohen-Solal A, Filippatos G, McMurray JJ, Ponikowski P, Poole-Wilson PA, Stromberg A, van Veldhuisen DJ, Atar D, Hoes AW, Keren A, Mebazaa A, Nieminen M, Priori SG, Swedberg K. ESC guidelines for the diagnosis and treatment of acute and chronic heart failure 2008: the Task Force for the Diagnosis and Treatment of Acute and Chronic Heart Failure 2008 of the European Society of Cardiology. Developed in collaboration with the Heart Failure Association of the ESC (HFA) and endorsed by the European Society of Intensive Care Medicine (ESICM). Eur Heart J. 2008;29(19):2388–442.

Diegoli M, Grasso M, Favalli V, Serio A, Gambarin FI, Klersy C, Pasotti M, Agozzino E, Scelsi L, Ferlini A, Febo O, Piccolo G, Tavazzi L, Narula J, Arbustini E. Diagnostic work-up and risk stratification in X-linked dilated cardiomyopathies caused by dystrophin defects. J Am Coll Cardiol. 2011;58(9):925–34.

Duboscq-Bidot L, Xu P, Charron P, Neyroud N, Dilanian G, Millaire A, Bors V, Komajda M, Villard E. Mutations in the Z-band protein myopalladin gene and idiopathic dilated cardiomyopathy. Cardiovasc Res. 2008;77(1):118–25.

Duboscq-Bidot L, Charron P, Ruppert V, Fauchier L, Richter A, Tavazzi L, Arbustini E, Wichter T, Maisch B, Komajda M, Isnard R, Villard E. Mutations in the ANKRD1 gene encoding CARP are responsible for human dilated cardiomyopathy. Eur Heart J. 2009;30(17):2128–36.

Duong NT, Morris GE, Lam LT, Zhang Q, Sewry CA, Shanahan CM, Holt I. Nesprins: tissue-specific expression of epsilon and other short isoforms. PLoS One. 2014;9(4):e94380.

Dweck D, Hus N, Potter JD. Challenging current paradigms related to cardiomyopathies. Are changes in the Ca2+ sensitivity of myofilaments containing cardiac troponin C mutations (G159D and L29Q) good predictors of the phenotypic outcomes? J Biol Chem. 2008;283(48):33119–28.

Egan KR, Ralphe JC, Weinhaus L, Maginot KR. Just sinus bradycardia or something more serious? Case Rep Pediatr. 2013;2013:736164.

Ehler E, Perriard JC. Cardiomyocyte cytoskeleton and myofibrillogenesis in healthy and diseased heart. Heart Fail Rev. 2000;5(3):259–69.

Ehlermann P, Weichenhan D, Zehelein J, Steen H, Pribe R, Zeller R, Lehrke S, Zugck C, Ivandic BT, Katus HA. Adverse events in families with hypertrophic or dilated cardiomyopathy and mutations in the MYBPC3 gene. BMC Med Genet. 2008;9:95.

El-Battrawy I, Zhao Z, Lan H, Cyganek L, Tombers C, Li X, Buljubasic F, Lang S, Tiburcy M, Zimmermann WH, Utikal J, Wieland T, Borggrefe M, Zhou XB, Akin I. Electrical dysfunc-

tions in human-induced pluripotent stem cell-derived cardiomyocytes from a patient with an arrhythmogenic right ventricular cardiomyopathy. Europace. 2018;20:f46–56.
Ellinor PT, MacRae CA, Thierfelder L. Arrhythmogenic right ventricular cardiomyopathy. Heart Fail Clin. 2010;6(2):161–77.
Elliott P, Andersson B, Arbustini E, Bilinska Z, Cecchi F, Charron P, Dubourg O, Kuhl U, Maisch B, McKenna WJ, Monserrat L, Pankuweit S, Rapezzi C, Seferovic P, Tavazzi L, Keren A. Classification of the cardiomyopathies: a position statement from the European Society of Cardiology Working Group on Myocardial and Pericardial Diseases. Eur Heart J. 2008;29(2):270–6.
Elliott P, O'Mahony C, Syrris P, Evans A, Rivera SC, Sheppard MN, Carr-White G, Pantazis A, McKenna WJ. Prevalence of desmosomal protein gene mutations in patients with dilated cardiomyopathy. Circ Cardiovasc Genet. 2010;3(4):314–22.
Elliott PM, Anastasakis A, Borger MA, Borggrefe M, Cecchi F, Charron P, Hagege AA, Lafont A, Limongelli G, Mahrholdt H, McKenna WJ, Mogensen J, Nihoyannopoulos P, Nistri S, Pieper PG, Pieske B, Rapezzi C, Rutten FH, Tillmanns C, Watkins H. 2014 ESC guidelines on diagnosis and management of hypertrophic cardiomyopathy: the Task Force for the Diagnosis and Management of Hypertrophic Cardiomyopathy of the European Society of Cardiology (ESC). Eur Heart J. 2014;35(39):2733–79.
Erdmann J, Hassfeld S, Kallisch H, Fleck E, Regitz-Zagrose V. Genetic variants in the promoter (g983G>T) and coding region (A92T) of the human cardiotrophin-1 gene (CTF1) in patients with dilated cardiomyopathy. Hum Mutat. 2000;16(5):448.
Fang X, Bogomolovas J, Wu T, Zhang W, Liu C, Veevers J, Stroud MJ, Zhang Z, Ma X, Mu Y, Lao DH, Dalton ND, Gu Y, Wang C, Wang M, Liang Y, Lange S, Ouyang K, Peterson KL, Evans SM, Chen J. Loss-of-function mutations in co-chaperone BAG3 destabilize small HSPs and cause cardiomyopathy. J Clin Invest. 2017;127(8):3189–200.
Fatkin D. Guidelines for the diagnosis and management of familial dilated cardiomyopathy. Heart Lung Circ. 2011;20(11):691–3.
Fatkin D, Graham RM. Molecular mechanisms of inherited cardiomyopathies. Physiol Rev. 2002;82(4):945–80.
Fatkin D, MacRae C, Sasaki T, Wolff MR, Porcu M, Frenneaux M, Atherton J, Vidaillet HJ, Spudich S, De Girolami U, Seidman JG, Seidman C, Muntoni F, Muehle G, Johnson W, McDonough B. Missense mutations in the rod domain of the lamin A/C gene as causes of dilated cardiomyopathy and conduction-system disease. N Engl J Med. 1999;341(23):1715–24.
Feldman AM, Begay RL, Knezevic T, Myers VD, Slavov DB, Zhu W, Gowan K, Graw SL, Jones KL, Tilley DG, Coleman RC, Walinsky P, Cheung JY, Mestroni L, Khalili K, Taylor MR. Decreased levels of BAG3 in a family with a rare variant and in idiopathic dilated cardiomyopathy. J Cell Physiol. 2014;229(11):1697–702.
Ferlini A, Galie N, Merlini L, Sewry C, Branzi A, Muntoni F. A novel Alu-like element rearranged in the dystrophin gene causes a splicing mutation in a family with X-linked dilated cardiomyopathy. Am J Hum Genet. 1998;63(2):436–46.
Finsterer J, Stollberger C. Non-compaction associated with an ACTC1 mutation. Acta Cardiol. 2015;70(1):88–9.
Franaszczyk M, Bilinska ZT, Sobieszczanska-Malek M, Michalak E, Sleszycka J, Sioma A, Malek LA, Kaczmarska D, Walczak E, Wlodarski P, Hutnik L, Milanowska B, Dzielinska Z, Religa G, Grzybowski J, Zielinski T, Ploski R. The BAG3 gene variants in Polish patients with dilated cardiomyopathy: four novel mutations and a genotype-phenotype correlation. J Transl Med. 2014;12:192.
Fratev F, Mihaylova E, Pajeva I. Combination of genetic screening and molecular dynamics as a useful tool for identification of disease-related mutations: ZASP PDZ domain G54S mutation case. J Chem Inf Model. 2014;54(5):1524–36.
Frustaci A, De Luca A, Guida V, Biagini T, Mazza T, Gaudio C, Letizia C, Russo MA, Galea N, Chimenti C. Novel alpha-actin gene mutation p.(Ala21Val) causing familial hypertrophic

cardiomyopathy, myocardial noncompaction, and transmural crypts. Clinical-pathologic correlation. J Am Heart Assoc. 2018;7(4):e008068.

Fu L, Luo S, Cai S, Hong W, Guo Y, Wu J, Liu T, Zhao C, Li F, Huang H, Huang M, Wang J. Identification of LAMP2 mutations in early-onset Danon disease with hypertrophic cardiomyopathy by targeted next-generation sequencing. Am J Cardiol. 2016;118(6):888–94.

Fukuzawa A, Lange S, Holt M, Vihola A, Carmignac V, Ferreiro A, Udd B, Gautel M. Interactions with titin and myomesin target obscurin and obscurin-like 1 to the M-band: implications for hereditary myopathies. J Cell Sci. 2008;121(11):1841–51.

Garcia-Pavia P, Syrris P, Salas C, Evans A, Mirelis JG, Cobo-Marcos M, Vilches C, Bornstein B, Segovia J, Alonso-Pulpon L, Elliott PM. Desmosomal protein gene mutations in patients with idiopathic dilated cardiomyopathy undergoing cardiac transplantation: a clinicopathological study. Heart. 2011;97(21):1744–52.

Gautel M. The sarcomeric cytoskeleton: who picks up the strain? Curr Opin Cell Biol. 2011;23(1):39–46.

Gerull B, Gramlich M, Atherton J, McNabb M, Trombitas K, Sasse-Klaassen S, Seidman JG, Seidman C, Granzier H, Labeit S, Frenneaux M, Thierfelder L. Mutations of TTN, encoding the giant muscle filament titin, cause familial dilated cardiomyopathy. Nat Genet. 2002;30(2):201–4.

Gianni D, Li A, Tesco G, McKay KM, Moore J, Raygor K, Rota M, Gwathmey JK, Dec GW, Aretz T, Leri A, Semigran MJ, Anversa P, Macgillivray TE, Tanzi RE, Del MF. Protein aggregates and novel presenilin gene variants in idiopathic dilated cardiomyopathy. Circulation. 2010;121(10):1216–26.

Girolami F, Iascone M, Tomberli B, Bardi S, Benelli M, Marseglia G, Pescucci C, Pezzoli L, Sana ME, Basso C, Marziliano N, Merlini PA, Fornaro A, Cecchi F, Torricelli F, Olivotto I. Novel alpha-actinin 2 variant associated with familial hypertrophic cardiomyopathy and juvenile atrial arrhythmias: a massively parallel sequencing study. Circ Cardiovasc Genet. 2014;7(6):741–50.

Givertz MM, Mann DL. Epidemiology and natural history of recovery of left ventricular function in recent onset dilated cardiomyopathies. Curr Heart Fail Rep. 2013;10(4):321–30.

Guo J, Zhang H, Xiao J, Wu J, Ye Y, Li Z, Zou Y, Li X. Monocyte chemotactic protein-1 promotes the myocardial homing of mesenchymal stem cells in dilated cardiomyopathy. Int J Mol Sci. 2013;14(4):8164–78.

Hamid MS, Norman M, Quraishi A, Firoozi S, Thaman R, Gimeno JR, Sachdev B, Rowland E, Elliott PM, McKenna WJ. Prospective evaluation of relatives for familial arrhythmogenic right ventricular cardiomyopathy/dysplasia reveals a need to broaden diagnostic criteria. J Am Coll Cardiol. 2002;40(8):1445–50.

Hanley A, Walsh KA, Joyce C, McLellan MA, Clauss S, Hagen A, Shea MA, Tucker NR, Lin H, Fahy GJ, Ellinor PT. Mutation of a common amino acid in NKX2.5 results in dilated cardiomyopathy in two large families. BMC Med Genet. 2016;17(1):83.

Hartmannova H, Kubanek M, Sramko M, Piherova L, Noskova L, Hodanova K, Stranecky V, Pristoupilova A, Sovova J, Marek T, Maluskova J, Ridzon P, Kautzner J, Hulkova H, Kmoch S. Isolated X-linked hypertrophic cardiomyopathy caused by a novel mutation of the four-and-a-half LIM domain 1 gene. Circ Cardiovasc Genet. 2013;6(6):543–51.

Hassel D, Dahme T, Erdmann J, Meder B, Huge A, Stoll M, Just S, Hess A, Ehlermann P, Weichenhan D, Grimmler M, Liptau H, Hetzer R, Regitz-Zagrosek V, Fischer C, Nurnberg P, Schunkert H, Katus HA, Rottbauer W. Nexilin mutations destabilize cardiac Z-disks and lead to dilated cardiomyopathy. Nat Med. 2009;15(11):1281–8.

Haywood NJ, Wolny M, Rogers B, Trinh CH, Shuping Y, Edwards TA, Peckham M. Hypertrophic cardiomyopathy mutations in the calponin-homology domain of ACTN2 affect actin binding and cardiomyocyte Z-disc incorporation. Biochem J. 2016;473(16):2485–93.

Herman DS, Lam L, Taylor MR, Wang L, Teekakirikul P, Christodoulou D, Conner L, DePalma SR, McDonough B, Sparks E, Teodorescu DL, Cirino AL, Banner NR, Pennell DJ, Graw S, Merlo M, Di Lenarda A, Sinagra G, Bos JM, Ackerman MJ, Mitchell RN, Murry CE, Lakdawala NK,

Ho CY, Barton PJ, Cook SA, Mestroni L, Seidman JG, Seidman CE. Truncations of titin causing dilated cardiomyopathy. N Engl J Med. 2012;366(7):619–28.
Hershberger RE, Siegfried JD. Update 2011: clinical and genetic issues in familial dilated cardiomyopathy. J Am Coll Cardiol. 2011;57(16):1641–9.
Hershberger RE, Parks SB, Kushner JD, Li D, Ludwigsen S, Jakobs P, Nauman D, Burgess D, Partain J, Litt M. Coding sequence mutations identified in MYH7, TNNT2, SCN5A, CSRP3, LBD3, and TCAP from 313 patients with familial or idiopathic dilated cardiomyopathy. Clin Transl Sci. 2008;1(1):21–6.
Hershberger RE, Norton N, Morales A, Li D, Siegfried JD, Gonzalez-Quintana J. Coding sequence rare variants identified in MYBPC3, MYH6, TPM1, TNNC1, and TNNI3 from 312 patients with familial or idiopathic dilated cardiomyopathy. Circ Cardiovasc Genet. 2010;3(2):155–61.
Hirtle-Lewis M, Desbiens K, Ruel I, Rudzicz N, Genest J, Engert JC, Giannetti N. The genetics of dilated cardiomyopathy: a prioritized candidate gene study of LMNA, TNNT2, TCAP, and PLN. Clin Cardiol. 2013;36(10):628–33.
Hobbiebrunken E, Stettner GM, Hehr U, Hübner A, Wilichowski E. Homozygote deletion of Fukutin gene (FKTN) exon 3 causes dilated cardiomyopathy and mild limb-girdle muscular dystrophy without brain involvement. Neuropediatrics. 2016;47(S 01):P4–14.
Horstkotte J, Perisic T, Schneider M, Lange P, Schroeder M, Kiermayer C, Hinkel R, Ziegler T, Mandal PK, David R, Schulz S, Schmitt S, Widder J, Sinowatz F, Becker BF, Bauersachs J, Naebauer M, Franz WM, Jeremias I, Brielmeier M, Zischka H, Conrad M, Kupatt C. Mitochondrial thioredoxin reductase is essential for early postischemic myocardial protection. Circulation. 2011;124(25):2892–902.
Houtkooper RH, Turkenburg M, Poll-The BT, Karall D, Perez-Cerda C, Morrone A, Malvagia S, Wanders RJ, Kulik W, Vaz FM. The enigmatic role of tafazzin in cardiolipin metabolism. Biochim Biophys Acta. 2009;1788(10):2003–14.
Ichida F, Hamamichi Y, Miyawaki T, Ono Y, Kamiya T, Akagi T, Hamada H, Hirose O, Isobe T, Yamada K, Kurotobi S, Mito H, Miyake T, Murakami Y, Nishi T, Shinohara M, Seguchi M, Tashiro S, Tomimatsu H. Clinical features of isolated noncompaction of the ventricular myocardium: long-term clinical course, hemodynamic properties, and genetic background. J Am Coll Cardiol. 1999;34(1):233–40.
Inagaki N, Hayashi T, Arimura T, Koga Y, Takahashi M, Shibata H, Teraoka K, Chikamori T, Yamashina A, Kimura A. Alpha B-crystallin mutation in dilated cardiomyopathy. Biochem Biophys Res Commun. 2006;342(2):379–86.
Jacoby D, McKenna WJ. Genetics of inherited cardiomyopathy. Eur Heart J. 2012;33(3):296–304.
Jamuar SS, Tan EC. Clinical application of next-generation sequencing for Mendelian diseases. Hum Genomics. 2015;9:10.
Kamisago M, Sharma SD, DePalma SR, Solomon S, Sharma P, McDonough B, Smoot L, Mullen MP, Woolf PK, Wigle ED, Seidman JG, Seidman CE. Mutations in sarcomere protein genes as a cause of dilated cardiomyopathy. N Engl J Med. 2000;343(23):1688–96.
Karkkainen S, Miettinen R, Tuomainen P, Karkkainen P, Helio T, Reissell E, Kaartinen M, Toivonen L, Nieminen MS, Kuusisto J, Laakso M, Peuhkurinen K. A novel mutation, Arg71Thr, in the delta-sarcoglycan gene is associated with dilated cardiomyopathy. J Mol Med (Berl). 2003;81(12):795–800.
Kaski JP, Syrris P, Burch M, Tome-Esteban MT, Fenton M, Christiansen M, Andersen PS, Sebire N, Ashworth M, Deanfield JE, McKenna WJ, Elliott PM. Idiopathic restrictive cardiomyopathy in children is caused by mutations in cardiac sarcomere protein genes. Heart. 2008;94(11):1478–84.
Kedar V, McDonough H, Arya R, Li HH, Rockman HA, Patterson C. Muscle-specific RING finger 1 is a bona fide ubiquitin ligase that degrades cardiac troponin I. Proc Natl Acad Sci U S A. 2004;101(52):18135–40.
Khan MA, Reckman YJ, Aufiero S, van den Hoogenhof MM, van der Made I, Beqqali A, Koolbergen DR, Rasmussen TB, van der Velden J, Creemers EE, Pinto YM. RBM20 regulates circular RNA production from the titin gene. Circ Res. 2016;119(9):996–1003.

Kimura S, Ikezawa M, Ozasa S, Ito K, Ueno H, Yoshioka K, Ijiri S, Nomura K, Nakamura K, Matuskura M, Miike T. Novel mutation in splicing donor of dystrophin gene first exon in a patient with dilated cardiomyopathy but no clinical signs of skeletal myopathy. J Child Neurol. 2007;22(7):901–6.

Kimura K, Takenaka K, Ebihara A, Uno K, Morita H, Nakajima T, Ozawa T, Aida I, Yonemochi Y, Higuchi S, Motoyoshi Y, Mikata T, Uchida I, Ishihara T, Komori T, Kitao R, Nagata T, Takeda S, Yatomi Y, Nagai R, Komuro I. Prognostic impact of left ventricular noncompaction in patients with Duchenne/Becker muscular dystrophy – prospective multicenter cohort study. Int J Cardiol. 2013;168(3):1900–4.

Knoll R, Hoshijima M, Hoffman HM, Person V, Lorenzen-Schmidt I, Bang ML, Hayashi T, Shiga N, Yasukawa H, Schaper W, McKenna W, Yokoyama M, Schork NJ, Omens JH, McCulloch AD, Kimura A, Gregorio CC, Poller W, Schaper J, Schultheiss HP, Chien KR. The cardiac mechanical stretch sensor machinery involves a Z disc complex that is defective in a subset of human dilated cardiomyopathy. Cell. 2002;111(7):943–55.

Knoll R, Postel R, Wang J, Kratzner R, Hennecke G, Vacaru AM, Vakeel P, Schubert C, Murthy K, Rana BK, Kube D, Knoll G, Schafer K, Hayashi T, Holm T, Kimura A, Schork N, Toliat MR, Nurnberg P, Schultheiss HP, Schaper W, Schaper J, Bos E, Den Hertog J, van Eeden FJ, Peters PJ, Hasenfuss G, Chien KR, Bakkers J. Laminin-alpha4 and integrin-linked kinase mutations cause human cardiomyopathy via simultaneous defects in cardiomyocytes and endothelial cells. Circulation. 2007;116(5):515–25.

Kontrogianni-Konstantopoulos A, Ackermann MA, Bowman AL, Yap SV, Bloch RJ. Muscle giants: molecular scaffolds in sarcomerogenesis. Physiol Rev. 2009;89(4):1217–67.

Lakdawala NK, Funke BH, Baxter S, Cirino AL, Roberts AE, Judge DP, Johnson N, Mendelsohn NJ, Morel C, Care M, Chung WK, Jones C, Psychogios A, Duffy E, Rehm HL, White E, Seidman JG, Seidman CE, Ho CY. Genetic testing for dilated cardiomyopathy in clinical practice. J Card Fail. 2012;18(4):296–303.

Lakdawala NK, Winterfield JR, Funke BH. Dilated cardiomyopathy. Circ Arrhythm Electrophysiol. 2013;6(1):228–37.

Lamont PJ, Wallefeld W, Hilton-Jones D, Udd B, Argov Z, Barboi AC, Bonneman C, Boycott KM, Bushby K, Connolly AM, Davies N, Beggs AH, Cox GF, Dastgir J, DeChene ET, Gooding R, Jungbluth H, Muelas N, Palmio J, Penttila S, Schmedding E, Suominen T, Straub V, Staples C, Van den Bergh PY, Vilchez JJ, Wagner KR, Wheeler PG, Wraige E, Laing NG. Novel mutations widen the phenotypic spectrum of slow skeletal/beta-cardiac myosin (MYH7) distal myopathy. Hum Mutat. 2014;35(7):868–79.

Landstrom AP, Weisleder N, Batalden KB, Bos JM, Tester DJ, Ommen SR, Wehrens XH, Claycomb WC, Ko JK, Hwang M, Pan Z, Ma J, Ackerman MJ. Mutations in JPH2-encoded junctophilin-2 associated with hypertrophic cardiomyopathy in humans. J Mol Cell Cardiol. 2007;42(6):1026–35.

Landstrom AP, Adekola BA, Bos JM, Ommen SR, Ackerman MJ. PLN-encoded phospholamban mutation in a large cohort of hypertrophic cardiomyopathy cases: summary of the literature and implications for genetic testing. Am Heart J. 2011;161(1):165–71.

Lazzarini E, Jongbloed JD, Pilichou K, Thiene G, Basso C, Bikker H, Charbon B, Swertz M, van Tintelen JP, van der Zwaag PA. The ARVD/C genetic variants database: 2014 update. Hum Mutat. 2015;36(4):403–10.

Levitas A, Muhammad E, Harel G, Saada A, Caspi VC, Manor E, Beck JC, Sheffield V, Parvari R. Familial neonatal isolated cardiomyopathy caused by a mutation in the flavoprotein subunit of succinate dehydrogenase. Eur J Hum Genet. 2010;18(10):1160–5.

Li D, Parks SB, Kushner JD, Nauman D, Burgess D, Ludwigsen S, Partain J, Nixon RR, Allen CN, Irwin RP, Jakobs PM, Litt M, Hershberger RE. Mutations of presenilin genes in dilated cardiomyopathy and heart failure. Am J Hum Genet. 2006;79(6):1030–9.

Li J, Goossens S, van Hengel J, Gao E, Cheng L, Tyberghein K, Shang X, De Rycke R, van Roy F, Radice GL. Loss of αT-catenin alters the hybrid adhering junctions in the heart and leads to dilated cardiomyopathy and ventricular arrhythmia following acute ischemia. J Cell Sci. Journal of Cell Science. 2012;125(4):1058–67.

Li RG, Li L, Qiu XB, Yuan F, Xu L, Li X, Xu YJ, Jiang WF, Jiang JQ, Liu X, Fang WY, Zhang M, Peng LY, Qu XK, Yang YQ. GATA4 loss-of-function mutation underlies familial dilated cardiomyopathy. Biochem Biophys Res Commun. 2013;439(4):591–6.

Li J, Liu WD, Yang ZL, Yuan F, Xu L, Li RG, Yang YQ. Prevalence and spectrum of GATA4 mutations associated with sporadic dilated cardiomyopathy. Gene. 2014;548(2):174–81.

Lin X, Ruiz J, Bajraktari I, Ohman R, Banerjee S, Gribble K, Kaufman JD, Wingfield PT, Griggs RC, Fischbeck KH, Mankodi A. Z-disc-associated, alternatively spliced, PDZ motif-containing protein (ZASP) mutations in the actin-binding domain cause disruption of skeletal muscle actin filaments in myofibrillar myopathy. J Biol Chem. 2014;289(19):13615–26.

Liu Z, Shan H, Huang J, Li N, Hou C, Pu J. A novel lamin A/C gene missense mutation (445 V > E) in immunoglobulin-like fold associated with left ventricular non-compaction. Europace. 2016;18(4):617–22.

Liu JS, Fan LL, Li JJ, Xiang R. Whole-exome sequencing identifies a novel mutation of Desmocollin 2 in a Chinese family with Arrhythmogenic right ventricular cardiomyopathy. Am J Cardiol. 2017;119(9):1485–9.

Lopes LR, Elliott PM. Genetics of heart failure. Biochim Biophys Acta. 2013;1832(12):2451–61.

Lopez-Ayala JM, Ortiz-Genga M, Gomez-Milanes I, Lopez-Cuenca D, Ruiz-Espejo F, Sanchez-Munoz JJ, Oliva-Sandoval MJ, Monserrat L, Gimeno JR. A mutation in the Z-line Cypher/ZASP protein is associated with arrhythmogenic right ventricular cardiomyopathy. Clin Genet. 2015;88(2):172–6.

Luxan G, Casanova JC, Martinez-Poveda B, Prados B, D'Amato G, MacGrogan D, Gonzalez-Rajal A, Dobarro D, Torroja C, Martinez F, Izquierdo-Garcia JL, Fernandez-Friera L, Sabater-Molina M, Kong YY, Pizarro G, Ibanez B, Medrano C, Garcia-Pavia P, Gimeno JR, Monserrat L, Jimenez-Borreguero LJ, de la Pompa JL. Mutations in the NOTCH pathway regulator MIB1 cause left ventricular noncompaction cardiomyopathy. Nat Med. 2013;19(2):193–201.

Maiellaro-Rafferty K, Wansapura JP, Mendsaikhan U, Osinska H, James JF, Taylor MD, Robbins J, Kranias EG, Towbin JA, Purevjav E. Altered regional cardiac wall mechanics are associated with differential cardiomyocyte calcium handling due to nebulette mutations in preclinical inherited dilated cardiomyopathy. J Mol Cell Cardiol. 2013;60:151–60.

Man E, Lafferty KA, Funke BH, Lun KS, Chan SY, Chau AK, Chung BH. NGS identifies TAZ mutation in a family with X-linked dilated cardiomyopathy. BMJ Case Rep 2013;2013.

Marian AJ, Braunwald E. Hypertrophic cardiomyopathy: genetics, pathogenesis, clinical manifestations, diagnosis, and therapy. Circ Res. 2017;121(7):749–70.

Maron BJ, Maron MS. Hypertrophic cardiomyopathy. Lancet. 2013;381(9862):242–55.

Maron BJ, Towbin JA, Thiene G, Antzelevitch C, Corrado D, Arnett D, Moss AJ, Seidman CE, Young JB. Contemporary definitions and classification of the cardiomyopathies: an American Heart Association Scientific Statement from the Council on Clinical Cardiology, Heart Failure and Transplantation Committee; Quality of Care and Outcomes Research and Functional Genomics and Translational Biology Interdisciplinary Working Groups; and Council on Epidemiology and Prevention. Circulation. 2006;113(14):1807–16.

Maron BJ, Maron MS, Semsarian C. Genetics of hypertrophic cardiomyopathy after 20 years: clinical perspectives. J Am Coll Cardiol. 2012;60(8):705–15.

Marston S, Montgiraud C, Munster AB, Copeland O, Choi O, Dos RC, Messer AE, Ehler E, Knoll R. OBSCN mutations associated with dilated cardiomyopathy and haploinsufficiency. PLoS One. 2015;10(9):e138568.

McKoy G, Protonotarios N, Crosby A, Tsatsopoulou A, Anastasakis A, Coonar A, Norman M, Baboonian C, Jeffery S, McKenna WJ. Identification of a deletion in plakoglobin in arrhythmogenic right ventricular cardiomyopathy with palmoplantar keratoderma and woolly hair (Naxos disease). Lancet. 2000;355(9221):2119–24.

McLaughlin HM, Kelly MA, Hawley PP, Darras BT, Funke B, Picker J. Compound heterozygosity of predicted loss-of-function DES variants in a family with recessive desminopathy. BMC Med Genet. 2013;14:68.

McNally EM, Golbus JR, Puckelwartz MJ. Genetic mutations and mechanisms in dilated cardiomyopathy. J Clin Invest. 2013;123(1):19–26.

Meder B, Ruhle F, Weis T, Homuth G, Keller A, Franke J, Peil B, Lorenzo BJ, Frese K, Huge A, Witten A, Vogel B, Haas J, Volker U, Ernst F, Teumer A, Ehlermann P, Zugck C, Friedrichs F, Kroemer H, Dorr M, Hoffmann W, Maisch B, Pankuweit S, Ruppert V, Scheffold T, Kuhl U, Schultheiss HP, Kreutz R, Ertl G, Angermann C, Charron P, Villard E, Gary F, Isnard R, Komajda M, Lutz M, Meitinger T, Sinner MF, Wichmann HE, Krawczak M, Ivandic B, Weichenhan D, Gelbrich G, El-Mokhtari NE, Schreiber S, Felix SB, Hasenfuss G, Pfeufer A, Hubner N, Kaab S, Arbustini E, Rottbauer W, Frey N, Stoll M, Katus HA. A genome-wide association study identifies 6p21 as novel risk locus for dilated cardiomyopathy. Eur Heart J. 2014;35(16):1069–77.

Memo M, Leung MC, Ward DG, Dos RC, Morimoto S, Zhang L, Ravenscroft G, McNamara E, Nowak KJ, Marston SB, Messer AE. Familial dilated cardiomyopathy mutations uncouple troponin I phosphorylation from changes in myofibrillar Ca(2)(+) sensitivity. Cardiovasc Res. 2013;99(1):65–73.

Mendes DAR, Tavares J, Martins S, Carvalho T, Enguita FJ, Brito D, Carmo-Fonseca M, Lopes LR. Whole gene sequencing identifies deep-intronic variants with potential functional impact in patients with hypertrophic cardiomyopathy. PLoS One. 2017;12(8):e182946.

Merlo M, Sinagra G, Carniel E, Slavov D, Zhu X, Barbati G, Spezzacatene A, Ramani F, Salcedo E, Di Lenarda A, Mestroni L, Taylor MR. Poor prognosis of rare sarcomeric gene variants in patients with dilated cardiomyopathy. Clin Transl Sci. 2013;6(6):424–8.

Mestroni L, Taylor MR. Genetics and genetic testing of dilated cardiomyopathy: a new perspective. Discov Med. 2013;15(80):43–9.

Meyer S, van der Meer P, van Tintelen JP, van den Berg MP. Sex differences in cardiomyopathies. Eur J Heart Fail. 2014;16(3):238–47.

Milano A, Vermeer AM, Lodder EM, Barc J, Verkerk AO, Postma AV, van der Bilt IA, Baars MJ, van Haelst PL, Caliskan K, Hoedemaekers YM, Le Scouarnec S, Redon R, Pinto YM, Christiaans I, Wilde AA, Bezzina CR. HCN4 mutations in multiple families with bradycardia and left ventricular noncompaction cardiomyopathy. J Am Coll Cardiol. 2014;64(8):745–56.

Milasin J, Muntoni F, Severini GM, Bartoloni L, Vatta M, Krajinovic M, Mateddu A, Angelini C, Camerini F, Falaschi A, Mestroni L, Giacca M. A point mutation in the 5′ splice site of the dystrophin gene first intron responsible for X-linked dilated cardiomyopathy. Hum Mol Genet. 1996;5(1):73–9.

Minoretti P, Arra M, Emanuele E, Olivieri V, Aldeghi A, Politi P, Martinelli V, Pesenti S, Falcone C. A W148R mutation in the human FOXD4 gene segregating with dilated cardiomyopathy, obsessive-compulsive disorder, and suicidality. Int J Mol Med. 2007;19(3):369–72.

Miszalski-Jamka K, Jefferies JL, Mazur W, Glowacki J, Hu J, Lazar M, Gibbs RA, Liczko J, Klys J, Venner E, Muzny DM, Rycaj J, Bialkowski J, Kluczewska E, Kalarus Z, Jhangiani S, Al-Khalidi H, Kukulski T, Lupski JR, Craigen WJ, Bainbridge MN. Novel genetic triggers and genotype-phenotype correlations in patients with left ventricular noncompaction. Cardiovasc Genet. 2017;10(4):e001763.

Mogensen J, Murphy RT, Shaw T, Bahl A, Redwood C, Watkins H, Burke M, Elliott PM, McKenna WJ. Severe disease expression of cardiac troponin C and T mutations in patients with idiopathic dilated cardiomyopathy. J Am Coll Cardiol. 2004;44(10):2033–40.

Mogensen J, van Tintelen JP, Fokstuen S, Elliott P, van Langen IM, Meder B, Richard P, Syrris P, Caforio AL, Adler Y, Anastasakis A, Gimeno JR, Klingel K, Linhart A, Imazio M, Pinto Y, Newbery R, Schmidtke J, Charron P. The current role of next-generation DNA sequencing in routine care of patients with hereditary cardiovascular conditions: a viewpoint paper of the European Society of Cardiology working group on myocardial and pericardial diseases and members of the European Society of Human Genetics. Eur Heart J. 2015;36(22):1367–70.

Mohapatra B, Jimenez S, Lin JH, Bowles KR, Coveler KJ, Marx JG, Chrisco MA, Murphy RT, Lurie PR, Schwartz RJ, Elliott PM, Vatta M, McKenna W, Towbin JA, Bowles NE. Mutations in the muscle LIM protein and alpha-actinin-2 genes in dilated cardiomyopathy and endocardial fibroelastosis. Mol Genet Metab. 2003;80(1–2):207–15.

Molina-Navarro MM, Rosello-Lleti E, Ortega A, Tarazon E, Otero M, Martinez-Dolz L, Lago F, Gonzalez-Juanatey JR, Espana F, Garcia-Pavia P, Montero JA, Portoles M, Rivera M. Differential gene expression of cardiac ion channels in human dilated cardiomyopathy. PLoS One. 2013;8(12):e79792.

Mook OR, Haagmans MA, Soucy JF, van de Meerakker JB, Baas F, Jakobs ME, Hofman N, Christiaans I, Lekanne DR, Mannens MM. Targeted sequence capture and GS-FLX Titanium sequencing of 23 hypertrophic and dilated cardiomyopathy genes: implementation into diagnostics. J Med Genet. 2013;50(9):614–26.

Morales A, Hershberger RE. Genetic evaluation of dilated cardiomyopathy. Curr Cardiol Rep. 2013;15(7):375.

Morita H, Seidman J, Seidman CE. Genetic causes of human heart failure. J Clin Invest. 2005;115(3):518–26.

Morrison AC, Felix JF, Cupples LA, Glazer NL, Loehr LR, Dehghan A, Demissie S, Bis JC, Rosamond WD, Aulchenko YS, Wang YA, Haritunians T, Folsom AR, Rivadeneira F, Benjamin EJ, Lumley T, Couper D, Stricker BH, O'Donnell CJ, Rice KM, Chang PP, Hofman A, Levy D, Rotter JI, Fox ER, Uitterlinden AG, Wang TJ, Psaty BM, Willerson JT, van Duijn CM, Boerwinkle E, Witteman JC, Vasan RS, Smith NL. Genomic variation associated with mortality among adults of European and African ancestry with heart failure: the cohorts for heart and aging research in genomic epidemiology consortium. Circ Cardiovasc Genet. 2010;3(3):248–55.

Moulik M, Vatta M, Witt SH, Arola AM, Murphy RT, McKenna WJ, Boriek AM, Oka K, Labeit S, Bowles NE, Arimura T, Kimura A, Towbin JA. ANKRD1, the gene encoding cardiac ankyrin repeat protein, is a novel dilated cardiomyopathy gene. J Am Coll Cardiol. 2009;54(4):325–33.

Muchtar E, Blauwet LA, Gertz MA. Restrictive cardiomyopathy: genetics, pathogenesis, clinical manifestations, diagnosis, and therapy. Circ Res. 2017;121(7):819–37.

Mueller II, Gawaz M, Linke RP, Zuern C, Steiner D, Altland K, Von Beckerath N, Weig HJ. Restrictive cardiomyopathy in inherited ATTR amyloidosis (TTR-Ser23Asn) in a patient of German-Italian extraction. BMJ Case Rep. 2010;2010.

Muller T, Krasnianski M, Witthaut R, Deschauer M, Zierz S. Dilated cardiomyopathy may be an early sign of the C826A Fukutin-related protein mutation. Neuromuscul Disord. 2005;15(5):372–6.

Muntoni F, Cau M, Ganau A, Congiu R, Arvedi G, Mateddu A, Marrosu MG, Cianchetti C, Realdi G, Cao A, et al. Brief report: deletion of the dystrophin muscle-promoter region associated with X-linked dilated cardiomyopathy. N Engl J Med. 1993;329(13):921–5.

Murakami T, Hayashi YK, Noguchi S, Ogawa M, Nonaka I, Tanabe Y, Ogino M, Takada F, Eriguchi M, Kotooka N, Campbell KP, Osawa M, Nishino I. Fukutin gene mutations cause dilated cardiomyopathy with minimal muscle weakness. Ann Neurol. 2006;60(5):597–602.

Murakami C, Nakamura S, Kobayashi M, Maeda K, Irie W, Wada B, Hayashi M, Sasaki C, Nakamaru N, Furukawa M, Kurihara K. Analysis of the sarcomere protein gene mutation on cardiomyopathy – mutations in the cardiac troponin I gene. Leg Med (Tokyo). 2010;12(6):280–3.

Murphy RT, Mogensen J, Shaw A, Kubo T, Hughes S, McKenna WJ. Novel mutation in cardiac troponin I in recessive idiopathic dilated cardiomyopathy. Lancet. 2004;363(9406):371–2.

Ng D, Johnston JJ, Teer JK, Singh LN, Peller LC, Wynter JS, Lewis KL, Cooper DN, Stenson PD, Mullikin JC, Biesecker LG. Interpreting secondary cardiac disease variants in an exome cohort. Circ Cardiovasc Genet. 2013;6(4):337–46.

Nguyen TP, Wang DW, Rhodes TH, George AJ. Divergent biophysical defects caused by mutant sodium channels in dilated cardiomyopathy with arrhythmia. Circ Res. 2008;102(3):364–71.

Norgett EE, Hatsell SJ, Carvajal-Huerta L, Cabezas JC, Common J, Purkis PE, Whittock N, Leigh IM, Stevens HP, Kelsell DP. Recessive mutation in desmoplakin disrupts desmoplakin-intermediate filament interactions and causes dilated cardiomyopathy, woolly hair and keratoderma. Hum Mol Genet. 2000;9(18):2761–6.

Norton N, Li D, Rieder MJ, Siegfried JD, Rampersaud E, Zuchner S, Mangos S, Gonzalez-Quintana J, Wang L, McGee S, Reiser J, Martin E, Nickerson DA, Hershberger RE. Genome-wide studies of copy number variation and exome sequencing identify rare variants in BAG3 as a cause of dilated cardiomyopathy. Am J Hum Genet. 2011;88(3):273–82.

Norton N, Li D, Rampersaud E, Morales A, Martin ER, Zuchner S, Guo S, Gonzalez M, Hedges DJ, Robertson PD, Krumm N, Nickerson DA, Hershberger RE. Exome sequencing and genome-wide linkage analysis in 17 families illustrate the complex contribution of TTN truncating variants to dilated cardiomyopathy. Circ Cardiovasc Genet. 2013;6(2):144–53.

Ntusi NA, Shaboodien G, Badri M, Gumedze F, Mayosi BM. Clinical features, spectrum of causal genetic mutations and outcome of hypertrophic cardiomyopathy in South Africans. Cardiovasc J Afr. 2016;27(3):152–8.

Ohno S, Omura M, Kawamura M, Kimura H, Itoh H, Makiyama T, Ushinohama H, Makita N, Horie M. Exon 3 deletion of RYR2 encoding cardiac ryanodine receptor is associated with left ventricular non-compaction. Europace. 2014;16(11):1646–54.

Olson TM, Michels VV, Thibodeau SN, Tai YS, Keating MT. Actin mutations in dilated cardiomyopathy, a heritable form of heart failure. Science. 1998;280(5364):750–2.

Olson TM, Kishimoto NY, Whitby FG, Michels VV. Mutations that alter the surface charge of alpha-tropomyosin are associated with dilated cardiomyopathy. J Mol Cell Cardiol. 2001;33(4):723–32.

Olson TM, Illenberger S, Kishimoto NY, Huttelmaier S, Keating MT, Jockusch BM. Metavinculin mutations alter actin interaction in dilated cardiomyopathy. Circulation. 2002;105(4):431–7.

Osio A, Tan L, Chen SN, Lombardi R, Nagueh SF, Shete S, Roberts R, Willerson JT, Marian AJ. Myozenin 2 is a novel gene for human hypertrophic cardiomyopathy. Circ Res. 2007;100(6):766–8.

Parent JJ, Towbin JA, Jefferies JL. Left ventricular noncompaction in a family with lamin A/C gene mutation. Tex Heart Inst J. 2015;42(1):73–6.

Parks SB, Kushner JD, Nauman D, Burgess D, Ludwigsen S, Peterson A, Li D, Jakobs P, Litt M, Porter CB, Rahko PS, Hershberger RE. Lamin A/C mutation analysis in a cohort of 324 unrelated patients with idiopathic or familial dilated cardiomyopathy. Am Heart J. 2008;156(1):161–9.

Parry TL, Melehani JH, Ranek MJ, Willis MS. Functional amyloid signaling via the inflammasome, necrosome, and signalosome: new therapeutic targets in heart failure. Front Cardiovasc Med. 2015;2:25.

Peled Y, Gramlich M, Yoskovitz G, Feinberg MS, Afek A, Polak-Charcon S, Pras E, Sela BA, Konen E, Weissbrod O, Geiger D, Gordon PM, Thierfelder L, Freimark D, Gerull B, Arad M. Titin mutation in familial restrictive cardiomyopathy. Int J Cardiol. 2014;171(1):24–30.

Perez-Serra A, Toro R, Campuzano O, Sarquella-Brugada G, Berne P, Iglesias A, Mangas A, Brugada J, Brugada R. A novel mutation in lamin a/c causing familial dilated cardiomyopathy associated with sudden cardiac death. J Card Fail. 2015;21(3):217–25.

Perez-Serra A, Toro R, Sarquella-Brugada G, de Gonzalo-Calvo D, Cesar S, Carro E, Llorente-Cortes V, Iglesias A, Brugada J, Brugada R, Campuzano O. Genetic basis of dilated cardiomyopathy. Int J Cardiol. 2016;224:461–72.

Piccolo P, Attanasio S, Secco I, Sangermano R, Strisciuglio C, Limongelli G, Miele E, Mutarelli M, Banfi S, Nigro V, Pons T, Valencia A, Zentilin L, Campione S, Nardone G, Lynnes TC, Celestino-Soper PB, Spoonamore KG, D'Armiento FP, Giacca M, Staiano A, Vatta M, Collesi C, Brunetti-Pierri N. MIB2 variants altering NOTCH signalling result in left ventricle hypertrabeculation/non-compaction and are associated with Menetrier-like gastropathy. Hum Mol Genet. 2017;26(1):33–43.

Pilotto A, Marziliano N, Pasotti M, Grasso M, Costante AM, Arbustini E. alphaB-crystallin mutation in dilated cardiomyopathies: low prevalence in a consecutive series of 200 unrelated probands. Biochem Biophys Res Commun. 2006;346(4):1115–7.

Ploski R, Rydzanicz M, Ksiazczyk TM, Franaszczyk M, Pollak A, Kosinska J, Michalak E, Stawinski P, Ziolkowska L, Bilinska ZT, Werner B. Evidence for troponin C (TNNC1) as a gene for autosomal recessive restrictive cardiomyopathy with fatal outcome in infancy. Am J Med Genet A. 2016;170(12):3241–8.

Protonotarios N, Tsatsopoulou A. Naxos disease and Carvajal syndrome: cardiocutaneous disorders that highlight the pathogenesis and broaden the spectrum of arrhythmogenic right ventricular cardiomyopathy. Cardiovasc Pathol. 2004;13(4):185–94.

Puckelwartz MJ, Kessler EJ, Kim G, Dewitt MM, Zhang Y, Earley JU, Depreux FF, Holaska J, Mewborn SK, Pytel P, McNally EM. Nesprin-1 mutations in human and murine cardiomyopathy. J Mol Cell Cardiol. 2010;48(4):600–8.

Pugh TJ, Kelly MA, Gowrisankar S, Hynes E, Seidman MA, Baxter SM, Bowser M, Harrison B, Aaron D, Mahanta LM, Lakdawala NK, McDermott G, White ET, Rehm HL, Lebo M, Funke BH. The landscape of genetic variation in dilated cardiomyopathy as surveyed by clinical DNA sequencing. Genet Med. 2014;16(8):601–8.

Rai TS, Ahmad S, Ahluwalia TS, Ahuja M, Bahl A, Saikia UN, Singh B, Talwar KK, Khullar M. Genetic and clinical profile of Indian patients of idiopathic restrictive cardiomyopathy with and without hypertrophy. Mol Cell Biochem. 2009;331(1–2):187–92.

Ramoglu MG, Ucar T, Ceylaner S, Atalay S, Tutar E. A novel mutation in the desmoplakin gene in two female siblings with a rare form of dilated cardiomyopathy: Carvajal syndrome. Anatol J Cardiol. 2017;18(6):435–6.

Rimessi P, Gualandi F, Duprez L, Spitali P, Neri M, Merlini L, Calzolari E, Muntoni F, Ferlini A. Genomic and transcription studies as diagnostic tools for a prenatal detection of X-linked dilated cardiomyopathy due to a dystrophin gene mutation. Am J Med Genet A. 2005;132A(4):391–4.

Rodriguez G, Ueyama T, Ogata T, Czernuszewicz G, Tan Y, Dorn GN, Bogaev R, Amano K, Oh H, Matsubara H, Willerson JT, Marian AJ. Molecular genetic and functional characterization implicate muscle-restricted coiled-coil gene (MURC) as a causal gene for familial dilated cardiomyopathy. Circ Cardiovasc Genet. 2011;4(4):349–58.

Roux-Buisson N, Gandjbakhch E, Donal E, Probst V, Deharo JC, Chevalier P, Klug D, Mansencal N, Delacretaz E, Cosnay P, Scanu P, Extramiana F, Keller D, Hidden-Lucet F, Trapani J, Fouret P, Frank R, Fressart V, Faure J, Lunardi J, Charron P. Prevalence and significance of rare RYR2 variants in arrhythmogenic right ventricular cardiomyopathy/dysplasia: results of a systematic screening. Heart Rhythm. 2014;11(11):1999–2009.

Rubattu S, Bozzao C, Pennacchini E, Pagannone E, Musumeci BM, Piane M, Germani A, Savio C, Francia P, Volpe M, Autore C, Chessa L. A next-generation sequencing approach to identify gene mutations in early- and late-onset hypertrophic cardiomyopathy patients of an Italian cohort. Int J Mol Sci. 2016;17(8):1239.

Sabater-Molina M, Perez-Sanchez I, Hernandez DRJ, Gimeno JR. Genetics of hypertrophic cardiomyopathy: a review of current state. Clin Genet. 2018;93(1):3–14.

Sala V, Gallo S, Leo C, Gatti S, Gelb BD, Crepaldi T. Signaling to cardiac hypertrophy: insights from human and mouse RASopathies. Mol Med. 2012;18:938–47.

San RI, Navarro M, Martinez F, Albert L, Polo L, Guardiola J, Garcia-Molina E, Munoz-Esparza C, Lopez-Ayala JM, Sabater-Molina M, Gimeno JR. Unclassifiable arrhythmic cardiomyopathy associated with Emery-Dreifuss caused by a mutation in FHL1. Clin Genet. 2016;90(2):171–6.

Schanzer A, Rupp S, Graf S, Zengeler D, Jux C, Akinturk H, Gulatz L, Mazhari N, Acker T, Van Coster R, Garvalov BK, Hahn A. Dysregulated autophagy in restrictive cardiomyopathy due to Pro209Leu mutation in BAG3. Mol Genet Metab. 2018;123(3):388–99.

Schmitt JP, Kamisago M, Asahi M, Li GH, Ahmad F, Mende U, Kranias EG, MacLennan DH, Seidman JG, Seidman CE. Dilated cardiomyopathy and heart failure caused by a mutation in phospholamban. Science. 2003;299(5611):1410–3.

Schonberger J, Wang L, Shin JT, Kim SD, Depreux FF, Zhu H, Zon L, Pizard A, Kim JB, Macrae CA, Mungall AJ, Seidman JG, Seidman CE. Mutation in the transcriptional coactivator EYA4 causes dilated cardiomyopathy and sensorineural hearing loss. Nat Genet. 2005;37(4):418–22.

Schweizer PA, Schroter J, Greiner S, Haas J, Yampolsky P, Mereles D, Buss SJ, Seyler C, Bruehl C, Draguhn A, Koenen M, Meder B, Katus HA, Thomas D. The symptom complex of familial sinus node dysfunction and myocardial noncompaction is associated with mutations in the HCN4 channel. J Am Coll Cardiol. 2014;64(8):757–67.

Sebillon P, Bouchier C, Bidot LD, Bonne G, Ahamed K, Charron P, Drouin-Garraud V, Millaire A, Desrumeaux G, Benaiche A, Charniot JC, Schwartz K, Villard E, Komajda M. Expanding the phenotype of LMNA mutations in dilated cardiomyopathy and functional consequences of these mutations. J Med Genet. 2003;40(8):560–7.

Sen-Chowdhry S, Syrris P, McKenna WJ. Genetics of restrictive cardiomyopathy. Heart Fail Clin. 2010;6(2):179–86.

Shah SN, Bhimji SS. Arrhythmogenic right ventricular cardiomyopathy. Treasure Island: StatPearls Publishing; 2018.

Shan L, Makita N, Xing Y, Watanabe S, Futatani T, Ye F, Saito K, Ibuki K, Watanabe K, Hirono K, Uese K, Ichida F, Miyawaki T, Origasa H, Bowles NE, Towbin JA. SCN5A variants in Japanese patients with left ventricular noncompaction and arrhythmia. Mol Genet Metab. 2008;93(4):468–74.

Shan S, He X, He L, Wang M, Liu C. Coexistence of congenital left ventricular aneurysm and prominent left ventricular trabeculation in a patient with LDB3 mutation: a case report. J Med Case Rep. 2017;11(1):229.

Sibbing D, Pfeufer A, Perisic T, Mannes AM, Fritz-Wolf K, Unwin S, Sinner MF, Gieger C, Gloeckner CJ, Wichmann HE, Kremmer E, Schafer Z, Walch A, Hinterseer M, Nabauer M, Kaab S, Kastrati A, Schomig A, Meitinger T, Bornkamm GW, Conrad M, von Beckerath N. Mutations in the mitochondrial thioredoxin reductase gene TXNRD2 cause dilated cardiomyopathy. Eur Heart J. 2011;32(9):1121–33.

Sikkema-Raddatz B, Johansson LF, de Boer EN, Almomani R, Boven LG, van den Berg MP, van Spaendonck-Zwarts KY, van Tintelen JP, Sijmons RH, Jongbloed JD, Sinke RJ. Targeted next-generation sequencing can replace Sanger sequencing in clinical diagnostics. Hum Mutat. 2013;34(7):1035–42.

Song XW, Yuan QN, Tang Y, Cao M, Shen YF, Zeng ZY, Lei CH, Li S, Zhao XX, Yang YJ. Conditionally targeted deletion of PSEN1 leads to diastolic heart dysfunction. J Cell Physiol. 2018;233(2):1548–57.

Subbarayan V, Mark M, Messadeq N, Rustin P, Chambon P, Kastner P. RXRalpha overexpression in cardiomyocytes causes dilated cardiomyopathy but fails to rescue myocardial hypoplasia in RXRalpha-null fetuses. J Clin Invest. 2000;105(3):387–94.

Subramanian K, Gianni D, Balla C, Assenza GE, Joshi M, Semigran MJ, Macgillivray TE, Van Eyk JE, Agnetti G, Paolocci N, Bamburg JR, Agrawal PB, Del MF. Cofilin-2 phosphorylation and sequestration in myocardial aggregates: novel pathogenetic mechanisms for idiopathic dilated cardiomyopathy. J Am Coll Cardiol. 2015;65(12):1199–214.

Tamargo J. TGFbeta3 mutations cause arrhythmogenic right ventricular dysplasia type 1 and open the door to understanding the biological role of TGFbeta3 (where there's a will, there's a way). Cardiovasc Res. 2012;96(2):188–90, 191–4

Taylor MR, Slavov D, Gajewski A, Vlcek S, Ku L, Fain PR, Carniel E, Di Lenarda A, Sinagra G, Boucek MM, Cavanaugh J, Graw SL, Ruegg P, Feiger J, Zhu X, Ferguson DA, Bristow MR, Gotzmann J, Foisner R, Mestroni L. Thymopoietin (lamina-associated polypeptide 2) gene mutation associated with dilated cardiomyopathy. Hum Mutat. 2005;26(6):566–74.

Taylor MR, Ku L, Slavov D, Cavanaugh J, Boucek M, Zhu X, Graw S, Carniel E, Barnes C, Quan D, Prall R, Lovell MA, Mierau G, Ruegg P, Mandava N, Bristow MR, Towbin JA, Mestroni L. Danon disease presenting with dilated cardiomyopathy and a complex phenotype. J Hum Genet. 2007;52(10):830–5.

Teekakirikul P, Kelly MA, Rehm HL, Lakdawala NK, Funke BH. Inherited cardiomyopathies: molecular genetics and clinical genetic testing in the postgenomic era. J Mol Diagn. 2013;15(2):158–70.

Theis JL, Sharpe KM, Matsumoto ME, Chai HS, Nair AA, Theis JD, de Andrade M, Wieben ED, Michels VV, Olson TM. Homozygosity mapping and exome sequencing reveal GATAD1 mutation in autosomal recessive dilated cardiomyopathy. Circ Cardiovasc Genet. 2011;4(6):585–94.

Thompson D, Patrick-Esteve J, Surcouf JW, Rivera D, Castellanos B, Desai P, Lilje C, Lacassie Y, Marble M, Zambrano R. RAF1 variants causing biventricular hypertrophic cardiomyopathy in two preterm infants: further phenotypic delineation and review of literature. Clin Dysmorphol. 2017;26(4):195–9.

Towbin JA. Left ventricular noncompaction: a new form of heart failure. Heart Fail Clin. 2010;6(4):453–69.

Towbin JA, Lowe AM, Colan SD, Sleeper LA, Orav EJ, Clunie S, Messere J, Cox GF, Lurie PR, Hsu D, Canter C, Wilkinson JD, Lipshultz SE. Incidence, causes, and outcomes of dilated cardiomyopathy in children. JAMA. 2006;296(15):1867–76.

Towe EC, Bos JM, Ommen SR, Gersh BJ, Ackerman MJ. Genotype-phenotype correlations in apical variant hypertrophic cardiomyopathy. Congenit Heart Dis. 2015;10(3):E139–45.

Tsatsopoulou A. Desmoplakin: highlights on a left ventricular arrhythmogenic disorder. Int J Cardiol. 2017;249:324–5.

Tse HF, Ho JC, Choi SW, Lee YK, Butler AW, Ng KM, Siu CW, Simpson MA, Lai WH, Chan YC, Au KW, Zhang J, Lay KW, Esteban MA, Nicholls JM, Colman A, Sham PC. Patient-specific induced-pluripotent stem cells-derived cardiomyocytes recapitulate the pathogenic phenotypes of dilated cardiomyopathy due to a novel DES mutation identified by whole exome sequencing. Hum Mol Genet. 2013;22(7):1395–403.

Tsubata S, Bowles KR, Vatta M, Zintz C, Titus J, Muhonen L, Bowles NE, Towbin JA. Mutations in the human delta-sarcoglycan gene in familial and sporadic dilated cardiomyopathy. J Clin Invest. 2000;106(5):655–62.

Tucker NR, McLellan MA, Hu D, Ye J, Parsons VA, Mills RW, Clauss S, Dolmatova E, Shea MA, Milan DJ, Scott NS, Lindsay M, Lubitz SA, Domian IJ, Stone JR, Lin H, Ellinor PT. Novel mutation in FLNC (Filamin C) causes familial restrictive cardiomyopathy. Circ Cardiovasc Genet. 2017;10(6):e001780.

Ucar SK, Mayr JA, Feichtinger RG, Canda E, Coker M, Wortmann SB. Previously unreported Biallelic mutation in DNAJC19: are sensorineural hearing loss and basal ganglia lesions additional features of Dilated Cardiomyopathy and Ataxia (DCMA) syndrome? JIMD Rep. 2017;35:39–45.

Valdes-Mas R, Gutierrez-Fernandez A, Gomez J, Coto E, Astudillo A, Puente DA, Reguero JR, Alvarez V, Moris C, Leon D, Martin M, Puente XS, Lopez-Otin C. Mutations in filamin C cause a new form of familial hypertrophic cardiomyopathy. Nat Commun. 2014;5:5326.

van Riet EE, Hoes AW, Wagenaar KP, Limburg A, Landman MA, Rutten FH. Epidemiology of heart failure: the prevalence of heart failure and ventricular dysfunction in older adults over time. A systematic review. Eur J Heart Fail. 2016;18(3):242–52.

van Waning JI, Caliskan K, Hoedemaekers YM, van Spaendonck-Zwarts KY, Baas AF, Boekholdt SM, van Melle JP, Teske AJ, Asselbergs FW, Backx A, du Marchie SG, Dalinghaus M, Breur J, Linschoten M, Verlooij LA, Kardys I, Dooijes D, Lekanne DR, IJpma AS, van den Berg MP, Hofstra R, van Slegtenhorst MA, Jongbloed J, Majoor-Krakauer D. Genetics, clinical features, and long-term outcome of noncompaction cardiomyopathy. J Am Coll Cardiol. 2018;71(7):711–22.

Vatta M, Mohapatra B, Jimenez S, Sanchez X, Faulkner G, Perles Z, Sinagra G, Lin JH, Vu TM, Zhou Q, Bowles KR, Di Lenarda A, Schimmenti L, Fox M, Chrisco MA, Murphy RT, McKenna W, Elliott P, Bowles NE, Chen J, Valle G, Towbin JA. Mutations in Cypher/ZASP in patients with dilated cardiomyopathy and left ventricular non-compaction. J Am Coll Cardiol. 2003;42(11):2014–27.

Villard E, Perret C, Gary F, Proust C, Dilanian G, Hengstenberg C, Ruppert V, Arbustini E, Wichter T, Germain M, Dubourg O, Tavazzi L, Aumont MC, DeGroote P, Fauchier L, Trochu JN, Gibelin P, Aupetit JF, Stark K, Erdmann J, Hetzer R, Roberts AM, Barton PJ, Regitz-Zagrosek V, Aslam U, Duboscq-Bidot L, Meyborg M, Maisch B, Madeira H, Waldenstrom

A, Galve E, Cleland JG, Dorent R, Roizes G, Zeller T, Blankenberg S, Goodall AH, Cook S, Tregouet DA, Tiret L, Isnard R, Komajda M, Charron P, Cambien F. A genome-wide association study identifies two loci associated with heart failure due to dilated cardiomyopathy. Eur Heart J. 2011;32(9):1065–76.

Waldmuller S, Schroeder C, Sturm M, Scheffold T, Imbrich K, Junker S, Frische C, Hofbeck M, Bauer P, Bonin M, Gawaz M, Gramlich M. Targeted 46-gene and clinical exome sequencing for mutations causing cardiomyopathies. Mol Cell Probes. 2015;29(5):308–14.

Walsh R, Buchan R, Wilk A, John S, Felkin LE, Thomson KL, Chiaw TH, Loong C, Pua CJ, Raphael C, Prasad S, Barton PJ, Funke B, Watkins H, Ware JS, Cook SA. Defining the genetic architecture of hypertrophic cardiomyopathy: re-evaluating the role of non-sarcomeric genes. Eur Heart J. 2017;38(46):3461–8.

Wang J, Wang Y, Zou Y, Sun K, Wang Z, Ding H, Yuan J, Wei W, Hou Q, Wang H, Liu X, Zhang H, Ji Y, Zhou X, Sharma RK, Wang D, Ahmad F, Hui R, Song L. Malignant effects of multiple rare variants in sarcomere genes on the prognosis of patients with hypertrophic cardiomyopathy. Eur J Heart Fail. 2014;16(9):950–7.

Wang X, Zhou K, Hua Y, Li Y. Approaching the facts between genetic mutation and clinical practice of hypertrophic cardiomyopathy: a case report with RAF1 770C>T mutant. Medicine (Baltimore). 2016;95(37):e4815.

Weintraub RG, Semsarian C, Macdonald P. Dilated cardiomyopathy. Lancet. 2017;390(10092):400–14.

Weir-McCall JR, Yeap PM, Papagiorcopulo C, Fitzgerald K, Gandy SJ, Lambert M, Belch JJ, Cavin I, Littleford R, Macfarlane JA, Matthew SZ, Nicholas RS, Struthers AD, Sullivan F, Waugh SA, White RD, Houston JG. Left ventricular noncompaction: anatomical phenotype or distinct cardiomyopathy? J Am Coll Cardiol. 2016;68(20):2157–65.

Wells QS, Veatch OJ, Fessel JP, Joon AY, Levinson RT, Mosley JD, Held EP, Lindsay CS, Shaffer CM, Weeke PE, Glazer AM, Bersell KR, Van Driest SL, Karnes JH, Blair MA, Lagrone LW, Su YR, Bowton EA, Feng Z, Ky B, Lenihan DJ, Fisch MJ, Denny JC, Roden DM. Genome-wide association and pathway analysis of left ventricular function after anthracycline exposure in adults. Pharmacogenet Genomics. 2017;27(7):247–54.

Williams T, Hundertmark M, Nordbeck P, Voll S, Arias-Loza PA, Oppelt D, Muhlfelder M, Schraut S, Elsner I, Czolbe M, Seidlmayer L, Heinze B, Hahner S, Heinze K, Schonberger J, Jakob P, Ritter O. Eya4 induces hypertrophy via regulation of p27kip1. Circ Cardiovasc Genet. 2015;8(6):752–64.

Xiong Q, Cao Q, Zhou Q, Xie J, Shen Y, Wan R, Yu J, Yan S, Marian AJ, Hong K. Arrhythmogenic cardiomyopathy in a patient with a rare loss-of-function KCNQ1 mutation. J Am Heart Assoc. 2015;4(1):e1526.

Xu L, Zhao L, Yuan F, Jiang WF, Liu H, Li RG, Xu YJ, Zhang M, Fang WY, Qu XK, Yang YQ, Qiu XB. GATA6 loss-of-function mutations contribute to familial dilated cardiomyopathy. Int J Mol Med. 2014;34(5):1315–22.

Xu JH, Gu JY, Guo YH, Zhang H, Qiu XB, Li RG, Shi HY, Liu H, Yang XX, Xu YJ, Qu XK, Yang YQ. Prevalence and spectrum of NKX2-5 mutations associated with sporadic adult-onset dilated cardiomyopathy. Int Heart J. 2017;58(4):521–9.

Yang Z, Bowles NE, Scherer SE, Taylor MD, Kearney DL, Ge S, Nadvoretskiy VV, DeFreitas G, Carabello B, Brandon LI, Godsel LM, Green KJ, Saffitz JE, Li H, Danieli GA, Calkins H, Marcus F, Towbin JA. Desmosomal dysfunction due to mutations in desmoplakin causes arrhythmogenic right ventricular dysplasia/cardiomyopathy. Circ Res. 2006;99(6):646–55.

Yang J, Zhu M, Wang Y, Hou X, Wu H, Wang D, Shen H, Hu Z, Zou J. Whole-exome sequencing identify a new mutation of MYH7 in a Chinese family with left ventricular noncompaction. Gene. 2015;558(1):138–42.

Yokoyama R, Kinoshita K, Hata Y, Abe M, Matsuoka K, Hirono K, Kano M, Nakazawa M, Ichida F, Nishida N, Tabata T. A mutant HCN4 channel in a family with bradycardia, left bundle branch block, and left ventricular noncompaction. Heart Vessel. 2018;33:802.

Yoshida Y, Hirono K, Nakamura K, Suzuki T, Hata Y, Nishida N. A novel ACTC1 mutation in a young boy with left ventricular noncompaction and arrhythmias. Heart Rhythm Case Rep. 2016;2(1):92–7.

Yoskovitz G, Peled Y, Gramlich M, Lahat H, Resnik-Wolf H, Feinberg MS, Afek A, Pras E, Arad M, Gerull B, Freimark D. A novel titin mutation in adult-onset familial dilated cardiomyopathy. Am J Cardiol. 2012;109(11):1644–50.

Yuan F, Qiu XB, Li RG, Qu XK, Wang J, Xu YJ, Liu X, Fang WY, Yang YQ, Liao DN. A novel NKX2-5 loss-of-function mutation predisposes to familial dilated cardiomyopathy and arrhythmias. Int J Mol Med. 2015;35(2):478–86.

Zhang XL, Dai N, Tang K, Chen YQ, Chen W, Wang J, Zhao CM, Yuan F, Qiu XB, Qu XK, Yang YQ, Xu YW. GATA5 loss-of-function mutation in familial dilated cardiomyopathy. Int J Mol Med. 2015a;35(3):763–70.

Zhang XL, Qiu XB, Yuan F, Wang J, Zhao CM, Li RG, Xu L, Xu YJ, Shi HY, Hou XM, Qu XK, Xu YW, Yang YQ. TBX5 loss-of-function mutation contributes to familial dilated cardiomyopathy. Biochem Biophys Res Commun. 2015b;459(1):166–71.

Zhao L, Xu JH, Xu WJ, Yu H, Wang Q, Zheng HZ, Jiang WF, Jiang JF, Yang YQ. A novel GATA4 loss-of-function mutation responsible for familial dilated cardiomyopathy. Int J Mol Med. 2014;33(3):654–60.

Zhao Y, Feng Y, Zhang YM, Ding XX, Song YZ, Zhang AM, Liu L, Zhang H, Ding JH, Xia XS. Targeted next-generation sequencing of candidate genes reveals novel mutations in patients with dilated cardiomyopathy. Int J Mol Med. 2015;36(6):1479–86.

Zhao CM, Bing-Sun, Song HM, Wang J, Xu WJ, Jiang JF, Qiu XB, Yuan F, Xu JH, Yang YQ. TBX20 loss-of-function mutation associated with familial dilated cardiomyopathy. Clin Chem Lab Med. 2016;54(2):325–32.

Zhao Y, Feng Y, Ding X, Dong S, Zhang H, Ding J, Xia X. Identification of a novel hypertrophic cardiomyopathy-associated mutation using targeted next-generation sequencing. Int J Mol Med. 2017;40(1):121–9.

Zhou W, Zhao L, Jiang JQ, Jiang WF, Yang YQ, Qiu XB. A novel TBX5 loss-of-function mutation associated with sporadic dilated cardiomyopathy. Int J Mol Med. 2015;36(1):282–8.

Zhou YM, Dai XY, Huang RT, Xue S, Xu YJ, Qiu XB, Yang YQ. A novel TBX20 lossoffunction mutation contributes to adultonset dilated cardiomyopathy or congenital atrial septal defect. Mol Med Rep. 2016;14(4):3307–14.

Zhou C, Li C, Zhou B, Sun H, Koullourou V, Holt I, Puckelwartz MJ, Warren DT, Hayward R, Lin Z, Zhang L, Morris GE, McNally EM, Shackleton S, Rao L, Shanahan CM, Zhang Q. Novel nesprin-1 mutations associated with dilated cardiomyopathy cause nuclear envelope disruption and defects in myogenesis. Hum Mol Genet. 2017;26(12):2258–76.

Zhou Y, Qian Z, Yang J, Zhu M, Hou X, Wang Y, Wu H, Zou J. Whole exome sequencing identifies novel candidate mutations in a Chinese family with left ventricular noncompaction. Mol Med Rep. 2018;17:7325–30.

Zimmerman RS, Cox S, Lakdawala NK, Cirino A, Mancini-DiNardo D, Clark E, Leon A, Duffy E, White E, Baxter S, Alaamery M, Farwell L, Weiss S, Seidman CE, Seidman JG, Ho CY, Rehm HL, Funke BH. A novel custom resequencing array for dilated cardiomyopathy. Genet Med. 2010;12(5):268–78.

Chapter 12
Warfarin and Its Pharmacogenomic Study

Shuiping Dai

Abstract Warfarin as an anticoagulant is clinically prescribed to prevent blood clot formation. Genetic variants affect Warfarin response through pharmacokinetic and pharmacodynamic effects, which could account partially for the interracial difference to Warfarin. Warfarin activity is determined partially by the polymorphism in two genes, VKOPC1 and CYP2C9. Although the clinical application of pharmacogenomic information is controversial, the future gene sequencing and functional characterization of common genetic variation based on individual's genetic information would provide a useful tool for precision medication.

Keyword Warfarin · Anticoagulant · Pharmacogenomics

1 History

"Late in December, he had lost two young heifers. In January one of his favorite old cows had developed a massive hematoma on a thigh and following a skin puncture, fatal bleeding set in rapidly. Finally, two young cows had died on Friday and the bull was oozing blood from the nose." As described in *Circulation*, Wisconsin chemist Dr. Karl Paul Link met a local farmer Ed Carlson and would like to figure out how the sweet clover disease happened in farmer's cattle in the chilly winter of 1933. The sweet clover disease was a cattle disease; when the cattle were ingesting moldy silage made from sweet clover which was a potent anticoagulant discovered in a comprehensive and thorough study made at the North Dakota Agricultural Experiment Station, the cattle were hemorrhaging to death spontaneously or after a minor procedure. At that time, Dr. Link worked as a faculty at the University of Wisconsin-Madison, and he started to characterize the hemorrhagic agent.

Dr. Link and his colleagues spent the next 5 years to isolate the anticoagulant dicoumarol and confirm that the dicoumarol was formed by plant molecule coumarin working with fungi in the rotten clover in 1939 (Link 1959). His research was

S. Dai (✉)
University of Wisconsin Hospital and Clinics, Madison, WI, USA

H. Jiang, M. Liu (eds.), *Heart Genomics*, Translational Bioinformatics 16,
https://doi.org/10.1007/978-981-13-1429-2_12

largely supported by the Wisconsin Alumni Research Foundation (WARF); the new molecule was named after WARF and coumarin as Warfarin. Chemically the drug Warfarin is composed of a mixture of two active enantiomers (R and S forms). R isomer is 2.5 times less potent than S isomer. Both isomers of Warfarin were metabolized through cytochrome CYP-mediated enzymes. Warfarin has a long half-life and needs to be taken once per day for patients while it remains effective for several days after patients discontinued it (Kresge et al. 2005).

2 Warfarin and Its Pharmacogenomics

Warfarin therapy decreases the clotting forming tendency but not prevents the clotting completely. Blood testing was used to monitor the effect of Warfarin; according to the blood testing, patient's daily dose of Warfarin must be adjusted to keep blood clotting time within a target range. As a standard blood test, prothrombin time test (PT) was reported as INR (international normalized ratio). If the INR is too low, blood clots will not be prevented; if the INR is too high, there is an increased risk of bleeding, so Warfarin must be taken exactly as prescribed.

Warfarin is commonly used to treat blood clots such as deep vein thrombosis and pulmonary embolism as well as to prevent stroke. It is also used after ST-segment elevation myocardial infarction (STEM) and orthopedic surgery. However, the common side effect of Warfarin usage after these procedures is bleeding.

Blood clots can cause a heart attack, stroke, deep vein thrombosis, or pulmonary embolism; Warfarin as an anticoagulant is clinically prescribed to prevent blood clot formation. Warfarin's major side effects are bleeding due to excessive anticoagulation, whereas larger clot could be generated due to the low dosage of Warfarin. Excessive bleeding can occur in any area of the body with most serious bleeding in gastrointestinal or intracerebral vessels (Kresge et al. 2005; Link 1959).

Regarding the mechanism, Warfarin decreases the formation of blood clots through the generation of vitamin K-dependent clotting factors. Vitamin K is required to make clotting factors and prevent bleeding. Warfarin blocks vitamin K epoxide reductase (VKOR) which allows the vitamin K epoxide to reactivate back to vitamin K. With sufficient active vitamin K, clotting factors II, VII, XI, and X can normally work to form clotting. Warfarin prevents the coagulation factors being carboxylated at certain glutamic acid residue which will lead the failure of glutamic acid residue binding to the endothelial surface of blood vessels.

Variation in drug effect was mainly determined by its pharmacokinetic (PK) and pharmacodynamic (PD). Metabolizing enzymes such as cytochrome (CYP P 450) could influence plasma drug concentration and time in achieving the therapeutic effect on the targets. Recently published guidelines suggest that the initial Warfarin doses are 5–10 mg per day although some patient required significant high doses (up to 145 mg/day).

Genetic variants affect drug response through their pharmacokinetic and pharmacodynamic effects. Drug response variation and incidence of the adverse event could be partially explained by genetic variants. Pharmacogenomics is the study of the role of inheritance in drug's response variation among individuals. The expression and functional alteration of drug metabolizing enzymes could produce high or low levels of enzyme activity and alter drug's effects.

Warfarin activity is determined partially by the polymorphism in two genes, VKOPC1 and CYP2C9. Approximately 30% of the dose variation among patients could be explained through VKORC1 polymorphisms.

Rieder et al. in 2005 identified ten common noncoding VKORC1 single-nucleotide polymorphisms and three major haplotypes (Rieder et al. 2005).

Among the three haplotype groups, about 2.7 mg per day for A/A, 4.9 mg per day for A/B, and 6.2 mg per day for B/B, VKORC1 haplotype groups A and B contribute approximately to 25% of the variances in dose. A higher percentage of A haplotype was found in Asian American, whereas African Americans present with a higher percentage of group B haplotype.

VKORC1 polymorphism explains why African Americans have a higher percentage of group A haplotype on average; Asian Americans are more resistant to Warfarin because Asian Americans have a higher percentage of group A haplotype. The average INR was about 2.5, but patients in VKORC1 polymphorphism make a more rapid achievement of a therapeutic INR, and a shorter time to reach an INR over 4 which are associated with bleeding.

CYP2C9 polymorphism represents 10% of the dose variation in patients, particularly among Caucasian patients, but this polymorphism is rare in African- and Asian Americans. Patients with CYP2C9 polymorphisms have shorter time to therapeutic INR over 4.

Zhou HH et al. firstly explored for the interracial difference in drug responsiveness by focusing on reversible interactions with plasma constituents (Zhou et al. 1990). Predictable pharmacokinetic analysis may find the differences of patients' responsiveness to various drugs (Liu et al. 2017)

Genetic variants could account partially for the interracial difference to Warfarin, 8–12% European patients are CYP2C9*2 and CYP2C9*3 alleles. Asian patients are more sensitive to Warfarin with 0–2% being CYP2C9*2 and CYP2C9*3 alleles. Veenstra et al. reported that Chinese patients need a 40–50% less maintenance dose of Warfarin compared with the European patients (Veenstra et al. 2005).

They found that haplotypes H1 and H2 were associated with a low-dose of Warfarin phenotype and haplotypes H7, H8, and H9 with a high-dose of Warfarin. The mean stable warfarin dose was 3.5 ± 1.6 mg/day. These patients exhibited a restricted set of VKORC1 haplotypes relative to the European Warfarin patients. Haplotypes H1 and H7 were the most common with frequencies of 86.3% and 13.0%, respectively.

The frequencies for the H1/H1, H1/H7, and H7/H7 genotypes were 75.4%, 20.3%, and 2.9%, respectively. Fifty-three patients (76.8%) that were homozygous for VKORC1 group A homozygous combinations (H1/H1, 76.8% of all patients) need a mean stable Warfarin dose of 2.9 mg/day. This dose was significantly lower than the mean stable dose of 4.9 mg/day of heterozygotes (H1/H7, 20.3%). Two patients homozygous for H7 (2.9%) need maintenance doses of 6.5 and 6.7 mg/day.

For the Chinese population in Hong Kong, Veenstra summarized after a multivariate analysis as below: "age explained 21.5% of the variability in warfarin dose, gender explained 0.4%, VKORC1 genotype explained 31% and CYP2C9 genotype explained

7.9%. No additional factors were considered in the analysis. All four factors combined explained 60.8% of the variability in Warfarin dose" (Veenstra et al. 2005).

The clinical application of pharmacogenomic information is controversial. The Centers for Medicare and Medicaid Services stated in 2009 that "the available evidence does not demonstrate that pharmacogenomic testing of CYP2C9 or VKORC1 alleles to predict Warfarin responsiveness improves health outcomes in Medicare beneficiaries" (Jensen et al. 2009).

In Stergiopoulos' study, a genotype-guided dosing strategy did not produce a greater percentage of the INR being within the therapeutic range, or reductions of major bleeding or thromboembolic incidents compared with clinical dosing algorithms (Stergiopoulos and Brown 2014).

The Food and Drug Administration (FDA) provided pharmacogenetic information to the Warfarin product label but did not alter a specific method for using genetic information to predict the dose requirement for individual patients.

The future gene sequencing and functional characterization of common genetic variation based on individual's genetic information would guide us for precision medication to further predict drug's therapeutic effect and side effect. The clinical implementation of pharmacogenomic techniques will be accessible and available for clinical providers, patients, and other colleagues in preventative medicine.

Acknowledgment The author would like to thank Dr. Xiaobo Wu and Dr. Ming Liu for constructive criticism and helps.

References

Jensen TS, Jacques LB, Ciccanti M, Long K, Eggleston L, Roche J. Decision memo for pharmacogenomic testing for warfarin response (CAG-00400N). Centers for Medicare and Medicaid Services. 2009.

Kresge N, Simoni RD, Hill RL. Hemorrhagic sweet clover disease, Dicumarol and warfarin: the work of Karl Paul Link. J Biol Chem. 2005;280:e5.

Link KP. The discovery of Dicumarol and its sequels. Circulation. 1959;19:97–107.

Liu J, Jiang HH, Wu DK, Zhou YX, Ye HM, Li X, Luo ZY, Guo Z, Zhang YL, Wang YC, Zhang W, Zhou HH, Wang LS. Effect of gene polymorphims on the warfarin treatment at initial stage. Pharmacogenomics. 2017;17(1):47–52.

Rieder MJ, Reiner AP, Gage BF, Nickerson DA, Eby CS, McLeod HL, Blough DK, Thummel KE, Veenstra DL, Rettie AE. Effect of VKORC1 haplotypes on transcriptional regulation and warfarin dose. N Engl J Med. 2005;352(22):2285–93.

Stergiopoulos K, Brown DL. Genotype-guided vs clinical dosing of warfarin and its analogues: meta-analysis of randomized clinical trials. JAMA Intern Med. 2014;174(8):1330–8.

Veenstra DL, You JH, Rieder MJ, Farin FM, Wilkerson HW, Blough DK, Cheng G, Rettie AE. Association of vitamin K epoxide reductase complex 1 (VKORC1) variants with warfarin dose in a Hong Kong Chinese patient population. Pharmacogenet Genomics. 2005;15(10):687–91.

Zhou HH, Adedoyin A, Wilkinson GR. Differences in plasma binding of drugs between Caucasians and Chinese subjects. Clin Pharmacol Ther. 1990;48(1):10–7.

Chapter 13
Gene Therapy and Genomic Application in Heart Disease

Feng Zhu and Kai Huang

1 Gene Therapy in Heart Disease

Heart disease is a leading cause of mortality all over the world. The mortality for all cardiovascular diseases was 262 per 100,000 persons, accounting for 1 of every 3 deaths in China (Weiwei et al. 2016). Pharmacologic drugs for heart diseases have multiple systemic side effects and limitations, and there is an unmet need for improving clinical outcomes without side effects. Interventional cardiovascular therapy including percutaneous coronary intervention (PCI), pacemaker implantation, and cardiac ablation has improved the prognosis for patients with heart diseases. However, large numbers of patients with heart diseases are still left with significant morbidity despite those therapies. This limitation in available therapies has prompted extensive investigation of new treatments. The improved understanding of the molecular basis of the heart diseases and the development of gene transferring vehicles and delivering systems have given researchers the tools to target-specific genes and pathways which play a role in heart diseases. The first clinical trial of gene therapy for heart disease was initiated more than two decades ago. Through the use of a vector, a new copy of a gene or relevant nucleotide sequence is inserted into a patient's heart tissue to either compensate or correct for the deleted or defective version that is causing the disease or enhance an existing version. The development of gene therapy in heart disease depends on the development of gene delivering systems, achievement of highly efficient, long-term and targeted expression to heart tissue, as well as design of gene transferring vectors that are safe for human administration.

F. Zhu · K. Huang (✉)
Clinic Center of Human Gene Research, Union Hospital, Tongji Medical College, Huazhong University of Science and Technology, Wuhan, China

Department of Cardiology, Union Hospital, Tongji Medical College, Huazhong University of Science and Technology, Wuhan, China

H. Jiang, M. Liu (eds.), *Heart Genomics*, Translational Bioinformatics 16,
https://doi.org/10.1007/978-981-13-1429-2_13

1.1 Vectors for Gene Therapy in Heart Disease

Successful delivery of a gene to heart tissue is paramount to therapeutic efficacy and safety. Current gene delivery vehicles, namely, vectors, are divided into two groups: non-viral vectors and viral vectors. Each of the vector systems applied for cardiac gene transfer has its own advantages and disadvantages (Table 13.1). The main vectors currently employed in gene therapy in heart disease will be discussed below.

1.1.1 Naked DNA Plasmid

Non-viral vectors include naked plasmid DNA, liposomal DNA complexes, polymer-carried DNA, polymeric nanoparticles, and oligonucleotides (Shim et al. 2017). Plasmids are double-stranded circular DNA molecules, which contains transgenes encoding target proteins, enhancer and promoter sequences. Compared with recombinant viral vectors, it can be easily produced by bacteria as plasmid DNA and is therefore cost-efficient in large-scale production. DNA plasmids can be delivered directly to targeted tissue creating a localized homing signal at the site of delivery, without significant expression of the therapeutic protein outside the targeted tissue. DNA plasmids do not integrate into the genome and do not result in permanent genetic modification. DNA plasmids have been shown to be non-immunogenic or tolerogenic in hosts and thus may be administered repeatedly. One major disadvantage of DNA plasmids is low efficiency of gene transfer to target tissue. However, plasmid modifications can improve transgene expression in target tissue, which include different combinations of the promoter, other enhancer elements in expression, nuclear localization signals, prokaryotic components, and CpG-dinucleotide content (Williams and Kingston 2011). And ultrasound-targeted microbubble destruction at the target site has been used to enhance the efficacy of cardiac gene transfer in the form of plasmid DNA (Williams and Kingston 2011). Plasmid-mediated angiogenesis with enhancing expression of vascular endothelial growth factor (VEGF) was applied in treatment of coronary artery disease (CAD). Clinical studies have demonstrated angiogenic activity of VEDF plasmids, not only

Table 13.1 The various vectors commonly used for cardiac gene therapy

	Naked DNA plasmid	Adeno-associated virus	Lentivirus	Adenovirus
Type	Non-viral	Viral	Viral	Viral
Diameter	NA	20 nm	90 nm	100 nm
Genome	DNA	Single-stranded DNA	RNA	Double-stranded DNA
Expression duration	Expression up to 2 months	Long-term cardiac expression	Long-term cardiac expression	Expression up to 2 weeks
Immune response	Moderate	Low	Mild	Strong

anatomically but also functionally (Khurana et al. 2005). DNA plasmids have also been used in phase I and II clinical trials of heart failure, which involved direct injection of DNA plasmids encoding human stromal cell-derived factor-1 (SDF-1) in to the peri-infarct area of patients with ischemic cardiomyopathy (Chung et al. 2015; Penn et al. 2013). Plasmid-based transferring SDF-1 aimed to recruit stem cells over a short period of time in an attempt to myocardial regeneration (Askari et al. 2003). Although patients treated with SDF-1 trended to an improvement in cardiac function, the results were not significant in the entire population in the phase II clinical trial (Chung et al. 2015).

1.1.2 Lentivirus

Four decades earlier, human immunodeficiency virus (HIV) had been identified as the cause of AIDS, and rapid studies of its biology led to the idea that this genus of retrovirus and lentivirus could be optimized for gene therapy (Dunbar et al. 2018). Lentiviral systems have been highly modified from HIV over several generations to make them safe to handle and useful for applications. Lentivirus (the complete virus particle) contains a protein capsid, lipid encapsulated, with diameter ranging from 80 to 130 nm. Lentivirus genome (3.5–10 kilobases or kB) has two copies of single-stranded molecule of RNA and contains a complete set of genetic instructions. Lentiviruses were the first viral vectors to be used in gene therapy because they offer the advantage of a potentially stable transgene expression due to integration of the target gene into host genome. The experimental use of such viral vectors in preclinical models is expanding in circulatory system. Lentiviruses harboring cardiac-specific promoters have been applied as a therapeutic method for treating mouse models of Fabry disease, an inherited disorder caused by deficiency of α-galactosidase A, which leads to progressive intracellular accumulation of globotriaosylceramide in various tissues and organs, including the heart, and causes heart failure (Lee et al. 2011). However, an important consideration with lentiviruses integrating in the host genome is the location of integration. If the integration of target gene does not occur at a predictable, specific site rather than at a random site, it could be oncogenic (Hayward et al. 2014). This complication was found that lentiviruses were used to treat severe combined immunodeficiency in children (Hacein-Bey-Abina et al. 2003). The safety issue concerns are still restraining their application for human cardiac gene therapy.

1.1.3 Adeno-Associated Virus

Adeno-associated virus (AAV) belongs to the parvovirus family of single-stranded small DNA viruses, which is naturally replication-defective (Hastie and Samulski 2015; Snyder 1999). The AAV genome consists of approximately 4.7 kB, which comprises two inverted terminal repeats (ITRs) flanking genes for replication and building the virus capsid. Wild-type AAV requires another virus such as an adenovirus or a herpesvirus to replicate (Asokan et al. 2012). To generate a recombinant AAV (rAAV), the entire sequence of virus except ITRs is replaced by the expression cassette of choice. rAAV comes in many serotypes and has been shown to able to target various tissues including the eye, brain, liver, and heart. Thirteen serotypes of AAV have thus far been identified, with the best characterized and most commonly used being AAV2 (Asokan et al. 2012). And AAV has been successful in treatment of blindness, hemophilia, and muscular dystrophy. In 2012, Europe approved the first human gene therapy product, namely, Glybera, a rAAV vector expressing lipoprotein lipase (LPL) for the treatment of LPL deficiency, representing a milestone in rAAV gene therapy (Yla-Herttuala 2012). AAV1, AAV6, AAV8, and AAV9 have been shown to transduce myocardium efficiently. AAV1-mediated sarcoplasmic reticulum calcium (Ca2+) ATPase 2a (SERCA2a) gene therapy (AAV1.SERCA2a) in the patients with heart failure have yielded very promising clinical results in phase I study (Hajjar et al. 2008; Jaski et al. 2009). However, the phase 2b clinical trial of intracoronary infusion to deliver AAV1.SERCA2a did not show efficacy in randomized testing (Greenberg et al. 2016). A major drawback of AAV vector is that its maximum cargo capacity is limited to less than 4.8 kB, which precludes delivering large therapeutic genes. Another drawback of AAVs is active immune responses against the AAV delivery vehicles by host and the existence of neutralizing antibodies in the host serum (Hayward et al. 2014). As a result, there are still a large number of obstacles to overcome, and clinical trials based on rAAV are still at an early stage in the development pathway.

1.1.4 Adenovirus

Adenovirus is a medium-sized (90–100 nm), non-enveloped icosahedral virus, with a nucleocapsid surrounding a double-stranded DNA genome (Rolland et al. 2009). There are over 51 different serotypes in humans, which cause upper respiratory infections, conjunctivitis, gastroenteritis, cystitis, and serious infections in neonates. Adenovirus enter cells mainly via clathrin-dependent, coxsackie-adenovirus receptor-mediated endocytosis. The double-stranded DNA is then transported to the nucleus, allowing it to efficiently transduce a wide range of dividing and nondividing cells such as cardiomyocytes. To date, adenovirus has been the most widely used vector in clinical gene transfer. In the heart, transgene expression after adenovirus transduction is robust but transient (1–2 weeks), which has been applied to deliver angiogenic genes to treat ischemic heart disease. An adenovirus carrying AC6 gene, which encoding the protein that catalyzes the conversion of adenosine triphosphate (ATP) to cyclic adenosine monophosphate (cAMP), resulted in increased of left ventricular function beyond standard heart failure therapy (Hammond et al. 2016). One major disadvantage of adenoviruses for cardiac gene therapy is the induction of a wide variety of immune responses both humoral and cellular, which compromise their safety and efficacy in clinical trials. Another limitation for the use of adenoviruses is the high prevalence of neutralizing antibodies in the host serum. About 97% of the population has neutralizing antibodies against type C adenoviruses, which include the frequently used adenovirus serotypes in gene therapy such as serotypes 2 and 5 (Nayak and Herzog 2010). Therefore, the application of adenovirus in cardiac gene therapy trials must be carefully evaluated and new generation adenoviral vectors need to be developed, which are less stimulation of the host immune system and more selectively targeted to cardiac tissue.

The various vector systems have different expression kinetics and tissue tropisms that must be taken into consideration in the design of human cardiac gene therapy trials. Vectors that can conveniently, efficiently, and safely deliver therapeutic genetic material to the heart tissue have been at the center stage of cardiac gene therapy. The continual development of non-viral vectors and viral vectors will no doubt improve the options available in the future clinic.

1.2 Genome-Editing Technologies

In contrast to non-viral vectors or viral vectors, which can mediate only gene addition, genome-editing technologies can mediate a variety of modifications to the targeted genome including gene addition, gene ablation, and gene correction. The field of genome editing was based on the discovery that targeted DNA double-strand breaks (DSBs) could be used to stimulate the endogenous cellular repair machinery (Maeder and Gersbach 2016). DSBs trigger DNA repair via one of two major cellular pathways: nonhomologous end joining (NHEJ) or homology-directed recombination (HDR) (Takata et al. 1998). NHEJ is the classical homology-independent pathway that repairs DSBs. It functions by simply rejoining the DNA ends of DSB without requiring any homologous sequence. In contrast to NHEJ, HDR requires a homologous DNA sequence to repair DNA lesions. HDR is the more accurate repair machinery for DSBs due to the requirement of a homologous DNA sequence between the damaged and intact donor strands of DNA. If the DNA template used for repair is identical to the original DNA sequence at the DSB, there is no error in the process. It also can introduce very specific mutations into the damaged DNA (Pardo et al. 2009). The basic process of genome editing is to create a specific DSB in the genome and then allow the cell's own endogenous repair machinery to repair the break, by either NHEJ or by HDR. Programmable nucleases enable precise genome editing by introducing DSBs at specific genomic loci. There are four basic nuclease platforms that include zinc finger nucleases (ZFNs), transcription activator-like effector nucleases (TALENs), engineered meganucleases, and clustered regularly interspaced short palindromic repeat (CRISPR)/CRISPR-associated endonuclease Cas9 (CRISPR/Cas9) (Table 13.2). All of these nucleases have been demonstrated to achieve efficient genome editing in a wide range of model organisms, and efforts are taken to develop these tools as therapeutics.

Table 13.2 Comparison of different genome-editing nuclease platforms

Name	Components	Mechanism of action	Specificity/ off-target effects	Recognition site	Ease of engineering
Zinc finger nucleases (ZFNs)	Fok1 restriction nuclease fused to multiple zinc finger peptides	Induces DSBs in target DNA	Have off-target effects	Typically 9–18 bp per ZFN monomer, 18–36 bp per ZFN pair	Difficult
Transcription activator-like effector nucleases (TALENs)	Non-specific DNA-cleaving nuclease fused to a DNA-binding domain specific for genomic locus	Induces DSBs in target DNA	Highly specific	Typically 14–20 bp per TALEN monomer, 28–40 bp per TALEN pair	Moderate
Meganucleases	Endonuclease with a large recognition site for DNA	Induces DSBs in target DNA	Highly specific	From 14 and 40 bp	Difficult
Clustered regularly interspaced short palindromic repeat-associated nuclease Cas9 (CRISPR/Cas9)	20 nt crRNA fused to tracrRNA and Cas9 endonuclease	Induces DSBs in target DNA or single-strand DNA nicks (Cas9 nickase)	Some off-target effects	22 bp (20 bp guide sequence +2 bp PAM sequence for S. pyogenes Cas9), up to 44 bp for double nicking	Easily

1.2.1 Zinc Finger Nucleases (ZFNs)

ZFNs are artificial restriction enzymes generated by fusing a zinc finger DNA-binding domain to a FokI nuclease DNA-cleavage domain (Chandrasegaran and Smith 1999). Zinc finger domains can be custom engineered to bind desired DNA sequences. FokI requires dimerization to cut DNA (Bitinaite et al. 1998). The binding of two heterodimers of designed ZFN-FokI hybrid molecules to two contiguous target sequences in each DNA strand separated by a 6 base-pair cleavage site results in FokI dimerization and subsequent DNA cleavage. ZFNs can be used to disable dominant mutations in heterozygous individuals by producing DSBs in the mutant allele, which will be repaired by NHEJ (Mittelman et al. 2009). ZFNs can also rewrite the sequence of an allele by inducing HDR to repair the DSB with the supplied DNA fragment as a template (Porteus and Carroll 2005). By this mechanism, genome-editing technologies based on ZFNs are used to modify genomic sequences to correct a mutation or create a mutation.

1.2.2 Transcription Activator-Like Effector Nucleases (TALENs)

Like ZFNs, TALENs are chimeric enzymes comprised a customizable DNA-binding domain fused to a non-specific FokI nuclease domain. The DNA-binding domain is composed of highly conserved repeats derived from transcription activator-like effectors (TALEs), which are secreted by the bacteria *Xanthomonas* and bind to sequences in the host plant genome, activating transcription (Joung and Sander 2013). The repeat domain of TALEs confers its sequence specificity. This domain is 33–35 amino acids long with divergent 12th and 13th amino acids. These two positions are known as the repeat variable diresidues (RVDs). Different combinations at the RVDs allow for specific nucleotide recognition, which determines TALE specificity (Joung and Sander 2013; Laufer and Singh 2015). Similar to ZFNs, TALEs are fused to the DNA-cleavage domain of FokI nuclease and shown to function as dimers to cleave their intended DNA target site. TALEs can be engineered to bind practically any given DNA sequence. By fusing such an engineered TALE with a DNA-cleavage domain, one can engineer chimeric enzymes that will specifically cut any desired DNA sequence. To introduce a DSB into DNA, engineered TALENs must work as pairs and bind to the forward and reverse strands with an optimal spacer sequence (12–25 bp) to allow a sufficient distance for dimerization of the FokI domains, resulting in the formation of DSBs (Mahfouz et al. 2011). Also similar to ZFNs, TALENs have been shown to efficiently induce both NHEJ and HDR in human somatic cells for the genome-editing purpose.

1.2.3 Meganucleases

Meganucleases, also named homing endonucleases, can be divided into five major classes based on their conserved structural motifs: LAGLIDADG, GIY-YIG, HNH, His-Cys box, and PD-(D/E)XK. The LAGLIDADG meganucleases, the most well-studied family, have become a valuable tool for the study of genome editing. With their DNA-cleavage site surrounded by the two parts of the DNA-binding domain, LAGLIDADG meganucleases are either homodimers (e.g., I-CreI) or monomers with an internal symmetry (e.g., I-SceI). LAGLIDADG meganucleases can be altered to deliver DSBs in a site-specific manner, which were extensively applied as genome-editing tools (Marcaida et al. 2010). To achieve the highly specificity of LAGLIDADG meganucleases in gene targeting, their long-range DNA recognition properties (14–40 bp) must be designed to cleave any desired sequence (Silva et al. 2011). Combinations of computational design, in vivo selection, artificial dimers and high-throughput screening procedures were used to generate meganucleases with tailored specificities. I-CreI, containing a single copy of LAGLIDADG motif, has been well-characterized and commonly used in the study of genomes and genome editing (Silva et al. 2011). I-CreI binds the DNA within its beta-ribbon region of each I-CreI monomer. The beta ribbons curve along the DNA major groove providing a long sequence recognition. When bound by I-CreI, DNA is bent and the minor groove compressed, bringing the two scissile phosphate groups at the active site into close proximity to each other. DNA cleavage occurs across the minor groove between the two scissile phosphate groups. The resulting DSB can serve as a trigger of homologous recombination that will ultimately transfer the open reading frame encoding I-CreI to the cleaved homing DNA site []. A variety of gene therapy applications can potentially make use of the properties of an endonuclease, if such an enzyme is introduced into or expressed in a cell, to drive a gene conversion process. Depending on the presence or absence of a corrective DNA template for break repair, and on the catalytic properties of the endonuclease, such applications can lead to different modification into the target gene.

1.2.4 Clustered Regularly Interspaced Short Palindromic Repeat (CRISPR)/CRISPR-Associated Endonuclease Cas9 (CRISPR/Cas9)

The clustered, regularly interspaced short, palindromic repeat (CRISPR) is a component of a defense system found in bacteria containing an array of short, conserved repeat sequences interspaced by unique DNA sequences of similar size called spacers. The CRISPR array together with Cas genes form the CRISPR/Cas9 system (Barrangou 2015; Horvath and Barrangou 2010). The CRISPR system cuts out small fragments from invading foreign DNA from a virus or plasmid DNA and incorporates them into the host CRISPR locus as spacers between CRISPR RNA (crRNA) repeats. Subsequently, the CRISPR array containing acquired spacers is transcribed, and the RNA fragments (crRNA) corresponding to the integrated foreign DNA fragments will be excised and integrated in the Cas9 endonuclease with the help of a bacterial trans-activating crRNA (tracrRNA). This activated Cas9 endonuclease can specifically recognize and cleave DNA sequences complimentary to the crRNA only if an additional 3–5 bp conserved sequence, the protospacer adjacent motif (PAM), is present at the 3′-end of the target. The PAM sequence is recognized only by the Cas9 protein and is not present in the repeats of the CRISPR array, preventing the cleavage of the bacterial DNA (Horvath and Barrangou 2010).

CRISPR/Cas9 system can be reduced to two components by a combination of the crRNA and tracrRNA to a single-guide RNA (gRNA). Re-targeting of the Cas9/gRNA complex to new sites for introducing site-specific DSBs can be easily done by changing the minor sequences of gRNA (Hsu et al. 2014). The only sequence limitation of the system derives from the necessity of PAM, which is specific to the species of Cas9. However, Cas9 variants with novel PAMs may be engineered by directed evolution, thus dramatically expanding the number of potential target sequences. Through an understanding of the basic mechanisms that regulate this process, scientists have devised strategies that can be used to edit genes in potentially any organism and any type of cell in a rapid and accurate manner. The CRISPR/Cas9 system has been widely adopted since its initial demonstration in 2012. To date, this approach has been used to generate genome edited animals in rat, pig, monkey, rabbit, zebra fish, *Drosophila*, *C. elegans*, as well as human (Pyzocha et al. 2014). Several research groups have now applied this method to introduce single point mutations (deletions or insertions) in a particular target gene, via a single gRNA (Nishimasu et al. 2014; Qi et al. 2013). Using a pair of gRNA-directed Cas9 nucleases instead, it is also possible to induce large deletions or genomic rearrangements, such as inversions or translocations (Ma et al. 2014). The CRISPR/Cas9 system requires only the redesign of the crRNA to change target specificity. This contrasts with other genome-editing tools, including ZFNs and TALENs, which require a redesign of protein-DNA interfaces. With the improvement of CRISPR/Cas systems, we now have powerful tools for genome editing, shifting the bottleneck on the way to the clinical applications to safe and effective evaluation.

1.2.5 RNA Interference (RNAi)

RNA interference (RNAi) is a biological process in which RNA molecules inhibit gene expression, typically by breaking down specific mRNA molecules. The discovery of RNAi allowed for a shift from gene therapy focused on gene augmentation to a focus on downregulation. Cells produce single-stranded RNA (ssRNA), which provide a template for the formation of double-stranded RNAs (dsRNA), which involves the activity of RNA-dependent RNA polymerases. The dsRNA is then cleaved by the RNAse III-like enzyme, called Dicer, into form small 21–23 nucleotide siRNAs (Zamore et al. 2000; Bernstein et al. 2001). These siRNAs form a complex known as the RNA-induced silencing complex (RISC), which leads to cognate target RNA destruction and/or inhibiting protein synthesis. The currently most important therapeutic application of RNAi in cardiovascular medicine is post-transcriptional silencing of genes and inhibiting protein synthesis involved in disease pathogenesis. The major obstacle against the application of RNAi therapy is delivery because they are rapidly cleared in vivo, are not tissue-specific, and their negative charge and size prevent passive endocytosis across the cellular membrane (Lorden et al. 2015). Although the clinical application of RNAi is promising, most of those fall outside the realm of regenerative medicine, focusing rather on disease remediation.

1.3 Strategies for Myocardial Gene Delivery

Methods of gene delivery are an important factor to determine the efficacy of gene transfer to myocardium. To increase gene transferring efficacy, it is necessary to consider the existing delivery barriers in bio-distribution, cellular uptake, and intracellular trafficking. Depending on the type of vector, various limitations have to be overcome. For efficient gene transfer, the naked DNA needs to be protected from serum DNases which would degrade the transgene. For viral vectors, they need to escape both humoral and cellular immune responses against the vector or the transgene. It is better to directly delivery target genes into the myocardium or coronary circulation. If not directly injected into the myocardium, viral vectors need to cross the endothelial barrier in the capillary wall of coronary circulation. Spreading of the vector throughout the myocardium and binding to the cell surface are further crucial steps of transduction. The regulatory sequence could improve cardiac transcription and be resistant to downregulation as it may occur with strong viral promoters (Paillard 1997; Brooks et al. 2004).

A number of gene therapy trials for heart diseases have been conducted or are currently ongoing. Although various delivery methods have been tested in the pre-clinical studies, antegrade intracoronary injection and direct intramyocardial injection are the most widely chosen methods in the clinical trials (Jaski et al. 2009; Hedman et al. 2003; Vale et al. 2001; Ripa et al. 2006; Fuchs et al. 2006; Grines et al. 2003; Rosengart et al. 1999). Figure 13.1 illustrates the gene delivery methods in cardiac gene therapy. There are two methods of direct injection into the

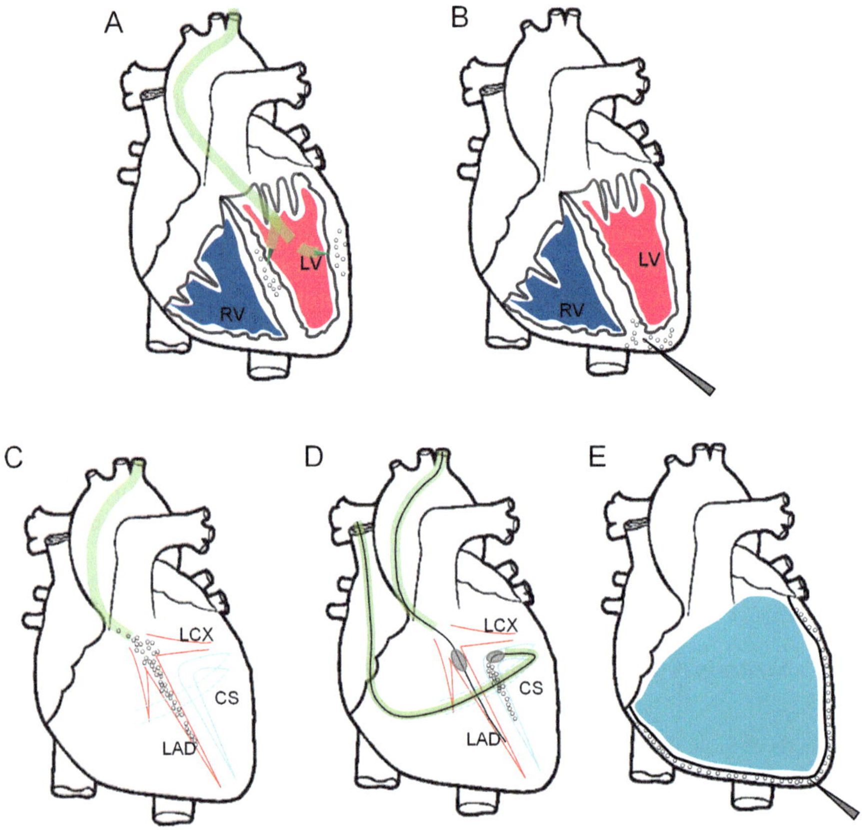

Fig. 13.1 The schematic presentation of gene delivery methods in cardiac gene therapy. (**a**) Direct myocardial injection: the vectors are injected into the endocardial myocardium through the left ventricle by a catheter-based approach; (**b**) Direct myocardial injection: the vectors are injected into the epicardial myocardium by a surgical approach; (**c**) Antegrade intracoronary injection: the vectors are injected through a catheter; (**d**) Retrograde injections through coronary sinus: the vectors are injected into the coronary sinus with simultaneous blockade of the antegrade flow; (**e**) Pericardial injection: the vectors are injected into the pericardial space

myocardium: a percutaneous catheter-based approach (Fig. 13.1a) and a surgical approach (Fig. 13.1b). Both of them allows for the application of a high concentration of vectors directly at the target site and avoid humoral and cellular immune responses against vectors. These approaches also avoid the challenge of vector penetrance through the endothelial barrier. In addition, these methods resulted in successful therapeutic myocardial angiogenesis and focal treatment of cardiac arrhythmias through effects on cellular electrophysiology (Koransky et al. 2002; Edelberg et al. 2001).

Selective coronary catheterization with antegrade intracoronary injection is one of the most attractive gene delivery method involving catheter-based, percutaneous infusion of genetic vector with cardiac selectivity (Fig. 13.1c) (Katz et al. 2012). As a result of the delicate the hemodynamic balance and severe cardiac dysfunction in heart failure for cardiac gene therapy, the minimally invasive nature of this method has a great advantage over other methods. The vectors go through the physiological pathway; thus homogeneous distribution is attained, and the vectors can be accurately delivered to specific coronary artery branches with this method. However, early research using simple bolus injection of vectors into the coronary artery achieved low transduction efficiency. There are many factors that influence uptake of transferring genes in myocardium. For example, the desired distribution may be impaired, if there is a coronary total occlusion. Higher speed coronary flow may lead limited exposure time of the vectors. In addition, the endothelium of coronary capillaries acts as a barrier for the vectors to reach myocardium. Thus, various agents have been deployed to increase the permeability of the coronary capillary bed. Drugs including adenosine, vascular endothelial growth factor (Vale et al. 1999), histamine (Kornowski et al. 2000), and nitroglycerine (Sylven et al. 2002) have been shown to successfully improve the transduction efficiency. Coronary artery blockade with the occlusion balloon, enabling controls the infusion rate and slow down the dilution of the injection of vectors, has been reported to achieve higher gene transduction in myocardium (Emani et al. 2003). And in this closed circuit, repeat circulation of the vectors results in longer vector exposure to the myocardium, thus providing high transduction efficiency.

Retrograde injection into the coronary sinus is another cardiac gene delivery method through target the coronary system. Retrograde infusion using the coronary sinus (Fig. 13.1d) enables access to the ischemic myocardium, regardless of the coronary occlusion, thus allowing homogeneous distribution even in patients with severe CAD (Ishikawa et al. 2011). This approach requires balloon occlusion of both the coronary artery and the sinus to prevent rapid washout of the genetic vectors. Percutaneous retrograde gene delivery by selective pressure-regulated retroinfusion of the coronary veins has shown to be more effective compared to direct myocardial injection and antegrade infusion (Boekstegers et al. 2000; Raake et al. 2004). It is expected that the level of gene expression will increase with the increase of occlusion time. However, the benefit of achieving higher efficiency needs to be balanced with the risk of causing complications associated occlusion time.

The pericardial cavity, comprising a closed space in between the pericardium and the myocardium, has been accessed for the delivery of drugs or genes (Ishikawa et al. 2011; Ayers et al. 1996). The pericardial space is accessible to both surgical and percutaneous delivery. Pericardial injection may be used to overcome some of the restrictions inherent in other gene transfer methods (Fig. 13.1e). First, access to the pericardial cavity can be obtained through rather simple procedure which is minimally invasive. Second, since the pericardium has a closed cavity, this may be used to maintain high concentration and long duration of the genetic vectors. The main drawback of this approach is gene transduction mostly in pericardial cells and only very limited in myocardial cells.

Although efficiency and safety of gene delivery systems and methods have been applied in clinical trials for heart diseases, many hurdles still have to be overcome.

1.4 Molecular Targets of Gene Therapy in Heart Diseases

1.4.1 Angiogenic Factors

Refractory angina classically occurs in patients with advanced, often diffuse, CAD that failed to be completely revascularized by percutaneous coronary intervention (PCI) and/or coronary artery bypass graft with remaining angina pectoris symptoms despite aggressive pharmacological therapy with β-blockers, ACE inhibitors, and statins. Angiogenic factors induce the formation of new vascular networks and make them alternative treatment for CAD. Well-developed coronary collateral network is associated with significant clinical benefits, including reduced infarct size after acute myocardial infarction, fewer future cardiovascular events, and reduced cardiovascular and all-cause mortality (Billinger et al. 2002; Steg et al. 2010; Hansen 1989; Meier et al. 2007). However, majority of patients with CAD have no or only poorly developed coronary collateral network. Therapeutic angiogenesis aims to facilitate collateral development in patients with CAD. Gene-delivering angiogenic growth factors, such as vascular endothelial growth factor (VEGF) or fibroblast growth factor (FGF), may lead to stimulation of coronary collateral vessel development by shifting the balance toward a proangiogenic/proarteriogenic microenvironment (Rubanyi 2013).

The first in-human studies of angiogenic gene therapy for treatment of CAD were reported in late 1990 (Schumacher et al. 1998). Published controlled clinical trials for gene therapy in ischemic heart disease included VEGF-165, VEGF-121, VEGF-C, and FGF-4 (Table 13.3). Adenovirus-mediated VEGF-165 (AdVEGF-165) gene therapy has been proven successful in KAT trial and encouraging in the Euroinject One trial. In the KAT trial, catheter-based intracoronary administration of AdVEGF-165 in patients with severe angina undergoing percutaneous transluminal coronary angioplasty (PTCA) showed a significant increase in myocardial perfusion (Hedman et al. 2003). In the Euroinject One trial, plasmid gene transfer of VEGF-165 (phVEGF-165) via intramyocardial injections in patients with stable severe angina showed little improvement in perfusion defects after 3 months (Kastrup et al. 2005). Similarly, the NORTHERN trial, involving phVEGF-165, did

Table 13.3 Controlled trials of angiogenic gene therapy

Trail	Reference	Treatment(s)	Route of administration	Patient characteristics	Follow-up primary	Primary endpoint(s)	Outcomes
	Laitinen et al. (2000)	Plasmid/liposome VEGF	Intracoronary infusion after PTCA	Patients with PTCA	6 months	Restenosis at 6 months	No change
	Losordo et al. (2002)	Plasmid VEGF2	Percutaneous intramyocardial injection	Class 3–4 angina patients with no other therapeutic options	12 weeks	Myocardial perfusion, CCS	Improved, improved
KAT	Hedman et al. (2003)	Adenoviral and plasmid liposome VEGF165	Intracoronary infusion after PCI	Class 2–3 angina patients undergoing PTCA	6 months	Myocardial perfusion (SPECT), CCS	Improved (AdV), No change (plasmid), no change (AdV and plasmid)
Euroinject One	Kastrup et al. (2005)	Plasmid VEGF165	NOGA-percutaneous—intramyocardial injection	Class 3–4 angina with no other therapeutic options	3 months	Myocardial perfusion; wall motion by NOGA Mapping and LV gram, CCS	No change, improved, no change, no change
REVASC	Stewart et al. (2006)	Adenoviral VEGF121	Intramyocardial injection	Patients with severe angina not amenable to revascularization	26 weeks	ETT time to 1 mm ST Depression, CCS	Improved, improved

(continued)

Table 13.3 (continued)

Trail	Reference	Treatment(s)	Route of administration	Patient characteristics	Follow-up primary	Primary endpoint(s)	Outcomes
	Ripa et al. (Rosengart et al. 1999)	Plasmid VEGF165 and G-CSF	NOGA-percutaneous endocardial injections	Patients with CABG or thoracotomy	3 months	Myocardial perfusion (SPECT and MRI), EDV (MRI), ESV (MRI)	No change and no change, no change, improved
EMAT	Ruel et al. (2008)	Plasmid VEGF165	Cardial injections	Patients with surgical triple-vessel coronary disease and a severely diffusely diseased left anterior descending artery	3 months	Differences in anterior myocardial territory blood flow reserve quantified by PET	Improved
NORTHERN	Stewart et al. (2009)	VEGF165plasmid DNA	NOGA-percutaneous endocardial injections	Class 3–4 angina patients who have reached maximal medical therapy	6 months	Myocardial perfusion (SPECT); CCS	No change, improved
KAT301	Hartikainen et al. (2017)	AdVEGF-DΔNΔC	NOGA catheter-mediated intramyocardial delivery	Patients with refractory angina	12 months	Myocardial perfusion was determined with PET; quality of life was assessed with a standardized 15 dimensions (15D) questionnaire, CCS	Improved, no change, improved
AGENT	Grines et al. (2002)	Ad5-FGF4	Intracoronary infusion	Patients with stable class 2–3 angina	12 weeks	Stress-induced wall motion, ETT	No change, improved

AGENT-2	Grines et al. (2003)	Ad5-FGF4	Intracoronary infusion	Patients with class 2–4 angina on maximal antianginal therapy and not optimal candidates for revascularization	12 months	Perfusion defect size, angina/week, NTG/week	No change, no change, no change
ASPIRE trial	Kaski et al. (Kaski and Consuegra-Sanchez 2013)	Ad5-FGF4	Intracoronary administration	Patients with stable angina pectoris being treated with chronic antianginal medication(s) at a stable dose	8 weeks	SPECT myocardial perfusion imaging	NA
AGENT-3 and AGENT-4	Henry et al. (2007)	AdV-FGF-4	Intracoronary infusion	Patients with class 2–4 angina on maximal anti-Angeline therapy who did not require immediate PCI or CABG	12 months	ETT, CCS, angina/week, NTG/week	No change, improved (gender bias), no change, no change

VEGF vascular endothelial growth factor, *PTCA* percutaneous transluminal coronary angioplasty, *PCI* percutaneous coronary intervention, *CCS* Canadian Cardiovascular Society functional classification of angina, *AdV* adenoviral vector, *LV* left ventricle, *NTG* nitroglycerine, *NOGA* non-fluoroscopic electromechanical mapping, *ETT* exercise tolerance time, *RCT* randomized controlled trial, *MRI* magnetic resonance imaging, *EDV* end diastolic volume, *ESV* end-systolic volume, *FGF* fibroblast growth factor, *PET* Positron-emission tomography, *SPECT* single photon emission computerized tomography, *CABG* coronary artery bypass grafting, *NA* no data not show any improvement in perfusion and exercise time in the Canadian Cardiovascular Society functional classification of angina class 3–4 chronic angina (Stewart et al. 2009). However, both the Euroinject One and NORTHERN trials clearly have demonstrated that safety of gene therapy and the lack of clear benefit were not detrimental and did not worsen clinical outcomes. Since plasmid DNA was used in both these studies, low gene transfection could be one of the reasons why these trials did not achieve clinical improvement. The viral vector-mediated gene transfer may be applied in future trials. Several other clinical trials have demonstrated symptom improvement with gene therapy. In the REVASC trial, adenovirus containing VEGF-121 particle units administered via intramyocardial injection in patients with severe angina not amenable to revascularization resulted in improved exercise time and symptom at 26 weeks of follow-up (Stewart et al. 2006). Plasmid VEGF-2 therapy in patients with CCS class 3–4 angina was safe and symptom improvement in initial pilot studies as well as in long-term follow-up (Losordo et al. 2002; Reilly et al. 2005). In the AGENT and AGENT-2 trials, fibroblast growth factor-4 (FGF-4) gene therapy administered to patients with CCS class 2–3 angina achieved the increase of exercise time at 4 weeks (Grines et al. 2002, 2003). But the AGENT-3 and AGENT-4 trials, trials of adenovirus-mediated FGF-4 gene therapy involving 532 patients with chronic angina, were halted when an interim analysis of the AGENT-3 trial showed that the primary end point (total exercise tolerance testing time) change from baseline did not reach significance at 12 weeks (Henry et al. 2007). However, subgroup analysis in these trials have shown exercise time in women significantly improved and may have been due to increased severity of CAD in women compared to men. Therefore, gender-based differences in response to gene therapy are an important factor, and further studies need to be performed to identify target populations that would benefit from gene therapies. There is an ongoing FGF gene therapy phase 3 trial in women with CAD (AWARE trial). Therapeutic angiogenesis based on delivering angiogenic growth factors genes to the heart have been tested in many clinical trials in the past two decades, but the potential of this new treatment paradigm has not been realized yet in late stage clinical trials (Lavu et al. 2011).

1.4.2 SERCA2a

Sarcoplasmic reticulum calcium(Ca2+) ATPase 2a (SERCA2a) is a critical ATPase responsible for Ca2+ reuptake during excitation-contraction coupling. SERCA2a leads to muscle relaxation by reducing the cytosolic Ca2+ and restores the Ca2+ reserves in the sarcoplasmic reticulum, which are necessary for muscle contraction (Shareef et al. 2014). A common cellular defect in heart failure is altered calcium transport in cardiac muscle cells, which is associated with reduced expression and activity of the calcium handling protein SERCA2a independent of the etiology of heart failure (Kranias and Hajjar 2012). Studies have shown that overexpression of SERCA2a in myocardium improves cardiac function in various animal models of heart failure (Sikkel et al. 2014). The Ca2+ upregulation by percutaneous administration of gene therapy in cardiac disease (CUPID) trials aimed to transfer the SERCA2A gene to fail human heart using a recombinant adeno-associated virus type 1(rAAV1) vector as a treatment for heart failure (Jaski et al. 2009). The rAAV1/SERCA2a was administrated by percutaneous intracoronary infusion, which was performed by using standard catheters after coronary angiography. Phase 1 trial was conducted to restore the levels of this key enzyme in patients with advanced heart failure by gene transfer to the human heart with the use of rAAV1/SERCA2a, demonstrating safety and feasibility (Jaski et al. 2009). Thirty-nine patients with class III/IV HF received intracoronary infusion of placebo or escalation dose rAAV1/SERCA2a vector via percutaneous intracoronary artery infusion. The rAAV1/SERCA2a in high-dose group met the pre-specified criteria for efficacy. Patients in high-dose group demonstrated improvement or stabilization in symptoms, functional class, NT-proBNP levels, and left ventricular end-systolic volumes. And patients in the treatment showed no increase in adverse events, disease-related events or arrhythmias, correlating well with the preclinical studies described earlier, and contrary to the perceived concerns. However, a potentially problematic issue of preexisting neutralizing antibodies against adeno-associated virus was identified in this trial. Patients with even the lowest titers of the antibodies appeared not to receive any therapeutic benefit. Therefore, before consideration for eligibility in the CUPID trial, a serum prescreening protocol was performed to assess baseline status of preexisting neutralizing antibodies against adeno-associated virus. Based on the encouraging results of CUPID, the pivotal CUPID 2 study was undertaken. The CUPID-2 study was a phase 2b randomized, double-blind, multicenter study that enrolled patients with stable heart failure (Greenberg et al. 2016). It is the largest gene transfer study done in patients with heart failure so far. A total of 250 patients were randomized 1:1 to receive intracoronary infusion of placebo or rAAV1/SERCA2a. The study product was delivered in proportion to myocardium perfused; targeting two-thirds of the dose to the anterolateral wall, intravenous nitroglycerin was infused during test article delivery. In a detailed analysis of time to recurrent heart failure events, time to first terminal event, and a separate analysis of all subgroups for the primary and secondary endpoints, there were no significant differences found between groups in any endpoints. Their investigation revealed that inadequate gene delivery was a potential cause of the negative trial and called for improvement in cardiac gene transferring in the future.

1.4.3 S100A1

S100 proteins are a family of Ca2+-modulated proteins implicated in intracellular regulatory activities. S100A1 belongs to the S100 protein family and is the most abundant isoform in cardiomyocytes. It is a regulator of myocardial contractility, which promotes cardiac contractile and relaxation through enhancing the activity of both ryanodine receptors and SERCA2a (Kettlewell et al. 2005). A key underpinning of heart failure is the depletion of S100A1 in cardiomyocytes, which results in abnormal intracellular Ca2+ cycling and poor contractility. Restoring S100A1 expression through gene therapy had beneficial effects on myocardial mechanical function, calcium handling and energetics in small and large animal models of heart failure, and more recently in failing human cardiomyocytes. Recombinant adeno-associated viral vector (rAAV)-mediated stable expression of S100A1 lead to increased myocardial contractile performance in a rat model of heart failure (Pleger et al. 2007). Intracoronary administration of adenovirus-mediated S100A1 gene therapy in a rat model of post-myocardial infarction normalized myocardial contractility and Ca2+ handling (Most et al. 2004). rAAV-mediated gene transfer of S100A1 by retrograde coronary venous delivery in a preclinical large animal model of ischemic cardiomyopathy resulted in improvements in contractile function (Pleger et al. 2011; Weber et al. 2014). All the preclinical data support the use of S100A1 as a gene therapy for the patients with heart failure.

1.4.4 Adenylyl Cyclase 6 (AC6)

Adenylyl cyclase type 6 (AC6), a dominant AC type in heart muscle cells, is a 130-kD membrane protein that catalyzes the conversion of adenosine triphosphate to cyclic adenosine monophosphate, a second messenger that is an important determinant of heart function(Phan et al. 2007). The amount and function of AC6 are reduced in failing hearts. Preclinical studies have shown benefits of increased cardiac AC6 content on cardiac myocytes and the failing heart (Gao et al. 1998; Lai et al. 2004; Lai et al. 2000). Adenovirus 5 encoding human adenylyl cyclase 6 (Ad5.hAC6) is being delivered by intracoronary injection to patients with heart failure. Onetime administration of Ad5.hAC6 in coronary arteries appears to be safe and that AC6 gene transfer provides a dose-related beneficial effect on cardiac function beyond standard heart failure therapy (Hammond et al. 2016).

1.4.5 Stromal Cell-Derived Factor-1 (SDF-1)

The stromal cell-derived factor-1 (SDF-1) is a chemokine protein that binds the G protein-coupled CXCR4 (Bleul et al. 1996). SDF-1 was originally shown to be an important regulator of endogenous tissue repair by recruiting bone marrow-derived stem cells to the site of myocardial injury. The overexpression of SDF-1 has been shown to induce homing of both bone marrows derived and cardiac stem cells to the site of ischemic injury, inhibit cardiac myocyte cell death, and lead to preventing and remodeling of myocardial scar. JVS-100 is a DNA plasmid encoding human SDF-1 designed to be delivered directly to the site of injured tissue. JVS-100 is non-immunogenic which provides for a strong safety profile and the ability to allow for repeat administrations. In a phase I study, JVS-100 delivered by an endomyocardial needle catheter improved heart failure symptoms in patients with ischemic cardiomyopathy (Penn et al. 2013). STOP-HF was a phase II, double-blind, randomized, placebo-controlled trial that evaluated the safety and efficacy of SDF-1 in patients with ischemic heart failure (Chung et al. 2015). JVS-100 was administered via 15 endomyocardial injections. Two different doses of JVS-100 were tested (15 and 30 mg) and compared with a placebo in a total of 93 patients. The primary endpoint, quality of life in heart failure, was not different between the two treated groups and the placebo group at 4 months follow-up. In patients with the lowest tertile of baseline left ventricular eject faction (LVEF), JVS-100 at the dose of 30 mg induced a significant increase in LVEF when compared with placebo. RETRO-HF was a similar study design as STOP-HF to a phase I/II study to evaluate the safety and efficacy of JVS-100 administered by retrograde delivery to cohorts of adults with ischemic heart failure. Patient recruitment for RETRO-HF has been completed in 2014, and phase I part of the data was reported to have preliminary signs of efficacy across several key parameters at 4 months posttreatment. However, the detailed information is not yet available. If RETRO-HF similarly presents better improvement in low LVEF patients, it may support the sub-analysis results of STOP-HF.

1.4.6 Transthyretin (TTR)

Transthyretin (TTR) is an evolutionarily conserved serum and cerebrospinal fluid protein that transports thyroxine and holo-retinol-binding protein (Buxbaum and Reixach 2009). It is a homotetrameric protein synthesized mainly in the liver, choroid plexus, retinal pigment epithelium, and pancreas. Mutant and wild-type TTR gives rise to various forms of amyloid deposition (amyloidosis), originally defined pathologically by the formation and aggregation of misfolded proteins which result in extracellular deposits that impair organ function. The clinical syndromes associated with TTR aggregation are familial amyloid polyneuropathy (FAP) and cardiac amyloidosis, in which mutant TTR protein aggregates in peripheral and autonomic nerves and the heart, respectively. Patients who suffer from cardiac amyloidosis could experience extreme postural hypotension, conduction system disturbances, left wall thickness, valve thickening, pericardial effusion, and heart failure mainly in a restrictive infiltrative pattern. The pattern of myocardial involvement and prognosis varies according to the specific mutation. For example, patients with the FAP V30 M mutation mainly manifest as a progressive peripheral sensorimotor and autonomic polyneuropathy and rarely present with heart failure (Rapezzi et al. 2010). In contrast, FAP T60A is a progressive disease dominated by cardiac amyloidosis associated with varying degrees of autonomic and peripheral polyneuropathy and has a poor outcome (Sattianayagam et al. 2012). Gene therapy based on RNAi has been developed for phenotypes related to TTR gene. The genetic medicines include patisiran and revusiran, which are responsible for targeting TTR amyloidosis within patients suffering from FAP. The activity of these agents is based on the delivery of a small interfering RNA (siRNA) that binds to a conserved sequence in the TTR mRNA common to both mutant and wild-type TTR gene and knocks down the disease-causing TTR protein (Coelho et al. 2013; Hawkins et al. 2015). Patisiran consists of an siRNA encapsulated in a lipid nanoparticle. Following intravenous administration, the LNPs are opsonized by apolipoprotein E and are delivered to the liver through interaction with ApoE receptors expressed on hepatocytes.

Patisiran was generally well tolerated and demonstrated strong knockdown (up to 94% achieved) in a phase I study in healthy volunteers (Coelho et al. 2013). This high level of TTR knockdown is comparable to the reduction in mutant TTR achieved following liver transplantation. In phase II, volunteers were subjected to TTR stabilizers, and this phase was directed more toward knowing the tolerance of the alternate doses of patisiran (Suhr et al. 2015). Phase II open-label extension helped further ensure the safety of the doses used and drug tolerance during the long run. Finally, phase III trial of patisiran has been initiated to prove how effective and safe the use of patisiran drug is in patients suffering from TTR amyloidosis (Adams et al. 2017). Revusiran is another investigational RNAi therapeutic agent for the treatment of TTR amyloidosis by subcutaneous administration. It consists of a TTR-targeting siRNA conjugated to a triantennary N-acetylgalactosamine ligand, which facilitates uptake via a cell surface receptor and has been explored for efficient drug delivery (Nair et al. 2014). Because of its abundant expression on hepatocytes, ability to support multiple rounds of uptake, and ligand specificity, the asialoglycoprotein receptor has been utilized for liver-specific drug and gene delivery. In a phase I study in healthy volunteers, revusiran was generally well tolerated. And revusiran elicited a significant reduction of serum TTR versus the placebo, with mean TTR reductions of approximately 90% observed with multiple dosing. The effects of revusiran in patients with TTR amyloidosis with cardiomyopathy are under assessment in a phase II pilot study and in a phase III study in patients with TTR amyloidosis with cardiomyopathy that was initiated. Provided that patisiran and revusiran meets all the safety guidelines for humans, the gene therapy based on RNAi is intended to support marketing authorization for TTR amyloidosis patients in the near future.

1.4.7 Myosin-Binding Protein C (MYBPC3)

Hypertrophic cardiomyopathy (HCM) is a genetic disorder that is characterized by unexplained and asymmetric left ventricular hypertrophy and a non-dilated left ventricle with preserved or increased ejection fraction. The epidemiology of HCM suggests that it is present in about 1 in 500 adults (Maron et al. 1995). HCM is the commonest cause of sudden death in otherwise healthy young athletes (Maron 2003). Mutations in over a dozen genes encoding sarcomere-associated proteins cause HCM. *MYH7* and *MYBPC3*, encoding β-myosin heavy chain and myosin-binding protein C, respectively, are the two most common genes involved, together accounting for about half of the HCM families (Authors/Task Force members 2014). Approximately two-thirds of *MYBPC3* mutations, including splice site, insertion/deletion, and nonsense mutations, are predicted to generate a truncated mRNA and protein. These mutant products can be degraded by nonsense-mediated mRNA decay and the ubiquitin-proteasome system, which results in an insufficient amount of normal full-length MYBPC3 protein and altered sarcomeric function (Tian et al. 2013). Due to delayed presentation of HCM, the mutations escape natural selection and are often inherited to the offspring leading to high frequency in some patient populations. Current treatments for HCM are mainly symptomatic remission without addressing genetic causation. One option to prevent the inheritance of certain mutation-associated HCM is to apply genetic testing during in vitro fertilization (IVF) fertility treatment. This allows for the selection of embryos for implantation without pathogenic mutation. It is hypothesized that precise genome-editing approach could enable an individual with HCM to increase the percentage of embryos without disease mutation available for IVF implantation (Ma et al. 2017). By applying CRISPR/Cas9, induced DSBs at the *MYBPC3* mutant paternal allele were repaired primarily based on the homologous wild-type maternal gene rather than a synthetic DNA template. Researchers did not identify off-target effects caused by introducing CRISPR/Cas9. And mosaicism can be diminished by coinjecting the sperm and the CRISPR complex at the earliest stage possible. Embryonic genome editing based on CRISPR/Cas9 has potential to be used for the correction of pathogenic mutations in human embryos by complementing preimplantation genetic diagnosis.

2 Genomic Applications in Heart Disease

2.1 Genome-Wide Association Studies(GWAS) in Heart Diseases

The sequencing of the human genome, together with an everexpanding catalog of variations mapped as single nucleotide polymorphisms (SNPs), has allowed large-scale genome-wide association studies (GWAS), which have the potential to fill the gap in our understanding of the genetic basis of heart diseases (Hirschhorn and Daly 2005).

GWAS have identified more than 30 loci associated with CAD. The most recent meta-analysis carried out by the CARDIoGRAMplusC4D consortium, which included about 185,000 cases and controls, revealed about 50 distinct loci that were associated with CAD at the genome-wide significance threshold (Nikpay et al. 2015). Meta-analysis of GWAS data in large numbers of subjects have also been carried out for numerous quantitative phenotypes clinically associated with CAD, including conventional risk factors, such as lipid levels and blood pressure, as well as less traditional biomarkers related to other biological mechanisms, such as inflammation, oxidation, coagulation, and amino acid metabolism (Hartiala et al. 2017). A number of GWAS have been also performed to identify and evaluate causal or susceptibility variants in heart failure syndromes, but as yet no consistent themes or common pathways are emerging (Villard et al. 2011; Meder et al. 2014; Stark et al. 2010). Susceptibility variants are located in genes of catecholamine secretion, cell signaling, and protein degradation, which suggesting complexity of the underlying mechanism of heart failure. Moreover, GWAS studies have identified variants in the susceptibility of developing the cardiac arrhythmias such as atrial fibrillation (AF), the most common arrhythmia associated with significant morbidity and mortality. The first GWAS for AF identified a susceptibility locus at chromosome 4q25 in a population of European descent and replicated this association in an Asian population (Gudbjartsson et al. 2007). So far genetic variants identified by GWAS seem to affect pathways in cardiac development, generation and propagation of electrical impulses, inflammation signaling pathways, atrial remodeling, and development of fibrosis (Thorolfsdottir et al. 2017). In general, conditions that slow conduction velocity and shorten action potential duration or atrial effective refractory period favor development of AF (Gudbjartsson et al. 2007; Schotten et al. 2011; Nielsen et al. 2018).

Although GWAS studies of heart diseases have provided important insights into the underlying genetic mechanisms and defined responses to therapies, translation of these discoveries to the bedside care of individual patients has been limited. There is a need for randomized controlled trials that are genotype-directed for the treatment of heart diseases.

2.2 *Next-Generation Sequencing in Heart Diseases*

Next-generation sequencing (NGS), also known the massively parallel sequencing, is the catch-all term used to describe a number of different modern sequencing technologies including Illumina sequencing, Roche 454 sequencing, Proton/PGM sequencing, and SOLiD sequencing (Mardis 2013). These recent technologies allow us to sequence DNA and RNA much more quickly and cheaply than the previously used Sanger sequencing and as such have revolutionized the study of genomics and molecular biology (Fig. 13.2). All NGS platforms can perform sequencing of millions of small fragments of DNA sample parallelly. Bioinformatics analyses are used to align these fragments by mapping the generated reads to the reference sequence. Each base of the DNA is sequenced multiple times, which provides high depth to deliver accurate data and an insight into unexpected genetic variation. Thus, NGS can easily be applied to genome sequencing, transcriptome profiling, epigenome profiling, and epitranscriptome characterization and broaden the capacity in which to study regulators of gene networks in heart diseases (de Magalhaes et al. 2010; Slagle and Conlon 2016). Figure 13.2 shows an overview of the main high-throughput technologies discussed here.

2.2.1 DNA Sequencing

The process of NGS of DNA starts with extraction of DNA of an individual, most commonly from peripheral leukocytes obtained from blood sample but can be from another tissue such as buccal swab or saliva (Jamuar and Tan 2015). The DNA is then broken down into short fragments and amplified using PCR- or hybridization-based approaches (Fig. 13.2). The regions that are amplified could include either a subset of genes or whole genome (Bamshad et al. 2011). Within cardiovascular

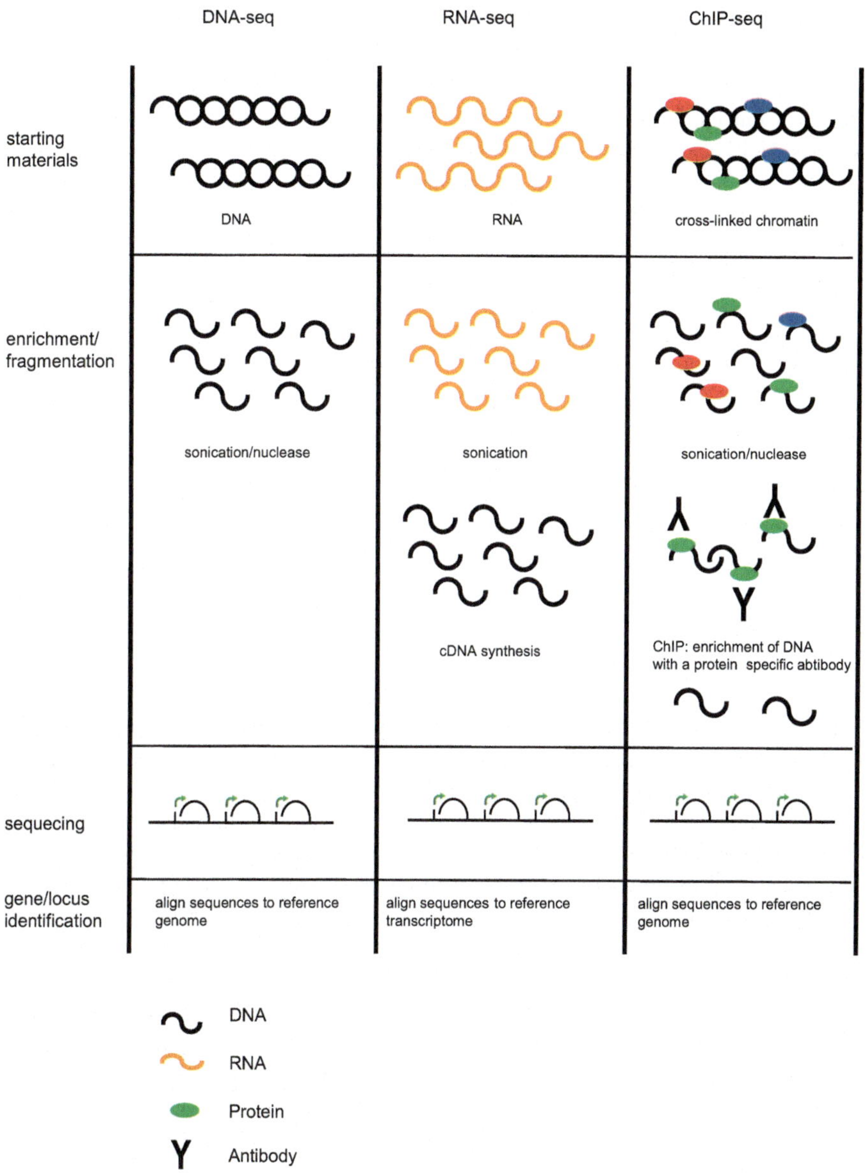

Fig. 13.2 The schematic presentation of the main high-throughput technologies

medicine, NGS is applied in the diagnosis of Mendelian diseases. Cardiovascular Mendelian diseases span multiple syndromes, such as long QT syndrome, familial hypercholesterolemia, familial cardiomyopathies, and Marfan syndrome. Characterizing the underlying genetic DNA variation of these cases reassures the patient, directs family screening and fertility planning, and can often actually guide therapy (Kathiresan and Srivastava 2012). For example, identification of mutation in a gene causing long QT syndrome in the proband can allow identification of other at-risk family members, who can then have regular Holter monitoring and, in some instances, implantation of a cardiac pacemaker before a catastrophic event occurs. NGS can capture a broader spectrum of mutations than Sanger sequencing. The spectrum of DNA variation in a human genome comprises small base changes (substitutions), insertions and deletions of DNA, large genomic deletions of exons or whole genes, and rearrangements such as inversions and translocations. All these DNA variations in the patient with inherited heart disease can be derived from NGS sequencing data directly, obviating the need for dedicated assays while harvesting the full spectrum of genomic variation in a single experiment. In the case of unrevealing genetic cause, whole-genome sequencing (WGS) or whole-exome sequencing (WES) can yield important information about potentially pathogenic genetic variation. NGS with or without additional positional information generated by genome-wide linkage analysis of the pedigree can identify novel genes mutated in the inherited heart diseases. In the future, continuing advances in NGS will increase the uptake of genetic testing when lower costs make it a routine technology in clinical practice.

2.2.2 RNA Sequencing

The transcriptome of a given tissue represents a complete set of the transcribed genome and encompasses all the transcripts, including the protein-coding mRNAs and the noncoding RNAs (ncRNAs) such as miRNAs and lncRNAs. Profiling the transcriptome by high-throughput RNA-sequencing (RNA-seq), also called whole-transcriptome shotgun sequencing, transcriptome complexity and its dynamics can be revealed at different levels in heart diseases (Wang et al. 2009). In RNA-seq, RNA is commonly converted to a more stable complementary DNA (cDNA). Then, library preparation and sequencing of cDNA follow the same sequencing procedure as DNA sequencing (Fig. 13.2). RNA-seq focused on mRNA, small RNA, ncRNA, or miRNAs can be achieved by including additional isolation or enrichment steps before cDNA synthesis. RNA-seq allows determination of the transcriptional structure of genes, in terms of their transcription start sites, splicing patterns, and other posttranscriptional modifications at different levels in heart development and diseases (Parkhomchuk et al. 2009). It provides a dramatic increase in our knowledge of the transcriptional landscape and regulation, including insight into functional pathways and regulation of biological processes of coordinated genes during heart development, under different conditions and in different heart diseases.

2.2.3 Methylation Sequencing (Methyl-seq) and Chromatin Immunoprecipitation Sequencing (ChIP-seq)

Epigenetics is the study of external factors besides DNA sequence variation that influence the process of gene transcription. In principle, this can encompass a diverse set of mechanisms ranging from DNA methylation to histone modifications, transcription factors, microRNAs, and long intervening noncoding RNAs (Perrino et al. 2017). The advances in NGS technologies have had a significant impact on epigenetic research.

DNA methylation, which most commonly occurs at the C5 position of cytosines within CpG dinucleotides, is frequently found clustered at gene regulatory sites such as promoter regions. Areas of the genome high in CpG concentration have increased methyl transferase activity and may be referred to as CpG islands. Methylation at CpG islands decreases the activity of promoters and generally decreases gene expression. DNA methylation at CpG islands has been shown to strongly suppress promoter activity and seems to occur as a function of age, causing loss-of-function phenotypes, and may be a target in many disorders. Bisulfite genomic sequencing is regarded as a gold-standard technology for detection of DNA methylation (Li and Tollefsbol 2011). Treatment of DNA with bisulfite converts cytosine residues to uracil but leaves 5-methylcytosine residues unaffected. After PCR amplification, uracil residues are converted to thymine. DNA methylation status can be determined by direct PCR sequencing or cloning sequencing. Bisulfite genomic sequencing, as a fundamental method of DNA methylation analysis, has been widely used in various research and clinical settings. Methyl-seq data analysis involves sequencing bisulfite-treated DNA and comparing this with the fraction that did not undergo bisulfite treatment, producing differences that can identify the regions that did not convert to uracil; an uracil is read as a thymidine during sequencing and is reported as a methylation site. The differences in methylation could be observed when analyzing the methylation pattern at the single gene level. By applying methyl-seq, increasing evidences were found that DNA methylation plays a causative role in cardiac hypertrophy and reducing cardiac contractility in human heart failure, suggesting that demethylation is a potential therapeutic strategy in the treatment of heart failure.

Chromatin is a complex of DNA and proteins in the nucleus of a cell. The primary protein components of chromatin are histones, including histone H1/H5, H2A, H2B, H3, and H4. Histones H2A, H2B, H3, and H4 are known as the core histones, while histones H1 and H5 are known as the linker histones. Histones and other regulatory proteins bind to the DNA and maintain its three-dimensional structure. There are many types of histone modifications, including acetylation, methylation, ubiquitination, citrullination, and phosphorylation of specific amino acids within the histone protein, usually toward the C-terminal end of the protein (Kimura 2013). These modifications can both positively and negatively regulate gene expression by changing the way in which histones bind to DNA. Genome-wide profiling histone modifications can be detected by chromatin immunoprecipitation followed by NGS (ChIP-seq). Antibodies specific to a histone modification are used to immunoprecipitate the histone-DNA complex. The bound DNA is then coprecipitated, purified, and sequenced by applying NGS (Fig. 13.2) (Park 2009). ChIP-seq can be used to interactions between specific proteins and DNA in the cell and determining their localization on a specific genomic locus (Mundade et al. 2014). It has played a significant role in our understanding of mechanisms and in exploring epigenetic changes in heart development and diseases.

2.2.4 N6-Methyladenosine Sequencing (m6A-seq) and Methylated RNA Immunoprecipitation Sequencing (MeRIP-seq)

Within the field of molecular biology, the epitranscriptome includes all the biochemical modifications of the RNA within a cell. All known RNA modified residues (over 150 altogether) are formed posttranscriptionally, by complex enzymatic mechanisms involving hundreds of different catalytic proteins and cofactors (Licht and Jantsch 2016). These modifications happen to all types of cellular RNA including, but not limited to, ribosomal RNA(rRNA), transfer RNA (tRNA), messenger RNA (mRNA), and small nuclear RNA (snRNA). Major mRNA modifications in the transcriptome of eukaryotic cells include N6-methyladenosine(m6A), N6, 2′-O-dimethyladenosine, 5-methylcytidine, 5-hydroxylmethylcytidine, inosine, pseudouridine, and N(1)-methyladenosine (Li et al. 2016). Discovered in 1974, m6A is the most abundant eukaryotic mRNA modification (Dominissini et al. 2012; Desrosiers et al. 1974). To profile the m6A, N6-methyladenosine sequencing (m6A-seq) and methylated RNA immunoprecipitation sequencing (MeRIP-seq) were developed (Dominissini et al. 2012). Both methods are very similar in the initial RNA preparation and immunoprecipitation steps. m6A-specific antibodies are used to immunoprecipitate RNA by binding with the m6A site. The unbound RNA fragments are removed. The RNA fragments in the elute are reverse-transcribed to cDNA and sequenced. Deep sequencing provides high-resolution reads of m6A-methylated RNA.

2.3 Pharmacogenetics Applications of Medications for Heart Diseases

Pharmacogenetics entails the use of genetic information to predict a person's response to drugs, both in terms of therapeutic effect and adverse effect. The ultimate objective is to deliver the right therapy at the right dose for the right patient and improve prescribing safety and efficacy (Johnson and Cavallari 2013). Genomic research has enabled a deeper understanding of a variety of phenotypes relevant to cardiovascular medicine including efficacy, toxicity, drug metabolism, and drug interactions. The DNA variants used in pharmacogenetic tests are identified in one of two ways: through analysis of DNA variants in candidate genes with biological links to drug activity or in an unbiased GWAS to find SNPs that are associated with a particular drug response or adverse effect. Table 13.4 lists cardiovascular drugs that the Food and Drug Administration (FDA) pharmacogenetic marker labels for genetic testing advisory.

Warfarin is an anticoagulant, which is widely prescribed for the prevention and treatment of thromboembolic disease. It is challenging to use because of highly

Table 13.4 Pharmacogenomic marker labels for cardiovascular drug

Drugs	Drug categories	Pharmacogenomic markers	Label sections
Carvedilol	β adrenergic receptor blocker and α adrenergic receptor blockers	CYP2D6	Drug interactions, clinical pharmacology
Clopidogrel	Platelet aggregation inhibitors	CYP2C19	Boxed warning, warnings and precautions, clinical pharmacology
Hydralazine	Smooth muscle relaxants	NAT1, NAT2	Clinical pharmacology
Isosorbide dinitrate	Nitrates	CYB5R	Overdosage
Isosorbide mononitrate	Nitrates	CYB5R	Overdosage
Metoprolol	β adrenergic receptor blockers	CYP2D6	Drug interactions, clinical pharmacology
Nebivolol	β adrenergic receptor blockers	CYP2D6	Dosage and administration, clinical pharmacology
Propranolol	β adrenergic receptor blocker	CYP2D6	Clinical pharmacology
Propafenone	Antiarrhythmic agents, sodium channel blockers	CYP2D6	Dosage and administration, warnings and precautions, drug interactions, clinical pharmacology
Prasugrel	Platelet aggregation inhibitors	CYP2B6, CYP3A5, CYP2C9, CYP2C19	Use in specific populations, clinical pharmacology, clinical studies
Quinidine	Antiarrhythmic agents, sodium channel blockers	CYP2D6	Precautions
Rosuvastatin	Cholesterol reducing medications	SLCO1B1	Clinical pharmacology
Ticagrelor	Platelet aggregation inhibitors	CYP2C19	Clinical pharmacology
Warfarin	Anticoagulants	CYP2C9, VKORC1, PROC, PROS1	Dosage and administration, drug interactions, clinical pharmacology, warnings, and precautions

Data consolidated from FDA website as of May, 2018

variable responses among patients and even within an individual patient. Warfarin is metabolized primarily via oxidation in the liver by cytochrome P450 2C9 (CYP2C9) and exerts its anticoagulant effect by inhibiting the protein vitamin K epoxide reductase complex, subunit 1 (VKORC1). Polymorphisms in two genes, *CYP2C9* and *VKORC1*, explain up to 40% of the interindividual variation in stable therapeutic dosing of warfarin, and FDA updated the label on warfarin to allow for genotype-specific dose ranges (Table 13.4) (Loebstein et al. 2007). Using a combination of genetic and clinical factors to predict the maintenance warfarin dose appears to be more accurate than a fixed-dose approach (International Warfarin Pharmacogenetics Consortium et al. 2009).

Platelet aggregation inhibitors are a group of drugs that inhibits platelets from aggregating to form a plug. Clopidogrel, an antiplatelet agent, is used to prevent heart attack and stroke in people who are at high risk of these events, including those with a history of myocardial infarction and other forms of acute coronary syndrome, stroke, and those with peripheral artery disease. Patients display variable responses to clopidogrel therapy because clopidogrel is not itself an active drug but must be converted into an active metabolite by the hepatic cytochrome P-450 2C19 enzyme (CYP2C19). Three landmark studies of mostly post-acute coronary syndrome or post-PCI on clopidogrel therapy found that carriers of loss-of-function (LOF) *CYP2C19* variants experienced significantly higher rates of cardiovascular death, myocardial infarction, and stroke (Mega et al. 2009; Simon et al. 2009; Collet et al. 2009). Subsequent studies with mostly lower-risk patients who did not undergo PCI did not find a difference in the effects of clopidogrel on LOF variant carriers versus noncarriers (Pare et al. 2010). Meta-analyses of numerous studies have reached conflicting conclusions over whether LOF variant carriers are at increased risk for adverse events when treated with clopidogrel especially after PCI (Mega et al. 2010; Holmes et al. 2011). The available data suggest that patients at highest risk for cardiovascular events may have worse outcomes on clopidogrel if they are LOF variant carriers. FDA has issued a black box warning advising practitioners to consider alternative treatment in patients identified as CYP2C19-poor metabolizers and note that these patients can be identified by genotyping. Polymorphisms of *CYP2C19* do not seem to affect prasugrel or ticagrelor pharmacokinetics, its effect on platelet aggregation, or clinical outcomes (Pereira et al. 2016). Therefore, these therapies may be useful alternatives to clopidogrel in the carriers of *CYP2C19* LOF genetic variants.

Evidence of other pharmacogenetic interactions that may be clinically important is accumulating. Variation in the β1 adrenergic receptor blockers, *ADRB1*, is associated with altered responsiveness to beta-blockade in heart failure (de Groote et al. 2005). A variant in *SLCO1B1* has been implicated in statin-related myopathy (SEARCH Collaborative Group et al. 2008). Given the complexity of heart physiology and the genetic complexity, the effect of genetic variation about pharmacogenetics was not well established. Deep sequencing of variants and genes related to drug absorption, distribution, metabolism, and excretion may identify specific variants that contribute to the heterogeneity of responsiveness to cardiovascular drugs.

2.4 The Challenge of Interpreting Genetic Information

Advances in high-throughput sequencing offer unique opportunities, as well as significant challenges, for the future of genetic testing in heart diseases. Genetic diagnostics require accurate and consistent classification of sequence variants. Detection of a pathogenic variant may influence a patient's diagnosis, prognosis, family screening, and lifelong therapy. Our understanding of the clinical significance of any given sequence variant falls along a gradient, ranging from those in which the variant is almost certainly pathogenic for a disorder to those that are almost certainly benign. To clarify the effect of a nucleotide change, the term "mutation," which has a negative connotation but does not actually infer a deleterious effect, has been replaced with the neutral term "variant" that is modified to define its association to disease phenotype, with descriptors such as pathogenic or benign.

In an attempt to broadly guide interpretation of variations identified in Mendelian disorders, the American College of Medical Genetics and Genomics (ACMG) and Association for Molecular Pathology (AMP) have published guidelines for classifying a variety of genetic variants along the spectrum of pathogenicity (Richards et al. 2015). The ACMG/AMP guidelines, which rely on original evidence of pathogenicity rather than previously classified data, include a semi-structured process that strives to unify the classification process so that conclusions reached by different laboratories using the same set of evidence are consistent. The guidelines classify the variants identified in the Mendelian disorders into five categories: pathogenic, likely pathogenic, likely benign, benign, and uncertain significance. The recommendation describes criteria for classifying variant, which is based on the population evidence, genetic evidence, functional evidence, bioinformatics software-based evidence, and literature evidence.

Publicly available large-scale exome/genome databases provide allele frequency data for the general population, which is applied as population evidence to assess variant pathogenicity. These databases include the Exome Aggregation Consortium (http://exac.broadinstitute.org/), Exome Variant Server (http://evs.gs.washington.edu/EVS), 1000 Genomes (http://browser.1000genomes.org), dbSNP (http://www.ncbi.nlm.nih.gov/snp), and dbVar (http://www.ncbi.nlm.nih.gov/dbvar) (Lek et al. 2016; Fu et al. 2013; Lappalainen et al. 2013; Sherry et al. 1999). In principle, when a variant is more common than the incidence of a disease, and disease penetrance is taken into consideration, the variant is unlikely to be a pathogenic variant. On the contrary, if a variant is absent or is present at a low frequency in the general population, the variant is likely to be disease-causing. However, there is no absolute frequency threshold to prove that a variant is likely benign or likely pathogenic. Several databases recording disease-causing or disease-associated variation are applied for the assessment of the variants' pathogenicity. These include the Human Gene Mutation Database (http://www.hgmd.cf.ac.uk), OMIM (http://www.omim.org/), ClinVar (http://www.ncbi.nlm.nih.gov/clinvar/), Locus/Disease/Ethnic/Other-Specific Databases (http://www.hgvs.org/dblist/glsdb.html), and DECIPHER (http://decipher.sanger.ac.uk). Disease databases often contain variants that are incorrectly classified including incorrect claims in published literature because many databases do not perform primary review of evidence (Stenson et al. 2014; Landrum et al. 2016; Greenblatt et al. 2008; Firth et al. 2009). When using disease databases, it is important to consider how patients were ascertained.

The importance of inheritance pattern, as well as family data for variant interpretation, cannot be overemphasized. De novo occurrence of a variant is considered strong support for pathogenicity, especially when paternity and maternity are confirmed. However, de novo variants can occur for any gene at a frequency of about 1 per exome, and most of an individual's de novo variants are expected to be clinically irrelevant (Samocha et al. 2014). Thus, additional evidence is required, which includes genotype-phenotype matching and ruling out of germ-line mosaicism. Although linkage evidence does not automatically translate into a causal relationship between variants and disease, the higher the level of cosegregation, the stronger the support for a variant-phenotype association. A high logarithm of the odds score supports a gene-level causal relationship, but linkage disequilibrium should be considered when assigning the causal relationship at the variant level.

Software-based prediction tools, both publicly and commercially available, facilitate analysis of single nucleotide variants at the level of annotation, interpretation, or both. Some of these tools, such as ANNOVAR, TREAT, VarioWatch, SeqAnt, and Anntools, have been specifically designed to handle the annotation task (Wang et al. 2010; Asmann et al. 2012; Cheng et al. 2012; Shetty et al. 2010; Makarov et al. 2012). Other tools focus on the interpretation step. A category of interpretation tools has been developed to assess whether a missense change is damaging to the resultant protein function or structure and those that predict if there is an effect on splicing. The pathogenicity of a missense change depends on the criteria such as the evolutionary conservation of an amino acid or nucleotide, the location in the protein structure and impact the overall protein, as well as the biochemical consequence of the amino acid substitution. These prediction tools are widely used but considered as low-weight evidence. Commonly used software-based prediction tools like PolyPhen-2 and SIFT are often contradictory among themselves and other non-computational evidence. In general, most tools for missense variant prediction are 65–80% accurate when examining known disease variants. Therefore, the use of software-based prediction tools in sequence variant interpretation should be used carefully and is not recommended to be used as the sole source of evidence to make a clinical assertion (Richards et al. 2015).

Evidence from functional studies is often considered the ultimate confirmation of causal relationships between variants and phenotypes. When introduced to cultured cells or laboratory animals, the suspect variant causes alterations of the normal phenotype, and these abnormalities are rescued by methods that recover the wild-type function. In reality, however, functional studies often fail to unequivocally demonstrate a causal relationship, because functional assays are often performed in nonhuman systems using proxy readouts rather than directly measuring or observing a specific disease phenotype. Therefore, a thorough understanding of the biology of the disease and mutation mechanisms is important.

Variant assessment also includes searching the scientific and medical literature. The immediate clinical use of the new information on genes and genetic variants from research studies, without further validation, may lead to premature conclusions and misinterpretation and might cause irreparable harm.

For the vast majority of identified susceptibility variants in GWAS, the underlying mechanism of their contribution to disease is unknown currently. Characterization of the biological mechanism underlying associations between genetic variants and diseases can provide a better understanding of disease pathogenesis and therefore lead to better clinical care of patients.

Eventually, all the lines of evidence described above, as well as other sources and types of related knowledge, need to be evaluated in order to achieve the most accurate insight on the potential role of a given variant in a given disease or symptom, in a given patient. Increasingly, public and commercial bioinformatics packages are being developed that exploit data from multiple sources, taking into account different lines of evidence in a combined way. They can carry out iterative operations to contrast hypothesis, identify relationship patterns, and generate metadata to facilitate biomedical interpretation. However, these machine learning systems are still in an early development stages for clinical applications.

Acknowledgment The authors would like to acknowledge financial support from the National Natural Science Foundation, China (No. 81570348).

References

Adams D, et al. Trial design and rationale for APOLLO, a phase 3, placebo-controlled study of patisiran in patients with hereditary ATTR amyloidosis with polyneuropathy. BMC Neurol. 2017;17(1):181.

Askari AT, et al. Effect of stromal-cell-derived factor 1 on stem-cell homing and tissue regeneration in ischaemic cardiomyopathy. Lancet. 2003;362(9385):697–703.

Asmann YW, et al. TREAT: a bioinformatics tool for variant annotations and visualizations in targeted and exome sequencing data. Bioinformatics. 2012;28(2):277–8.

Asokan A, Schaffer DV, Samulski RJ. The AAV vector toolkit: poised at the clinical crossroads. Mol Ther. 2012;20(4):699–708.

Authors/Task Force members, et al. ESC guidelines on diagnosis and management of hypertrophic cardiomyopathy: the task force for the diagnosis and management of hypertrophic cardiomyopathy of the European Society of Cardiology (ESC). Eur Heart J., 2014. 2014;35(39):2733–79.

Ayers GM, et al. Amiodarone instilled into the canine pericardial sac migrates transmurally to produce electrophysiologic effects and suppress atrial fibrillation. J Cardiovasc Electrophysiol. 1996;7(8):713–21.

Bamshad MJ, et al. Exome sequencing as a tool for Mendelian disease gene discovery. Nat Rev Genet. 2011;12(11):745–55.

Barrangou R. The roles of CRISPR-Cas systems in adaptive immunity and beyond. Curr Opin Immunol. 2015;32:36–41.

Bernstein E, et al. Role for a bidentate ribonuclease in the initiation step of RNA interference. Nature. 2001;409(6818):363–6.

Billinger M, et al. Physiologically assessed coronary collateral flow and adverse cardiac ischemic events: a follow-up study in 403 patients with coronary artery disease. J Am Coll Cardiol. 2002;40(9):1545–50.
Bitinaite J, et al. FokI dimerization is required for DNA cleavage. Proc Natl Acad Sci U S A. 1998;95(18):10570–5.
Bleul CC, et al. A highly efficacious lymphocyte chemoattractant, stromal cell-derived factor 1 (SDF-1). J Exp Med. 1996;184(3):1101–9.
Boekstegers P, et al. Myocardial gene transfer by selective pressure-regulated retroinfusion of coronary veins. Gene Ther. 2000;7(3):232–40.
Brooks AR, et al. Transcriptional silencing is associated with extensive methylation of the CMV promoter following adenoviral gene delivery to muscle. J Gene Med. 2004;6(4):395–404.
Buxbaum JN, Reixach N. Transthyretin: the servant of many masters. Cell Mol Life Sci. 2009;66(19):3095–101.
Chandrasegaran S, Smith J. Chimeric restriction enzymes: what is next? Biol Chem. 1999;380(7–8):841–8.
Cheng YC, et al. VarioWatch: providing large-scale and comprehensive annotations on human genomic variants in the next generation sequencing era. Nucleic Acids Res. 2012;40(Web Server issue):W76–81.
Chung ES, et al. Changes in ventricular remodelling and clinical status during the year following a single administration of stromal cell-derived factor-1 non-viral gene therapy in chronic ischaemic heart failure patients: the STOP-HF randomized phase II trial. Eur Heart J. 2015;36(33):2228–38.
Coelho T, et al. Safety and efficacy of RNAi therapy for transthyretin amyloidosis. N Engl J Med. 2013;369(9):819–29.
Collet JP, et al. Cytochrome P450 2C19 polymorphism in young patients treated with clopidogrel after myocardial infarction: a cohort study. Lancet. 2009;373(9660):309–17.
de Groote P, et al. Association between beta-1 and beta-2 adrenergic receptor gene polymorphisms and the response to beta-blockade in patients with stable congestive heart failure. Pharmacogenet Genomics. 2005;15(3):137–42.
de Magalhaes JP, Finch CE, Janssens G. Next-generation sequencing in aging research: emerging applications, problems, pitfalls and possible solutions. Ageing Res Rev. 2010;9(3):315–23.
Desrosiers R, Friderici K, Rottman F. Identification of methylated nucleosides in messenger RNA from Novikoff hepatoma cells. Proc Natl Acad Sci U S A. 1974;71(10):3971–5.
Dominissini D, et al. Topology of the human and mouse m6A RNA methylomes revealed by m6A-seq. Nature. 2012;485(7397):201–6.
Dunbar CE, et al. Gene therapy comes of age. Science. 2018;359(6372):pii: eaan4672.
Edelberg JM, et al. Molecular enhancement of porcine cardiac chronotropy. Heart. 2001;86(5):559–62.
Emani SM, et al. Catheter-based intracoronary myocardial adenoviral gene delivery: importance of intraluminal seal and infusion flow rate. Mol Ther 2003;8:306–313.
Firth HV, et al. DECIPHER: database of chromosomal imbalance and phenotype in humans using Ensembl resources. Am J Hum Genet. 2009;84(4):524–33.
Fu W, et al. Analysis of 6,515 exomes reveals the recent origin of most human protein-coding variants. Nature. 2013;493(7431):216–20.
Fuchs S, et al. A randomized, double-blind, placebo-controlled, multicenter, pilot study of the safety and feasibility of catheter-based intramyocardial injection of AdVEGF121 in patients with refractory advanced coronary artery disease. Catheter Cardiovasc Interv. 2006;68(3):372–8.
Gao M, et al. Increased expression of adenylylcyclase type VI proportionately increases beta-adrenergic receptor-stimulated production of cAMP in neonatal rat cardiac myocytes. Proc Natl Acad Sci U S A. 1998;95(3):1038–43.
Greenberg B, et al. Calcium upregulation by percutaneous administration of gene therapy in patients with cardiac disease (CUPID 2): a randomised, multinational, double-blind, placebo-controlled, phase 2b trial. Lancet. 2016;387(10024):1178–86.

Greenblatt MS, et al. Locus-specific databases and recommendations to strengthen their contribution to the classification of variants in cancer susceptibility genes. Hum Mutat. 2008;29(11):1273–81.

Grines CL, et al. Angiogenic Gene Therapy (AGENT) trial in patients with stable angina pectoris. Circulation. 2002;105(11):1291–7.

Grines CL, et al. A randomized, double-blind, placebo-controlled trial of Ad5FGF-4 gene therapy and its effect on myocardial perfusion in patients with stable angina. J Am Coll Cardiol. 2003;42(8):1339–47.

Gudbjartsson DF, et al. Variants conferring risk of atrial fibrillation on chromosome 4q25. Nature. 2007;448(7151):353–7.

Hacein-Bey-Abina S, et al. LMO2-associated clonal T cell proliferation in two patients after gene therapy for SCID-X1. Science. 2003;302(5644):415–9.

Hajjar RJ, et al. Design of a phase 1/2 trial of intracoronary administration of AAV1/SERCA2a in patients with heart failure. J Card Fail. 2008;14(5):355–67.

Hammond HK, et al. Intracoronary gene transfer of adenylyl cyclase 6 in patients with heart failure: a randomized clinical trial. JAMA Cardiol. 2016;1(2):163–71.

Hansen JF. Coronary collateral circulation: clinical significance and influence on survival in patients with coronary artery occlusion. Am Heart J. 1989;117(2):290–5.

Hartiala J, et al. The genetic architecture of coronary artery disease: current knowledge and future opportunities. Curr Atheroscler Rep. 2017;19(2):6.

Hartikainen J, et al. Adenoviral intramyocardial VEGF-DDeltaNDeltaC gene transfer increases myocardial perfusion reserve in refractory angina patients: a phase I/IIa study with 1-year follow-up. Eur Heart J. 2017;38(33):2547–55.

Hastie E, Samulski RJ. Adeno-associated virus at 50: a golden anniversary of discovery, research, and gene therapy success – a personal perspective. Hum Gene Ther. 2015;26(5):257–65.

Hawkins PN, et al. Evolving landscape in the management of transthyretin amyloidosis. Ann Med. 2015;47(8):625–38.

Hayward C, Patel H, Lyon A. Gene therapy in heart failure. SERCA2a as a therapeutic target. Circ J. 2014;78(11):2577–87.

Hedman M, et al. Safety and feasibility of catheter-based local intracoronary vascular endothelial growth factor gene transfer in the prevention of postangioplasty and in-stent restenosis and in the treatment of chronic myocardial ischemia: phase II results of the Kuopio Angiogenesis Trial (KAT). Circulation. 2003;107(21):2677–83.

Henry TD, et al. Effects of Ad5FGF-4 in patients with angina: an analysis of pooled data from the AGENT-3 and AGENT-4 trials. J Am Coll Cardiol. 2007;50(11):1038–46.

Hirschhorn JN, Daly MJ. Genome-wide association studies for common diseases and complex traits. Nat Rev Genet. 2005;6(2):95–108.

Holmes MV, et al. CYP2C19 genotype, clopidogrel metabolism, platelet function, and cardiovascular events: a systematic review and meta-analysis. JAMA. 2011;306(24):2704–14.

Horvath P, Barrangou R. CRISPR/Cas, the immune system of bacteria and archaea. Science. 2010;327(5962):167–70.

Hsu PD, Lander ES, Zhang F. Development and applications of CRISPR-Cas9 for genome engineering. Cell. 2014;157(6):1262–78.

International Warfarin Pharmacogenetics Consortium, et al. Estimation of the warfarin dose with clinical and pharmacogenetic data. N Engl J Med. 2009;360(8):753–64.

Ishikawa K, et al. Gene delivery methods in cardiac gene therapy. J Gene Med. 2011;13(10):566–72.

Jamuar SS, Tan EC. Clinical application of next-generation sequencing for Mendelian diseases. Hum Genomics. 2015;9:10.

Jaski BE, et al. Calcium upregulation by percutaneous administration of gene therapy in cardiac disease (CUPID trial), a first-in-human phase 1/2 clinical trial. J Card Fail. 2009;15(3):171–81.

Johnson JA, Cavallari LH. Pharmacogenetics and cardiovascular disease – implications for personalized medicine. Pharmacol Rev. 2013;65(3):987–1009.

Joung JK, Sander JD. TALENs: a widely applicable technology for targeted genome editing. Nat Rev Mol Cell Biol. 2013;14(1):49–55.
Kaski JC, Consuegra-Sanchez L. Evaluation of ASPIRE trial: a phase III pivotal registration trial, using intracoronary administration of Generx (Ad5FGF4) to treat patients with recurrent angina pectoris. Expert Opin Biol Ther. 2013;13(12):1749–53.
Kastrup J, et al. Direct intramyocardial plasmid vascular endothelial growth factor-A165 gene therapy in patients with stable severe angina pectoris A randomized double-blind placebo-controlled study: the Euroinject One trial. J Am Coll Cardiol. 2005;45(7):982–8.
Kathiresan S, Srivastava D. Genetics of human cardiovascular disease. Cell. 2012;148(6):1242–57.
Katz MG, et al. Gene delivery technologies for cardiac applications. Gene Ther. 2012;19(6):659–69.
Kettlewell S, et al. S100A1 increases the gain of excitation-contraction coupling in isolated rabbit ventricular cardiomyocytes. J Mol Cell Cardiol. 2005;39(6):900–10.
Khurana R, et al. Role of angiogenesis in cardiovascular disease: a critical appraisal. Circulation. 2005;112(12):1813–24.
Kimura H. Histone modifications for human epigenome analysis. J Hum Genet. 2013;58(7):439–45.
Koransky ML, Robbins RC, Blau HM. VEGF gene delivery for treatment of ischemic cardiovascular disease. Trends Cardiovasc Med. 2002;12(3):108–14.
Kornowski R, et al. Electromagnetic guidance for catheter-based transendocardial injection: a platform for intramyocardial angiogenesis therapy. Results in normal and ischemic porcine models. J Am Coll Cardiol. 2000;35(4):1031–9.
Kranias EG, Hajjar RJ. Modulation of cardiac contractility by the phospholamban/SERCA2a regulatome. Circ Res. 2012;110(12):1646–60.
Lai NC, et al. Intracoronary delivery of adenovirus encoding adenylyl cyclase VI increases left ventricular function and cAMP-generating capacity. Circulation. 2000;102(19):2396–401.
Lai NC, et al. Intracoronary adenovirus encoding adenylyl cyclase VI increases left ventricular function in heart failure. Circulation. 2004;110(3):330–6.
Laitinen M, et al. Catheter-mediated vascular endothelial growth factor gene transfer to human coronary arteries after angioplasty. Hum Gene Ther. 2000;11(2):263–70.
Landrum MJ, et al. ClinVar: public archive of interpretations of clinically relevant variants. Nucleic Acids Res. 2016;44(D1):D862–8.
Lappalainen I, et al. DbVar and DGVa: public archives for genomic structural variation. Nucleic Acids Res. 2013;41(Database issue):D936–41.
Laufer BI, Singh SM. Strategies for precision modulation of gene expression by epigenome editing: an overview. Epigenetics Chromatin. 2015;8:34.
Lavu M, Gundewar S, Lefer DJ. Gene therapy for ischemic heart disease. J Mol Cell Cardiol. 2011;50(5):742–50.
Lee CJ, et al. Promoter-specific lentivectors for long-term, cardiac-directed therapy of Fabry disease. J Cardiol. 2011;57(1):115–22.
Lek M, et al. Analysis of protein-coding genetic variation in 60,706 humans. Nature. 2016;536(7616):285–91.
Li Y, Tollefsbol TO. DNA methylation detection: bisulfite genomic sequencing analysis. Methods Mol Biol. 2011;791:11–21.
Li X, Xiong X, Yi C. Epitranscriptome sequencing technologies: decoding RNA modifications. Nat Methods. 2016;14(1):23–31.
Licht K, Jantsch MF. Rapid and dynamic transcriptome regulation by RNA editing and RNA modifications. J Cell Biol. 2016;213(1):15–22.
Loebstein R, et al. A coding VKORC1 Asp36Tyr polymorphism predisposes to warfarin resistance. Blood. 2007;109(6):2477–80.
Lorden ER, Levinson HM, Leong KW. Integration of drug, protein, and gene delivery systems with regenerative medicine. Drug Deliv Transl Res. 2015;5(2):168–86.
Losordo DW, et al. Phase 1/2 placebo-controlled, double-blind, dose-escalating trial of myocardial vascular endothelial growth factor 2 gene transfer by catheter delivery in patients with chronic myocardial ischemia. Circulation. 2002;105(17):2012–8.

Ma Y, et al. Generating rats with conditional alleles using CRISPR/Cas9. Cell Res. 2014;24(1):122–5.
Ma H, et al. Correction of a pathogenic gene mutation in human embryos. Nature. 2017;548(7668):413–9.
Maeder ML, Gersbach CA. Genome-editing technologies for gene and cell therapy. Mol Ther. 2016;24(3):430–46.
Mahfouz MM, et al. De novo-engineered transcription activator-like effector (TALE) hybrid nuclease with novel DNA binding specificity creates double-strand breaks. Proc Natl Acad Sci U S A. 2011;108(6):2623–8.
Makarov V, et al. AnnTools: a comprehensive and versatile annotation toolkit for genomic variants. Bioinformatics. 2012;28(5):724–5.
Marcaida MJ, et al. Homing endonucleases: from basics to therapeutic applications. Cell Mol Life Sci. 2010;67(5):727–48.
Mardis ER. Next-generation sequencing platforms. Annu Rev Anal Chem (Palo Alto, Calif). 2013;6:287–303.
Maron BJ. Sudden death in young athletes. N Engl J Med. 2003;349(11):1064–75.
Maron BJ, et al. Prevalence of hypertrophic cardiomyopathy in a general population of young adults. Echocardiographic analysis of 4111 subjects in the CARDIA study. Coronary artery risk development in (young) adults. Circulation. 1995;92(4):785–9.
Meder B, et al. A genome-wide association study identifies 6p21 as novel risk locus for dilated cardiomyopathy. Eur Heart J. 2014;35(16):1069–77.
Mega JL, et al. Cytochrome p-450 polymorphisms and response to clopidogrel. N Engl J Med. 2009;360(4):354–62.
Mega JL, et al. Reduced-function CYP2C19 genotype and risk of adverse clinical outcomes among patients treated with clopidogrel predominantly for PCI: a meta-analysis. JAMA. 2010;304(16):1821–30.
Meier P, et al. Beneficial effect of recruitable collaterals: a 10-year follow-up study in patients with stable coronary artery disease undergoing quantitative collateral measurements. Circulation. 2007;116(9):975–83.
Mittelman D, et al. Zinc-finger directed double-strand breaks within CAG repeat tracts promote repeat instability in human cells. Proc Natl Acad Sci U S A. 2009;106(24):9607–12.
Most P, et al. Cardiac adenoviral S100A1 gene delivery rescues failing myocardium. J Clin Invest. 2004;114(11):1550–63.
Mundade R, et al. Role of ChIP-seq in the discovery of transcription factor binding sites, differential gene regulation mechanism, epigenetic marks and beyond. Cell Cycle. 2014;13(18):2847–52.
Nair JK, et al. Multivalent N-acetylgalactosamine-conjugated siRNA localizes in hepatocytes and elicits robust RNAi-mediated gene silencing. J Am Chem Soc. 2014;136(49):16958–61.
Nayak S, Herzog RW. Progress and prospects: immune responses to viral vectors. Gene Ther. 2010;17(3):295–304.
Nielsen JB, et al. Genome-wide study of atrial fibrillation identifies seven risk loci and highlights biological pathways and regulatory elements involved in cardiac development. Am J Hum Genet. 2018;102(1):103–15.
Nikpay M, et al. A comprehensive 1,000 genomes-based genome-wide association meta-analysis of coronary artery disease. Nat Genet. 2015;47(10):1121–30.
Nishimasu H, et al. Crystal structure of Cas9 in complex with guide RNA and target DNA. Cell. 2014;156(5):935–49.
Paillard F. Promoter attenuation in gene therapy: causes and remedies. Hum Gene Ther. 1997;8(17):2009–10.
Pardo B, Gomez-Gonzalez B, Aguilera A. DNA repair in mammalian cells: DNA double-strand break repair: how to fix a broken relationship. Cell Mol Life Sci. 2009;66(6):1039–56.
Pare G, et al. Effects of CYP2C19 genotype on outcomes of clopidogrel treatment. N Engl J Med. 2010;363(18):1704–14.
Park PJ. ChIP-seq: advantages and challenges of a maturing technology. Nat Rev Genet. 2009;10(10):669–80.

Parkhomchuk D, et al. Transcriptome analysis by strand-specific sequencing of complementary DNA. Nucleic Acids Res. 2009;37(18):e123.
Penn MS, et al. An open-label dose escalation study to evaluate the safety of administration of nonviral stromal cell-derived factor-1 plasmid to treat symptomatic ischemic heart failure. Circ Res. 2013;112(5):816–25.
Pereira NL, et al. Pharmacogenetics of clopidogrel: an unresolved issue. Circ Cardiovasc Genet. 2016;9(2):185–8.
Perrino C, et al. Epigenomic and transcriptomic approaches in the post-genomic era: path to novel targets for diagnosis and therapy of the ischaemic heart? Position paper of the European Society of Cardiology Working Group on cellular biology of the heart. Cardiovasc Res. 2017;113(7):725–36.
Phan HM, et al. New signaling pathways associated with increased cardiac adenylyl cyclase 6 expression: implications for possible congestive heart failure therapy. Trends Cardiovasc Med. 2007;17(7):215–21.
Pleger ST, et al. Stable myocardial-specific AAV6-S100A1 gene therapy results in chronic functional heart failure rescue. Circulation. 2007;115(19):2506–15.
Pleger ST, et al. Cardiac AAV9-S100A1 gene therapy rescues post-ischemic heart failure in a preclinical large animal model. Sci Transl Med. 2011;3(92):92ra64.
Porteus MH, Carroll D. Gene targeting using zinc finger nucleases. Nat Biotechnol. 2005;23(8):967–73.
Pyzocha NK, et al. RNA-guided genome editing of mammalian cells. Methods Mol Biol. 2014;1114:269–77.
Qi LS, et al. Repurposing CRISPR as an RNA-guided platform for sequence-specific control of gene expression. Cell. 2013;152(5):1173–83.
Raake P, et al. Myocardial gene transfer by selective pressure-regulated retroinfusion of coronary veins: comparison with surgical and percutaneous intramyocardial gene delivery. J Am Coll Cardiol. 2004;44(5):1124–9.
Rapezzi C, et al. Transthyretin-related amyloidoses and the heart: a clinical overview. Nat Rev Cardiol. 2010;7(7):398–408.
Reilly JP, et al. Long-term (2-year) clinical events following transthoracic intramyocardial gene transfer of VEGF-2 in no-option patients. J Interv Cardiol. 2005;18(1):27–31.
Richards S, et al. Standards and guidelines for the interpretation of sequence variants: a joint consensus recommendation of the American College of Medical Genetics and Genomics and the Association for Molecular Pathology. Genet Med. 2015;17(5):405–24.
Ripa RS, et al. Intramyocardial injection of vascular endothelial growth factor-A165 plasmid followed by granulocyte-colony stimulating factor to induce angiogenesis in patients with severe chronic ischaemic heart disease. Eur Heart J. 2006;27(15):1785–92.
Rolland M, Kerlan C, Jacquot E. The acquisition of molecular determinants involved in potato virus Y necrosis capacity leads to fitness reduction in tobacco plants. J Gen Virol. 2009;90(Pt 1):244–52.
Rosengart TK, et al. Angiogenesis gene therapy: phase I assessment of direct intramyocardial administration of an adenovirus vector expressing VEGF121 cDNA to individuals with clinically significant severe coronary artery disease. Circulation. 1999;100(5):468–74.
Rubanyi GM. Mechanistic, technical, and clinical perspectives in therapeutic stimulation of coronary collateral development by angiogenic growth factors. Mol Ther. 2013;21(4):725–38.
Ruel M, et al. Concomitant treatment with oral L-arginine improves the efficacy of surgical angiogenesis in patients with severe diffuse coronary artery disease: the endothelial modulation in angiogenic therapy randomized controlled trial. J Thorac Cardiovasc Surg. 2008;135(4):762–70. 770 e1
Samocha KE, et al. A framework for the interpretation of de novo mutation in human disease. Nat Genet. 2014;46(9):944–50.
Sattianayagam PT, et al. Cardiac phenotype and clinical outcome of familial amyloid polyneuropathy associated with transthyretin alanine 60 variant. Eur Heart J. 2012;33(9):1120–7.

Schotten U, et al. Pathophysiological mechanisms of atrial fibrillation: a translational appraisal. Physiol Rev. 2011;91(1):265–325.

Schumacher B, et al. Induction of neoangiogenesis in ischemic myocardium by human growth factors: first clinical results of a new treatment of coronary heart disease. Circulation. 1998;97(7):645–50.

SEARCH Collaborative Group, et al. SLCO1B1 variants and statin-induced myopathy – a genome-wide study. N Engl J Med. 2008;359(8):789–99.

Shareef MA, Anwer LA, Poizat C. Cardiac SERCA2A/B: therapeutic targets for heart failure. Eur J Pharmacol. 2014;724:1–8.

Sherry ST, Ward M, Sirotkin K. dbSNP-database for single nucleotide polymorphisms and other classes of minor genetic variation. Genome Res. 1999;9(8):677–9.

Shetty AC, et al. SeqAnt: a web service to rapidly identify and annotate DNA sequence variations. BMC Bioinformatics. 2010;11:471.

Shim G, et al. Therapeutic gene editing: delivery and regulatory perspectives. Acta Pharmacol Sin. 2017;38(6):738–53.

Sikkel MB, et al. SERCA2a gene therapy in heart failure: an anti-arrhythmic positive inotrope. Br J Pharmacol. 2014;171(1):38–54.

Silva G, et al. Meganucleases and other tools for targeted genome engineering: perspectives and challenges for gene therapy. Curr Gene Ther. 2011;11(1):11–27.

Simon T, et al. Genetic determinants of response to clopidogrel and cardiovascular events. N Engl J Med. 2009;360(4):363–75.

Slagle CE, Conlon FL. Emerging field of cardiomics: high-throughput investigations into transcriptional regulation of cardiovascular development and disease. Trends Genet. 2016;32(11):707–16.

Snyder RO. Adeno-associated virus-mediated gene delivery. J Gene Med. 1999;1(3):166–75.

Stark K, et al. Genetic association study identifies HSPB7 as a risk gene for idiopathic dilated cardiomyopathy. PLoS Genet. 2010;6(10):e1001167.

Steg PG, et al. Impact of collateral flow to the occluded infarct-related artery on clinical outcomes in patients with recent myocardial infarction: a report from the randomized occluded artery trial. Circulation. 2010;121(25):2724–30.

Stenson PD, et al. The human gene mutation database: building a comprehensive mutation repository for clinical and molecular genetics, diagnostic testing and personalized genomic medicine. Hum Genet. 2014;133(1):1–9.

Stewart DJ, et al. Angiogenic gene therapy in patients with nonrevascularizable ischemic heart disease: a phase 2 randomized, controlled trial of AdVEGF(121) (AdVEGF121) versus maximum medical treatment. Gene Ther. 2006;13(21):1503–11.

Stewart DJ, et al. VEGF gene therapy fails to improve perfusion of ischemic myocardium in patients with advanced coronary disease: results of the NORTHERN trial. Mol Ther. 2009;17(6):1109–15.

Suhr OB, et al. Efficacy and safety of patisiran for familial amyloidotic polyneuropathy: a phase II multi-dose study. Orphanet J Rare Dis. 2015;10:109.

Sylven C, et al. Catheter-based transendocardial myocardial gene transfer. J Interv Cardiol. 2002;15(1):7–13.

Takata M, et al. Homologous recombination and non-homologous end-joining pathways of DNA double-strand break repair have overlapping roles in the maintenance of chromosomal integrity in vertebrate cells. EMBO J. 1998;17(18):5497–508.

Thorolfsdottir RB, et al. A missense variant in PLEC increases risk of atrial fibrillation. J Am Coll Cardiol. 2017;70(17):2157–68.

Tian T, et al. Progress in the molecular genetics of hypertrophic cardiomyopathy: a mini-review. Gerontology. 2013;59(3):199–205.

Vale PR, et al. Catheter-based myocardial gene transfer utilizing nonfluoroscopic electromechanical left ventricular mapping. J Am Coll Cardiol. 1999;34(1):246–54.

Vale PR, et al. Randomized, single-blind, placebo-controlled pilot study of catheter-based myocardial gene transfer for therapeutic angiogenesis using left ventricular electromechanical mapping in patients with chronic myocardial ischemia. Circulation. 2001;103(17):2138–43.

Villard E, et al. A genome-wide association study identifies two loci associated with heart failure due to dilated cardiomyopathy. Eur Heart J. 2011;32(9):1065–76.

Wang Z, Gerstein M, Snyder M. RNA-Seq: a revolutionary tool for transcriptomics. Nat Rev Genet. 2009;10(1):57–63.

Wang K, Li M, Hakonarson H. ANNOVAR: functional annotation of genetic variants from high-throughput sequencing data. Nucleic Acids Res. 2010;38(16):e164.

Weber C, et al. Therapeutic safety of high myocardial expression levels of the molecular inotrope S100A1 in a preclinical heart failure model. Gene Ther. 2014;21(2):131–8.

Weiwei C, et al. Outline of the report on cardiovascular diseases in China, 2014. Eur Heart J Suppl. 2016;18(Suppl F):F2–F11.

Williams PD, Kingston PA. Plasmid-mediated gene therapy for cardiovascular disease. Cardiovasc Res. 2011;91(4):565–76.

Yla-Herttuala S. Endgame: glybera finally recommended for approval as the first gene therapy drug in the European union. Mol Ther. 2012;20(10):1831–2.

Zamore PD, et al. RNAi: double-stranded RNA directs the ATP-dependent cleavage of mRNA at 21 to 23 nucleotide intervals. Cell. 2000;101(1):25–33.

Printed by Printforce, the Netherlands